ENVIRONMENTAL GEOLOGY

ENVIRONMENTAL GEOLOGY

Fourth Edition

Edward A. Keller

University of California/Santa Barbara

With assistance from
E. M. Burt
Williams & Works
Grand Rapids, Michigan

Charles E. Merrill Publishing Company
A Bell & Howell Company
Columbus   Toronto   London   Sydney

Published by
Charles E. Merrill Publishing Company
A Bell & Howell Company
Columbus, Ohio 43216

Cover Designer: Cathy Watterson
Text Designer: Cynthia Brunk
Production Editor: Molly Kyle
Art Coordination: Marilyn Reeves

Cover photo: Pismo Dunes State Vehicular Recreation Area, south
of Pismo Beach, California. Use of off-road vehicles in this area has
damaged the environment and created controversy over coastal land
use.

Title page photo: Courtesy of George B. Cleveland, California
Division of Mines and Geology; Ian Campbell, California Academy
of Sciences; and U.S. Geological Survey at Menlo Park.

Library of Congress Catalog Card Number: 84-62584
International Standard Book Number: 0-675-20373-2
Printed in the United States of America
1 2 3 4 5 6 7 8 9—90 89 88 87 86 85

For Denny and Kathy

Preface

When I began work on the first edition of *Environmental Geology* as a graduate student at Purdue University in 1971, it never occurred to me that thirteen years later I might be working on the fourth edition, not far from where I grew up in southern California. Nevertheless, I am pleased to have the opportunity to better present some old ideas and introduce some new concepts, case histories, and facts related to the ever-changing field of environmental geology.

Basically, environmental geology is applied geology, and as such, focuses on the entire spectrum of possible interactions between people and the physical environment. The fourth edition of *Environmental Geology,* like the previous three, is intended as an introduction to the study of applied geology. Students who become particularly interested in the study of applied geology may then go on to take courses in applied hydrology and engineering geology.

The study of environmental geology is facilitated by previous exposure either to physical geology or geography; however, I recognize that students, unless they are majoring in geology, seldom have the latitude in their undergraduate studies to take more than a single geology course. Therefore, *Environmental Geology* is designed to allow for the study of applied geology without previous geologic training. I have selected case histories and other materials that are most relevant to a wide variety of students, including those in professional schools such as engineering and architecture and planning; traditional scientific disciplines such as geology, physics, chemistry, biology, and physical geography; and liberal arts students majoring in disciplines such as economics, literature, art, sociology, and human geography.

The organization of the fourth edition is essentially the same as that of the previous three. Part One introduces philosophy and fundamental principles important in the study of environmental geology. The objective is to unite the cultural and physical environments and introduce important geologic concepts and terminology that are useful in understanding the remainder of the book.

Part Two introduces the major natural processes and geologic hazards, including flooding, landslides, earthquakes, volcanoes, and coastal phenomena that continue to take human lives and inflict property damage. New material in Part Two includes consideration of the snow avalanche hazard in mountainous areas; discussion of the Long Valley caldera in California that is providing valuable information about catastrophic caldera-forming eruptions and a subsequent volcanic hazard; new methods and ideas about long-term earthquake predictions and the relevance of such predictions to earthquake hazard reduction programs; and discussion of the tidal flood hazard in London, England, and the new mechanical barrier that has been constructed to alleviate such floods.

Part Three discusses various aspects of human interaction with the environment. Particularly important are hydrology, especially as it relates to supply, water use, water pollution, and channelization; waste disposal as it relates to topics such as solid-waste disposal, hazardous chemical waste, radioactive waste management, and waste-water treatment; and geologic aspects of environmental health that are important in relationships among people, environment, and chronic disease. Changes and additions in Part Three include an expanded discussion of hazardous chemical waste and monitoring of toxic materials through the biosphere; revision of material dealing with water supply and use; and the addition of several case histories, including the first major effort to undo a channelization project in Florida.

Part Four discusses minerals, energy, and environmental issues associated with resource utilization and management. New material in Part Four includes expanded treatment of environmental impact of mineral development; discussion of the newly discovered sulphide mineral deposits on the floor of the ocean; and the principles of petroleum production.

Part Five discusses issues related to land use and decision making and draws together much of the material from earlier parts of the text. We discuss environmental impact analysis and regulations and consider emergency planning following large-scale environmental disruption from natural events such as hurricanes, volcanic eruptions, and earthquakes. Of particular interest is the update of the Cape Hatteras National Seashore case history which, during the four editions of *Environmental Geology,* has proceded from an initial environmental impact analysis to development of a comprehensive general management plan for the Seashore.

Successful completion of a textbook that includes numerous photographs, figures, and case histories is not feasible without the cooperation of many individuals, companies, and agencies. In particular, I wish to thank the U.S. Geological Survey and individual state Geological Surveys, which have provided much of the information and concepts important in the development of applied geology. To all those individuals who were so helpful in this endeavor, I offer my sincere thanks and appreciation.

Reviews of chapters by Roger J. Bain, Douglas G. Brookins, Thomas L. Davis, John G. Drost, Richard V. Fisher, Stanley T. Fisher, Cal Janes, Donald L. Johnson, Harold L. Krivoy, Robert M. Norris, James R. Lauffer, John S. Pomeroy, James Dennis Rash, Derek J. Rust, Samuel E. Swanson, and William S. Wise are acknowledged and appreciated. And for the fourth edition, I wish to thank Ernst Kastning, P. Thompson Davis, Roger Bain, and Robert Mathews. Special thanks goes to Wakefield Dort, who made suggestions for improvement of the figures and captions.

I am indebted to my editors at Charles E. Merrill, Kathy Nee and Molly Kyle, for coordinating the production and suggesting improvements to the text.

The Environmental Studies program and Department of Geological Sciences at the University of California, Santa Barbara and the Department of Geography and Earth Sciences at the University of North Carolina at Charlotte provided encouragement and a stimulating environment for writing. I would like to thank the staff members at both institutions who readily gave of their time in the thoughtful discussion that is so necessary in explaining and presenting complex environmental issues.

Special thanks is offered to my wife, Jacqueline J. Keller, who has remained very understanding during the times when writing seems to take up all the available time. Jacqueline also found the time near Christmas to complete the index for the fourth edition.

Edward A. Keller
Santa Barbara, California

Contents

COLOR PLATES

ENVIRONMENTAL GEOLOGY

ONE

Philosophy and Fundamental Principles

Everything has a beginning and an end. Our earth began approximately 5 billion years ago when a cloud of interstellar gas known as a *solar nebula* collapsed, forming protostars and planetary sytems; life on earth began about 2 billion years later, or 3 billion years ago. Since then, a multitude of different types of organisms have emerged, prospered, and died out, leaving only their fossils to mark their place in earth's history. Several million years ago, on one of the more recent pages in earth history, our ancestors set the stage for the eventual dominance of the human race. As certainly as our sun will eventually die, we, too, will disappear. The impact of humanity on earth history may not be significant, but to us living now, our children and theirs, our environment is significant indeed.

Environmental geology is applied geology. Specifically, it is the application of geologic information to solving conflicts, minimizing possible adverse environmental degradation, or maximizing possible advantageous conditions resulting from our use of the natural and modified environment. This includes evaluation of *natural hazards* such as floods, landslides, earthquakes, and volcanic activity to minimize loss of human life and property damage; evaluation of the *landscape* for site selection, land-use planning, and *environmental impact analysis*; and evaluation of *earth materials* (such as elements, minerals, rocks, soils, and water) to determine their potential use as resources or waste disposal sites and the effects on human health, and to assess the need for conservation practices. In a broader sense, environmental geology is that branch of earth science that emphasizes the entire spectrum of human interactions with the physical environment.

Photograph courtesy of the California Department of Water Resources.

Environment may be considered as the total set of circumstances that surround an individual or a community. It may be defined to include two parts: first, physical conditions such as air, water, gases, landforms, and so on that affect the growth and development of an individual or a community; and second, social and cultural aspects such as ethics, economics, aesthetics, and so on that affect the behavior of an individual or a community. Therefore, a complete introduction to environmental geology involves consideration of philosophical and cultural aspects that influence how we perceive and react to our landscape, as well as the physical earth processes, resources, and landforms that may be more readily recognized by the observant earth scientist.

Chapters 1 through 3 provide the philosophical framework for the remainder of the book. Chapters 1 and 2 integrate the influence of cultural and physical activities into our total environment, and Chapter 3 introduces the physical environment through the geological cycle. The term *cycle* emphasizes that most earth materials, such as air, water, soil, minerals, and rock, although changed physically and chemically and transported from place to place, are constantly being reworked, conserved, and renewed by natural earth processes. Chapter 3 also introduces basic earth science terminology and engineering properties of earth materials necessary to understand the remainder of the book.

Chapter 1

Cultural Basis for the Environmental Crisis

The cultural aspect of the environmental crisis involves the entire way of life that we have transmitted from one generation to another. Therefore, to uncover the roots of our present condition, we must look to the past and consider various functional categories and social institutions that have developed. The functional categories of society that are especially significant in environmental studies are ethical, economic, political, aesthetic, and, perhaps, religious. The interaction between individuals and the institutions responsible for maintaining these functions are intimately associated with the way we perceive and respond to our physical environment.

ENVIRONMENTAL ETHICS

What started as the "quiet crisis" of the 1960s has evolved into what Stewart Udall, statesman and conservationist, refers to as the "crisis of survival" (1). More important than the certainty of a crisis is whether society believes there is a crisis. In other words, is there a new awareness that is destined to change our life-style, morals, ethics, and institutions, or is the en-

vironmental revolution just another prestigious fad that interests the intellectual community?

The evolution of ethics (Figure 1.1) is an important environmental trend. Aldo Leopold emphasizes the lack of ethics regarding property through the story of Odysseus, who, upon returning from Troy, hanged a dozen slave women for suspected misbehavior during his absence. The hangings involved no question of ownership: the women were property, and the disposal of property was a matter of expediency, much as it is today. Although concepts of right and wrong were present in Greece three thousand years ago, these ethical values did not extend to slaves (2). Since that time, ethical values have been extended to many other areas of human behavior; but, apparently, only within this century has the relationship between civilization and its physical environment begun to emerge as a relationship with moral considerations.

Ecological ethics involve limitations on social as well as individual freedom of action in the struggle for existence in our stressed environment (2). A land ethic assumes that we are ethically responsible not only to other individuals and society but also to the total environment, that larger community consisting of plants, animals, soil, atmosphere, and so forth. The environ-

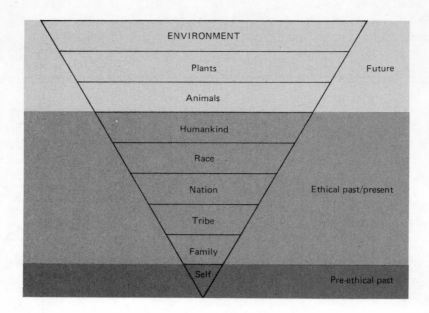

mental ethic proposed by Leopold affirms the right of all resources, including plants, animals, and earth materials, to continued existence and, at least in certain locations, continued existence in a natural state. This ethic effectively changes our role from that of conqueror of the land to that of citizen and protector of the environment. This role change obviously requires us to revere and love our land and not, for instance, to allow economics to determine all land use.

A possible dichotomy or source of confusion exists between an ideal and a realistic land ethic. To give rights to the plants, animals, and landscape might be interpreted as granting to individual plants and animals the fundamental right to live. If we are to be part of the environment, however, we must extract the energy necessary to survive. Therefore, although the land ethic assigns rights for animals such as deer, cattle, or chickens to survive as a *species*, it does not necessarily assign rights to an *individual* deer, cow, or chicken. The same argument may be given to justify the use of stream gravel for construction material, or to mine and use the other resources necessary for our well-being. However, unique landscapes with high aesthetic value, like endangered species, are in need of complete protection within our ethical framework.

Environmental ethics and moral responsibility are re-restated by Stewart Udall. Each generation has its own rendezvous with the land, for despite our fee titles and claims of ownership, we are all brief tenants on this planet. By choice or by default, we will carve out a land for our heirs. We can misuse the land and diminish the usefulness of resources, or we can create a

world in which physical affluence and spiritual affluence go hand in hand (1).

The resounding message is that humanity is an integral part of the environment. A person is no more than any other being, and has a moral obligation to those beings who will follow. This obligation is to insure that they will also have the opportunity to experience the pleasure of belonging to and cooperating with the entire land community.

ECONOMIC AND POLITICAL SYSTEMS

Arriving in late fall of 1620, after two months on the stormy North Atlantic, 73 men and 29 women from the *Mayflower* confronted what they considered a wild and savage land. The colonists were not equipped with the skills and knowledge necessary to adapt quickly to their new environment. Regardless of these shortcomings and despite their fear of the wilderness, they brought three things that assured their success in the New World. First, they brought a new technology. Reportedly, when the Pilgrims landed, they did not even have a saw, but they did have Iron Age skills necessary to insure relentless subjugation of the land and its earliest inhabitants, the Indians. In the long run, the ax, gun, and wheel asserted their supremacy. Second, the colonists brought with them the blueprints to remake the New World. They knew how to organize work, use work animals, and sell their surplus to overseas markets. Third, they brought with them a concept of land ownership completely different from that of the Indians, whose bonds to the land were religious and held by kinship with nature rather than exclusive possession. The Indians had a deep affection for the land and had a notion of ownership different from that of the colonists, whose idea of ownership involved an absolute title to land regardless of who worked the land or how far away the owner was. After the Indians were displaced, land use or abuse depended entirely on the attitude of the owner.*

America now, as in its early years, suffers greatly from the Myth of Superabundance. This myth assumes that the land and resources in America are inexhaustible and that management of resources is therefore unnecessary. Management of the people and their society, however, continues to reflect the deep roots transposed from the Old World.

Stewart Udall writes that the land myth was instrumental in environmental degradation from the ''birth of land policy'' in the eighteenth century throughout the ''raid on resources'' that lasted into the twentieth century. Even young Thomas Jefferson, who in later life was to become aware of the value of conservation, said there was such a great deal of farmland that it could be wasted as he pleased. However, the real raid on resources probably began with the mountain men and their beaver trapping in the 1820s. This was only the beginning; there followed the invention of machines that were capable of large-scale removal of resources and landscape alteration. The inventions included the sawmills that precipitated the destruction of the American forests, and the ''Little Giant'' hose nozzle that could tear up an entire hillside in the search for California gold.

The ''Great Giveaway'' of land which resulted in the destruction of forests and consequent soil erosion eventually ended, and in 1884, hydraulic mining was outlawed. However, the effects of these repugnant land-use practices are still visible today. Similar examples of the raid on resources are the plights of the fur seal, the buffalo, and the passenger pigeon, and the Dust Bowl of the 1930s (1).

The seeds of conservation were planted in the latter part of the nineteenth century by men such as Carl Schurz, Secretary of the Interior, and John Wesley Powell, geologist and explorer. Their messages concerning conservation of resources and land-use planning, although largely ignored when first introduced, today stand as landmarks in perceptive and innovative conservation (1).

The historical roots of our landscape heritage, while not a pretty picture, are full of lessons to be learned. For example, we painstakingly learned that our resources are not infinite and that land and water management is necessary for meaningful existence. This conclusion has become even more significant over the years, as American society continues to urbanize and consume resources at an ever-increasing rate.

The convergence of available resources with the needs of society, along with an ever-growing production of waste, has produced what is popularly referred to as the *environmental crisis*. This impending crisis in America, according to Lewis W. Moncrief, is a result of individual and institutional inability in our democratic system to organize technology, conservation, urbanization, and the capitalistic mission to the betterment of our landscape (3). Moncrief further contends that the present condition is characterized by three features that tend to prevent a quick solution to environmental problems: first, the absence of individual and personal moral direction concerning the way we treat our natural resources; second, the inability of our social institutions to make adjustments to reduce environmental stress; and third, an abiding faith in technology.

Overpopulation, urbanization, and industrialization, combined with little ethical regard for our land and inadequate institutions (or perhaps too many institutions stumbling over one another) to cope with environmental stress, may well be the immediate source of the crisis. The overpopulation, urbanization, and industrial factors, although part of the political and economic systems, currently tend to transcend those systems.

Political and economic theorists are often surprised to learn that disruption of the environment is as serious a problem in the USSR as in the United States. Goldman reports that the Soviets have greatly misused their natural resources. For example, most major Soviet cities have air pollution problems, and water pollution there has resulted in massive fish kills, in turn resulting in an increase in mosquitoes and malaria peril. The oil-covered Iset River was accidentally ignited in 1966, as was the Cuyahoga River in Ohio. A large percent of the factories in Russia discharge their waste without treatment, and a large percent of the cities and suburbs have no water treatment facilities. Land-use problems in Russia include a system of dams, reservoirs, and canals that have diverted so much water that there is serious concern for the future of the Caspian Sea and its famous caviar fisheries. The reservoirs have also increased evaporation, which is disrupting natural moisture patterns and changing rainfall cycles. Furthermore, seepage from unlined canals has caused the water table to rise in normally dry areas, facilitating the deposition of harmful salts in the soil. Mining of beach deposits for construction material in conjunction with a decreased supply of sediment to the beaches (the

*Stewart Udall, *The Quiet Crisis*, pp. 25–27.

reservoirs are holding back the natural flow of sediment) has resulted in serious coastal erosion; without the sand and gravel to impede the impact of waves, the coastline is subject to rapid erosion.

The factors responsible for environmental problems in the USSR, as in America, are population explosion and rapid industrialization. Furthermore, the Soviet government, as sole owner of the productive resources, has been no more successful than other countries in regulating or controlling environmental degradation. In fact, the national commitment to centralized control of industry has given rise to unique environmental problems resulting from uniform regulations for industry regardless of local conditions.

Goldman notes several other disadvantages of an urbanizing, socialized system. First, no private interests can challenge government proposals. Thus, a well-intentioned program leading to possible unanticipated degradation to the environment might be introduced. Such a program would more likely be terminated in a society that allows public criticism. Second, environmental disruption occurs because of lack of private ownership. For example, the removal of gravel and subsequent erosion of the Black Sea coast is a result of the absence of private control. Because no one owns the gravel, it is free to anyone who can carry it away. Consequently, contractors who want gravel need not worry about infringing on the rights of coastal property owners worried about aesthetic degradation, erosion, and loss of tourist trade. Third, because the Soviet government is the sole owner of the country's industry, it is generally unable to take the role of impartial referee between consumer and industry. This condition is not likely to change as long as the USSR, which has only recently become industrialized, stresses rapid growth and production.

On the other hand, Goldman notes some aspects of socialism in the USSR that have, by design, or perhaps by fortuitous circumstances, resulted in lesser environmental disruption. For example, Russia has de-emphasized the production of consumer goods, resulting in less prolific production of easily disposed of items and less waste material. Furthermore, relatively low labor costs, which are not necessarily a planned characteristic of their system, facilitate trash collection and sewage disposal despite the general lack of a sewer system. This results because the cost of collecting sewage "nite soil" is apparently offset by its value as fertilizer. The central power of the state may also be an advantage in situations where expediency is sought. For example, establishment of national parks may be a relatively quick and inexpensive process compared to countries with private ownership of property.*

A recent study stated that the 1970s was a period of environmental awareness in both Russia and the U.S. The Soviets passed pollution abatement and other environmental laws, as did the United States. However, the Russian emphasis on high production rates coupled with institutional inflexibility has reduced their effectiveness, as is the case, to a lesser extent, in America (5).

The conclusion from consideration of political systems is that not private enterprise, but rather industrialization, urbanization, economic consideration, and lack of a land ethic are primarily responsible for environmental degradation. Therefore, the salvation of the landscape community involves social, economic, and ethical behavior on the part of individuals rather than by political systems as they exist today.

Optimistically, it appears that the emerging environmental ethics inherent in the spirit of individual actions and new legislation will facilitate the types of changes needed. These changes are possible because our democratic system, with private ownership of resources and free enterprise, has the flexibility necessary to allow meaningful change. We emphasize, however, that the system cannot be expected to react until individual citizens are willing to practice environmental ethics and the political, economic, and legal institutions and processes are modernized and streamlined in keeping with the original constitutional intent. There appears to be a tendency to solve problems by "legislating" them out of existence, while "insuring" no real progress by keeping antiquated or conflicting laws or institutions *ad infinitum*, sometimes by simply changing their names.

AESTHETIC PREFERENCE AND JUDGMENT

Environmental intangibles, such as the pleasures of private outdoor experiences in nature, are extremely difficult to evaluate. The hunter in a blind on a crisp autumn morning; fishermen in icy mountain streams as the day gives way to darkness; the nature photographer about to culminate the search for an elusive subject on a lonely mountaintop; hikers showing their three-year-old child a snail shell; picnickers relaxing; or a motorist out for a Sunday drive in the country—all perceive to a lesser or greater extent various aspects of the landscape and react to it with various types of be-

*M.I. Goldman, *Environmental Disruption in the Soviet Union*, pp. 63–65.

havior. Their experiences and memories cannot readily be equated to economic value, but to the individual they may be priceless.

An understanding of beauty was originally studied in the branch of philosophy known as *aesthetics*. Today, little attention is given to philosophers, and beauty is defined by artists and art critics (6). There appears to be an important distinction between the philosopher who studies aesthetics to establish evaluative judgments, similar to verdicts and findings, and the artist and art critic who may be more concerned with appreciative aesthetic judgments that express preferences such as affection or antipathy (7). Given a distinction between evaluation and preference, the former constitutes a more objective approach.

One perplexing problem of aesthetic evaluation is the impact of personal preference. For example, one person may appreciate a meandering river in an isolated swamp; another may prefer a bubbling mountain stream; a third would rather visit a public park in an urban area. Regardless of such preferences, if we are going to consider aesthetic factors in local, regional, and national land-use planning, we must develop a method of aesthetic evaluation for landscapes that is easy to understand and is quantitative, credible, and predictive.

Three basic criteria necessary to judge aesthetic quality have been recognized: unity, vividness, and variety (8). *Unity* refers to the quality of wholeness of the perceived landscape, not as an assemblage but as a single harmonious unit. *Vividness* refers to that quality of the landscape that reflects a visually striking scene. This is nearly synonymous with *intensity*, *novelty*, or *clarity*. *Variety* refers to how different one landscape is from another. *Diversity* and *uniqueness* also have similar meanings. Greater diversity, however, is not necessarily an indication of higher aesthetic value.

IMPACT OF RELIGION

The role of religion in causing, perpetuating, or condoning environmental disruption and degradation is vigorously debated. One school of thought holds that the Judeo-Christian heritage of Western civilization is responsible for the way we treat the environment. A second school refutes this and argues that human treatment of the land is a characteristic that transcends religious and cultural teaching.

Arguing that the Judeo-Christian heritage is responsible for Western attitudes and behavior with respect to the environment, Lynn White, Jr., cites the following evidence. First, Christianity is the most anthropocentric religion the world has seen. It establishes a dualism of humanity and nature, insisting that it is God's will that the human race exploit nature. Second, by destroying pagan animism, Christianity, which previously tended to unite humanity with nature, made it possible for us to degrade the environment in a way completely indifferent to the rights or feelings of natural objects. Third, Western science, technology, and industrialization are a natural result of Judeo-Christian dogma of creation, which teaches that humankind was created after plants, animals, and fishes as their rightful monarch. Fourth, great Western scientists from the thirteenth century to the eighteenth explained their motivation in religious terms, indicating that science was seeking to understand God's mind by discovering how creation operates rather than through the previous effort to decode the physical symbols of God's communication with people. In other words, before the thirteenth century, our predecessors studied nature because God created nature and consequently nature revealed the divine mentality. In the early church, therefore, nature was conceived as a symbolic system through which God communicated. In the West, this changed in the thirteenth century when scientists, in the name of religious progress, began investigating physical processes of light and matter (9).

The second school argues that the environmental crisis is not a religious problem. They criticize White's thesis on several grounds. First, prehistoric people, through the use of fire and water, also caused considerable environmental disruption. White also recognizes this. Second, before the birth of Christ, the early Greeks and Romans both imposed their will on the environment. Third, the triumph of Christianity over paganism brought no revolutionary change to the relation between society and nature. Fourth, although the ideals of some cultures may suggest that land is sacred, there is a considerable hiatus between ethical ideals and actual land-use practices (3, 10).

One can conclude that religious attitudes and beliefs are not a primary cause of the environmental crisis. This does not suggest that religious activities are not responsible for considerable environmental disruption. On the contrary, numerous examples, such as timber shortages in China caused by Buddhists' cremation of the dead and deforestation in Japan associated with construction of huge wooden Buddhist halls and temples, suggest that religious activity can promote environmental problems.

The implication that one religion, one culture, or one political system is responsible for the way people

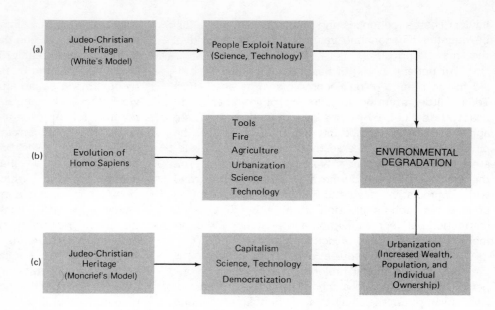

FIGURE 1.2
Models showing possible paths leading to environmental degradation.

(a) Judeo-Christian Heritage (White's Model) → People Exploit Nature (Science, Technology) → ENVIRONMENTAL DEGRADATION

(b) Evolution of Homo Sapiens → Tools / Fire / Agriculture / Urbanization / Science / Technology → ENVIRONMENTAL DEGRADATION

(c) Judeo-Christian Heritage (Moncrief's Model) → Capitalism / Science, Technology / Democratization → Urbanization (Increased Wealth, Population, and Individual Ownership) → ENVIRONMENTAL DEGRADATION

treat the land cannot be rigorously defended. Therefore, White's model (Figure 1.2a) is not entirely correct; likewise, Moncrief's modification of White's model (Figure 1.2c) to include capitalism and democracy, with or without the Judeo-Christian start, is not a complete and accurate progression. Since environmental degradation apparently transcends both religious belief and political system, we must look further for the primary cause of our present condition. A simpler explanation (Figure 1.2b) assumes that our environmental problems result from a pattern of human development that began when the earliest people attempted to use tools to better their chances for survival. A product of harsh times, early Homo sapiens, like other animals, extended their niche as far as restraints allowed. Therefore, each innovation not only asserted the individual but also assured that everyone who followed had

an easier time. As a result, increased populations created greater demand on resources as well as demanded more innovations. This spiral has continued to the present, when there are signs that we may be on a collision course with our environment. In a small way, the condition may be analogous to what sometimes happens to deer when, through our artificial management of animals, their numbers exceed the carrying capacity of the land. Deprived of their natural enemies, the deer herds increase until these ''artificial deer'' eat all available food, insuring a serious shortage of winter feed and little reproduction of food plants for the following spring. Everything from wildflowers to trees is gradually impoverished, and the deer either become dwarfed from malnutrition or starve to death (2). Are we, with our increasingly artificial environment, doing to ourselves what we have done to the deer?

SUMMARY AND CONCLUSIONS

Functional categories of society that are significant in environmental studies and are the cultural bases for environmental degradation are ethical, economic, political, aesthetic, and, perhaps, religious.

Our ethical framework appears to be slowly expanding and will eventually include the total environment in a land ethic. This ethic affirms the right of all resources, including plants, animals, and earth materials, to continued existence, and, at least in certain locations, continued existence in a natural state (2).

The immediate cause of environmental degradation is overpopulation, urbanization, and industrialization combined with, as yet, little ethical regard to our land and inadequate institutions to cope with environmental stress. These problems are not unique to a particular political system; and, therefore, we conclude that the salvation of the landscape community necessitates changing social, economic, and ethical behavior that transcends political systems.

Aesthetic factors are now being considered in local, regional, and national land-use planning, and scenery is considered a natural resource. A problem remains in finding a method of aesthetic evaluation that is easy to understand and is quantitative, credible, and predictive. Until we do have a satisfactory methodology, it will remain difficult to balance aesthetic and economic costs and benefits.

The role of religion in causing, perpetuating, or condoning environmental degradation remains a much debated issue. Some authors argue that the Judeo-Christian heritage is responsible for Western man's attitudes and behavior toward the environment. The argument is that the Judeo-Christian teachings and practices destroyed pagan animism, which had previously tended to unite nature and humanity, and thereby made it possible for humans to degrade the environment with complete indifference. This argument cannot be rigorously defended. Prehistoric humans and modern peoples who believe in both Eastern and Western religions have exploited and disrupted their land. One can conclude that religious institutions have indeed been responsible for some environmental problems, but that the general tendency for degradation of the environment is a more universal problem that transcends religious teachings.

REFERENCES

1 UDALL, S. L. 1963. *The quiet crisis*. New York: Avon Books.

2 LEOPOLD, A. 1949. *A Sand County almanac*. New York: Oxford University Press.

3 MONCRIEF, L. W. 1970. The cultural basis for our environmental crisis. *Science 170:* 508–12.

4 GOLDMAN, M. I. 1971. Environmental disruption in the Soviet Union. In *Man's impact on environment*, ed. T. R. Detwyler, pp. 61–75. New York: McGraw-Hill.

5 PRYDE, P. R. 1983. The ''Decade of the environment'' in the U.S.S.R. *Science* 220: 274–279.

6 FLORMAN, S. C. 1968. *Engineering and the liberal arts*. New York: McGraw-Hill.

7 ZUBE, E. H. 1973. Scenery as a natural resource. *Landscape Architecture* 63: 126–32.

8 LITTON, R. B. 1973. Aesthetic dimensions of the landscape. In *Natural Environments*, ed. J. V. Kantilla, pp. 262–91. Baltimore: Johns Hopkins University Press.

9 WHITE, L., JR. 1967. The historical roots of our ecological crisis. *Science* 155: 1203–7.

10 YI-FU, T. 1970. Our treatment of the environment in ideal and actuality. *American Scientist* 58: 244–249.

Chapter 2

Fundamental Concepts

In this chapter we will discuss concepts basic to the understanding and study of environmental geology. Although these concepts probably do not constitute a complete list, they provide the philosophical framework of this book. They are not to be memorized. An understanding of the general thesis of each concept will be a significant help in comprehending and evaluating philosophical and technical material throughout the remainder of the text.

CONCEPT ONE

The earth is essentially a closed system.

A *system* may be considered any part of the universe that is isolated in thought or in fact for the purpose of studying or observing changes that take place under various imposed conditions (1). Examples of systems might include a planet, a volcano, or an ocean basin. Most systems contain various component parts which mutually adjust, and each part exerts partial control on the others. For example, the earth may be considered a system with four parts: the *atmosphere*; the *hydrosphere*; the *biosphere*; and the *lithosphere*. The mutual interaction of these parts is responsible for the surface features of the earth today. Furthermore, any change in

the magnitude or frequency or processes in one part will affect the other parts. The propensity for change in the various parts of the environment is known as the *principle of environmental unity*—that is, everything affects everything else. For example, a change in the magnitude of the processes that produce mountains may affect the atmosphere by releasing volcanic gas and causing regional changes in precipitation patterns as a new rain shadow is produced. This in turn affects the local hydrosphere as greater or less runoff reaches the ocean basins. Biospheric changes as a result of changes in the environment also can be expected, and eventually, the steeper slopes will also affect the lithosphere by facilitating increased erosion, which in turn will change the rate and types of sediments produced and, thus, the types of rocks produced from the sediments. These interactions among the variables in systems are not random and may be understood by examining each variable to determine how it interacts with other variables and how it varies spatially over a site, area, or region. In the hydrosphere, for example, the spatial distribution of the oceans with respect to the sun affects the evaporation process of ocean waters, which in turn affects atmospheric conditions by increasing or decreasing the amount of water in the atmosphere.

11

We know that the earth is not static; rather, it is a dynamic, evolving system in which material and energy are constantly changing. Such dynamics might be considered evidence that the earth is an open system with no boundaries of energy or material. This interpretation is applicable as long as the sun continues to impart energy to the earth. However, considering natural earth cycles such as the water and rock cycles in which there is a continual recycling of earth materials, we can best think of the earth as a closed system or, in reality, a coalition of a large number of closed systems (2). For example, the rain that falls today will eventually return to the atmosphere, and the sediment deposited yesterday will be transformed into solid rock. Therefore, although the earth is currently, and seemingly forever will be, an open system in terms of energy and material, it is essentially a closed system in terms of natural earth cycles.

As more and more demands are made on the earth and its limited resources, it becomes increasingly important for us to understand the magnitude and frequency of the processes that maintain earth cycles. For example, if we hope to manage the water resources of a region, we must know the nature and extent to which natural processes will supply groundwater and surface water. Or, if we are concerned with disposal of dangerous chemicals in a disposal well, we must know how the disposal procedure will interact with natural cycles to insure that we or our heirs will not be exposed to hazardous chemicals. This becomes especially critical in dealing with radioactive wastes which must be contained from several centuries to as long as a quarter million years. Therefore, it is exceedingly important to recognize earth cycles and determine the length of time involved in various parts of specific cycles. Tables 2.1 and 2.2 list the residence times of selected earth materials and rates of some natural processes.

CONCEPT TWO

The earth is the only suitable habitat we have, and its resources are limited.

The place of humanity in the universe is well stated in the *Desiderata*: "You are a child of the universe, no less than the trees and the stars; you have a right to be here. And whether or not it is clear to you, no doubt the universe is unfolding as it should" (3).

Leo F. Laporte, senior author of *The Earth and Human Affairs*, believes the context of Concept Two includes two fundamental truths: first, that this earth is indeed the only place to live that is now accessible to us; and second, that our resources are limited, and

TABLE 2.1
Residence times of some selected materials.

Earth Materials	Some Typical Residence Times
Atmosphere circulation	
Water vapor	10 days (lower atmosphere)
Carbon dioxide	5 to 10 days (with sea)
Aerosol particles	
Stratosphere (upper atmosphere)	Several months to several years
Troposphere (lower atmosphere)	One week to several weeks
Hydrosphere circulation	
Atlantic surface water	10 years
Atlantic deep water	600 years
Pacific surface water	25 years
Pacific deep water	1,300 years
Terrestrial groundwater	150 years [above 760 m depth]
Biosphere circulation[a]	
Water	2,000,000 years
Oxygen	2,000 years
Carbon Dioxide	300 years
Seawater constituents[a]	
Water	44,000 years
All salts	22,000,000 years
Calcium ion	1,200,000 years
Sulfate ion	11,000,000 years
Sodium ion	260,000,000 years
Chloride ion	Infinite

[a]Average time it takes these materials to recycle with the atmosphere and hydrosphere.

Source: *The Earth and Human Affairs* by the National Academy of Sciences. Copyright © 1972 by the National Academy of Sciences (Canfield Press). By permission of Harper & Row, Publishers.

TABLE 2.2
Rates of some natural processes.

Earth Processes	Some Typical Rates
Erosion	
Average U. S. erosion rate[a]	6.1 cm per 1,000 years
Colorado River drainage area	16.5 cm per 1,000 years
Mississippi River drainage area	5.1 cm per 1,000 years
N. Atlantic drainage area	4.8 cm per 1,000 years
Pacific slope (Calif.)	9.1 cm per 1,000 years
Sedimentation[b]	
Colorado River	281 million metric tons per year
Mississippi River	431 million metric tons per year
N. Atlantic coast of U.S.	48 million metric tons per year
Pacific slope (Calif.)	76 million metric tons per year
Tectonism	
Sea-floor spreading	
N. Atlantic	2.5 cm per year
E. Pacific	7 to 10 cm per year
Faulting	
San Andreas (Calif.)	1–5 cm per year
Mountain uplift	
Cajon Pass, San Bernardino Mts. (Calif.)	1 cm per year

[a]Thickness of the layer of surface of the continental United States eroded per 1,000 years.
[b]Includes solid particles and dissolved salts.
Source: *The Earth and Human Affairs* by the National Academy of Sciences. Copyright © 1972 by the National Academy of Sciences (Canfield Press). By permission of Harper & Row, Publishers.

while some resources are renewable, many are not. Therefore, we eventually will need large-scale recycling of many materials, and a large part of our solid and liquid waste-disposal problems could be alleviated if these wastes were recycled. In other words, many things that are now considered pollutants could be considered resources out of place.

There are at least two dichotomous views on natural resources. One school holds that finding resources is not so much a problem as is finding ways to use them. In other words, the entire earth, including the ocean and atmosphere, has raw materials that can be made useful if we can develop the necessary ingenuity and skill (4). The basic assumption is that as long as there is freedom to think and innovate, we will be able to produce sufficient energy and locate sufficient resources to meet our needs. There is evidence to support this line of reasoning: first, efficient and intelligent use of materials has historically been a successful venture; second, we know more about extracting minerals and fuel than we did in the past and so can find new resources faster and mine lower-grade mineral deposits; and third, new work with atomic power and recycling of resources can help us meet the needs of the future.

The second school holds that "cornucopian premises" such as that outlined above are fallacious on grounds that an exponential increase of people and mineral products on a finite resource base is impossi-

ble. Furthermore, Preston Cloud claims that we are in a resource crisis because of improvements in medical technology contributing to overpopulation of the earth; second, an unrealistic view of the necessity of an ever-increasing gross national product based on obsolescence and waste; third, the finite nature of the earth's accessible minerals; and fourth, increased risk of irreversible damage to the environment as a result of overpopulation, waste, and the necessity of larger and larger mining operations to obtain ever smaller proportions of useful minerals (5).

The history of Homo sapiens can be traced back only several million years. Geologically, this is a very short time. Dinosaurs, for example, ruled the land for more than 100 million years. Evidence from earth history suggests that more species have become extinct than have survived! What then will be our history, and who will write it? We can hope that we will be something more in the geologic record than a good index fossil indicating a brief time in earth history when the human race flourished.

CONCEPT THREE

Today's physical processes are modifying our landscape and have operated throughout much of geologic time. However, the magnitude and frequency of these processes are subject to natural and artificially-induced change.

The concept that understanding the present processes that form and modify our landscapes will facilitate the development of inferences concerning the geologic history of a landscape is known as the doctrine of *uniformitarianism*. Stated simply as "the present is the key to the past," uniformitarianism was first suggested by James Hutton in 1785, elegantly restated by John Playfair in 1802, and popularized by Charles Lyell in the early part of the nineteenth century. Today it is heralded as a fundamental concept of the earth sciences.

Uniformitarianism does not demand or even suggest that the magnitude and frequency of natural processes remain constant with time. Furthermore, it is obvious that the principle cannot be extended back throughout all of geologic time, because the processes operating in the oxygen-free environment of the first 2 billion years of earth history were quite different from today's processes. However, as long as past continents, oceans, and atmosphere were similar to those of today, we can infer that the present processes also operated in the past. For example, if we have studied present alpine glaciers and the characteristic erosional and depositional landforms associated with alpine glaciation, we can then infer that valleys with similar landforms were at one time glaciated even if no glacial ice is present today. Similarly, if one finds ancient gravel deposits with all the characteristics of stream gravel on the top of a mountain, then uniformitarianism can be used to suggest that a stream must have flowed there at one time. In other words, what was originally a stream valley has been changed by differential erosion and/or uplift to a mountain top, known as *inversion of topography*. Phenomena such as this would be difficult to determine correctly if it were not for the principle of uniformitarianism.

We must understand the effects of human activity on increasing or decreasing the magnitude and frequency of natural earth processes. For example, rivers will flood regardless of human activities, but the magnitude and frequency of flooding may be greatly increased or decreased because of human activities. Therefore, to predict the long-range effects of a certain process such as flooding, we must be able to determine how our future activities will change the rate of the process. In this case, the present may have to be the key to the future. We can assume that the same processes will operate but that rates will vary as the environment adjusts to human activity. Furthermore, we must conclude that ephemeral landforms such as beaches and lakes will appear and disappear in response to natural processes, and human influence may be small in comparison.

Although the effects of human activity may be small on a global scale, they are very pronounced in a local area. One year of erosion at a construction site may exceed many decades of erosion from an equivalent tract of woodland or even agricultural land (6). This erosion results from exposure of the soil following the removal of vegetation. Therefore, to maximize the value of geologic knowledge in land-use planning, we must be able to use our understanding of natural earth processes in both a historical and a predictive mode. For example, when environmental geologists examine recent mudflow (flowage of saturated heterogeneous debris) deposits in an area designated to become a housing development, they must use uniformitarianism to infer where there will be future mudflows, as well as to predict what effects urbanization will have on the magnitude and frequency of future flows.

CONCEPT FOUR

There have always been earth processes that are hazardous to people. These natural hazards must be recognized and avoided where possible, and their threat to human life and property must be minimized.

Our discussion of uniformitarianism established that present processes have been operating a good deal longer than humankind has been on the earth. Therefore, we have always been obligated to contend with processes that tend to make our lives difficult. Surprisingly, however, Homo sapiens appear to be a product of the Ice Age, one of the harshest of all environments.

Early in the history of the human race, its struggle with natural earth processes was probably a day-to-day experience. However, its numbers were neither great nor concentrated, and, therefore, losses from hazardous earth processes were not very significant. As people developed and learned to produce and maintain a constant food supply, both population and (we can guess) the effects of hazardous earth processes increased, because population centers probably became local centers of pollution and disease. The concentration of population and resources also increased the impact of periodic earthquakes, floods, and other natural disasters. This trend has continued until many people today live in areas that are likely to be damaged by hazardous earth processes or are susceptible to the adverse impact of such processes in adjacent areas.

Natural earth processes are *exogenetic* if they operate at or near the surface of the earth and *endogenetic* if they operate within or below the earth's crust. Exogenetic processes include weathering, mass wasting, and either erosion or deposition by such agents as running water, wind, or ice. Volcanic activity and diastrophism (processes that produce mountains, continents, ocean basins, and so on) are common endogenetic processes. The work of organisms, including people, are primarily exogenetic processes; however, we are now able to cause some endogenetic processes, such as earthquakes.

Many processes continue to cause loss of life and property damage, including flooding, earthquakes, volcanic activity, mass-wasting phenomena such as landslides and mudflows, and weathering. The magnitude and frequency of these processes depend on such factors as a region's climate, geology, and vegetation. For example, the effects of running water as an erosional or depositional process depend on the intensity of rainfall; the frequency of storms; how much and how fast the rainwater is able to infiltrate rock or soil; the rate of evaporation and transpiration of water back into the atmosphere; the nature and extent of the vegetation; and topography. Considering these factors, Peltier concluded that the present rate of erosion by running water is minimized in three areas: first, mid-latitude desert regions of low rainfall; second, arctic and subarctic regions of low rainfall; and third, tropical regions characterized by abundant plant cover (7). The areas of maximum erosion by running water are the mid-latitude regions with more abundant rainfall and less vegetation cover than is found in tropical regions (Figures 2.1 and 2.2). Similar reasoning can be used to determine areas of maximum weathering (Figure 2.3) and maximum mass wasting (Figure 2.4). The endogenetic processes associated with volcanoes, earthquakes, and tsunamis (very large sea waves, usually incorrectly referred to as *tidal waves*) are located primarily in re-

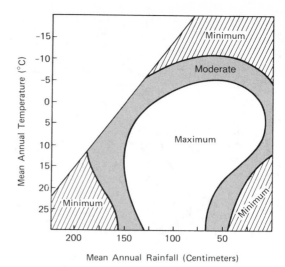

FIGURE 2.2
Intensity of erosion by running water. (After L. C. Peltier. Reproduced by permission from *ANNALS of the Association of American Geographers*, vol. 40, 1950.)

sponse to geologic conditions. For example, the "ring of fire" consisting of the circum-Pacific area is well known for its high incidence of earthquakes and volcanic activity resulting from dynamic, global geologic processes originating deep within the earth.

From our discussion of natural earth processes, we can conclude that many processes can be recognized and predicted by considering climatic, biologic, and geologic conditions. After earth scientists have identified potentially hazardous processes, they should

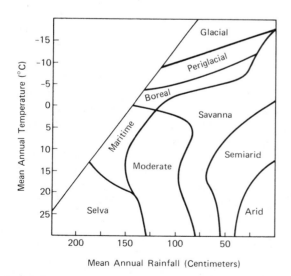

FIGURE 2.1
Morphogenetic regions. (After L. C. Peltier. Reproduced by permission from *ANNALS of the Association of American Geographers*, vol. 40, 1950.)

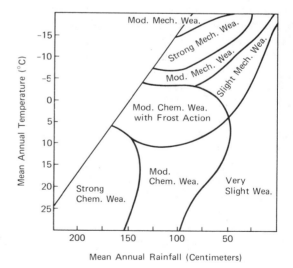

FIGURE 2.3
Intensity of weathering. (After L. C. Peltier. Reproduced by permission from *ANNALS of the Association of American Geographers*, vol. 40, 1950.)

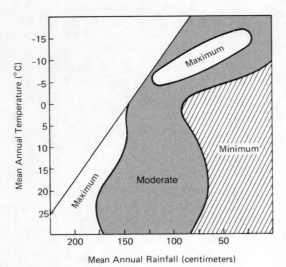

FIGURE 2.4
Intensity of mass wasting. (After L. C. Peltier. Reproduced by permission from *ANNALS of the Association of American Geographers*, vol. 40, 1950.)

make the information available to planners and decision makers who can then formulate various alternatives to avoid or minimize any threat to human life or property.

CONCEPT FIVE

Land- and water-use planning must strive to obtain a balance between economic considerations and the less tangible variables such as aesthetics.

Scenery is now considered a natural resource, and the aesthetic evaluation of a site or landscape before modification has become an important part of the "environmental impact" statement. We find it refreshing that consideration is now given to the less tangible variables such as aesthetics, as well as the more traditional benefits-cost analysis. Until this revolution, justification for a project was weighed by comparing the financial benefits over a period of time with the cost. The assumption now is that there are varying *scenic* values, just as there are varying economic values, associated with a landscape and proposed modification (8).

Most of the research on landscape aesthetics has been concerned with unique landscape nearly untouched by human activity. Since most of us now live in an urban environment, however, we should evaluate the aesthetic resources of these areas. Evaluation would facilitate land use by identifying various alternatives. Evaluation might temporarily interfere with the supply and demand aspects of urbanization, but the result might be worth the extra trouble.

Balancing economic criteria with aesthetic criteria is ambitious and optimistic as well as difficult. The problem is, on what basis can the two be compared? The one logical solution is a hierarchical ranking of the economic alternatives compared to a similar ranking of the aesthetic evaluation. Trade-offs and compromise can then be considered. Pragmatically, the problem of comparing economic considerations with aesthetic evaluation is not particularly difficult, provided we can develop a generally agreed-upon rating scale for aesthetic evaluation; develop a reliable quantitative method to analyze the data and hierarchically rank the alternatives; and develop techniques to map our scenic resources.

CONCEPT SIX

The effects of land use tend to be cumulative, and, therefore, we have an obligation to those who follow.

Several million years ago, when early hominids roamed the grasslands, marshy deltas, and adjacent forests of ancient Lake Rudoff along the Great Rift Valley system of East Africa, these prehistoric people were completely dependent upon their immediate environment. Their effect on that environment was probably insignificant as they hunted game and were in turn hunted by predators. This relationship between people and the environment probably existed until about 800,000 years ago, when they developed skill in the use of fire.

The use of fire brought new effects that differed from earlier impacts on the environment. First, fire was capable of affecting large areas of forest or grasslands. Second, it was a repetitive process capable of damaging the same area at rather frequent intervals. Third, it was a rather selective process, in that certain species were locally exterminated while other species that exhibited a resistance to or rapid recovery from fire were favored (9). Early use of fire for protection and hunting probably had a significant effect on the environment, and as people became more and more dependent on an increasing variety of resources for clothing, lodging, and hunting, they also increased their capacity to observe and test the environment. This early experimentation probably led to the use of plants and primitive agriculture about 7,000 B.C.

The emergence of agriculture was the first instance of an artificial land use capable of modifying the natural environment. It also set the stage for the development of a more or less continuously occupied site or cluster of sites that introduced further modification of the environment, such as shelter for living space, primitive latrines, and protective barriers against predators and other people. Furthermore, these early sites probably became the first areas to experience pollution problems resulting from disposal of waste and soil erosion problems resulting from removal of indigenous vegetation (9). The innovation of agriculture also supplied the necessary nutrients for an increasing population that necessitated the clearing of additional land. This activity certainly influenced an area's ecological balance as some species were domesticated or cultivated and others were removed as pests, so it is not surprising to note that the increase in human population is paralleled by an increase in the number of extinctions among birds and mammals (Figure 2.5).

The significant point of the entire developmental process of the human race through time is that as cities and farms increase, demand for diversification of land use increases, and the effects tend to be cumulative with time (10). If this is the case, then from an ethical and moral standpoint, we need to examine the effects of land use in a historical framework, if only to insure that our children and their children can survive in the environment they inherit. This is especially critical since it has been determined that at least since the beginning of civilization, 6,000 years ago, and perhaps as far back as 15,000 years ago, the entire surface of the earth has been altered by human activity. In other words, little, if any, land can be considered original or untouched (2). Furthermore, our ability to cause further changes is increasing at a rapid rate. In defense of human activity, it can be noted that, compared to the energy of mountain building, volcanic activity, and erosion power of streams, human impact is small. It is not insignificant, however, especially in the large metropolitan areas where there are many negative aspects of urbanization.

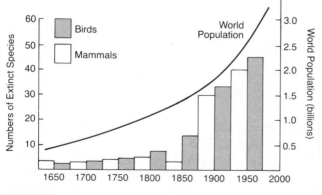

FIGURE 2.5
Increase in the human population paralleled by increase in the extinction of birds and animals. (Reproduced, by permission, from V. Ziswiler, *Extinct and Vanishing Species* [New York: Springer-Verlag, 1967].)

The lessons in land use from the Old World are explicit. Where sound conservation practices were used, there were successful adjustments of population; but, where wasteful exploitation of resources was practiced, the results varied from gullied fields and alluvial plains on rocky hills and steep slopes, to silted-up irrigation reservoirs and canals, to ruins of prosperous cities. Three examples from Lowdermilk's study will emphasize this point (10).

Ancient Phoenicia and Slope Farming

About 5,300 years ago, the Phoenicians migrated from the desert to settle along the eastern coast of the Mediterranean Sea and establish the coastal towns of Tyre and Sidon, Beyrouth and Byblos. The land is mountainous, with a relief of about 3,000 meters, and at that time was heavily forested with the famous cedars of Lebanon. These trees became the timber supply for the alluvial plains of the Nile and Mesopotamia. As the limited, flat land along the coast was populated to its carrying capacity, people moved to the slopes, and as the slopes were cleared and cultivated, they were subject to soil erosion. Today a large number of terrace walls, in various states of repair, indicate that the ancient Phoenician farmers attempted to control erosion with rock walls across the slope as many as 40 to 50 centuries ago.

The cedars of Lebanon retreated under the ax, until today very little of the original forest of approximately 2,600 square kilometers remains. Evidence suggests that given present climatic conditions, the forests would grow where soil has escaped the process of erosion. Today the bare limestone slopes strewn with remnants of former terrace walls are testimony to the results of erosion and the decline and loss of a country's resources.

Dead Cities of Syria

In northern Syria are a number of formerly prosperous cities that are now practically dead. Before the invasion of the Persians and Arabs, cities such as El Bare prospered from the conversion of forest to farmland, which resulted in exports of olive oil and wine to Rome. After the invasion, which destroyed the agriculture, as much as two meters of soil eroded from the slopes, and today the 13 centuries of neglect can be seen in the nearly complete destruction of the land. What is left of the formerly productive land is an artificial desert generally lacking vegetation, water, and soil.

Palestine Story

The Promised Land described by Moses on Mount Neba approximately 3,000 years ago was a land of

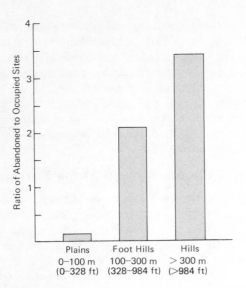

FIGURE 2.6
Relationship between abandoned village sites and topography in the Wadi Musrara watershed. (Data from W. C. Lowdermilk, *Lessons from the Old World to the Americans in Land Use* [Smithsonian Report, 1943, pp. 413–28].)

streams and springs, a land of wheat, barley, and vines, a land of abundant resources. The Promised Land at the time of Lowdermilk's investigation was a sad commentary on human use of the land. Soil erosion and accelerated runoff from barren slopes had caused many of the hills to become greatly depopulated. The Wadi Musrara watershed, which drains the western slope from Jerusalem to Tel Aviv, can be divided into three altitude zones: plains (0–100 m); foothills (100–300 m); and hills (over 300 m). The data in Figure 2.6 show the reduction in population of ancient village sites with topography since the seventh century. The breakdown of old terrace walls on steeper slopes and subsequent erosion of the soil are probably sufficient to explain the greater frequency of abandoned sites in the hills.

Although the land in Palestine cannot be restored to its original productivity, efforts to redeem the land in recent years show what can be done at great cost. Lowdermilk reported that as of 1960, Israel had more than doubled its cultivated land to about 4,000 square kilometers, drained 178 square kilometers of marshland, and established range cover on vast amounts of uncultivated land for its livestock industry. The results of their efforts are significant in that the country had at that time nearly attained agricultural self-sufficiency with an export-import balance in food (11).

Histories of long populated areas suggest that soil erosion is a serious problem that has destroyed land and retarded the progress of civilization. Therefore,

conservation of our soils must remain a national interest, for, as stated by Lowdermilk, "One generation of people replaces another, but productive soils destroyed by erosion are seldom restorable and never replaceable."

Potential for Catastrophe and Future Generations

A *catastrophe* is an event that causes great loss of life, suffering, and/or destruction. Increased population with stress on global resources in developing countries, along with industrialization and the arms race in others, is increasing the potential for and chances of a human-induced regional or even global catastrophe. On one front are problems of overgrazing, deforestation, and desertification; on another are hazardous chemical wastes and acid precipitation. However, the expected magnitude of disruption from these, while large, seems almost insignificant compared to the projected environmental consequences of large-scale nuclear war. The potential global impact of multiple nuclear explosions would be to produce a new "dark age" in human history. That event would destroy a significant part, if not all, of the biological and physical support systems necessary for civilization in the Northern Hemisphere. Those people, animals, and plants who were to survive the initial blast, fire, and radiation would be immediately subjected to the "nuclear winter," a period of extreme cold and darkness lasting several months or more that would severely alter global ecosystems for years (12, 13). The anticipated effects of nuclear war are so terrible that its prevention must be given top priority over all other potential environmental problems.

CONCEPT SEVEN

The fundamental component of every person's environment is the geologic factor, and understanding of this environment requires a broad-based comprehension and appreciation of the earth sciences and other related disciplines.

Concept Seven arises from the fact that all geology is environmental, and since we all live on the surface of the earth, we are both directly and indirectly affected by geologic processes (14). Therefore, an understanding of our complex environment requires considerable knowledge of such disciplines as *geomorphology*, the study of landforms and surface processes; *petrology*, the study of rocks and minerals; *sedimentology*, the study of environments of depositions of sediments; *tectonics*, the study of processes that produce continents, ocean basins, mountains, and other large structural features; *hydrogeology*, the study of surface and subsurface water; *pedology*, the study of soils; *economic geology*, the application of geology to locating and evaluating mineral materials; and *engineering geology*, the application of geologic information to engineering problems. Beyond this, the serious earth scientist should also be aware of the contributions to environmental research from areas such as biology, conservation, atmospheric science, chemistry, environmental law, architecture, and engineering, as well as physical, cultural, economic, and urban geography. Environmental geology is the domain of the generalist with strong interdisciplinary interest. This in no way refutes the significant contributions of specialists in various aspects of environmental studies, or the importance of the generalist's consideration of specific problems or specialty areas for research. It merely suggests that, although our research interest may be specialized, we should be generalists in terms of awareness of other disciplines and their contribution to environmental geology. Also, many projects may be studied best by an interdisciplinary team of scientists—but teams must be well disciplined.

The importance of the interdisciplinary nature of environmental geology becomes apparent when we explore the nature of environmental problems. Most projects are complex and involve many different facets that may be generalized into three categories: physical, bi-

ological, and of human use and interest. These are essentially the same categories used by Leopold to evaluate river valleys, and their extension to other areas of environmental research seems appropriate. (15). The category of physical factors includes such considerations as physical geography, geologic processes, hydrologic processes, rock and soil types, and climatology. Biologic factors include consideration of the nature of plant and animal activity, changes in biologic conditions or processes, and spatial analysis of biologic information. Human use and interest factors include such attributes as land use, economics, aesthetics, interaction between human activity and the physical and biological realms, and environmental law.

Obviously, no one project or research interest will involve all possible factors in each of the three categories, and there is considerable interaction among the categories. Projects such as waste-disposal operations, highway construction, mass transit systems, urban land-use planning, and mining of resources may be concerned with all of the categories. For example, the planning, construction, and operation of a sanitary landfill site is concerned with physical factors such as physical location, topography, soil type, and hydrologic conditions; biologic processes, which determine the rate of decay of the organic refuse, as well as any contamination of the biologic realm in the vicinity of the site; and human interests such as compliance with laws, regulations, and good engineering practice.

SUMMARY AND CONCLUSIONS

There are seven fundamental concepts basic to the understanding of environmental geology. The earth is essentially a closed system; the earth is the only suitable habitat we have, and its resources are limited; today's physical processes are modifying our landscape and have operated throughout much of geologic time, but the magnitude and frequency of these processes are subject to natural and artificially induced change. Earth processes that are hazardous to people have always existed. These natural hazards must be recognized and avoided where possible, and their threat to human life and property must be minimized. Land- and water-use planning must strive to obtain a balance between economic considerations and the less tangible variables such as aesthetics. The effects of land use tend to be cumulative, and, therefore, we have an obligation to those who follow. The fundamental component of every person's environment is the geologic factor, and understanding of this environment requires a broad-based comprehension and appreciation of the earth sciences and other related disciplines.

Although they probably do not constitute a complete list, these concepts establish a philosophical framework from which to investigate and discuss environmental geology.

REFERENCES

1 EHLERS, E. G. 1968. *Phase equilibria: Laboratory studies in mineralogy*. San Francisco: W. H. Freeman.

2 NATIONAL RESEARCH COUNCIL. 1971. *The earth and human affairs*. San Francisco: Canfield Press.

3 EHRMANN, M. 1927. *Desiderata*. Terre Haute, Indiana.

4 HOLMAN, E. 1952. Our inexhaustible resources. *Bulletin of the American Association of Petroleum Geologists* 6: 1323–29.

5 CLOUD P. E., JR. 1968. Realities of mineral distribution. In *Man and his physical environment*, ed. G. D. McKenzie, and R. O. Utgard, pp. 194–207. Minneapolis, Minnesota: Burgess.

6 WOLMAN, M. G., and SCHICK, A. P. 1967. Effects of construction on fluvial sediment, urban and suburban areas of Maryland. *Water Resources Research* 3: 451–64.

7 PELTIER, L. C. 1950. The geographic cycle in periglacial regions as it is related to climatic geomorphology. *Annals of the Association of American Geographers* 40: 214–36.

8 ZUBE, E. H. 1973. Scenery as a natural resource. *Landscape Architecture* 63: 126–32.

9 NICHOLSON, M. 1970. Man's use of the earth: historical background. In *Man's impact on environment*, ed. T. R. Detwyler, pp. 10–21. New York: McGraw-Hill.

10 LOWDERMILK, W. C. 1943. *Lessons from the Old World of the Americans in land use*. Smithsonian Report for 1943, pp. 413–28.

11 _____. 1960. The reclamation of a man-made desert. *Scientific American* 202: 54–63.

12 TURCO, R. P., TOON, O. B., ACKERMAN, T. P., POLLACK, J. B., and SAGAN, C. 1983. Nuclear winter: Global consequences of multiple nuclear explosions. *Science* 222: 1203–1292

13 EHRLICH, P. R., HARTE, J., HARWELL, M. A., RAVEN, P. H., SAGAN, C., WOODWELL, G. M., AYENSU, E. S., EHRLICH, A. H., EISNER, T., GOULD, S. J., GROVER, H. D., HERRERA, R., MAY, R. M., MAYR, E., MCKAY, C. P., MOONEY, H. A., MYERS, N., PIMENTAL, D., and TEAL, J. M. 1983. Long-term biological consequences of nuclear war. *Science* 222: 1293–1300.

14 OAKESHOTT, G. B. 1970. Controlling the geologic environment for human welfare. *Journal of Geological Education* 18: 193.

15 LEOPOLD, L. B. 1969. *Quantitative comparison of some aesthetic factors among rivers*. U.S. Geological Survey Circular 620.

Chapter 3

Earth Materials and Processes

GEOLOGIC CYCLE

Throughout the 4.5 billion years of earth history, the materials on or near the earth's surface have been created, maintained, and destroyed by numerous physical, chemical, and biochemical processes. Except during the early history of our planet, the processes that produce the earth materials necessary for our survival have periodically reproduced new materials. Collectively, the processes are referred to as the *geologic cycle* (Figure 3.1), which is really a group of subcycles (1). Two of the more important subcycles are the *tectonic* and *hydrologic* cycles.

The hydrologic cycle is the movement of water from the oceans, to the atmosphere, and back to the oceans, by way of evaporation, runoff in streams and rivers, and groundwater flow. Only a very small amount of the total water in the ocean is active in the hydrologic cycle at any one time, and yet this small amount of water is tremendously important in facilitating the movement and sorting of chemical elements in solution (geochemical cycle), sculpturing the landscape, weathering rocks, transporting and depositing sediments, and providing our water resources. Therefore, we should learn more about the hydrologic cycle to better understand and use this important resource.

Tectonic processes are driven by forces deep within the earth. They deform the earth's crust, producing external forms such as ocean basins, continents, and mountains. These processes are collectively known as the *tectonic cycle*. We now know that the outer layer of the earth, containing the continents and oceans, is about 100 kilometers thick. Called the *lithosphere*, it is not a continuous, uniform layer. Rather, the lithosphere is broken into several large parts called *plates* that move relative to one another (Figure 3.2) (2). As the lithospheric plates move over the asthenosphere, which is thought to be a more or less continuous layer of little strength below the lithosphere, the continents also move (Figure 3.3) (3). This moving of continents is called *continental drift*. It is believed that the most recent episode of drift started about 200 million years ago, when a supercontinent called *Pangaea* broke up.

The boundaries between plates are geologically active areas where most earthquakes and volcanic activities occur. The three types of boundaries are divergent, convergent, and transform fault (4). *Divergent boundaries* occur at oceanic ridges where plates are moving away from each other and producing new lithosphere. *Convergent boundaries (subduction zones)* occur when one plate dives beneath the leading edge

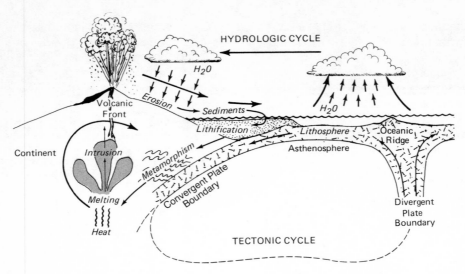

FIGURE 3.1
The geologic cycle is composed of subcycles, including the hydrologic and tectonic cycles.

of another plate. However, if both leading edges are composed of relatively light continental material, it is more difficult for subduction to start, and a special type of convergent plate boundary called a *collision boundary* may develop. This produces linear mountain systems such as the Alps and the Himalayas. *Transform fault* boundaries occur where one plate slides past another, as, for example, the San Andreas fault in California (Figure 3.2).

The importance of the tectonic cycle (Figure 3.1) or plate tectonics to environmental studies cannot be overstated. For the first time, earth scientists have a global, unifying theory to better explain how the earth works. The discovery of plate tectonics is to geology what Darwin's origin of the species was to biology. We now have an understanding, in the discovery of DNA, of the mechanism of biology. In geology we are still seeking the exact mechanism that drives plate tectonics. Nevertheless, everything living on the earth is affected by plate tectonics. As the plates slowly move a few centimeters per year, so do the continents and ocean basins, producing zones of resources (oil, gas, and minerals), earthquakes, and volcanoes. Furthermore, plate motion over long periods of time (millions of years) changes or modifies flow patterns in the oceans, influencing global climate and regional-local variation in precipitation, and thus the productivity of the land and where people wish to live.

Two other subcycles of the geologic cycle are the *geochemical* and *rock* cycles. Geochemistry is the study of the distribution and migration of elements in earth processes, and the geochemical cycle is the mi-

gratory path of elements during geologic changes. This cycle involves the chemistry of the lithosphere, asthenosphere, hydrosphere, and biosphere. The rock cycle (Figure 3.4) is a sequence of processes that produce the three rock families: igneous, sedimentary, and metamorphic. The geochemical and rock cycles are closely related to one another and intimately related to the hydrologic cycle which provides the water necessary for many chemical and physical processes. The tectonic cycle is also intimately related to the other cycles as it provides water from volcanic processes as well as heat and energy to form and change many earth materials.

Our discussion of the geologic cycle has established that earth materials such as minerals, rocks, soil, and water are constantly being created, changed, and destroyed by internal and external earth processes in numerous subcycles. In the remainder of this chapter we will discuss the different earth materials and various aspects of their occurrence and geology that have environmental significance.

MINERALS

Minerals are naturally occurring, solid, crystalline substances with physical and chemical properties that vary within known limits of selected minerals. (See color plate.) Although there are over 2,000 minerals, only a few are necessary to identify most rocks. Nearly 75 percent by weight of the earth's crust is oxygen and silicon. These two elements in combination with a few other elements (aluminum, iron, calcium, sodium, potassium, and magnesium) account for the chemical composition of minerals that make up about 95 percent of the earth's crust. Minerals that include the elements silicon and oxygen in their chemical composition are called *silicates*, and are the most abundant of the rock-forming minerals. The three most important rock-forming silicate minerals or mineral groups are quartz, feldspar, and ferromagnesium.

Quartz, one of the most abundant minerals in the crust of the earth, is a hard, resistant mineral composed of silicon and oxygen. It is often white, but, because of impurities, may also be rose, purple, black, or another color. It is usually recognized by its hardness, which is greater than that of glass, and the characteristic way it fractures—conchoidally (like a clam shell). Because it is very resistant to natural processes that lead to the breakdown of most minerals, quartz is the common mineral in river and most beach sands.

Feldspars, the most abundant and perhaps the most important group of rock-forming minerals in the crust of the earth, are aluminosilicates of potash, soda, and lime. They are generally white, gray, or pink and are fairly hard. Feldspars are very abundant and are important commercial minerals in the ceramics and glass industries. Feldspars weather chemically to form clays, which are hydrated aluminosilicates. This process has important environmental implications. All rocks are fractured, and water, which facilitates chemical weathering, enters the fractures. The feldspars and other minerals then weather. If clay forms along the fracture, it can greatly reduce the strength of the rock and increase the chance of a landslide.

Ferromagnesians are a group of silicates in which the silicon and oxygen combine with iron and magnesium. These are the dark minerals in most rocks. They are not very resistant to weathering and erosional processes. Thus, they tend to be altered or removed relatively quickly. Because they weather quickly to oxides such as limonite (rust), clays, and soluble salts, these minerals, when abundant, may produce weak rocks. Caution must be used when evaluating construction sites, such as for highways, tunnels, and reservoirs, that contain rocks that are high in ferromagnesians.

Other groups of minerals important in environmental studies include the oxides, carbonates, sulfides, and native elements.

Iron and aluminum are probably the most important metals in our industrial society. The most important iron ore (hematite) and the most important aluminum ore (bauxite) are both oxides. Magnetite, also an iron oxide but economically less important than hematite, is common in many rocks. It is a natural magnet (lodestone) that will attract and hold iron particles.

Where particles of magnetite are abundant, they may produce a black sand in streams or beach deposits.

Environmentally, the most important carbonate mineral is calcite, which is calcium carbonate. This mineral is the major constituent of limestone and marble, two very important rock types. Weathering by solution of this mineral from such rocks often produces *caverns, sinkholes* (surface pits), and other unique features. Cavern systems carry groundwater, and water pollution problems in urban areas over limestone bedrock are well known. In addition, construction of highways, reservoirs, and other engineering structures are a problem where caverns or sinkholes are likely to be encountered.

The sulfide minerals, such as pyrite, or iron sulfide (fool's gold), are sometimes associated with environmental degradation. This occurs most often when roads, tunnels, or mines cut through rocks such as coal containing sulfide minerals. The minerals oxidize to form compounds such as ferric hydroxide and sulfuric acid. This is a major problem in the coal regions of Appalachia.

Native elements such as gold, silver, copper, and diamonds have long been sought as valuable minerals. They usually occur in rather small accumulations but occasionally are found in sufficient quantities to justify mining. As we continue to mine these minerals in ever lower-grade deposits, the environmental impact will continue to increase.

ROCKS

Rocks are aggregates of one or more minerals (see color plate), and the rock cycle is the largest of the earth cycles. For this cycle to operate, the tectonic cycle is required for energy, the geochemical cycle for materials, and the hydrologic cycle for water. Water is used in the processes of weathering, erosion, transportation, deposition, and lithification of sediments.

The general model of the rock cycle is that the three rock families—igneous, metamorphic, and sedimentary—are involved in a worldwide recycling process. The elements of the cycle are shown in Figure 3.4. Internal heat from the tectonic cycle drives the rock cycle and with crystallization produces igneous rocks from molten materials. These may crystallize beneath or on the earth's surface. Rocks at or near the surface break down chemically and physically by weathering processes to form sediments that are transported by wind, water, and ice. The sediments accumulate in depositional basins, such as the ocean, where they are eventually transformed into sedimentary rocks. After the sedimentary rocks are buried to

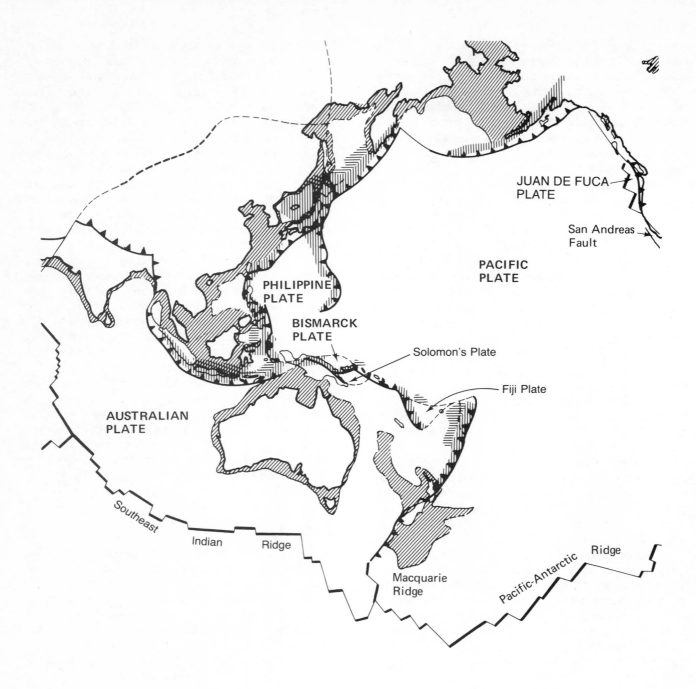

FIGURE 3.2

Lithospheric plates that form the earth's outer layer. Three types of plate junctions are
shown: *oceanic ridges*, forming divergent boundaries; *subduction zones*, forming convergent plate boundaries; and more rarely, *transform fault plate boundaries*, such as the San
Andreas Fault in California, where one plate is sliding by another. (From "Plate Tectonics"
by John F. Dewey. Copyright © May 1972 by Scientific American, Inc. All rights reserved.)

24

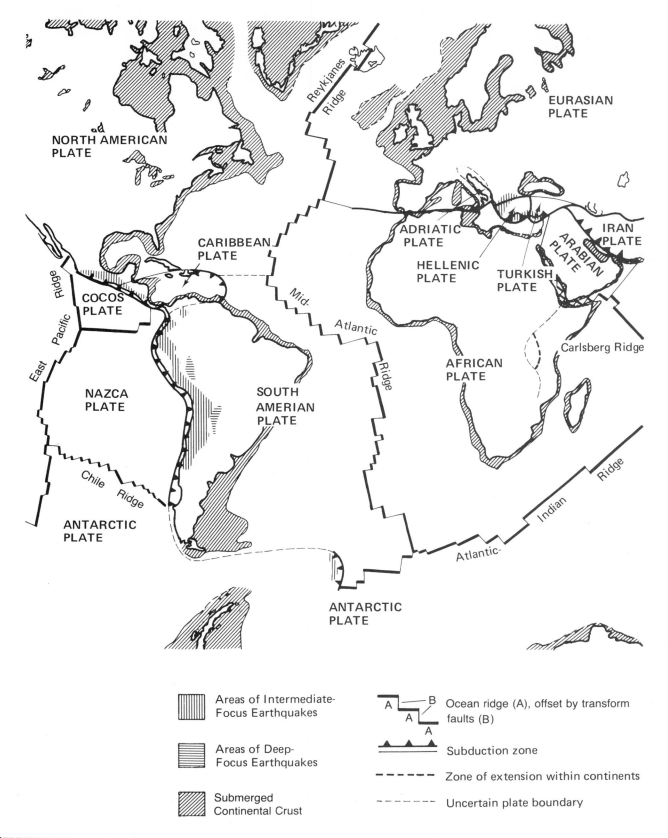

NORTH AMERICAN PLATE

Reykjanes Ridge

EURASIAN PLATE

CARIBBEAN PLATE

ADRIATIC PLATE

HELLENIC PLATE

TURKISH PLATE

ARABIAN PLATE

IRAN PLATE

COCOS PLATE

East Pacific Ridge

NAZCA PLATE

SOUTH AMERIAN PLATE

Mid-Atlantic Ridge

AFRICAN PLATE

Carlsberg Ridge

Chile Ridge

ANTARCTIC PLATE

Indian Ridge

Atlantic-

ANTARCTIC PLATE

Areas of Intermediate-Focus Earthquakes

Areas of Deep-Focus Earthquakes

Submerged Continental Crust

Ocean ridge (A), offset by transform faults (B)

Subduction zone

Zone of extension within continents

Uncertain plate boundary

(FIGURE 3.2, continued)

25

FIGURE 3.3

Idealized diagram showing the model of sea floor spreading which is thought to drive the movement of the lithospheric plates. New lithosphere is being produced at the oceanic ridge (divergent plate boundary). The lithosphere then moves laterally and eventually returns down to the interior of the earth at a convergent plate boundary (subduction zone). This process produces ocean basins and provides a mechanism that moves continents.

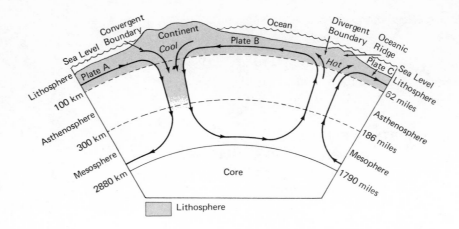

sufficient depth, they may be altered by heat, pressure, or chemically active fluids to produce metamorphic rocks, which may then melt to begin the cycle again.

Possible variations of the idealized sequence are indicated by the arrows in Figure 3.4. For example, either igneous or metamorphic rocks may be altered by metamorphism into a new metamorphic rock without ever being exposed to weathering or erosion processes.

The tectonic cycle provides several environments for the rock cycle (Figure 3.1). Generally, new lithosphere is produced at the oceanic ridges. This material moves laterally away from the ridges, which are divergent plate boundaries, until it dives beneath another plate to form a convergent plate boundary. These environments and associated rock-forming processes are shown in Figure 3.1. Transform fault and collision boundaries also produce rock-forming processes that fit in the generalized rock cycle. Recycling of rock and mineral material is the most important aspect of the

rock cycle, while the processes that drive and maintain the cycle are important in determining the properties of resulting rocks. Therefore, interest is more than academic because it is upon these earth materials that we build our homes, industries, roads, and other structures. Understanding of the various aspects of this cycle, such as the igneous-rock-forming processes or the soil-forming processes, facilitates the best use of these resources.

Other aspects of the rock cycle are responsible for concentrating as well as dispersing materials. These aspects are extremely important in mining minerals. If it were not for igneous, sedimentary, and metamorphic processes that concentrate minerals, it would be difficult indeed to extract resources. Thus, we take resources that are concentrated by one aspect of the rock cycle, transform these resources through industrial activities, and then return them to the cycle in a diluted form where they are further dispersed by continuing earth processes (5). Once this process has taken place, the resource cannot be concentrated again within a useful frame of time. Take, for example, the lead in automobile fuel. Lead is mined in a concentrated form; transformed and diluted in fuel; and further dispersed by traffic patterns, air current, and other processes. Eventually, lead may become abundant enough to contaminate soil and water but never sufficiently concentrated to be recycled. Similar examples may be cited for many other resources used in paints, solvents, and other industrial products.

Important aspects of human use of earth materials produced by the rock cycle include properties that affect engineering design. Therefore, the terminology of applied geology is somewhat different from that used in traditional geologic investigations. For example, the term *rock* can be reserved for earth materials that cannot be removed without blasting, and *soil* can be defined as those earth materials that can be excavated with normal earth-moving equipment. Therefore, a very

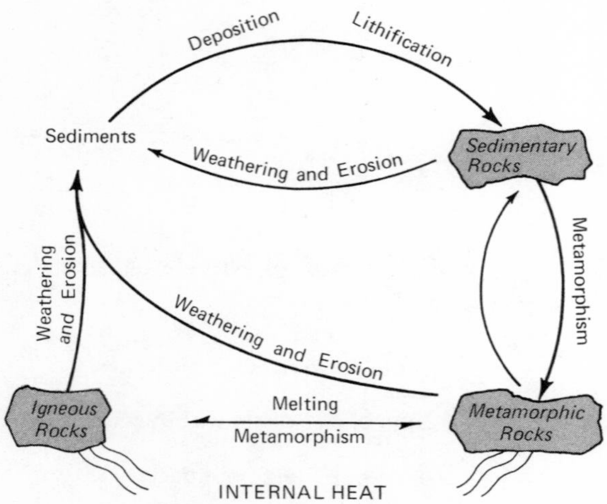

FIGURE 3.4

Idealized diagram showing the rock cycle.

friable (loosely compacted or cemented) sandstone may be considered a soil, but a well-compacted clay may be called a rock. Although confusing at first, this pragmatic approach has advantages over conventional terminology in that more useful information is conveyed to planners and designers. Clay, for example, is generally unconsolidated and easily removed. Thus, a contractor might assume, without further information, that clay is a soil and bid low for an excavation job on the belief that it can be removed without blasting. If the clay turns out to be well compacted, however, he may have to blast it, which is much more expensive. This kind of error is avoided if a proper preliminary investigation is made and the contractor forewarned that such a clay should be considered rock.

Strength of Rocks

The strength of earth materials varies with composition, texture, and location. Thus, weak rocks, such as those containing many altered ferromagnesium minerals, may creep or nearly flow under certain conditions and be very difficult to tunnel through. On the other hand, granite, a very common igneous rock, is generally a strong rock that needs little or no support. If granite were placed deep within the earth, however, it might also flow and be deformed. Hence, the strength of a rock may be quite different under different types and different amounts of stress.

Stress is defined as a force per unit area generally measured in kilograms (force) per square centimeter, or kgf/cm^2 (previously, pounds per square inch or lb/in^2). Three types of stress, *compressive, tensile,* and *shear,* are shown in Figure 3.5. *Strain* is defined as deformation induced by a stress. A material under stress may deform in an *elastic* or a *plastic* manner. Elastic deformation is like a rubber band; that is, the deformed material returns to its original shape after the stress is removed. Plastic deformation is characterized by permanent strain; that is, the material does not return to its original shape after the stress is removed. Materials that rupture before any plastic deformation are *brittle,* and those that rupture after elastic and plastic deformation are *ductile* (Figure 3.6) (6).

The strength of an earth material is usually described as the *compressive, shear,* or *tensile stress* necessary to break a sample of it. We must remember, however, that rocks are neither homogeneous nor isotropic, and both conditions are necessary before completely reliable strength values can be assigned to any material. Thus, we assign a safety factor (SF) to insure that a rock will perform as expected; that is, the rock is loaded only to a fraction of its assumed strength. For example, if a rock has a compressive strength of 700

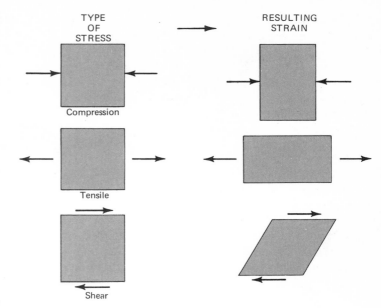

FIGURE 3.5
Common types of stress and resulting strain.

kgf/cm^2, then with a safety factor of 10, the rock should be loaded only to 70 kgf/cm^2. The testing necessary to establish strengths of rocks can be extremely expensive, so this information is usually obtained only for large engineering structures. For small structures, the experienced earth scientist can render a judgment consistent with good engineering practice.

Avoiding extensive testing to determine the strength of rocks at a site does not extend to a similar treatment of soils. The cost of a good soil survey to determine the strength characteristic of soils is well

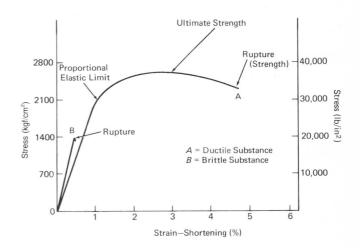

FIGURE 3.6
Typical stress-strain diagram. (After M.P. Billings, *Structural Geology* [New York: Prentice Hall, 1954]. Reproduced by permission.)

worth the initial cost compared to the possible damages and costs that can result from not considering these properties. Even small structures, such as an apartment building with a swimming pool and retaining wall, are subject to considerable structural damage (cracking of walls or steps pulling away from the buildings) by differential settling on unstable soils. With proper use of a soil survey before buildings are designed, however, damage caused by adverse soil conditions can be minimized.

The strength of rocks is greatly affected by the frequency and orientation of fractures, joints, and shear zones, which can range in size from small hairline fractures to huge shear zones such as the San Andreas fault zone (Figure 3.7) which scars the California landscape for hundreds of kilometers. When large structures are planned, these shear zones and fracture systems in rocks must be considered. One active shear zone in the proposed immediate vicinity of the foundation of a large masonry dam may well be sufficient reason to look for another site. It has been concluded that the Baldwin Hills Reservoir failure in 1963 in Southern California, which claimed five lives and caused $11 million in damages, resulted from gradual movement along shear zones (faults). The movement fractured the tile drain, foundation, and asphaltic membrane of the dam and reservoir floor. After the reservoir had drained, long cracks in the asphaltic concrete pavement were observed. The cracks extended continuously the length of the reservoir and in line with the deep gash in the dam (Figure 3.8). The dam was 71 meters high and 198 meters long. When construction began in 1947, it incorporated the most advanced knowledge available. The fracture, along with several others, was discovered during construction and was examined and judged not dangerous to the stability of the dam. Because of a previous experience with shear zones (the 1928 failure of the St. Francis dam), the design of the Baldwin Hills dam included some provisions for the fractures in the foundation material. In addition, a series of periodic inspections was initiated to insure the safety of the operation. Even with these precautions, however, the failure came quite suddenly. Were it not for fortunate circumstances that gave several hours warning and enabled officials to act quickly, the loss of life might have been much greater (7).

The problems and dangers associated with fractures, shear zones, and faults are not confined to "active" systems. Once a fracture has developed, it is subject to weathering, which may produce clay minerals. These minerals may be unstable and may become a lubricated surface facilitating landslides, or the clays may wash out and create open conduits for water to move through the rocks. Therefore, although active faulting is obviously a hazard to the stability of engineered structures, even small, inactive fractures, faults, and other shear zones must be inspected and carefully evaluated to maximize structural safety.

Types of Rocks

Our discussion of the generalized rock cycle established that there are three rock families: igneous, sedimentary, and metamorphic. Table 3.1 lists the com-

FIGURE 3.7
Aerial view of the San Andreas fault zone in the vicinity of Point Reyes, California. (Photo courtesy of the California Division of Mines and Geology.)

(a)

(b)

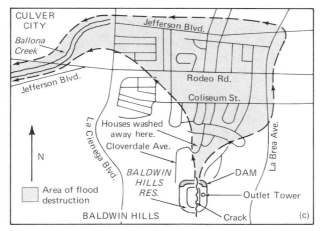

FIGURE 3.8
Failure of the Baldwin Hills Reservoir, aerial and ground views (a, b) and map showing
area flooded (c). (Photograph [a] courtesy of the California Department of Water Resources.
Drawing [c] from Walter E. Jessup, "Baldwin Hills Dam Failure." Reprinted with permission from the February 1964 issue of *Civil Engineering—ASCE*, official monthly publication
of the American Society of Civil Engineers, vol. 34, 1964. Photo [b] courtesy of Los Angeles
Department of Water and Power.)

mon rock types. Although rocks are primarily classified according to both mineralogy and texture, texture is most significant in environmental geology. Rock texture—the size, shape, and arrangement of grains—along with fractures, shear zones, and faults, determines the strength and utility of rock. This should not

be interpreted to mean that the mineralogy is not important. On the contrary, in specific cases it can be extremely significant. In considering groups of rocks such as the intrusive igneous rocks, however, mineralogy is of secondary importance to the texture and structure. Granitic rocks, for example, can be given a number of different names based upon their mineralogy, but the engineering properties of all of them will be nearly the same. More important in practical applications is the texture and nature of fractures and related alteration zones.

Igneous rocks are rocks that have crystallized from a naturally occurring, mobile mass of quasi-liquid earth material known as *magma*. If magma crystallizes below the surface of the earth, the resulting igneous rock is called *intrusive*. Therefore, whenever intrusive igneous rocks such as granite are exposed at the surface of the earth, we may conclude that erosional processes have removed the original cover material. Magma is probably generated in the upper asthenosphere or the lithosphere. As it moves up, it displaces the rock it intrudes, often breaking off portions of the rocks into which the magma moves, and incorporates them into the mass of moving magma. These foreign blocks, known as *inclusions*, are good evidence of forcible intrusion. A few varieties of granite, the most common intrusive igneous rock, do not show evidence of being intruded and may form in place by a metasomatism that involves replacement rather than magmatic processes.

Whether metasomatic or magmatic, intrusive igneous rocks are generally strong and have a relatively high unconfined compressive strength varying from about 700 to 2,800 kgf/cm^2 (8). The strength of an individual granite depends upon the grain size and the degree of fracturing. In general, fine-grained granites are stronger than coarse-grained. Fresh, unweathered intrusive rocks, unless extensively fractured, are usually satisfactory for all types of engineering construction and operations (9). If they can be economically quarried, they are often good sources of construction material—crushed stone for concrete aggregate, dimension stone for facing or veneer, broken stone for fill material, or riprap to protect slopes against erosion. Even with the general favorable physical properties of intrusive rocks, however, it is not safe to assume that

TABLE 3.1
Common rocks (engineering geology terminology)

Type	Texture	Materials
Igneous		
Intrusive		
Granitic[a]	Coarse[b]	Feldspar, quartz
Ultrabasic	Coarse	Ferromagnesians, ±quartz
Extrusive		
Basaltic[c]	Fine[d]	Feldspar, ±ferromagnesians, ±quartz
Volcanic breccia	Mixed—coarse and fine	Feldspar, ±ferromagnesians, ±quartz
Welded tuff	Fine volcanic ash	Glass, feldspar, ±quartz
Metamorphic		*Parent Material*
Foliated		
Slate	Fine	Shale or basalt
Schist	Coarse	
Gneiss	Coarse	Shale, basalt, or granite
Nonfoliated		
Quartzite	Coarse	Sandstone
Marble	Coarse	Limestone
Sedimentary		*Materials*
Detrital		
Shale	Fine	Clay
Sandstone	Coarse	Quartz, feldspar, rock fragments
Conglomerate	Mixed—very coarse and fine	Quartz, feldspar, rock fragments
Chemical		
Limestone	Coarse to fine	Calcite, shells, calcareous algae
Rocksalt	Coarse to fine	Halite

[a]Textural name used by engineers for a group of coarse-grained, intrusive igneous rocks, including granite, diorite, and gabbro.
[b]Individual mineral grains can be seen with naked eye.
[c]Textural name used by engineers for a group of fine-grained, extrusive igneous rocks, including rhyolite, andesite, and basalt.
[d]Individual mineral grains cannot be seen with naked eye.

all these rocks are satisfactory for their intended purpose. This is especially true for large, heavy structures whose stability depends upon a solid, safe foundation. In this case, an entire project may be jeopardized when the site is studied and evaluated, as is good engineering practice, and a shear zone or an alteration zone is identified. Therefore, although most intrusive rock *is* satisfactory for nearly all purposes, one should not *assume* that it is. Field evaluation, including mapping, drilling, and laboratory tests, is necessary before large structures are designed.

Extrusive igneous rocks form when magma reaches the surface and is blown out of a volcano as pyroclastic debris or flows out as lava. In either case, the resulting igneous rock is extrusive. The variety and composition of extrusive rocks are considerable, and, therefore, generalizations concerning their suitability for a specific purpose are difficult. Again, when one is working with these rocks, careful field examination, including detailed mapping and drilling, along with laboratory testing, is always necessary before large engineering structures are designed (9).

Extrusive rocks crystallized from lava flows may be stronger than granite (with an unconfined compressive strength exceeding 2,800 kgf/cm^2) (8). However, if the lava flows are mixed with volcanic breccia (angular fragments of broken lava and other material) or with thick vesicular or scoriaceous zones (produced as gas escapes from cooling lava), then the strength of the resulting rocks may be greatly reduced. Furthermore, after cooling and solidifying, lava flows often exhibit extensive columnar jointing (Figure 3.9) which may lower the strength of the rock. Solidified lava flows also may have subterranean voids known as *lava tubes*, which may either collapse from the weight of the overlying material or carry large amounts of groundwater, either of which may cause problems during the planning, design, or construction phases of a project.

Pyroclastic debris ejected from a volcano produces a variety of extrusive rocks. The most common material is *tephra*, a comprehensive term for any clastic material ejected from a volcano. Volcanic ash consists of rock fragments and glass shards less than 4 mm in diameter. When it is compacted, cemented, or welded together, it is called *tuff*. Although the strength of a tuff depends upon how well cemented or welded it is, generally it is a soft, weak rock that may have an unconfined compressive strength of less than 350 kgf/cm^2 (8). Some tuff may be altered to a clay known as *bentonite*, an extremely unstable material. When it is wet, bentonite expands to many times its original volume. Pyroclastic activity also produces larger fragments which, when mixed with ash and cemented together, form *volcanic breccia* or *agglomerate*. The strength of this

FIGURE 3.9
Devil's Postpile. Characteristic columnar joints that have formed because of contraction during cooling of lava. (Photo by Cecil W. Stoughton, courtesy of the U.S. Department of the Interior, National Park Service.)

material may be comparable to that of intrusive rocks or may be quite weak, depending upon how the individual particles are held together. In general, however, volcanic breccia and agglomerates make poor concrete aggregates because of the large and variable amount of fine materials produced when the rock is crushed.

From the discussion of extrusive igneous rocks, and especially pyroclastic debris, we can see that planning, design, and construction of engineering projects in or on these rocks can be complicated and risky (9). This was tragically emphasized on June 5, 1976, when the Teton Dam in Idaho failed, killing 14 people and inflicting approximately $1 billion in property damage. The causes of the failure had strong geologic aspects; namely, highly fractured volcanic rocks over which the dam was constructed and highly erodible wind-deposited clay-silts used in construction of the dam interior or core. Open fractures in the volcanic rocks were probably not completely filled with a cement slurry (grout) during construction, and while the reservoir rwas filling, water began moving under the foundation area of the dam. When the moving water came into contact with the highly erodible material of the core, it quickly eroded a tunnel through the base of the dam, explaining the observed whirlpool several meters across that formed in the reservoir just prior to failure. In other words, development of a vortex of water draining out of the reservoir near the dam strongly suggested the presence of a subsurface tunnel of free-flowing water below the dam. The final failure of the dam came minutes later, sending a wall of water up to 20 meters high downstream, destroying homes, farms, equipment, animals, and crops along a 160-kilometer reach of the Teton and Snake Rivers.

Sedimentary rocks form when sediments are transported, deposited, and then lithified by natural cement, compression, or other mechanism. There are two types of sedimentary rocks: *detrital* sedimentary rocks, which form from broken parts of previously existing rocks; and *chemical* sedimentary rocks, which form from chemical or biochemical processes that remove material carried in chemical solution.

Detrital sedimentary rocks include shale, sandstone, and conglomerate. Of the three, shale includes about 50 percent of all sedimentary rocks and is by far the most abundant. Shale also causes considerable environmental problems, and the existence of shale is a red flag to the applied earth scientist.

There are two types of shale: *compaction* shale and *cementation* shale. Compaction shale is held together primarily by molecular attraction of the fine clay particles. It is a very weak rock with the following environmental problems. First, depending upon the type of clay (mineralogy), it may have a high potential to absorb water and swell. Second, depending upon the bonding between the depositional layers (bedding planes), it may have a high potential to slide even on gentle slopes. Third, it has a high potential to slake; that is, contact with water will cause the surface to break away and curl up. This problem seriously retards the making of a firm bond between the rock and engineering structures such as foundations of buildings and dams. Fourth, because of the elastic nature of clays, these rocks tend to rebound if stress conditions change, making the rock a very poor foundation material. Fifth, these rocks have an unconfined compressive strength as low as 1.8 kgf/cm^2.

Depending upon the degree and type of cementing material, cemented shales can be very stable, strong rock suitable for all engineering purposes; but the presence of shale is a danger sign, calling for a close look to determine the type and extent of the rock and its physical properties.

Sandstones and conglomerates are coarse-grained and make up about 25 percent of all sedimentary rock. Depending on the type of cementing material, these rocks may be very strong and stable for engineering purposes. Common cementing materials are silica, calcium carbonate, and clay. Of these, silica is the strongest; calcium carbonate tends to dissolve in weak acid; and clay may be unstable and tend to wash away. It is always advisable to evaluate carefully the strength and stability of cementing materials in the detrital sedimentary rocks.

Limestones make up about 25 percent of all sedimentary rocks and are by far the most abundant of the chemical sedimentary rocks. Limestone is commonly composed of the mineral calcite. Human use and activity generally do not mix well with limestone. Although it may have sufficient strength, it weathers easily to form subsurface cavern systems and solution pits. In such limestone areas, most of the streams may be diverted to subterranean routes. These routes can easily become polluted by contaminated runoff that enters the groundwater. The mineralogy of limestone is such that little or no natural purification of contaminated waters in contact with them occurs. In addition, construction may be hazardous in areas with abundant caverns and surface solution pits called *sinkholes*.

Another important chemical sedimentary rock is *rocksalt,* which is composed primarily of halite (sodium chloride). Rocksalt forms when shallow seas of lakes dry up. As water evaporates, a series of salts, one of which is halite, are precipitated. The salts may later be covered up with other types of sedimentary rocks as the area again becomes a center of deposition of sediments. It is from these sedimentary basins that salt is mined, often by solution mining, a process in which water is pumped down into the salt and flows or is pumped out again supersaturated with salt. The process leaves huge holes in the salt deposits. Because rocksalt is abundant, usually completely isolated from circulating groundwater, and located in areas with low earthquake risk, it has been suggested as an acceptable place to dispose of radioactive wastes. Rocksalt is abundant in Kansas, New York, Michigan, Utah, Colorado, New Mexico, Texas, and the Gulf Coast. Specific sites in some of these areas are being given serious consideration as disposal sites if pilot studies are successful (10).

Geologic structures, such as *folds* and *unconformities,* have important relationships to human interest and activity. Folds or bends (Figure 3.10) are produced when rocks are deformed under stress. Geologic events that fold rocks are usually associated with the tectonic cycle. Unconformities are buried erosion surfaces (Figure 3.11). Our present landscape is an erosion surface produced by the action of running water, wind, and other processes. If the sea were to rise and cover part of the landscape, sediments would be deposited over the surface and bury it, and an unconformity would be formed.

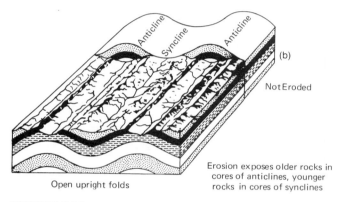

FIGURE 3.10
(a) Series of folds in the Rocky Mountains of southern British Columbia, and (b) idealized diagram showing the effects of erosion of folded rocks. (Photo courtesy of the Geological Survey of Canada.)

Folds and unconformities are important in environmental work for several reasons. First, they tend to influence the flow of groundwater in the rocks; that is, water usually migrates down the limbs of folds and along unconformities. Second, petroleum reservoirs are often found in folded sedimentary rocks and in association with unconformities. Third, folding produces typical fracture systems characterized by open tension fractures on the crest of anticlines and closed compression fractures in the troughs of synclines (Figure 3.12). This may be significant in the siting of reservoirs in areas of folded sedimentary rocks such as limestone, where solutional weathering enlarges the fractures. Synclines are likely to hold the water better than anticlines and, therefore, may be favored sites (9). Fourth, erosion of folded rocks may expose a variety of rock

types at different inclinations and hence affect the design of an engineering structure for that area.

Metamorphic rocks are changed rocks. Heat, pressure, and chemically active fluids produced in the tectonic cycle may change the mineralogy and texture of rocks, in effect producing new rocks. The two types of metamorphic rocks are *foliated,* those in which the elongated or flat mineral grains have a preferential parallel alignment or banding of light and dark minerals; and *nonfoliated,* those without preferential alignment or segregation.

Foliated metamorphic rocks, such as slate, schist, and gneiss, have a variety of physical and chemical properties, so it is difficult to generalize about the usefulness of such rocks for engineering projects. Slate is generally an excellent foundation material. Schist,

FIGURE 3.11
Unconformity (buried surface of erosion) near Morro Bay, California. The light rock is a steeply inclined shale that has been eroded and covered with more recent sediments that are nearly horizontal.

when composed of soft minerals, is a poor foundation material for large structures. Gneiss is usually a hard, tough rock suitable for most engineering purposes.

Foliation planes of metamorphic rocks are potential planes of weakness. The strength of the rock, its potential to slide, and the movement of water through the rock all vary with the orientation of the foliation. Consider, for example, the construction of road cuts and dams in terrain where foliated metamorphic rocks are common. In the case of road cuts, foliation planes inclined toward the road result in unstable blocks that might fall or slide downslope toward the road. Also, groundwater will tend to flow down the foliation, causing a drainage problem at the road. The preferred ori-

entation is with the foliation planes dipping away from the road cut (Figure 3.13). For construction of dams, the preferred orientation is with nearly vertical foliation planes parallel to the axis of the structure (Figure 3.14) (8). This position minimizes the chance of leaks and unstable blocks.

Important nonfoliated metamorphic rocks include *quartzite* and *marble*. Quartzite is a metamorphosed sandstone. It is a hard, strong rock suitable for many engineering purposes. The engineering properties of marble are similar to those of its parent rock (limestone), and cavern systems and surface pits should be expected.

On the night of March 12, 1928, more then 500 lives were lost and $10 million in property damage was done as ravaging flood waters raced down the San Francisquito Canyon near Saugus, California. This disaster did more than any previous event to focus public attention on the need for geologic investigation as part of siting reservoirs. The St. Francis Dam, 63 meters high with a main section 214 meters long and holding 47 million cubic meters of water, had failed. The cause was clearly geologic. Adverse geologic conditions (Figure 3.15) included these: first, the east canyon wall is metamorphic rock (schist) with foliation planes parallel to the wall. Before the failure, both recent and ancient landslides indicated the instability of the rock. Second, the rocks of the west side of the dam are sedimentary and form prominent ridges, suggesting that they were strong and resistant. Under semiarid conditions, this was true; when the rocks became wet, however, they

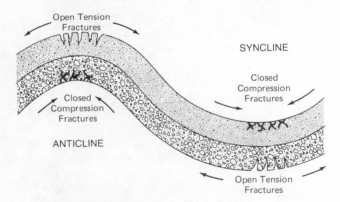

FIGURE 3.12
Idealized diagram showing an anticline and a syncline and the accompanying tension and compression fractures formed during folding.

One of the many clay minerals. When present in soils, clays may exhibit undesirable properties such as low strength, high water content, poor drainage, and high shrink-swell potential.

Mica minerals, biotite (black) and muscovite (light), both of which are important rock-forming minerals. It is unusual to find both together in one sample.

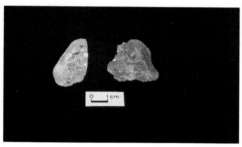

Clear and impure varieties of halite (salt). Halite is the main mineral in rocksalt, a potential host rock for high-level radioactive wastes.

Several blue and/or green copper minerals are present in this rock.

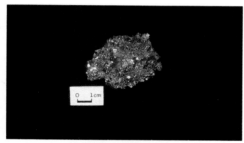

Pyrite (fool's gold) is iron sulfide, a common mineral associated with coal, which reacts with water and oxygen to form sulfuric acid.

Two varieties of feldspar, the most common rock-forming mineral in the earth's crust.

Two of the many varieties of quartz.

Calcite, the abundant mineral in limestone and marble. Limestone terrain is associated with caverns, sinkholes, subsidence, and potential water pollution and construction problems.

PLATE 1   Minerals

(Photographs by Ricardo Zepeda.)

Three samples of granite (intrusive igneous rock). White and pink mineral grains are feldspar, dark grains are ferromagnesian minerals, and quartz is clear or gray.

Basalt (extensive igneous rock). Note the fine-grain size compared to granite.

Two specimens of limestone (biochemical or chemical sedimentary rock). The sample on the left contains numerous fragments of shell material; the sample on the right is composed mostly of massive impure calcite.

Two samples of sandstone (detrital sedimentary rock). Red banding (left sample) and red color of the relatively coarse-grained sandstone (right sample) indicate the presence of iron oxides.

Serpentine, a weak rock that causes slope stability problems when encountered in excavations for roads or other structures.

Schist, or foliated metamorphic rock. Parallel alignment of mineral grains, in this case mica, demonstrates foliation.

Gneiss, a foliated metamorphic rock, with minerals segregated into white bands of feldspar and dark bands of ferromagnesian.

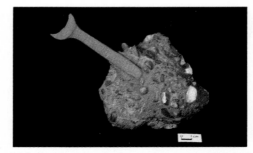

This iron-cemented conglomerate (detrital sedimentary rock) obviously formed quite recently.

PLATE 2   Rocks

(Photographs by Ricardo Zepeda.)

PLATE 3
This is the flood channel constructed after a flash flood through the small village of Lynmouth, England, killed 34 people and destroyed 93 houses in August, 1952. A flood wall protects the village on the left bank of the Lyn River and a rock berm with steps to the river protects the left bank. Tourists seeking recreation use the flat berm. A repeat of the 1952 event is unlikely.

PLATE 4
Landslides in Santa Barbara, California. Five shallow soil slips (top) developed during a period of heavy precipitation, and the large complex landslide that destroyed two homes (bottom) involved reactivation of an older slide in part because of urban runoff and erosion of the toe of the slope by waves. (Lower photograph courtesy of Don Weaver.)

PLATE 5
Large landslide associated with clear-cut timber harvesting in Oregon. The landslide started near a logging road (top right) and completely blocked Drift Creek, producing a small lake.

PLATE 6
San Gabriel Mountains, California. Stream erosion (bank failure) along Little Rock Creek has washed out this road.

PLATE 7
Homes in Anchorage, Alaska, destroyed by landslides associated with the 1964 earthquake. (Ward's Natural Science Establishment, Inc. photo.)

PLATE 8
Two large landslides near Carpinteria, California that occurred on slopes shortly after the land was converted from native vegetation to orchards.

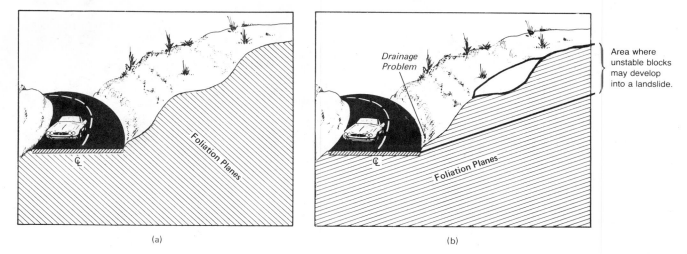

(a)

(b)

FIGURE 3.13
Idealized diagram showing two possible orientations of foliation in metamorphic rock and the effect on highway stability. Where the foliation is inclined away from the road (a), there is less likelihood that unstable blocks will fall on the roadway. Where the foliation is inclined toward the road (b), unstable blocks of rock above the road may produce a landslide hazard.

disintegrated. This characteristic was not discovered and tested until after the dam had failed. Third, the contact between the two rock types is a fault with approximately a 1.5-meter-thick zone of crushed and altered rock. The fault was on California's 1922 fault map, and was either not recognized or ignored. The causes of the dam failure were a combination of slip-ping of the metamorphic rock, disintegration and sliding of the sedimentary rock, and leakage of water along the fault zone that washed out the crushed rock (7,8). These three causes together destroyed the bond between the concrete and the rock and precipitated failure. This tragic event clearly illustrated the need to investigate carefully the properties of earth materials

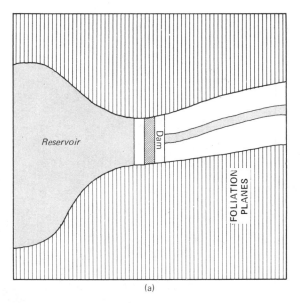

(a)

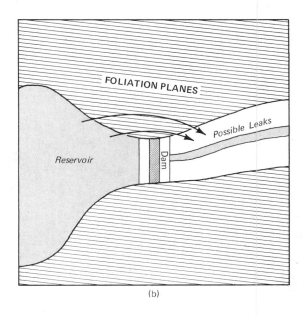

(b)

FIGURE 3.14
Idealized diagram showing two possible orientations of foliation in metamorphic rocks at a dam and reservoir site. The most favorable orientation of the foliation is shown in part (a), where the foliation is parallel to the axis of the dam. The least favorable orientation is where the foliation planes are perpendicular to the axis of the dam, as shown in (b).

FIGURE 3.15
St. Francis Dam (a) prior to failure; (b) geology along the axis of the dam; and (c) after failure. (Photos courtesy of Los Angeles Department of Water and Power.)

(a)

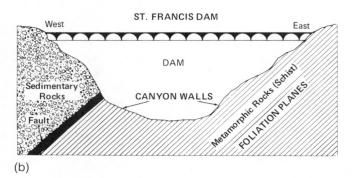

(b)

(c)

before construction of large engineering structures. Such investigations are now standard procedure.

SOILS

Intimate interactions of processes between the rock and hydrologic cycles produce weathered rock materials that are basic ingredients of soils. Weathering is the physical and chemical breakdown of rocks and the first step in the soil-forming process. The more insoluble weathered material may remain essentially in place and be further modified by organic processes to form a residual soil, as, for example, the red soils of the Piedmont in the southeastern United States. If the weathered material is transported by water, wind, or ice (glaciers), then deposited and further modified by organic processes, a transported soil forms. The fertile soils formed from glacial deposits in the Midwest are an example.

Soils may be defined in two ways. Soil scientists define soil as solid earth material that has been altered by physical, chemical, and organic processes, such that it can support rooted plant life. Engineers define soil as any solid earth material that can be removed without blasting. Both definitions are important in environmental studies.

Soil can be considered an open system. As such, soils are a function of climate, topography, parent material (material from which the soil is formed), maturity (time), and organic activity. Many of the differences we see in soils are effects of climate and topography, but the type of parent rock, the organic processes, and the amount of time the soil-forming processes have operated are also important.

Vertical and horizontal movement of material in the soil system produces a distinct soil layering or soil profile divided into zones or horizons (Figure 3.16). An important process is *leaching* (downward movement of soluble material) from the upper zone, called the *A horizon*, to the intermediate zone, or *B horizon*, where the leached material is deposited. In different climates, this may produce a "hardpan" layer of compacted clay, or calcium carbonate cemented *subsoil materials*, known as *caliche*. In addition, materials may also move upward in the soil in response to natural variations of the water table or when extensive irrigation raises the level of groundwater. When accompanied by extensive evaporation, this upward movement causes salts to move up and be deposited in the upper soil layers or on the surface of the ground. This happened in the Indus River Valley in Pakistan, where about 93,000 square kilometers were irrigated. The water table rose

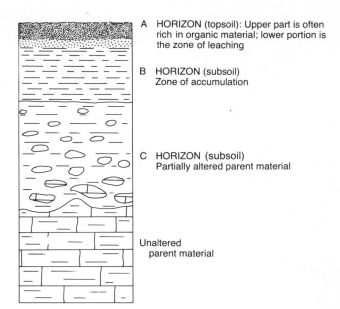

A HORIZON (topsoil): Upper part is often rich in organic material; lower portion is the zone of leaching

B HORIZON (subsoil) Zone of accumulation

C HORIZON (subsoil) Partially altered parent material

Unaltered parent material

FIGURE 3.16
Idealized diagram showing soil horizons and associated zones.

and extensive surface evaporation deposited salts, making the soil unsuitable for crops (11). The problem can be, and in some places has been, corrected by installing a drainage system for the land and then providing an abundance of water to leach the salts out. This procedure is carried out in the Great Valley of California, where salts are a problem.

Soil Classification

Terminology and classification of soils for environmental studies are a problem because in many cases we are interested in both the soil processes and the materials as they relate to human use and activity. A genetic classification that includes engineering properties is therefore most appropriate—but none exists. So, we will discuss the concept of *zonal soils*, which groups soils in a genetic mode, *Soil Taxonomy (the 7th approximation)*, and the *engineering classification*, which groups soils by material types and engineering properties.

The concept of zonal soils is based on the observation that many soil profiles are closely related to the climatic zone in which they are found. Hence, soils in which the profile is an adjustment to the climatic zone are called *zonal soils*. Recent surface materials such as sand dunes or flood deposits that do not have a distinctive soil layering are referred to as *azonal soils*. In addition, some soils have characteristics determined primarily by local conditions, such as excessive water as a result of flooding, high groundwater table or poor drainage, or perhaps an abundance of salts or lime. These soils are referred to as *intrazonal soils* (12).

Soil scientists have developed a comprehensive and systematic classification of soils known as *Soil Taxonomy* (informally called *the 7th approximation*). This classification is not as dependent on climate and vegetation as earlier schemes; rather, physical and chemical properties of the soil profile are emphasized by generic terminology. There is a six-fold hierarchy in the classification: Order, Suborder, Great Group, Subgroup, Family, and Series. There are ten Orders (Table 3.2) that are based on gross soil morphology, nutrient status, organic content, color, and general climatic considerations. With each step down the hierarchy, more information about a specific soil becomes known.

The 7th approximation is especially useful for agricultural and related land-use purposes, but it is too complex and lacks sufficient textural and engineering information to be of much use in specific site evaluation for engineering purposes. Nevertheless, it behooves the serious earth scientist to have knowledge of this classification because it is commonly used by soil scientists and Quaternary geologists.

The unified soil classification system is widely used in engineering practice. Because all natural soils are mixtures of coarse particles such as gravel and sand, fine particles such as silt and clay, and organic material, this classification is divided into these three groups. Each group is based on the predominant size or the abundance of organic material (Table 3.3). Coarse soils are those in which greater than 50 percent of the particles (by weight) are larger than 0.074 milli-

TABLE 3.2
General properties of soil orders used with the 7th approximation by soil scientists. (After Soil Survey Staff, 1960, *Soil Classification—A Comprehensive System*.)

Order	General Properties
Entisols	No horizon development (azonal); many are recent alluvium; synthetic soils are included; are often young soils.
Vertisols	Include swelling clays (greater than 35 percent) that expand and contract with changing moisture content. Generally form in regions with a pronounced wet and dry season.
Inceptisols	One or more of horizons have developed quickly; horizons are often difficult to differentiate; most often found in young but not recent land surfaces, have appreciable accumulation of organic material; most common in humid climates but range from the Arctic to Tropics; native vegetation is most often forest.
Aridisols	Desert soils; soils of dry places; low organic accumulation; have subsoil horizon where gypsum, caliche (calcium carbonate), salt, or other materials may accumulate.
Mollisols	Soils characterized by black, organic rich "A" Horizon (prairie soils), surface horizons are also rich in bases. Commonly found in semi-arid or subhumid regions.
Spodosols	Soils characterized by ash-colored sands over subsoil, accumulations of amorphous iron-aluminum sesquioxides and humus. They are acid soils that commonly form in sandy parent materials. Are found principally under forests in humid regions.
Alfisols	Soils characterized by: a brown or gray-brown surface horizon, an argillic (clay rich) subsoil accumulation with an intermediate to high base saturation (greater than 35 percent as measured by the sum of cations, such as calcium, sodium, magnesium, etc.). Commonly form under forests in humid regions of the mid-latitudes.
Ulfisols	Soils characterized by an argillic horizon with low base saturation (less than 35 percent as measured by the sum of cations); often have a red-yellow or reddish-brown color; restricted to humid climates and generally form on older landforms or younger, highly weathered parent materials.
Oxisols	Relatively featureless, often deep soils, leached of bases, hydrated, containing oxides of iron and aluminum (laterite) as well as kaolinite clay. Primarily restricted to tropical and subtropical regions.
Histosols	Organic soils (peat, muck, bog).

TABLE 3.3
Unified soil classification system used by engineers.

Major Divisions					Group Symbols	Soil Group Names
COARSE-GRAINED SOILS (Over half of material larger than 0.074 mm)	Gravels	Clean Gravels	Less than 5% fines		GW	Well-graded gravel
					GP	Poorly graded gravel
		Dirty Gravels	More than 12% fines		GM	Silty gravel
					GC	Clayey gravel
	Sands	Clean Sands	Less than 5% fines		SW	Well-graded sand
					SP	Poorly graded sand
		Dirty Sands	More than 12% fines		SM	Silty sand
					SC	Clayey sand
FINE-GRAINED SOILS (Over half of material smaller than 0.074 mm)	Silts	Non-plastic			ML	Silt
					MH	Micaceous silt
					OL	Organic silt
	Clays	Plastic			CL	Silty clay
					CH	High plastic clay
					OH	Organic clay
Predominantly Organics					PT	Peat and muck

meter in diameter. Fine soils are those with less than 50 percent of the particles greater than 0.074 millimeter (8). Organic soils have a high organic content and are identified by their black or gray color and sometimes by an odor of hydrogen sulfide, which smells like rotten egg. A useful way of estimating the size of the soil particles in the field is as follows: it is sand or larger (coarse soil) if you can see the individual grains; silt (fine soil) if you can see the grains with a 10X hand lens; and clay (fine soil) if you cannot see the grains with such a hand lens.

Engineering Properties of Soils

All soils above the water table have three distinct phases: the solid material, the liquid (mostly water), and the gas (mostly air). The usefulness of a soil is greatly affected by the variations in the proportions and structure of the three phases. The types of solid materials, the size of soil particles, and the water content are probably the most significant variables that determine engineering properties. For fine-grained soils, properties such as the *liquid limit* (LL) and the *plastic limit* (PL) are used to classify soils for engineering purposes. The liquid limit is defined by the water content at which the soil behaves as a liquid; and the plastic limit is defined by the water content below which the soil no longer behaves as a plastic material. The numerical difference between the liquid and plastic limits is the *plasticity index* (PI), the range in moisture content within which a soil behaves as a plastic material.

Soils greatly affect the determination of the best use of land, and a soil survey is an important part of planning for nearly all engineering projects. A soil survey should include a soil description; soil maps showing the horizontal and vertical extent of soils; and tests to determine grain size, moisture content, and strength. The purpose of the survey is to provide necessary information for identifying potential problem areas before construction.

The more important engineering properties of soils for planning are strength, sensitivity, compressibility, erodability, permeability, corrosion potential, ease of excavation, and shrink-swell potential.

Strength of soils is difficult to generalize, and numerical averages are often misleading, because earth materials are often composed of several different mixtures, zones, or layers of materials with different physical and chemical properties. The strength of a particular soil type is a function of cohesive and frictional forces. *Cohesion* is a measure of the ability of soil particles to stick together. The attraction of particles to each other in fine-grained soils is primarily the result of the presence of molecular and electrostatic forces acting between the particles, and is the significant factor determining the strength of the soil. The presence of moisture films between coarse grains of partly saturated soils may cause an apparent cohesion, explaining the ability of wet sand (which is cohesionless when dry) to stand in vertical walls in children's sand castles on the beach (Figure 3.17) (13). *Frictional forces* that contribute to the strength of a soil are the result of interactions among the soil's individual grains. The total frictional force is a function of the density, size, and shape of the soil particles as well as the weight of overlying particles forcing the grains together. Frictional

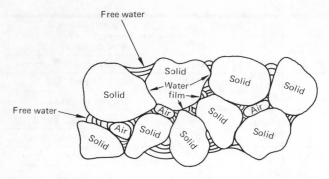

FIGURE 3.17
Partly saturated soil showing particle-water-air relationships. Particle size is greatly magnified. Attraction (apparent cohesion) between the water and soil particles (surface tension) develops a stress that holds the grains together. The cohesion is destroyed if the soil dries out or becomes completely saturated. (After R. Pestrong, *Slope Stability,* American Geological Institute, 1974.)

forces are most significant in the coarse-grained soils rich in sand and gravel. Since most soils are a mixture of coarse and fine particles, the strength is usually the result of a mixture of cohesion and internal friction. Although it is dangerous to generalize, clay and organic-rich soils to tend to have lower strengths than coarser soils.

Vegetation may also play an important role in soil strength. For example, tree roots may provide considerable apparent cohesion through the binding characteristics of a continuous root mat or by anchoring individual roots to bedrock beneath thin soils on steep slopes.

Sensitivity of soils reflects changes in the soil strength resulting from disturbances such as vibrations or excavations. Sand and gravel soils with no clay are the least sensitive. As fine material becomes abundant, soils become more and more sensitive. Some clay soils may lose 75 percent or more of their strength following disturbance (8).

Compressibility of soils is a measure of the tendency to consolidate, or decrease in volume. It is partly a function of the elastic nature of the soil particles and is directly related to settlement of structures (9). Excessive settlement will crack foundations and walls. Coarse materials such as gravels and sands have a low compressibility, and settlement will be considerably less than that of fine-grained or organic soils which have a high compressibility.

Erodability of soils refers to the ease with which soil materials can be removed by wind or water. Easily eroded materials (soils with a high erosion factor) include unprotected silts, sands, and other loosely consolidated materials. Cohesive soils, greater than 20 percent clay or naturally cemented soils, are not easily moved and therefore have a low erosion factor.

Permeability is a measure of the ease with which water moves through a material. Clean gravels or sands have the highest permeabilities and may transmit several hundred cubic meters per day per square meter ($m^3/day/m^2$). Mixtures of clean gravel and sand have permeabilities of about one hundred cubic meters per day per square meter. As fine particles in such a mixture increase, permeability decreases. Clays generally have a permeability of less than one tenth of a cubic meter per day per square meter.

Corrosion is a slow weathering or chemical decomposition that proceeds from the surface into the ground. All objects buried in the ground—pipes, cables, anchors, fenceposts—are subject to corrosion. The corrosiveness of a particular soil depends upon the chemistry of both the soil and the buried material, and on the amount of water available. It has been observed that the more easily a soil carries an electrical current (low resistivity), the higher its corrosion potential. Therefore, measurement of soil resistivity is one way to estimate the corrosion hazard (14).

Ease of excavation pertains to the procedures, and hence equipment, required to remove soils during construction. There are three general categories of excavation techniques. *Common excavation* is accomplished with an earth mover, backhoe, or dragline. This equipment essentially removes the soil without having to scrape it first. *Rippable excavation* requires breaking the soil up with a special ripping tooth or teeth before it can be removed. *Blasting* or *rock cutting* is the third, and often most expensive, category.

Shrink-swell potential in soil is its tendency to gain or lose water. Soils that tend to increase or decrease in volume with water content are *expansive.* The swelling is caused by chemical attraction and addition of layers of water molecules between the flat submicroscopic clay plates of certain clay minerals with high plasticity indices. The plates are primarily composed of silica, aluminum, and oxygen atoms, and layers of water are added between the plates as the clay expands or swells (Figure 3.18a) (15).

Expansive soils in the United States cause significant environmental problems and, as our most costly natural hazard, are responsible for more than $2 billion in damages annually to highways, buildings, and other structures—more than floods, earthquakes, hurricanes, and tornados combined. Every year more than 250,000 new houses are constructed on expansive soils. Of these, about 60 percent will experience some minor damage (cracks in foundation, walls, or walkways) (Figure 3.18b), but 10 percent will be seriously damaged—some beyond repair (16,17).

Montmorillonite is the common clay mineral associated with most expansive soils. With sufficient water, pure montmorillonite may expand up to 15 times

FIGURE 3.18
Expansive soils: idealized diagram showing an expansive clay (montmorillonite) as layers of water molecules are incorporated between clay plates on a microscopic scale (a); and effects of soil's shrinking and swelling at a home site (b). (After Mathewson and Castleberry, *Expansive Soils: Their Engineering Geology* [Texas A. & M University].) Cracked wall resulting from expansion of clay soil under the foundation (c). (Photo courtesy of U.S. Department of Agriculture.)

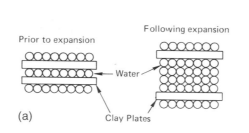

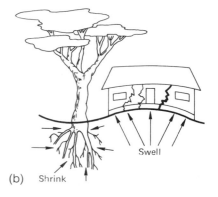

its original volume, but fortunately most soils contain limited amounts of the clay, so it is unusual for an expansive soil to swell beyond 25 to 50 percent. However, an increase in volume of greater than 3 percent is considered potentially hazardous (15).

Damages to structures on expansive soils are caused by volume changes in the soil in response to changes in moisture content. Factors that tend to determine the moisture content of an expansive soil at a site include climate, vegetation, topography, drainage, site control, and quality of construction (16). Regions in which most of the rainfall occurs in a pronounced wet season followed by a dry season (for example, the southwestern United States) are more likely to experience an expansive soil problem than regions where precipitation is more evenly distributed throughout the year. This results because the wet and dry seasonal pattern allows for a regular shrink-swell sequence. Vegetation can cause changes in the moisture content of

soils. Especially during the dry season, large trees draw and use a lot of local soil moisture, facilitating soil shrinkage (Figure 3.18b). Therefore, in areas with expansive soil, trees should not be planted close to foundations of light structures (such as homes). Topography and drainage are significant because adverse topographic and drainage conditions cause ponding of water around or near structures, increasing the swelling of expansive clays. Site control and construction procedures are aspects of the expansive soil problem that the owner and contractor can advantageously manipulate. Proper design of subsurface drains, rain gutters, and foundations can minimize damages associated with expansive soils by improving drainage and allowing the foundation to accommodate some shrinking and swelling of the soil (16).

It should be apparent that some soils are more desirable than others for specific uses. Planners concerned with land use will not make soil tests to evaluate engineering properties of soils, but they will be better prepared to design with nature and take advantage of geologic conditions if they understand the basic terminology and principles of earth materials. Our discussion of engineering properties established two general principles. First, because of their low strength, high sensitivity, high compressibility, low permeability, and variable shrink-swell potential, clay soils should be avoided in projects involving heavy structures, structures with minimal allowable settling, or projects needing well-drained soils. Second, soils that have high corrosive potentials or that require other than common excavation should be avoided if possible. If such soils cannot be avoided, extra care, special materials and/or techniques, and higher-than-average initial costs (planning, design, and construction) must be expected. The second costs (operation and maintenance) may also be greater. Tables 3.4 and 3.5 further summarize engineering properties of soils in terms of the unified soil classification.

WATER

The hydrologic or water cycle (Figure 3.19) is driven by solar energy and supplies nearly all our water resources. Of the water on earth, approximately 97 percent is in the oceans. A small percentage is locked up in glacier icecaps, and only a small fraction of 1 percent is in the entire atmosphere. Nevertheless, the freshwater phase of the hydrologic cycle is dependent upon this small portion of total water resources.

Of the many types of water than can be described, we will consider six. The first is *meteoric* water, the water in or derived from the atmosphere. The second is *connate* water, water that is no longer in circulation or contact with the present water cycle. Connate water is usually saline water trapped during the deposition of sediments and may be considered as "fossil water." The third is *juvenile* water, water derived from the interior of the earth that has not previously existed as atmospheric or surface water. The slow release of juvenile water, usually by volcanic activity today, is believed to be the source of most of the water in the hydrologic cycle. The fourth is *surface* water, the waters above the surface of the lithosphere (river, lake, ocean waters). The fifth is *subsurface* water, all the waters within the lithosphere. These include soil, capillary, and connate waters, and groundwater. *Groundwater*, the sixth, is that part of the subsurface waters within the zone of saturation. These waters may be fresh or saline. Fresh water is an important part of our potable water supply.

On land, meteoric and surface water processes, and, to a lesser extent, groundwater processes, are responsible for erosion and deposition of earth materials that form most of our landscape. Even in arid areas, running water produces most of the landforms. In addition, the same water processes interact and provide the water resources necessary for our existence.

The hydrologic cycle shown in Figure 3.19 is that of a humid region. In humid climates, processes of precipitation balance with those of evaporation, transpiration, runoff, and groundwater flow, so the entire system forms a water budget. The situation is simplified in arid areas, where evaporation may exceed precipitation, and the flow of water is essentially one way from the land up to the atmosphere (1).

Processes involved with the movement of surface waters and groundwaters are intimately related. In the contiguous United States, approximately 30 percent of the precipitation enters into the surface-subsurface flow system. Of this, approximately 1 percent reaches the ocean by way of groundwater flow (1). This is grossly oversimplified, however, because there are so many possible interactions where surface water enters into the groundwater flow, and conversely. The rate of surface runoff or infiltration into the groundwater system depends upon the distribution and amount of precipitation; the types of soils and rocks; the slope of the land; the amount and type of vegetation; and the amount of rejected recharge, or water that cannot enter the ground because the soil is already saturated.

Water that infiltrates the ground first enters the belt of soil moisture in the zone of aeration (Figure 3.20). It may then move through the intermediate belt, which is seldom saturated, and enter the groundwater system in the zone of saturation. The upper surface of this zone is the water table. The capillary fringe just

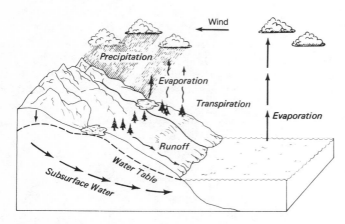

FIGURE 3.19
Idealized diagram of the water cycle.

TABLE 3.4

Generalized sizes, descriptions and properties of soils. (After Wagner, "The Use of the Unified Soil Classification System by the Bureau of Reclamation," International Conference on Soil Mechanics and Foundation Engineering, Proceedings [London], 1957.)

Soil	Soil Component	Symbol	Grain Size Range and Description	Significant Properties
Coarse-grained components	Boulder	None	Rounded to angular, bulky, hard, rock particle; average diameter more than 25.6 cm.	Boulders and cobbles are very stable components used for fills, ballast, and rip-rap. Because of size and weight, their occurrence in natural deposits tends to improve the stability of foundations. Angularity of particles increases stability.
	Cobble	None	Rounded to angular, bulky, hard, rock particle; average diameter 6.5–25.6 cm.	
	Gravel	G	Rounded to angular, bulky, hard, rock particles greater than 2 mm in diameter.	Gravel and sand have essentially the same engineering properties, differing mainly in degree. They are easy to compact, little affected by moisture, and not subject to frost action. Gravels are generally more pervious and less stable and resistant to erosion and piping than are sands. The well-graded sands and gravels are generally less pervious and more stable than those which are poorly graded and of uniform gradation. Irregularity of particles increases the stability slightly; finer, uniform sand approaches the characteristics of silt.
	Sand	S	Rounded to angular, bulky, hard, rock particles 0.074–2 mm in diameter.	
Fine-grained components	Silt	M	Particles 0.004–0.074 mm in diameter; slightly plastic or nonplastic regardless of moisture; exhibits little or no strength when air dried.	Silt is inherently unstable, particularly with increased moisture, and has a tendency to become quick when saturated. It is relatively impervious, difficult to compact, highly susceptible to frost heave, easily erodible, and subject to piping and boiling. Bulky grains reduce compressibility, whereas flaky grains (such as mica) increase compressibility, producing an elastic silt.
	Clay	C	Particles smaller than 0.004 mm in diameter; exhibits plastic properties within a certain range of moisture; exhibits considerable strength when air dried.	The distinguishing characteristic of clay is cohesion or cohesive strength, which increases with decreasing moisture. The permeability of clay is very low; it is difficult to compact when wet and impossible to drain by ordinary means; when compacted is resistant to erosion and piping; not susceptible to frost heave; and subject to expansion and shrinkage with changes in moisture. The properties are influenced not only by the size and shape (flat or platelike), but also by their mineral composition. In general, the montmorillonite clay mineral has the greatest and kaolinite the least adverse effect on the properties.
	Organic matter	O	Organic matter in various sizes and stages of decomposition.	Organic matter present even in moderate amounts increases the compressibility and reduces the stability of the fine-grained components. It may decay, causing voids, or change the properties of a soil by chemical alteration; hence, organic soils are not desirable for engineering purposes.

Note: The Unified Soil Classification does not recognize cobbles and boulders with symbols. The size range for these as well as the upper limit for clay are according to Wentworth (1922).

TABLE 3.5

Selected engineering properties of soils. (Modified after Wagner, 1957, and Detwyler and Marcus, 1972.)

				Properties			
	Unified Soil Classification Symbol	Strength	Frost Susceptibility	Permeability (after compaction)	Compressibility (after compaction and saturation)	Erodibility (compacted and saturated)	General Use As Construction Material
Coarse Soils — Gravels	GW: well-graded gravel	Very High	Negligible	High	Insignificant	Low-High	Excellent
	GP: poorly graded gravel	High	Very low	Very high	Insignificant	Low-High	Good
	GM: silty gravel	High	Medium-High	Low	Insignificant	Medium-High	Good
	GC: clayey gravel	High-Medium	Low	Very low	Very low	Medium-High	Good
Coarse Soils — Sands	SW: well-graded sand	Very high	Negligible	High	Insignificant	Very high	Excellent
	SP: poorly graded sand	High	Low	High	Very low	High	Fair
	SM: silty sand	High	Medium-High	Low	Low	High	Fair
	SC: clayey sand	High-Medium	Low-Medium	Very low	Low	Medium-High	Good
Fine Soils — Silts	ML: silt	Medium	High	Low	Medium	High	Fair
	MH: micaceous silt	Medium-Low	Very high	Low	High	High	Poor
	OL: organic silt	Low	High	Low	Medium	High	Fair
Fine Soils — Clays	CL: silty clay	Medium	Medium-High	Very low	Medium	Medium	Good-Fair
	CH: high plastic clay	Low	Medium	Very low	High	Low	Poor
	OH: organic clay	Low	Medium	Very low	High	Low	Poor

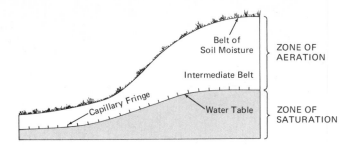

FIGURE 3.20
Idealized diagram showing the zones in which groundwater occurs.

above the water table is a narrow belt where water is drawn up by repulsion because of surface tension of the water.

The rate of movement of groundwater depends upon the nature and extent of the primary (or inter-granular) and secondary (or fractural) permeability of the soil or rocks and upon the hydraulic (pressure) gradient, which is related to the slope of the water table. The percentage of void or empty space in soil and rock is called *porosity*. The measure of the ability of water to move through a particular material is called *permeability*. Some of the most porous materials, such as clay, have very low permeability (Table 3.6) because the small, flat clay particles can have a great deal of pore space, but the individual small openings tenaciously hold water.

A zone of earth material capable of producing groundwater at a useful rate from the well is called an *aquifer*. A zone of earth material that will hold water but not transmit it fast enough to be pumped from a well is called an *aquiclude*. These zones may form an essentially impermeable layer through which little water moves. Clay soils, shale, and igneous or metamor-

TABLE 3.6
Porosity and permeability of selected earth materials.

Material	Porosity (%)	Permeability (m^3/day/m^2)
Unconsolidated		
Clay	45	0.041
Sand	35	32.8
Gravel	25	205.0
Gravel and sand	20	82.0
Rock		
Sandstone	15	28.7
Dense limestone or shale	5	0.041
Granite	1	0.0041

Source: Modified after Linsley, Kohler, and Paulhus, *Hydrology for Engineers* (New York: McGraw-Hill, 1958. Copyright © 1958 by McGraw-Hill Book Company. Used by permission of McGraw-Hill Book Company).

phic rock with little interconnected porosity and/or fractures are likely to form aquicludes. On the other hand, gravel, sand soils, and sandstone, as well as granite and metamorphic rocks with high porosity due to connected open fractures, will be quite good aquifers if groundwater is available.

Aquifers are said to be *unconfined* if there is no impermeable layer restricting the upper surface of the zone of saturation. If a *confining* layer is present, then the water below may be under pressure, forming artesian conditions. Water in artesian systems tends to rise to the height of the recharge zone. This condition is analogous to the function of a water tower's producing water pressure for houses (Figure 3.21). Both confined and unconfined aquifers may be found in the same area (Figure 3.22).

When water is pumped from a well, a cone of depression forms in the water table or artesian pressure gradient (Figure 3.23). Overpumping of an aquifer causes the water level to lower continuously with time. This necessitates lowering the pump settings or drilling deeper wells. These adjustments may work, depending on the hydrologic conditions, but are not always successful. For instance, continued deepening to correct for overpumping of wells that tap igneous and metamorphic rocks is limited. Water from these wells is pumped from open fracture systems that tend to close or diminish in number and size with increasing depth.

The two types of streams are *effluent* and *influent* (Figure 3.22). Effluent streams tend to be perennial; during the dry season, groundwater seepage into the channel maintains stream flow. Influent streams are everywhere above the groundwater table and flow in direct response to precipitation. Water from the channel moves down through the water table, forming a recharge mound. These streams may be intermittent or ephemeral in that they flow only part of the year.

Streams and rivers are the basic transportation systems of the part of the rock cycle involved with erosion and deposition of sediments. They are also a primary erosion agent in the sculpture of landscape. Streams and rivers are open systems that generally maintain a dynamic equilibrium between the work done and the load imposed (18). To accomplish this, the stream must maintain a delicate balance between the flow of water and movement of sediment. Since the stream cannot increase or decrease the amount of water or sediment it receives, adjustments are made in terms of channel slope and cross-sectional shape, which effectively change the velocity of the water. The change of velocity may, in turn, increase or decrease the amount of sediment carried in the system. The stream tends to have a slope to provide just the veloc-

FIGURE 3.21

Diagram showing how an artesian well system may develop. In (a), water rises in homes due to pressure created by water level in the tower. If friction in pipes is small, there will be little drop in pressure. As shown in (b), the pressure surface in natural systems declines away from the source because of friction in the flow system, but water may still rise above the surface of the ground if an impervious layer such as clay is present to cap the groundwater.

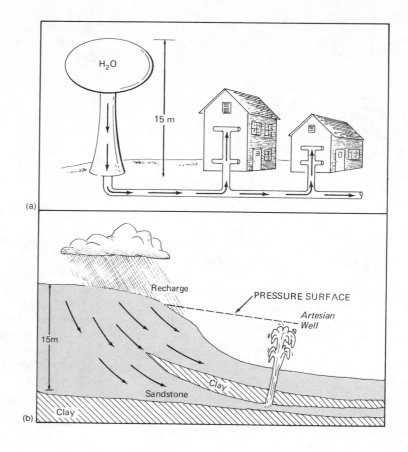

ity of water necessary to do the work of moving the sediment load (19). Since this is a delicate balance, any change in the sediment load or discharge will initiate slope changes to bring this system into balance again. Take, for example, a land-use change from forest to agricultural row crops. This change will cause increased soil erosion and an increase in the load supplied to the stream. The stream will be initially unable to transport the entire load, and sediment will be deposited, increasing the slope, which will in turn increase the velocity of water and allow the stream to move more sediment. Slope will increase by deposition in the channel until the velocity increases sufficiently to carry the new load. A new dynamic equilibrium may be reached, pro-

FIGURE 3.22

Idealized diagram showing an unconfined aquifer, a local (perched) water table, and influent and effluent streams.

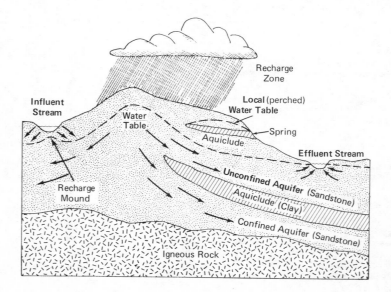

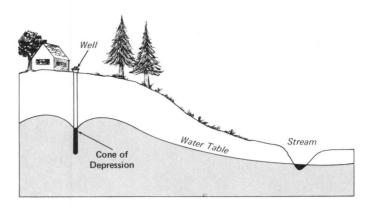

FIGURE 3.23
Cone of depression in water table resulting from pumping water from a well.

vided the rate of sediment increase levels out and the channel can adjust the slope and channel shape before another land-use change. If the reverse situation occurs, that is, if farmland is converted to forest, the sediment load will decrease and the stream will react by eroding the channel to lower the slope, which in turn will lower the velocity of the water. This will continue until an equilibrium between the load imposed and work done is achieved again.

This sequence of change is occurring in parts of the Piedmont of the southeastern United States. There, land that was forest in the early history of the country was cleared for farming, producing accelerated soil erosion and subsequent deposition of sediment in the stream (Figure 3.24). The land is now reverting to pine forests, and this, in conjunction with soil conservation measures, has reduced the sediment load delivered to streams. Thus, formerly muddy streams choked with sediment are now clearing slightly and eroding their channels. Whether this trend continues depends on future conservation measures and land use.

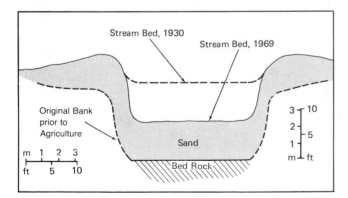

FIGURE 3.24
Accelerated sedimentation and subsequent erosion resulting from land-use changes (natural forest to agriculture and back to forest) at the Mauldin Millsite on the Piedmont of middle Georgia. (After S. W. Trimble, "Culturally Accelerated Sedimentation on the Middle Georgia Piedmont," Master's thesis [Athens: University of Georgia, 1969]. Reproduced by permission.)

Consider now the effect of constructing a dam on a stream. Considerable changes will take place both upstream and downstream of the reservoir. Upstream, the effect will be to slow down the stream, causing deposition of sediment. Downstream, the water coming out below the dam will have little sediment. (Most sediment is trapped in the reservoir.) As a result, the stream will erode its channel, picking up sediment, thus decreasing the slope until new equilibrium conditions are reached (Figure 3.25).

The Grand Canyon of the Colorado River provides a good example of a river's adjustment to the impact of a large dam. In 1963, the Glen Canyon Dam was built upstream from the Grand Canyon. Construction of the dam has drastically altered the pattern of flow and channel process downstream. From a hydrologic viewpoint, the Colorado River was tamed. Before the Glen Canyon Dam, the river reached maximum flow in May or June during the spring snow melt, then flow receded during the remainder of the year, except for occasional flash floods caused by upstream rainstorms. During periods of high discharge, the river had a tremendous capacity to transport sediment (mostly sand and silt) and vigorously scoured the channel. As the summer low flow approached, the stream was able to carry less sediment, so deposition along the channel formed large bars and terraces known as beaches to people who rafted the river. The high floods also moved large boulders off the rapids that formed because of shallowing of the river where it flows over alluvial fan deposits delivered from tributary canyons. After the dam was built, the mean annual flood (the average of the highest flow each year for a period of years) was reduced from approximately 2,500 cubic meters per second to 800 cubic meters per second, and the ten-year flood (the flood that can be expected on an average of every ten years) was reduced from about 3,500 cubic meters per second to 860 cubic meters per second. On the other hand, the dam did control the flow of water to such an extent that the median discharge actually increased from about 210 cubic meters per second to 350 cubic meters

CHAPTER 3 | EARTH MATERIALS AND PROCESSES **47**

FIGURE 3.25
Upstream deposition and downstream erosion from construction
of a dam and a reservoir.

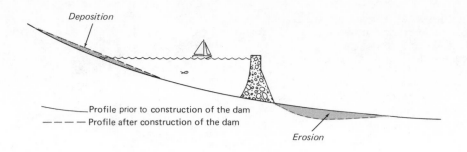

Deposition

Profile prior to construction of the dam
———— Profile after construction of the dam

Erosion

per second. The flow is highly unstable, however, because of fluctuating needs to generate power, and the
level of the river may vary by as much as 5 meters per
day with a mean daily high discharge of about 570 cubic meters per second and daily low of 130 cubic meters per second. The dam also greatly affected the sediment load of the Colorado River through the Grand
Canyon, and the median suspended sediment concentration was reduced by a factor of about 200 times immediately downstream from the dam. There was a
lesser reduction in sediment load further downstream,
because tributary channels continued to add sediment
to the channel (20).

The change in hydrology of the Colorado River in
the Grand Canyon has greatly changed the river's morphology. The rapids may be becoming more dangerous, because large floods no longer occur, and these
flows had previously moved some of the large boulders
further downstream (Figure 3.26). In addition, some of
the large sand bars or beaches are eroding because the
river is deficient in sediment below the dam and is
picking up sediment and thus causing erosion. The

(a)

(b)

FIGURE 3.26
Crystal Rapid of the Colorado River in the Grand Canyon before and after the construction
of the Glen Canyon Dam. Photograph (a) taken in 1963 shows the rapid as it existed prior
to the building of the dam. (Bureau of Reclamation photograph by Al Turner.) Photograph
(b) taken in 1967 shows the effect of a 1965 flash flood. The debris and rock delivered from
Crystal Creek (left) forms a fan-shaped deposit in the Colorado River. This deposit has
made the rapid more difficult to negotiate, and the debris will remain in the river for a
relatively long time as large floods that would normally remove some of the debris are di-
minished by the presence of the dam and reservoir. (Bureau of Reclamation photograph by
Mel Davis.)

changes in the river's flow (mainly deleting the large floods) have also resulted in vegetational shifts. Before the dam, yearly scouring of the channel and high bars somewhat restricted plant growth. With the now lower flows, some of these channels and bars are becoming vegetated by exotic plants that are displacing indigenous species. One last impact, of course, has been an increase in the number of people who raft through the Grand Canyon. Although this is now restricted to about 10,000 persons per year, the long-range impact of people, along with potential erosion of beaches that must serve as campsites, is a problem yet to be resolved. The Colorado in the 1970s and 1980s is a changed river, and cannot be expected to return to what it was before the construction of the dam (20).

Channel pattern is the configuration of the channel in plan view, as from an airplane. Two main patterns (Figure 3.27) are *braided* channels, characterized by numerous bars and islands that divide and reunite the channel, and those that do not braid. Because long, straight reaches are relatively rare, channels that do not braid are designated as *sinuous;* however, sinuous channels often contain relatively short, straight

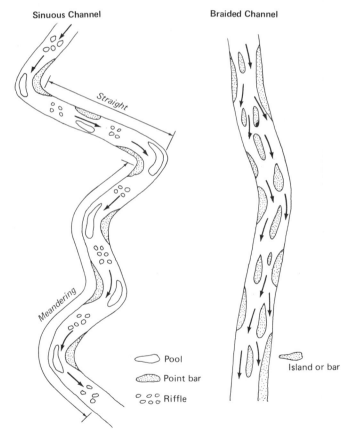

FIGURE 3.27
Channel patterns.

reaches. Individual bends in sinuous channels are called *meanders,* which migrate back and forth across the stream valley, depositing sediment on the inside of bends, forming point bars, and eroding the outside of bends. This process is prominent in constructing and maintaining floodplains. Overbank deposition during floods causes vertical accretion that is also important in development of floodplains.

Sinuous channels often contain a series of regularly spaced *pools* and *riffles* (Figure 3.28). Pools at low flow are deep areas characterized by relatively slow movement of water. They are produced by scour at high flow. Riffles at low flow are shallow areas recognized by relatively fast water. Riffles are produced by depositional processes at high flow. With respect to erosion and deposition of sediment, we therefore conclude that pools scour at high flow and fill at low flow, whereas riffles fill at high flow and scour at low flow (Figure 3.29). It is this pattern of scour and fill that maintains pools and riffles. The pool-riffle sequence is repeated approximately every five to seven times the channel width. Streams with well-developed pools and riffles tend to have considerable gravel in the streambed and a relatively low slope. Streams with finer bed material or steep slopes tend to lack regularly spaced pools and riffles. Bedrock channels also may develop pools and riffles. These forms are important in environmental studies because they provide a variety of flow conditions characterized by deep, slow-moving water alternating with shallow, fast-moving water which facilitates desirable biologic activity. They also provide visual and other sensory variety that increase the aesthetic amenity as well as the recreational potential by providing better fishing and boating conditions. In some instances, pools and riffles also help to stabilize the channel. For example, a straight reach with pools and riffles in which the deep part of the channel alternates from bank to bank may be morphologically more stable than a straight channel that lacks these forms.

WIND AND ICE

Wind- and ice-related processes are responsible for the erosion, transport, and deposition of tremendous quantities of surficial earth materials. Furthermore, these processes both modify and create a substantial number of landforms in environmentally sensitive areas— coastal, desert, arctic, and subarctic.

Wind

Windblown deposits are generally subdivided into two groups: *sand deposits,* mainly dunes; and *loess* (wind-

FIGURE 3.28
Well-developed pool-riffle sequence in Sims Creek near Blowing Rock, North Carolina. A deep pool is apparent in the middle distance and shallow riffles can be seen in the far distance and the foreground.

blown silt), further divided into *primary loess,* which is essentially unaltered since deposition, and *secondary loess,* which has been transported and reworked over a short distance by water or has been intensely weathered in place (8).

Extensive deposits of windblown sand and silt cover thousands of square kilometers in the United States (Figure 3.30). Sand dunes and related deposits are found along the coasts of the Atlantic and Pacific oceans and the Great Lakes. Inland sand is found in areas of Nebraska, southern Oregon, southern California, Nevada, and northern Indiana, and along large rivers flowing through semiarid regions, as, for example, the Columbia and Snake rivers in Oregon and Wash-

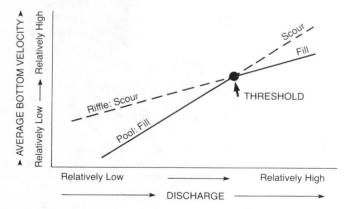

FIGURE 3.29
Idealized drawing showing the scour-fill pattern characteristic of a pool-riffle sequence. The threshold is a critical discharge at which change in process (scour-to-fill or fill-to-scour) occurs.

ington. The majority of loess is located adjacent to the Mississippi Valley, but some is also found in the Pacific Northwest and Idaho.

Sand dunes are constructed from sand moving close to the ground. They have a variety of sizes and shapes and develop under a variety of conditions (Figures 3.31, 3.32, and 3.33). Regardless of where they are located, how they form, or whether they are active or relic or otherwise stabilized, they tend to cause environmental problems. Migrating sand is particularly troublesome, and stabilization of sand dunes is a major problem in construction and maintenance of highways and railroads that cross sandy areas of deserts. The complex group of sand dunes shown in Figure 3.34a are encroaching on Highway 95 near Winnemucca, Nevada. The rate of movement is about 12 meters per year. The sand is removed about three times a year (Figure 3.34b), and approximately 1,500 to 4,000 cubic meters are removed each time. Attempts to stabilize the dunes by planting several varieties of grass in test plots have not been successful because there is simply not enough precipitation to support the grass.

Building and maintaining reservoirs in sand dune terrain are even more troublesome and tend to be extremely expensive. These reservoirs should be constructed only if very high water loss can be tolerated. Canals in sandy areas should be lined to hold water and control erosion (8).

In contrast to sand, which seldom moves more than a meter off the ground, windblown silt and dust can be carried in huge dust clouds thousands of meters in altitude (Figure 3.35). A typical dust storm 500 to 600 kilometers in diameter may carry over 100 million

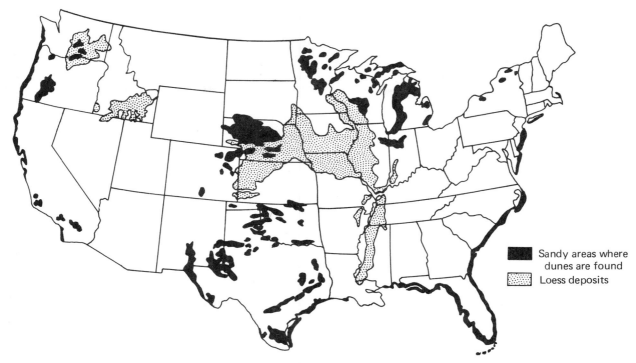

FIGURE 3.30
Distribution of windblown deposits within the contiguous United States. (Reproduced, by permission, from Douglas S. Way, *Terrain Analysis* [Stroudsburg, Pennsylvania: Dowden, Hutchinson & Roth, 1973].)

tons of silt and dust, sufficient to form a pile 30 meters high and 3 kilometers in diameter (21). Terrible dust storms in the 1930s probably exceeded even this, perhaps carrying over 58 thousand tons of dust per square kilometer.

An extremely high magnitude wind storm occurred on December 20, 1977, in the southern San Joaquin Valley in California. Winds that may locally have reached velocities as high as 300 km/hr caused mod-

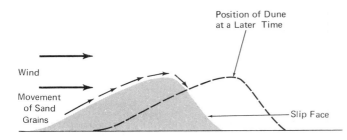

FIGURE 3.31
Idealized diagram showing the movement of a sand dune. The wind moves the individual sand grains along the surface of the gentle windward side of the dune until they fall down the steeper leeward face. In this manner, the dune slowly migrates in the direction of the wind. (From Robert J. Foster, *Physical Geology,* 2nd ed. [Columbus: Charles E. Merrill, 1975].)

erate to heavy damage to buildings, crops, automobiles, wildlife, and soils over an area of about 2000 km^2. Five people were killed in automobile accidents caused by impaired visibility and soil erosion released "valley fever" spores that caused significant increase in the incidence of the disease in distant downwind locations. The winds mobilized about forty-two thousand tons/km^2 of soil over 600 km^2 of grazing lands in a 24-hour period. The wind-scoured land was vulnerable to erosion and subsequent rainstorms during the following February caused accelerated erosion and flooding (22).

Loess, or windblown silt, in the United States derived primarily during the Pleistocene Ice Ages from glacial outwash in the vicinity of major streams that carried the glacial meltwater from the ice front. Retreat of the ice left large, unvegetated areas adjacent to rivers; these areas were highly susceptible to wind erosion. We know this because loess generally decreases rapidly in thickness with distance from the major rivers.

Loess is a mixture of fine sand, silt, and clay in which the grains are arranged in an open framework. Loess is porous, with a vertical permeability greater than its horizontal. This difference results in part from the presence of long, vertical tubes in the loess that are

Dune Type	Sketch Cross Section			Remarks

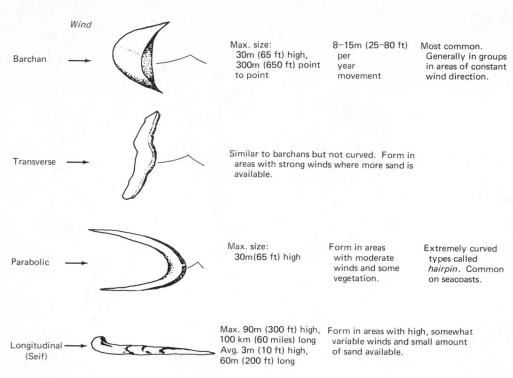

Barchan → Max. size: 30m (65 ft) high, 300m (650 ft) point to point | 8–15m (25–80 ft) per year movement | Most common. Generally in groups in areas of constant wind direction.

Transverse → Similar to barchans but not curved. Form in areas with strong winds where more sand is available.

Parabolic → Max. size: 30m (65 ft) high | Form in areas with moderate winds and some vegetation. | Extremely curved types called *hairpin.* Common on seacoasts.

Longitudinal (Seif) → Max. 90m (300 ft) high, 100 km (60 miles) long Avg. 3m (10 ft) high, 60m (200 ft) long | Form in areas with high, somewhat variable winds and small amount of sand available.

FIGURE 3.32
Types of sand dunes. (After Foster, *Physical Geology, 3d ed.* [Columbus: Charles E. Merrill, 1979].)

probably casts of plant roots that developed as the loess accumulated. Although loess may form nearly vertical slopes (Figure 3.36), it rapidly consolidates when subjected to a load (such as a building) and wetted, a process called *hydrocompaction,* which results as clay films or calcium carbonate cement around the silt grains washes away. Loess may therefore be a dangerous foundation material. Settling and cracking of a house reportedly took place overnight when water from a hose was accidentally left on. On the other hand, when loess is properly compacted and remolded, it acquires considerable strength and resistance to erosion, and, therefore, a reliable platform for building footings may be constructed (8).

Ice

The cold-climate phenomenon of ice has been an important environmental topic for several years. As more people live and work in the higher latitudes, we will

FIGURE 3.33
Barchan dunes in the Columbia River Valley, Oregon. (Photo by G. K. Gilbert, courtesy of the U. S. Geological Survey.)

(a)

(b)

FIGURE 3.34
Complex group of sand dunes encroaching on Highway 95 near Winnemucca, Nevada (a). Removal of the sand from the highway is a continuous problem (b). (Photos courtesy of J. O. Davis [a] and D. T. Trexler [b].)

have to learn more about how to insure the best use of these sometimes fragile environments.

Only a few thousand years ago, the most recent continental glaciers retreated from the Great Lakes region of the U.S. Figure 3.37 shows the maximum extent of the Pleistocene ice sheets. Several times in the last 2 to 3 million years, the ice advanced southward, and scientists are still speculating as to whether it will advance once again. We may, indeed, still be in the Ice Age. The effects of recent glacial events are easily seen in the landscape. The flat, nearly featureless ground moraine or till plains of central Indiana are composed of material carried and deposited by continental glaciers, *till,* which buried preglacial river valleys. It is hard to believe that beneath the glacial deposits is a topography formed by running water much like the hills and valleys of southern Indiana where glaciers never reached.

Continental and mountain glaciers produced a variety of erosional and depositional landforms. Because of the variable types of forms and deposits, the environmental geology in a recently glaciated area may be complex. Glacial-related deposits include sands and gravels from streams in, on, under, and in front of the ice, as well as heterogeneous material (till) deposited directly by the ice. In addition, numerous lakes formed when ice blocks buried by glacial deposits melted. These lakes often contained peat and other organic material forming high-organic soils. The wide variety of possibilities of earth materials in glaciated areas requires that detailed evaluation of the physical properties of surficial and subsurface materials be conducted before planning, designing, and building structures such as dams, highways, and large buildings.

In the higher latitudes, *permafrost,* permanently frozen ground, is a widespread natural phenomenon

FIGURE 3.35
Dust storm caused by cold front at Manteer, Kansas, 1935. (Photo courtesy of Environmental Science Services Administration.)

FIGURE 3.36
Nearly vertical exposure of loess near Chester, Illinois. (Photo courtesy of James Brice.)

that still underlies about 20 percent of the land area in the world (Figure 3.38). Two main types of permafrost are defined by the areal extent of the phenomenon: *discontinuous* permafrost and *continuous* permafrost. At higher latitudes, discontinuous permafrost is characterized by scattered islands of thawed ground that exist in a predominantly frozen area. Toward the southern border of the permafrost, however, the percentage of unfrozen ground increases until all the ground is unfrozen. In continuous permafrost areas, the only ice-free areas are beneath deep lakes or rivers. The distribution of these types of permafrost for Alaska, where about 85

percent of the state is underlain by frozen ground, is shown in Figure 3.39. Known thickness of the permafrost in Alaska varies from about 400 meters in northern Alaska to less than 0.3 meters at the southern margin of the frozen ground (23).

A cross section through permafrost (Figure 3.40) shows an upper active layer that thaws during the summer, and unfrozen layers within the permafrost below the active layer. The thickness of the active layer depends upon such factors as exposure, slope, amount of water, and particularly the presence or absence of a vegetation cover which greatly affects the thermal conductivity of the soil.

Special engineering problems are associated with the design, construction, and maintenance of structures such as roads, railroads, airfields, pipelines, and buildings in permafrost areas. Lack of knowledge about permafrost has led to very high maintenance costs and relocation or abandonment of highways, railroads, and other structures. Figure 3.41 shows a gravel road with severe differential subsidence caused by thawing of permafrost. Figure 3.42 shows a tractor trail constructed by bulldozing off the vegetation cover over permafrost on the North Slope of Alaska. The small ponds on the abandoned trail formed during the first summer following the trail construction. They will continue to grow deeper and wider as the permafrost continues to thaw (23).

The entire range of engineering problems in permafrost areas is extensive and beyond the scope of our discussion. Specific types of problems occur with

FIGURE 3.37
Maximum extent of ice sheets during the Pleistocene glaciation. (From Foster, *Physical Geology*, 3d ed. [Columbus: Charles E. Merrill, 1979].)

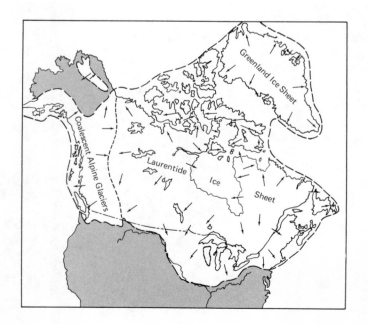

FIGURE 3.38
Extent of permafrost zones in the Northern Hemisphere. (From Ferrians, Kachadoorian, and Greene, *Permafrost and Related Engineering Problems in Alaska,* U.S. Geological Survey Professional Paper 678, 1969.)

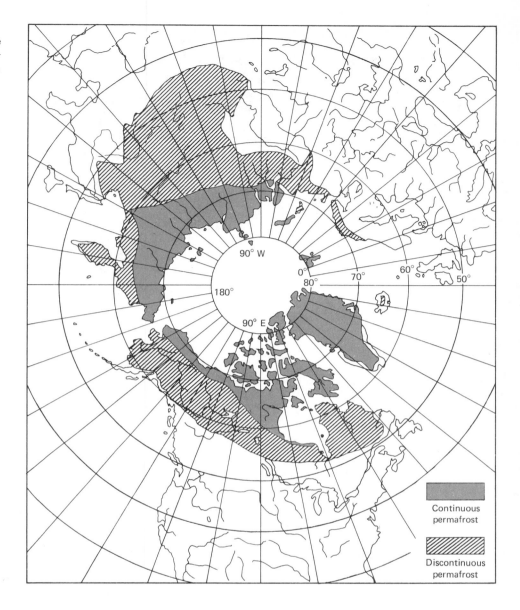

90° W

0°
80°

180°

70°

60°

50°

90° E

Continuous permafrost

Discontinuous permafrost

different types of earth materials. In general, the major problems are associated with permafrost that occurs in fine-grained, poorly drained, frost-susceptible materials. In these materials, a lot of ice melts if the thermal regime changes. Melting produces unstable materials, resulting in settling, subsidence, landslides, and lateral or downslope flowage of saturated sediment. This thawing of permafrost and subsequent frost heaving and subsidence caused by freezing and thawing of the active layer are responsible for most of the engineering problems in the arctic and subarctic regions (23).

Experience has shown that two basic methods of construction can be used on permafrost: the *active* method and the *passive* method. The active method is used where the permafrost is thin or discontinuous or

contains a relatively small amount of ice. The method consists of thawing the permafrost, and if the thawed material has sufficient strength, convential construction is used. This works best in coarse-grained, well-drained soils that are not particularly frost-susceptible. The passive method is used where it is impractical to thaw the permafrost. The basic principle is to keep the permafrost frozen and not upset the natural quasi-equilibrium among environmental factors that are so sensitive that even the passage of a tracked vehicle that destroys the vegetation cover will upset the equilibrium and initiate melting which, once started, is nearly impossible to reverse. Special design of structures and foundations to minimize melting of the permafrost is the key to the passive method (23).

*Mountainous areas, generally underlain
by bed rock at or near the surface.*

Underlain by continuous permafrost.

Underlain by discontinuous permafrost.

Underlain by isolated masses of permafrost.

*Lowland areas, generally underlain by
thick, unconsolidated deposits.*

Underlain by thick permafrost in areas
of either fine-grained or coarse-
grained deposits

Underlain by moderately thick to thin
permafrost in areas of fine-grained
deposits, and by discontinuous or isolated masses of permafrost in areas of
coarse-grained deposits.

Underlain by isolated masses of permafrost in areas of fine-grained deposits,
and generally free of permafrost in
areas of coarse-grained deposits.

AREAS OUTSIDE PERMAFROST
REGION

Generally free of permafrost, but a few
small isolated masses of permafrost
occur at high altitudes and in lowland
areas where ground insulation is high
and ground insolation is low, especially near the border of the permafrost region.

FIGURE 3.39
Extent and distribution of permafrost in Alaska. (From Ferrians, Kachadoorian, and Greene,
Permafrost and Related Engineering Problems in Alaska, U.S. Geological Survey Profes-
sional Paper 678, 1969.)

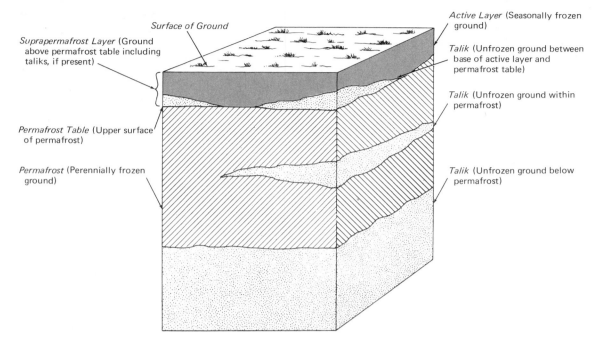

FIGURE 3.40
Block diagram of permafrost morphology. (From Ferrians, Kachadoorian, and Greene, *Permafrost and Related Engineering Problems in Alaska,* U.S. Geological Survey Professional Paper 678, 1969.)

FIGURE 3.41
Gravel road in Alaska showing severe differential subsidence caused by thawing of permafrost. (Photo by O. J. Ferrians, Jr., courtesy of the U.S. Geological Survey.)

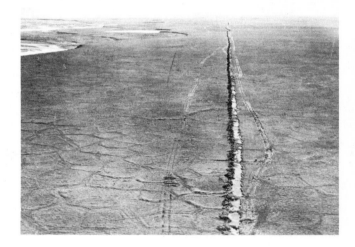

FIGURE 3.42
Tractor trail on the North Slope of Alaska. The small ponds (light patches above the straight road) are created by thawing of the permafrost in the roadway. The polygons are a type of "pattern ground" formed by vertical ice wedges in frost-susceptible materials. (Photo by O. J. Ferrians, Jr., courtesy of the U.S. Geological Survey.)

SUMMARY AND CONCLUSIONS

Processes that create, maintain, change, and destroy earth materials (water, minerals, rocks, and soil) are collectively referred to as the *geologic cycle*. More correctly, the geologic cycle is a group of subcycles, including the *tectonic, hydrologic, rock,* and *geochemical* cycles (1). In various activities, people use earth materials found in different parts of these cycles. The materials may be found in a concentrated or "pure" state; but after various artificial processes, they are returned to the earth and are dispersed, contaminated, or polluted in another part of the geologic cycle (5). Once dispersed or otherwise altered, these materials cannot be concentrated or made available for human use in any reliable timetable.

Application of geological information to environmental problems requires an understanding of both geological and engineering properties of earth materials. Different minerals, rocks, and soils behave predictably, but they behave differently for various land uses. Thus, clay soils and compaction shales are generally poor foundation materials for large engineering structures, whereas granite with few fractures is satisfactory for most purposes. In addition, the strength of rocks is greatly influenced by active fracture systems, or old fractures that have been weathered and altered to unstable minerals.

Most continental landforms are produced by running water, and an understanding of both surface and subsurface hydrogeologic processes is necessary to work on many environmental problems. Streams and groundwater systems are basically open systems in which a dynamic equilibrium between various parts of the system is established. Any change, artificial or natural, will cause the other parts to readjust to compensate and produce a new equilibrium.

Windblown sand, silt, and dust cover many thousands of square kilometers. Major deposits of sand are concentrated along coastal and interior areas, and windblown silt, or loess deposits, are concentrated along major rivers that carried the meltwater from continental glaciers during the Pleistocene glaciation.

Engineering problems caused by migrating and stabilized sand dunes involve expensive construction and maintenance costs for highways, buildings, and hydraulic structures.

Loess deposits, unless properly compacted, make hazardous foundations for engineering purposes because loess may hydroconsolidate and settle when it becomes wet.

Geologically, recent continental glaciation has produced a variety of different types of earth materials such as glacial till, highly organic soils, and water transported sediment, all of which may be located in a given area. As a result, careful evaluation of surface and subsurface deposits is necessary before planning, designing, and building.

Today permafrost underlies about 20 percent of the world's land area, producing a fragile and sensitive environment. Special engineering procedures are necessary to minimize adverse effects from artificial thawing of fine-grained, poorly drained, frozen ground.

REFERENCES

1 LONGWELL, C. L.; FLINT, R. F.; and SANDERS, J. E. 1969. *Physical geology.* New York: John Wiley & Sons.

2 LE PICHON, X. 1968. Sea-floor spreading and continental drift. *Journal of Geophysical Research* 73:3661–97.

3 ISACKS, B.; OLIVER, J.; and SYKES, L. 1968. Seismology and the new global tectonics. *Journal of Geophysical Research* 73:5855–99.

4 DEWEY, J. F. 1972. Plate tectonics. *Scientific American* 22:56–68.

5 Committee on Geological Sciences. 1972. *The earth and human affairs.* San Francisco: Canfield Press.

6 BILLINGS, M. P. 1954. *Structural geology.* 2nd ed. Englewood Cliffs, New Jersey: Prentice-Hall.

7 JESSUP, W. E. 1964. Baldwin Hills dam failure. *Civil Engineering* 34:62–64.

8 KRYNINE, D. P., and JUDD, W. R. 1957. *Principles of engineering geology and geotechnics.* New York: McGraw-Hill.

9 SCHULTZ, J. R., and CLEAVES, A. B. 1955. *Geology in engineering.* New York: John Wiley & Sons.

10 HOLDEN, C. 1971. Nuclear waste: Kansas riled by AEC plans for atom dump. *Science* 172:249–50.

11 DASMANN, R. F. 1972. *Environmental conservation.* 3rd ed. New York: John Wiley & Sons.

12 HUNT, C. B. 1972. *Geology of soils.* San Francisco: W. H. Freeman.

13 PESTRONG, R. 1974. *Slope stability.* American Geological Institute. New York: McGraw-Hill.

14 FLAWN, P. T. 1970. *Environmental Geology.* New York: Harper & Row.

15 HART, S. S. 1974. Potentially swelling soil and rock in the Front Range Urban Corridor. *Environmental Geology* 7. Colorado Geological Survey.

16 MATHEWSON, C. C.; CASTLEBERRY, J. P., II; and LYTTON, R. L. 1975. Analysis and modeling of the performance of home foundations on expansive soils in central Texas. *Bulletin of the Association of Engineering Geologists* 17, no. 4:275–302.

17 JONES D. E., Jr., and HOLTZ, W. G. 1973. Expansive soils—the hidden disaster. *Civil Engineering,* August:49–51.

18 LEOPOLD, L. B., and MADDOCK, T., JR. 1953. *The hydraulic geometry of stream channels and some physiographic implications.* U.S. Geological Professional Paper 252.

19 MACKIN, J. H. 1948. Concept of the graded river. *Geological Society of America Bulletin* 59:463–512.

20 DOLAN R., HOWARD, A., and GALLENSON, A., 1974. Man's impact on the Colorado River and the Grand Canyon. American Scientist, V. 62:392–401.

21 WAY, D. S. 1973. *Terrain analysis.* Stroudsburg, Pennsylvania: Dowden Hutchinson & Ross.

22 WILSHIRE, H. G., NAKATA, J. K., and HALLET, B. 1981. Field observations of the December 1977 windstorm, San Joaquin Valley, California. Geological Society of America, Special Paper 186:233–251.

23 FERRIANS, O. J., JR.; KACHADOORIAN, R.; and GREENE, G. W. 1969. *Permafrost and related engineering problems in Alaska.* U. S. Geological Survey Professional Paper 678.

TWO

Hazardous Earth Processes

The earth is a dynamic, evolving system with complex interactions of internal and external processes. Internal processes are responsible for moving giant lithospheric plates and thus the continents. Interactions at plate junctions generate internal stress that causes rock deformation, resulting in earthquakes, volcanic activity, and tectonic creep (slow surface movement resulting from movement along a fault zone). These processes trigger numerous external events such as landslides, mudflows, and tsunamis (giant sea waves). Other external activities, such as running water or moving waves, are the result of the interaction among the hydrosphere, atmosphere, and lithosphere. For example, tremendous flooding may result from hurricane activity. The nature and extent of an external event, however, is affected by internal processes which produce the uplifted land surface that running water and waves erode.

People generally do not perceive the full significance of processes that periodically cause loss of life and property. The casual attitude toward natural hazards is a complex manifestation of an internal optimism people have about their lives and their homes, combined with a lack of understanding of physical relations that control the magnitude (severity) and frequency of hazardous processes.

Chapters 4 through 8 will discuss our interaction with natural geological events that have damaged and destroyed, and will continue to damage and destroy, human life and property. These events include river floods, landslides, earthquakes, volcanic eruptions, and coastal processes. Especially important are the natural and artificial aspects of the magnitude and frequency of hazardous earth processes and how they relate to human use and interest. These will be discussed in detail for each hazard.

Chapter 4

River Flooding

More than any other surface process, running water is responsible for sculpturing the landscape. Rivers erode their valleys and, in the process, provide a system of drainage that is delicately balanced with the climate, topography, rocks, and soil. Rivers are also basic transportation systems that carry rock and soil material delivered to the channel by the tributaries and the slope processes on the valley sides. Much of the sediment transported in rivers is periodically stored by deposition in the channel and on the adjacent floodplain (Figure 4.1). These areas, collectively called the *riverine environment,* are the natural domain of the river. Lateral migration of bends of rivers and over-bank flow combine to produce the floodplain, which is periodically inundated by water and sediment. This natural process of over-bank flow is termed *flooding.* Channel discharge (the volume of water per unit time flowing past a particular location; cubic meters per second, m^3/s) at the point where water overflows the channel is called the *flood discharge* and may or may not coincide with property damage. The term *flood stage* frequently connotes that the elevation of the water surface has reached a high water condition likely to cause damage to personal property on the floodplain. This is obviously based on human perception of the event and therefore changes or varies as human use of the floodplain changes (1).

A summary of flood damage in the United States is shown in Figure 4.2, and a few of the more severe floods are listed in Table 4.1. These data emphasize the significant impact of this most universally experienced natural hazard. In the United States, the lives lost to flooding number about 80 persons per year, with property damage greater than $300 million per year. The loss of life is relatively low compared to loss in preindustrial societies that lack monitoring and warning systems before a flood and disaster relief afterward. During one twenty-year period (1947–1967), Asia, excluding Russia, reported 154,000 lives lost to floods, compared with about 1,300 lives lost in the United States during the same period. Although preindustrial societies with dense populations on floodplains lose a larger proportion of lives, they have a relatively lower amount of property damage than industrial societies (1, 2).

Most river flooding is a function of the total amount and distribution of precipitation and the rate at which it infiltrates the rock or soil and the topography; however, some floods result from rapid melting of ice and snow in the spring and, on rare occasions, from the failure of a dam.

FIGURE 4.1
Block diagram showing flood-plain and river channel.

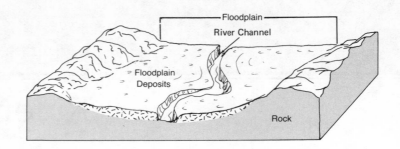

MAGNITUDE AND FREQUENCY OF FLOODS

Flooding is intimately related to the amount and intensity of precipitation. Catastrophic floods reported on television and in newspapers are produced by infrequent, large, intense storms. Smaller floods or *flows* are produced by less intense storms, which occur more frequently. All flow events that can be measured or estimated from stream-gauging stations (Figure 4.3) can be arranged in order of their magnitude of discharge, generally measured in cubic meters per second (m³/s). The list of flows so arranged is called an *array* and can be plotted on a discharge-frequency curve (Figure 4.4)

by deriving the recurrence interval R for each flow from the relationship

$$R = \frac{N + 1}{M}$$

where R is a recurrence interval in years, N is the number of years of record, and M is the rank of the individual flow in the array (3). The highest flow for nine years of data for the stream shown in Figure 4.4 is almost 283 m³/s and so has a rank M equal to one (4). Therefore, the recurrence interval of this flood is

$$R = \frac{N + 1}{M} = \frac{9 + 1}{1} = 10$$

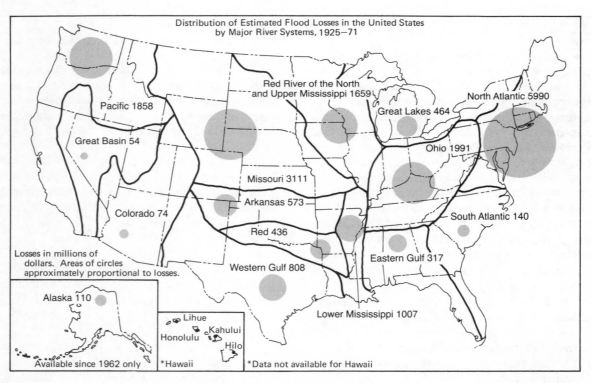

FIGURE 4.2
Distribution of estimated flood losses in the United States by major river systems from 1925 to 1976. (U.S. Department of Commerce, *Climatological Data, National Summary,* 1977.)

TABLE 4.1

Severe river floods in the United States.

Year	Month	Location	Lives Lost	Property Damage (In millions of dollars)
1937	Jan.–Feb.	Ohio and lower Mississippi river basins	137	417.7
1938	March	Southern California	79	24.5
1940	Aug.	Southern Virginia and Carolinas, and Eastern Tennessee	40	12.0
1947	May–July	Lower Missouri and middle Mississippi river basins	29	235.0
1951	June–July	Kansas and Missouri	28	923.2
1955	Dec.	West Coast	61	154.5
1963	March	Ohio River basin	26	97.6
1964	June	Montana	31	54.3
1964	Dec.	California and Oregon	40	415.8
1965	June	Sanderson, Texas (flash flood)	26	2.7
1969	Jan.–Feb.	California	60	399.2
1969	Aug.	James River basin, Virginia	154	116.0
1971	Aug.	New Jersey	3	138.5
1972	June	Black Hills, South Dakota	242	163.0
1972	June	Eastern United States	113	3,000.0
1973	March–June	Mississippi River	—	1,200.0
1976	July	Big Thompson River, Colorado	139	35.0
1977	July	Johnstown, Pennsylvania	76	330.0
1977	Sept.	Kansas City, Missouri and Kansas	25	80.0
1979	April	Mississippi and Alabama	10	500.0
1983	Sept.	Arizona	13	416.0

SOURCES: NOAA, *Climatological Data, National Summary,* 1970, 1972, 1973, and 1977; U.S. Geol. Survey.

which means that a flood with a magnitude equal to or exceeding 283 m^3/s can be expected about every ten years. Studies of many streams and rivers show that channels are formed and maintained by bank-full discharge with a recurrence interval of 1.5 to 2 years (28 m^3/s on Figure 4.4). Therefore, we can expect a stream to emerge from its banks and cover part of the floodplain with water and sediment once every year or so.

The longer flow records are collected, the more accurate the prediction of floods is likely to be. However, designing structures for a 10-year, 25-year, 50-year, or even 100-year flood, or in fact any flow below possible maximum, is itself a calculated risk because predicting a flood of a certain magnitude is based on probability, and therefore has an element of chance.

Theoretically, a 25-year flood should happen on the average of every 25 years, but two 25-year floods could occur in any one year (5). We can thus conclude that as long as we continue to build dams, highways, bridges, homes, and other facilities on flood-prone areas, we can expect continued loss of lives and property.

In discussing the concepts of magnitude and frequency of floods, we should distinguish between *upstream* floods and *downstream* floods (Figure 4.5). Upstream floods are in the upper parts of drainage areas and are generally produced by intense rainfall of short duration over a relatively small area. Although these floods can be severe over a relatively small area, they generally do not cause floods in the larger streams they

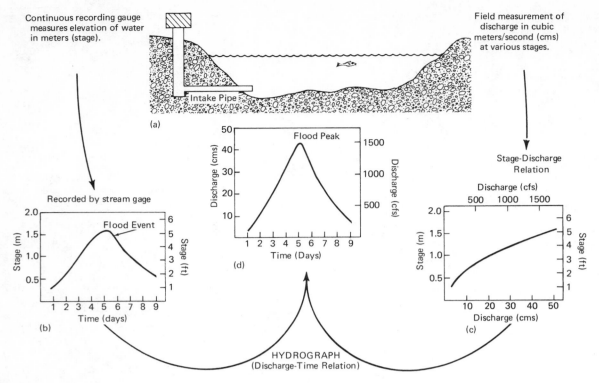

FIGURE 4.3
Field data (a) consist of a continuous recording of the water level (stage), which is used to produce a stage-time graph (b). Field measurements at various flows also produce a stage-discharge graph (c). Then graphs (b) and (c) are combined to produce the final hydrograph (d).

join downstream. Downstream floods, on the other hand, cover a wide area. They are usually produced by storms of long duration that saturate the soil and produce increased runoff. Flooding on small tributary basins is limited, but the contribution of increased runoff from thousands of tributary basins may cause a large flood downstream, which can be recognized by the downstream migration of an ever increasing flood wave with large rise and fall of discharge (6). The example of

the Chattanooga-Savannah River system in Figure 4.6a shows the downstream migration with time of a flood crest. It illustrates that a progressively longer time is necessary for the rise and fall of water as the flood wave proceeds downstream, and shows dramatically the tremendous increase in discharge from low-flow conditions to over 1,700 m³/s in five days and 257 kilometers of downstream flow (7). It is the large downstream floods that usually make television and news-

FIGURE 4.4
Example of a flood frequency curve. Each circle represents a flow event with recurrence interval plotted on probability paper. (After L. B. Leopold, U.S. Geological Survey Circular 559, 1968.)

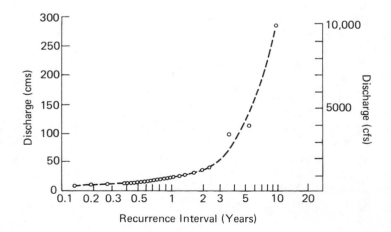

FIGURE 4.5
Idealized diagram comparing upstream flood (a) to downstream flood (b). Upstream floods generally cover relatively small areas and are caused by intense local storms, whereas downstream floods cover wide areas and are caused by regional storms or spring runoff. (Modified after U.S. Department of Agriculture drawing.)

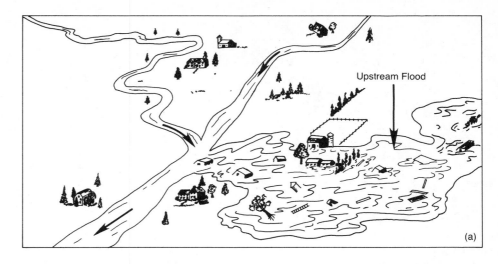

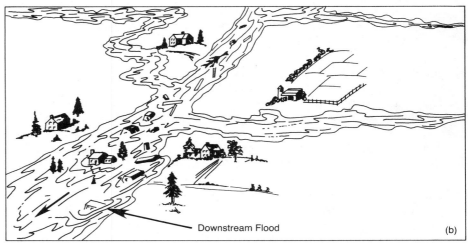

paper headlines. Figure 4.6b illustrates the same flood in terms of discharge per unit area, thus eliminating the effect of downstream increase in discharge. This better illustrates the shape and form (sharpness of peaking) of the flood wave as it moves downstream (7).

A few upstream floods of very high magnitude have been caused directly by structural failure. For example, the most destructive flood in West Virginia's history was caused by the failure of a coal-waste dam on the middle fork of Buffalo Creek. On the morning of February 26, 1972, that flood, which lasted only three hours, cost at least 118 lives, destroyed 500 homes, left 4,000 persons homeless, and caused more than $50 million of property damage. A wall of water 3 to 6 meters high swept through the Buffalo Creek Valley at an average rate of about 2.1 meters per second, or 8 kilometers per hour. For three days before the disaster, nearly 10 centimeters of rain fell in the area, producing a 10-year flood in local streams similar to Buffalo Creek. The volume of water released by the collapse of the

coal-waste dam was approximately 500 thousand cubic meters, producing a flood about 40 times greater than a naturally occurring 50-year flood. The U.S. Geological Survey concluded that the failure of the dam contributed almost all of the peak flood flow and that direct runoff from other sources was not significant. It was further concluded that there were several causes for the dam failure. First, the dam was neither designed nor constructed to withstand the amount and depth of water it impounded. It was, in fact, a waste pile that grew as more and more material was dumped. Second, there was no spillway or other adequate water level control in the dam to provide a means of removing water. Third, sludge waste from the coal mining operation was an inadequate foundation material for the dam, and seepage through the foundation caused openings to form through which water flowed, greatly reducing the dam's stability. Fourth, stability of the dam was further reduced because of its great thickness, and the dam became saturated and somewhat bouyant. Fifth,

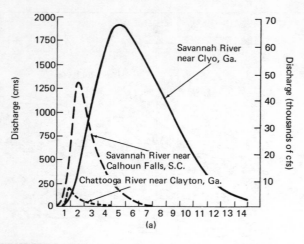

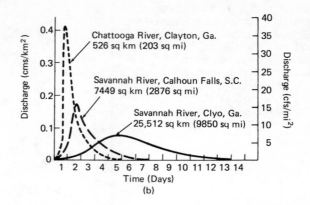

FIGURE 4.6
Downstream movement of a flood wave on the Savannah River, South Carolina and Georgia. (See text for further explanation.) (After William G. Hoyt and Walter B. Langbein, *Floods* [copyright 1955 by Princeton University Press] Fig. 8, p. 39. Reprinted by permission of Princeton University Press.)

the dam itself was constructed of coal waste, including fine coal, shale, clay, and mine rubbish, all of which disintegrate rapidly and are very unstable. A safe, economical dam cannot be constructed of this material alone (8).

After the failure of the dam, the impounded water emptied into Buffalo Creek in 15 minutes. At the time, the flow in the creek was well below bank-full stage. The flood wave traveled three hours before reaching the mouth of Buffalo Creek at the Town of Man. Along

the way, the peak flattened and the flood took longer to pass a particular point (Figure 4.7). When the wave hit the Guyandotte River at 11:00 A.M., it produced a sudden high peak in the already rising river (Figure 4.8).

Because the Buffalo Creek flood did not occur naturally, it is not strictly valid to compare it to other natural floods. Nevertheless, this flood was a serious warning of the possible effects of human activities on even small streams (8).

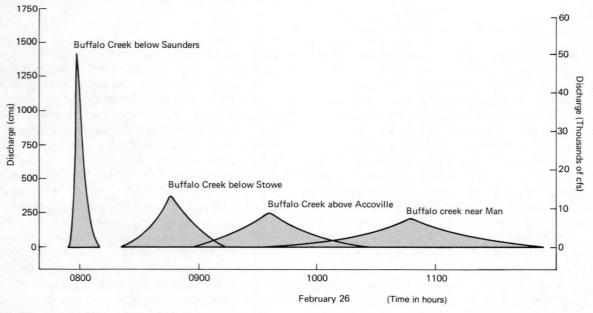

FIGURE 4.7
Estimated flood hydrographs for Buffalo Creek, West Virginia, on February 26, 1972. (After Davies, Bailey, and Donovan, U.S. Geological Survey Circular 667, 1972.)

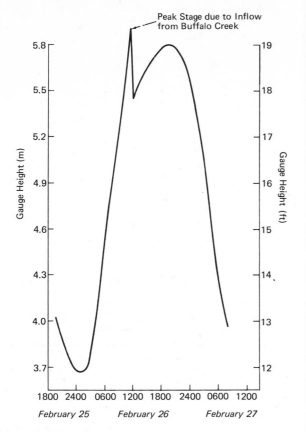

Peak Stage due to Inflow
from Buffalo Creek

FIGURE 4.8
Hydrograph of the Guyandotte River at Man, West Virginia, during the period February 25 to February 27, 1972. The flood inflow from Buffalo Creek produced a sudden high peak of 5.89 meters at the gauging station. Peak discharge was 2,950 cubic meters per second. (After Davies, Bailey, and Donovan, U.S. Geological Survey Circular 667, 1972.)

URBANIZATION AND FLOODING

Human use of land in the urban environment has increased both the magnitude and frequency of floods in small drainage basins of a few square kilometers. The rate of increase is a function of the percentage of the land that is covered with roofs, pavement, and cement (referred to as *impervious cover*) and the percentage of area served by storm sewers. Storm sewers are important because they allow urban runoff from impervious surfaces to quickly reach stream channels. Therefore, impervious cover and storm sewers are collectively a measure of the degree of urbanization. The graph in Figure 4.9 shows that an urban area with 40 percent impervious surface and 40 percent served by storm sewers can expect to have about three times as many overbank flows as before urbanization. This relation holds for floods of small and intermediate frequency, but as the size of the drainage basin increases, floods

of high magnitude with frequencies of 50 years or so are not much affected by urbanization (Figure 4.10).

Floods are a function of rainfall-runoff relations, and urbanization causes a tremendous number of changes in these relations. One study showed that urban runoff is 1.1 to 4.6 times preurban runoff (5). (See Figure 4.11.) The greatest differences correspond to larger storms. Estimates of discharge for different recurrence intervals at different degrees of urbanization are shown in Figure 4.12. The estimates dramatically indicate the tremendous increase of runoff with increasing impervious areas and storm sewer coverage.

The increase of runoff with urbanization occurs because less water infiltrates the ground, as suggested by the significant reduction in time between the majority of rainfall and the flood peak (lag-time) for urban versus rural conditions (Figure 4.13). Short lag-times, referred to as *flashy discharge,* are characterized by rapid rise and fall of flood water. Since little water infiltrates the soil, the low water or dry season flow in urban streams, sustained by groundwater seepage into the channel, is greatly reduced. This effectively concentrates any pollutants present and generally lowers the aesthetic amenities of the stream (4).

Urbanization generally increases runoff and thus flooding; however, in specific instances, relationships between land use and flooding for small drainage basins may be quite complex. One study concludes that not all types of urbanization increase all runoff and flood events (9). When row crops (corn and soybeans) are replaced by low density residential development, the predicted runoff and flood peaks for low magnitude events with recurrence intervals of two to four years increase (as expected). For events with recurrence intervals exceeding four years, however, the predicted runoff and flood peaks for the agricultural land may exceed that for residential development (Figure 4.14). The reason for this change in hydrologic character is that, as row crops are replaced by paved areas and grass, the runoff from the paved areas is greater, but the grass produces less runoff than the agricultural land. Therefore, the effect of the land-use change on runoff and flooding depends on the nature and extent of urbanization and in particular on the proportion of paved and grass-covered areas. Thus, for the land uses specified in Table 4.2 and shown in Figure 4.14, the predicted runoff and flood from a 2-year storm are greater for residential areas than for agricultural land. This results because, for low magnitude (light) rainfall events, the runoff from the small paved areas under residential development is a significant contribution to the total runoff, whereas the runoff from the row crop areas is low because of high infiltration. With higher magnitude (more intense) rainfall events, however, the runoff from

FIGURE 4.9
Relationship between the ratio of
overbank flows (after urbanization to before urbanization) and
measure of urbanization. This
figure shows that as the degree
of urbanization increases, the
number of over-bank flows per
year also increases. (From L. B.
Leopold, U.S. Geological Survey
Circular 559, 1968.)

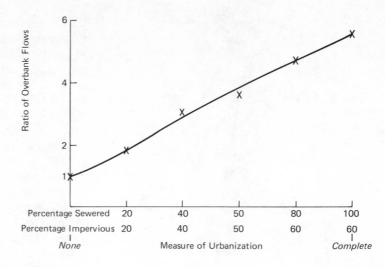

agricultural lands increases, and in comparison, the
runoff from the small paved areas in residential developments is not very significant. Furthermore, grass
areas have higher infiltration rates than do row crop
areas. When urbanization consists of apartments or
commercial development, the amount of paved area increases rapidly and the grassed areas decrease (Table
4.2); thus, the predicted runoff and floods are larger (at
all recurrence intervals) than for agricultural land (Figure 4.14) (9). Therefore we conclude that, because most
urbanization (except for very small drainage basins)
tends to be a mixture of various types of development,
the concept that urbanization increases runoff and
flooding remains valid in most situations.

TABLE 4.2
Relative cover (pavement or grass) as a function of different
types of urbanization for small basins in Illinois. (Data from
Terstriep, et al., "Conventional Urbanization and Its Effect
on Storm Runoff." Illinois State Water Survey, 1976.)

Type of Urbanization	Paved Area (%)	Grass (%)
Residential	23	77
Apartments	60	40
Commercial	75	25

FIGURE 4.10
Graph showing the variation of
flood frequency with percentage
of impervious area. The mean
annual flood is the average (over
a period of years) of the largest
flow that occurs each year. The
mean annual flood in a natural
river basin with no urbanization
has a recurrence interval of 2.33
years. Note that the smaller
floods with recurrence intervals
of just a few years are much
more affected by urbanization
than the larger floods. The 50-
year flood is little affected by the
amount of area that is rendered
impervious. (From L.A. Martens,
U.S. Geological Survey Water
Supply Paper 1591C, 1968.)

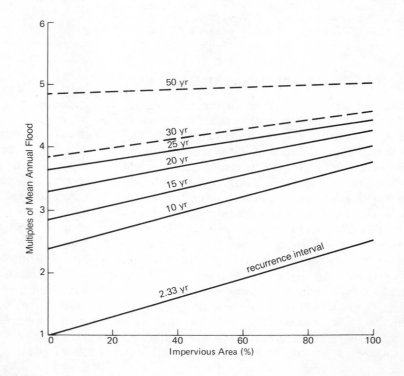

FIGURE 4.11

Comparison of the rainfall–runoff relationships for preurban and urban conditions. Data are for individual storms in one subarea of Nassau County, New York. (After G. E. Seaburn, U.S. Geological Survey Professional Paper 627B, 1969.)

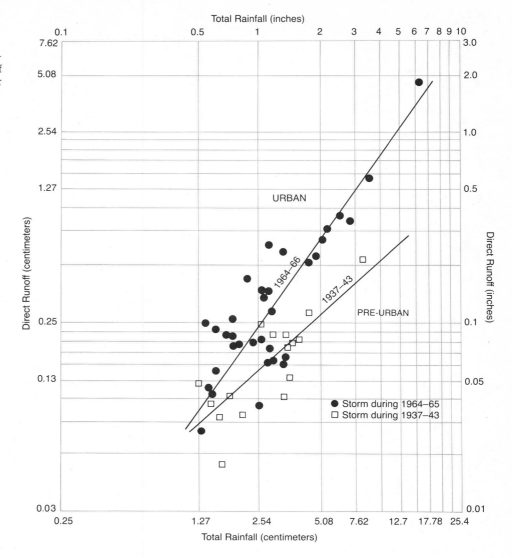

NATURE AND EXTENT OF FLOOD HAZARD

The nature and extent of flood hazard include factors that control flood damage, preventive measures and adjustments, and perception of the hazard.

Factors that control the damage caused by floods include (1) land use on the floodplain; (2) magnitude (depth and velocity of the water and frequency of flooding); (3) rate of rise and duration of flooding; (4) season; (5) sediment load deposited; and (6) effectiveness of forecasting, warning, and emergency systems.

Effects of the flooding may be *primary*, that is, caused by the flood, or *secondary*, caused by disruption and malfunction of services and systems associated with the flood (9). Primary effects include injury and loss of life, and damage caused by swift currents, debris, and sediment to farms, homes (Figures 4.15 and

4.16), buildings, railroads, bridges (Figure 4.17), roads, and communication systems. Erosion and deposition of sediment in the rural and urban landscape may also involve a loss of considerable soil and vegetation (Figure 4.18). Secondary effects may include short-term pollution of rivers, hunger and disease, and displacement of persons who have lost their homes. In addition, fires may be caused by short circuits or broken gas mains (10).

Preventive and adjustment measures for floods involve different approaches. *Prevention* includes engineering structures and projects such as levees and flood walls to serve as physical barriers against high water; reservoirs to store water for later release at safe rates; on-site storm water detention systems or retention basins (Figures 4.19 and 4.20); channel improvements *(channelization)* to increase channel size and move water off the land quickly; and channel diver-

FIGURE 4.12

Flood frequency curve for a 2.6-square-kilometer (one-square-mile) basin in various states of urbanization. Note 100–60 means basin is 100 percent sewered and 60 percent of surface area is impervious. Dashed line shows increase in mean annual flood with increasing urbanization. (After L. B. Leopold, U.S. Geological Survey Circular 559, 1968.)

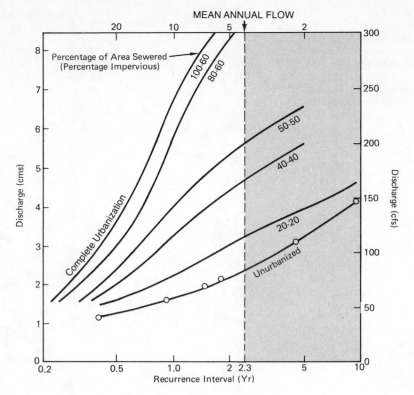

FIGURE 4.13

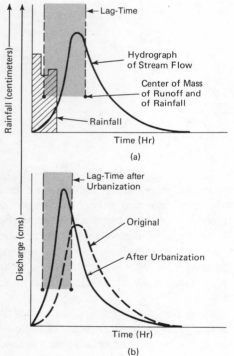

Generalized hydrographs. Hydrograph (a) shows the typical lag between the time when most of the rainfall occurs and the time when the stream floods. Hydrograph (b) shows the decrease in lag-time because of urbanization. (After L. B. Leopold, U.S. Geological Survey Circular 559, 1968.)

sions to route flood waters around areas requiring protection. A major problem with a structural approach to flood hazard is that upstream reservoirs, levees, and flood walls tend to produce a false sense of security that leads to floodplain development. *Adjustment* measures include floodplain regulation and flood insurance (10).

From an environmental view, the best approach to minimizing flood damage in urban areas is floodplain regulation. This is not to say that physical barriers, reservoirs, and channel works are not desirable. In areas developed on floodplains, they will be necessary to protect lives and property (Figure 4.21). We need to recognize, however, that the floodplain belongs to the river system, and encroachment that reduces the cross-sectional area of the floodplain increases flooding (Figure 4.22). Therefore, an ideal solution would be to discontinue development of areas subject to flooding that necessitates new physical barriers. In other words, we must "design with nature." Realistically, in most cases, the most effective and practical solution is a combination of physical barriers and floodplain regulations that will result in less physical modification of the river system. For example, reasonable floodplain zoning in conjunction with a diversion channel project or upstream reservoir may result in a smaller diversion channel or reservoir than would be necessary if no floodplain regulations were used.

FIGURE 4.14

Runoff volume-frequency curves of a small tributary of the Fox River, Illinois, for different land uses. (After Terstriep, et al., *Conventional Urbanization and Its Effect on Storm Runoff*, Illinois State Water Survey Publication, 1976.)

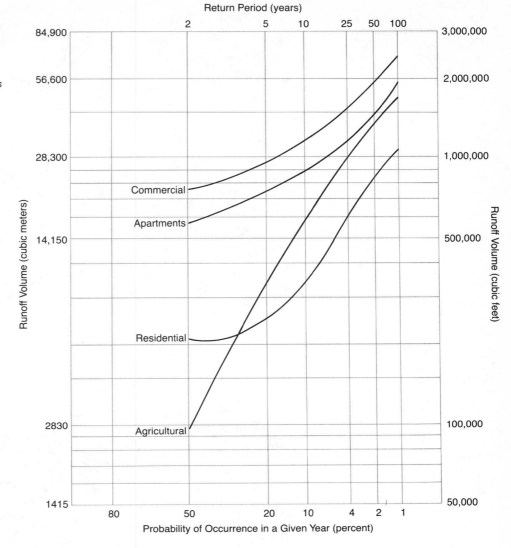

The purpose of floodplain regulation is to obtain the most beneficial use of floodplains while minimizing flood damage and cost of flood protection (11). It is a compromise between indiscriminate use of floodplains, which results in loss of life and tremendous property damage, and complete abandonment of floodplains, which gives up a valuable natural resource. A preliminary step to floodplain regulation is flood-hazard mapping, which is a means of providing floodplain information for land-use planning (12). These maps may delineate past floods or floods of a particular frequency, say, the 100-year flood, and are useful in deriving regulations for private development, purchasing of land for public use as parks and recreational facilities, and providing guidelines for future land use on floodplains.

Developing flood hazard maps for a particular drainage basin can be difficult and expensive. The maps are generally prepared by analyzing stream flow data from gaging stations over a period of years; however, flow data are not available in many cases, especially for small streams, so alternative data sources must be used to assess the flood hazard. Methods of "upstream" flood hazard evaluation may involve estimations of flood peak discharges based on physical properties of the drainage basin. A study of streams in central Texas characterized by periodic intense upstream flooding produced a preliminary empirical model to estimate flood peak discharges by measuring the number of source streams (*stream magnitude*) of a drainage basin and the *drainage density* (total length of all streams in a basin divided by the area of the basin) (13). In other words, the Texas work produced a statistically valid relationship between measured flood peak discharge and two measured physical parameters, stream magnitude and drainage density, that can be used to predict floods in basins where hydrologic

FIGURE 4.15
Flooding of farmland and buildings during the Missouri River floods of 1952. (Photo by Forsythe, courtesy of U.S. Department of Agriculture.)

FIGURE 4.16
Damage caused by flooding of the Missouri River in western Iowa, 1952. The upper photograph shows disruption of transportation systems (railroad) and the lower photograph shows a close-up of the flood damages to government-stored grain. (Photos by Forsythe, courtesy of U.S. Department of Agriculture.)

FIGURE 4.17
Flood damage to state highway
in Ohio. (Photo by N. Richard
Mowrey, courtesy of U.S. Department of Agriculture, Soil Conservation Service.)

information is unavailable or insufficient for detailed evaluation.

Flood hazard evaluation for "downstream" areas may also be accomplished in a general way by direct observation and measurement of physical parameters. For example, extensive flooding of the Mississippi River Valley during the spring of 1973 shows clearly on images produced from satellite-collected data (Figure 4.23).

Floods can also be mapped from aerial photographs taken during flood events or estimated from high water lines, flood deposits, scour marks, and

FIGURE 4.18
Aerial view of flooding of farm-
lands during the Missouri River
floods of 1952. (Photo by Forsythe, courtesy of U.S. Department of Agriculture.)

FIGURE 4.19
Comparison of runoff from a paved area through a storm drain with runoff from a paved area through a temporary storage site (retention pond). Notice that the paved area drained by way of the retention pond produces a lesser peak discharge and therefore is less likely to contribute to flooding of the stream. (Modified after U.S. Geological Survey Prof. Paper 950.)

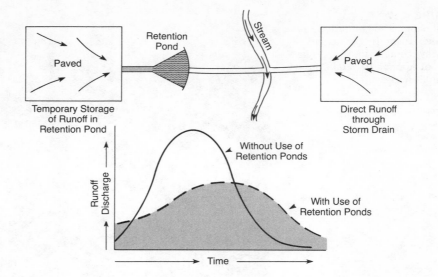

trapped debris on the floodplain, measured in the field after the water has receded (Figure 4.24).

Careful mapping of soils and vegetation can also help evaluate the downstream flood hazard. Soils on floodplains are often different from upland soils, and, with favorable conditions, certain soils can be correlated with flooding of known frequency. Figure 4.25 shows a fair correlation between soils and the 100-year flood (13). Vegetation type may facilitate flood hazard assessment because there is often a rough zonation of vegetation in river valleys that may correlate with flood zones. Some types of trees with shallow roots require an abundant supply of water and benefit from frequent submergence. These trees are often found near the banks of perennial streams that frequently flood. Other species of trees are more restricted to well-drained soils that are not subjected to prolonged or frequent flooding. While zonation of vegetation is helpful in evaluating flood-prone areas, the cause of the zones is complex and not directly caused by flooding. Therefore, use of vegetation, as with soils, should be combined with other methods of flood hazard evaluation such as satellite data, aerial photographs, historical records, and floodplain features (13).

The primary advantages of evaluating both upstream and downstream flood hazards from direct observation or properties of the drainage basin and river valley are expediency and cost. A recent study of the

FIGURE 4.20
Photograph of a storm water retention pond in Charlotte, N.C. This pond functions by draining water from a complex of buildings and parking lots, holding the water and discharging it into a nearby stream at a relatively slow rate.

FIGURE 4.21
Dike on the righthand side of the river protecting commercial property from flooding of the Missouri River in 1973. (Photo courtesy of Department of the Army, U.S. Corps of Engineers.)

Colorado River Valley near Austin, Texas, showed that these methods could easily distinguish areas with a frequent flood hazard (1- to 4-year recurrence interval) from areas with an intermediate (10- to 30-year) and infrequent (greater than 100-year) hazard (13). The only disadvantage is that of accuracy. Flood hazard evaluation based on sufficient hydrologic (stream flow) data generally provides more accurate prediction of flood events.

In urbanizing areas, the accuracy of flood-hazard mapping based entirely on stream flow data is questionable. A better map may be produced by assuming projected urban conditions with estimated percentage of impervious areas. A theoretical 100-year flood map can then be produced. The U.S. Geological Survey in Charlotte, North Carolina, is producing such maps for use with the Mecklenburg County Floodway Regulations (Figure 4.26). Two districts are mapped in the flood hazard area, the Floodway District and the Floodway Fringe District (Figure 4.27).

The *Floodway District* is the portion of the channel and floodplain of a stream designated to provide passage of the 100-year regulatory flood without increasing the elevation of the flood by more than 0.3 meters. On this land, permitted uses include farming, pasture, outdoor plant nurseries, wildlife sanctuaries

FIGURE 4.22
Idealized diagram showing the effect of floodplain encroachment on increasing flood heights. (From Water Resources Council, *Regulation of Flood Hazard Areas*, vol. 1, 1971.)

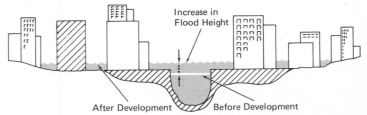

Floodplain encroachments increase flood heights.

Increase in Flood Height

After Development   Before Development

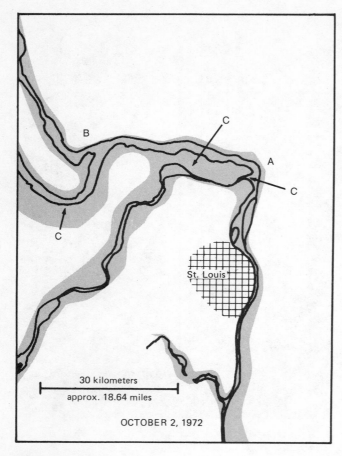

FIGURE 4.23

The spring 1973 flood of the Mississippi River as interpreted from satellite (Landsat) images. Point A is the confluence of the Missouri and Mississippi Rivers, and Point B is the confluence of the Illinois and Mississippi Rivers. Notice that, during the flood, the river is still very narrow through St. Louis. This is a result of the city's flood protection works. Points labelled C are large areas inundated by the flood waters, in part resulting from the flood control protection works that constrict the flow at St. Louis, causing upstream ponding of water. (After N.M.. Short, et al., *Mission to Earth: Landsat Views the World*, NASA, 1976.)

and game farms; loading areas, parking areas, rotary aircraft ports and similar uses, provided they are further than 8 meters from the stream channel; golf courses, tennis courts, archery ranges, and hiking and riding trails; streets, bridges, overhead utility lines, and storm drainage facilities; temporary facilities for certain specified activities such as circuses, carnivals, and plays; boat docks, ramps, and piers; and dams, if they are constructed in accordance with approved specifications. Other uses of the Floodway District, such as open storage of equipment or material, structures designed for human habitation, or underground storage of fuel or flammable liquids, require special permits.

The second district is the *Floodway Fringe District,* which consists of the land located between the Floodway District and the maximum elevation subject to flooding by the 100-year regulatory flood. Permitted uses in this area include any uses permitted in the Floodway District; residential accessory structures, provided they are firmly anchored to prevent flotation; fill material that is graded to a minimum of 1 percent grade and protected against erosion; and structural foundations if firmly anchored to prevent flotation. Aboveground storage or processing of any material that is flammable or explosive or that could cause injury to human, animal, or plant life in time of flooding is prohibited on the Floodway Fringe District.

Figure 4.28 shows a typical zoning map before and after establishment of floodplain regulations.

FLOODS OF THE 1970s AND EARLY 1980s

Floods recorded in the United States from 1972 through 1983 took hundreds of lives and caused several billion dollars of property damage. Floodplain residents will certainly long remember these years both for the high and dangerous waters and for the incalculable suffering of the survivors. Floods in June of 1972 rank as two of the worst flood disasters in the history of the United States (2). The first occurred on June 9 and 10 in the Black Hills area of South Dakota, and the second occurred in the eastern United States from June 21 through 24.

The South Dakota event claimed 242 lives in a catastrophic flash flood. Damages were estimated at $163 million. Most of the deaths and property losses were along Rapid Creek near Rapid City, South Dakota. The flood resulted from torrential rains that struck the Black Hills, delivering, at places, 38 centimeters of rain in five hours. Total rainfall from the 7-hour storm nearly equaled the yearly average. A storm of this magnitude is likely to strike only every several thousand years (2).

The floods in the eastern United States were produced by tremendous rainstorms associated with Hurricane Agnes. In terms of property damage (exceeding $3 billion), this was the most devastating flood ever to strike this country. One hundred thirteen persons lost their lives. Torrential rains and floods were recorded everywhere east of the Appalachians from North Carolina to New York State (2).

Total loss of property from flood damage in 1973 (Figure 4.23) was second only to that in 1972. The largest flood of the year was the spring flooding of the Mississippi River, which resulted in approximately $1.2 billion in property damage. Although tens of thousands of

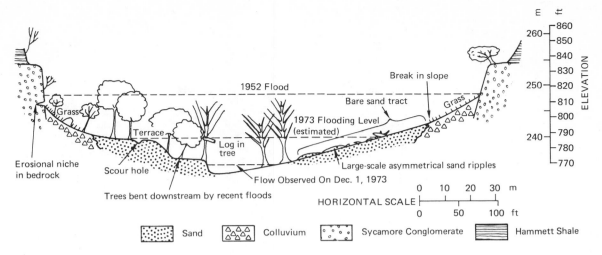

FIGURE 4.24
Schematic cross section of the Pedernales River Valley, Texas, illustrating floodplain features useful in estimating floods. (After V. R. Baker, "Hydrogeomorphic Methods for the Regional Evaluation of Flood Hazards," *Environmental Geology,* vol. 1, pp. 261–81, 1976.)

persons had to be evacuated and suffered appreciably as thousands of square kilometers of farmland were inundated throughout the Mississippi River Valley, fortunately, few deaths were caused directly by the flooding (14).

Violent flash floods, nourished by a complex system of thunderstorms delivering up to 25 centimeters of rain, swept through several canyons west of Love-land, Colorado, on the night of July 31–August 1, 1976, killing 139 people and inflicting over $35 million in damages to highways, roads, bridges, homes, and small businesses. Most of the damage and loss of life was in the Big Thompson Canyon, where hundreds of residents, campers, and tourists were caught in the flood with little or no warning. Although the storm and flood were rare events (several times the 100-year

FIGURE 4.25
Relationship between soils and flooding for a reach of the Colorado River near Austin, Texas. The intermediate frequency floods with a recurrence interval of 10 to 30 years are primarily associated with the Lincoln, Yahola, Norwood, and Bergstrom soils (on the floodplain). The 100-year flood tends to roughly correlate with the lower topographic boundary of the upland soils. (After V. R. Baker, "Hydrogeomorphic Methods for the Regional Evaluation of Flood Hazards," *Environmental Geology,* vol. 1, pp. 261–81, 1976.)

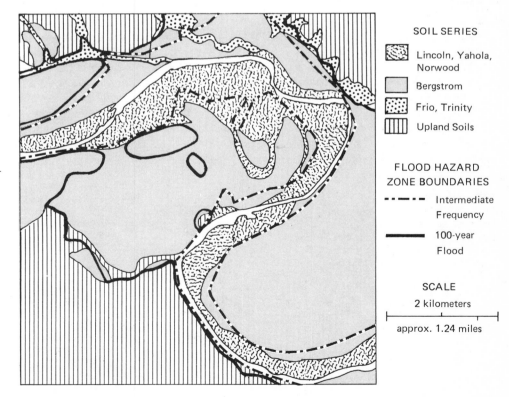

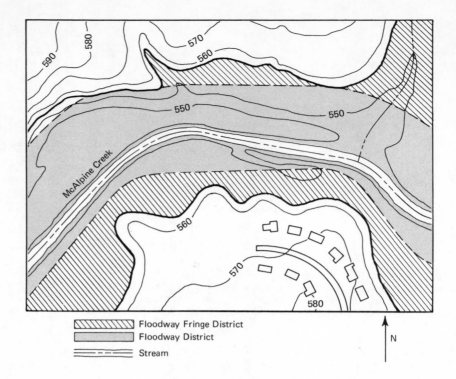

FIGURE 4.26
Example of a flood hazard map showing the Floodway Fringe District and the Floodway District. (From County Engineer, Mecklenburg County, North Carolina. (550 to 580 ft = 168 to 177 m)

flood), comparable floods have occurred in the past and others can be expected in the future for similar canyons along the Front Range of Colorado. This prediction is possible because many streams draining the Front Range, including Boulder Creek with substantial urban development in Boulder, Colorado, contain geologically very young flood deposits analogous to those transported and deposited by the Big Thompson Canyon flood (15, 16, 17).

The September, 1983 floods in Arizona killed at least 13 people and inflicted more than $416 million in damages to homes (over 1,300 destroyed or damaged), highways, roads, and bridges. Interestingly, officials complained that donated food, clothing, and money were so abundant that at times they actually made disaster relief efforts more difficult. Disaster planners

should learn to live with such difficulties—congratulations are in order to Arizona's ''good neighbor policy.''

PERCEPTION OF FLOODING

At the institutional level, sensitivity to the perception of flooding appears to be adequate; however, on the individual level, the situation is not as clear. Several conclusions have been drawn from perception studies. Accurate knowledge of flood hazard does not inhibit all persons from moving onto a floodplain. Flood hazard maps are not effective as a form of communication. Upstream development is frequently the scapegoat for downstream floodplain residents threatened by floods. People are tremendously variable in their knowledge of

FIGURE 4.27
Idealized diagram showing a valley cross section and its Flood Hazard Area, Floodway District, and Floodway Fringe District. (From Water Resources Council, *Regulation of Flood Hazard Areas*, vol. 1, 1971.)

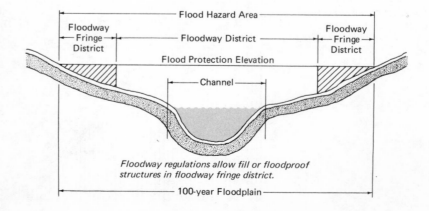

FIGURE 4.28
Typical zoning map before and after the addition of flood regulations. (From Water Resources Council, *Regulation of Flood Hazard Areas,* vol. 1, 1971.)

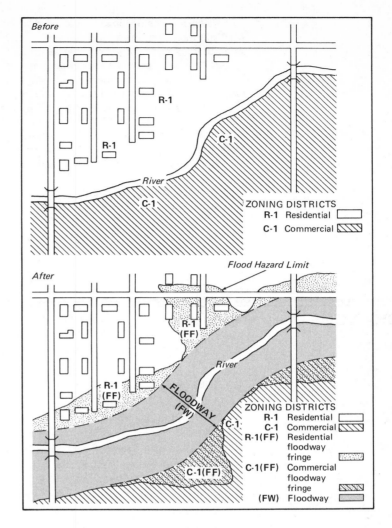

flooding, anticipation of future flooding, and willingness to accept adjustments caused by the hazard (1).

Recent progress at the institutional level includes the mapping of **(1)** flood-prone areas (over 13,000 maps have been prepared); **(2)** areas with a flash flood potential downstream from dams; and **(3)** areas where urbanization is likely to cause problems in the near future. In addition, the U.S. Flood Insurance Administration program has encouraged states and local communities to adopt floodplain management plans and has sponsored preparation of flood insurance plans in more than 6,000 of the country's nearly 21,000 communities with a history of flood problems (18).

SUMMARY AND CONCLUSIONS

River flooding is the most universally experienced natural hazard. Loss of life is relatively low in developed countries, with adequate monitoring and warning systems, compared to that in preindustrial societies; but property damage is much greater in industrial societies where floodplains are often extensively developed.

The magnitude and frequency of flooding are a function of the intensity and distribution of precipitation and the rate of infiltration of water into the soil, rock, and topography. Human use and interest in urban areas has increased the flood hazard in small drainage basins by increasing the amount of ground covered with buildings, roads, parking lots, and so on, which forces storm water to run off the land rather than infiltrate it.

Two types of floods are recognized: *upstream* floods produced by intense rainfall of short duration over a relatively small area; and *downstream* floods produced by storms of long duration over a large area that saturate the soil, causing increased runoff from thousands of tributary basins and resulting in large floods in the major rivers.

Factors that control damage caused by flooding include land use on the floodplain, magnitude and frequency of the flooding, amount of sediment deposited, and effectiveness of forecasting, warning, and emergency systems. From an environmental view, the best solution to minimizing flood damage is floodplain regulation. In highly urban areas, however, it will remain necessary to use physical barriers, reservoirs, and channel works to protect existing development. The realistic solution to minimizing flood damage involves a combination of floodplain regulation and engineering techniques.

An adequate perception of flood hazards exists at the institutional level; however, on the individual level, more public-awareness programs are needed to help people perceive the hazard of living in flood-prone areas.

REFERENCES

1 BEYER, J. L. 1974. Global response to natural hazards: Floods. In *Natural hazards,* ed. G. F. White, pp. 265–74. New York: Oxford University Press.

2 U.S. DEPARTMENT OF COMMERCE, 1972. *Climatological data, national summary* 23, no. 13.

3 LINSLEY, R. K., Jr.; KOHLER, M. A.; and PAULHUS, J. L. 1958. *Hydrology for engineers.* New York: McGraw-Hill.

4 LEOPOLD, L. B. 1968. *Hydrology for urban land planning.* U.S. Geological Survey Circular 559.

5 SEABURN, G. E. 1969. *Effects of urban development on direct runoff to East Meadow Brook, Nassau County, Long Island, New York.* U.S. Geological Survey Professional Paper 627B.

6 AGRICULTURAL RESEARCH SERVICE. 1969. *Water intake by soils.* Miscellaneous Publication No. 925.

7 STRAHLER, A. N., and STRAHLER, A. H. 1973. *Environmental geoscience.* Santa Barbara, California: Hamilton Publishing.

8 DAVIES, W. E.; BAILEY, J. F.; and KELLY, D. B. 1972. *West Virginia's Buffalo Creek flood: A study of the hydrology and engineering geology.* U.S. Geological Survey Circular 667.

9 TERSTRIEP, M.L.; VOORHEES, M. L.; and BENDER, G. M. 1976. *Conventional urbanization and its effect on storm runoff.* Illinois State Water Survey Publication.

10 OFFICE OF EMERGENCY PREPAREDNESS. 1972. *Disaster preparedness* 1, 3.

11 BUE, C. D. 1967. *Flood information for floodplain planning.* U.S. Geological Survey Circular 539.

12 SCHAEFFER, J. R.; ELLIS, D. W.; and SPIEKER, A. M. 1970. *Flood-hazard mapping in metropolitan Chicago.* U.S. Geological Survey Circular 601C.

13 BAKER, V. R. 1976. Hydrogeomorphic methods for the regional evaluation of flood hazards. *Environmental Geology* 1:261–281.

14 U.S. DEPARTMENT OF COMMERCE. 1973. *Climatological data, national summary* 24, no. 13.

15 McCAIN, J. F., HOXIT, L. R., MADDOX, R. A., CHAPPELL, C. F. and CARACENA, F. 1979. Storm and flood of July 31–August 1, 1976., in the Big Thompson River and Cache la Poudre River Basins, Larimer and Weld Counties, Colorado. U.S. Geological Survey Professional Paper 1115A.

16 SHROBA, R. R., SCHMIDT, P. W., CROSBY, E. J., and HANSEN, W. R. 1979. Storm and flood of July 31–August 1, 1976, in the Big Thompson River and Cache la Poudre River Basins, Larimer and Weld Counties. Colorado. U.S. Geological Survey Professional Paper 1115B.

17 BRADLEY, W. C., and MEARS, A. I. 1980. Calculations of flows needed to transport coarse fraction of Boulder Creek alluvium at Boulder, Colorado. Geol. Soc. Amer. Bull. Part II, v. 91:1057–1090.

18 EDELEN, G. W., Jr., 1981. Hazards from floods. In *Facing geological and hydrologic hazards, earth-science considerations,* ed. W. W. Hays. U.S. Geological Survey Professional Paper 1240-B:39–52.

Chapter 5

Landslides and Related Phenomena

Slopes are the most common landforms, and although most slopes appear stable and static, they are actually dynamic, evolving systems. Slope processes are one significant reason that stream valleys are much wider than the stream channel and adjacent floodplain. Material on most slopes is constantly moving down the slope at rates that vary from imperceptible creep of soil and rock to thundering avalanches and rockfalls that move at tremendous speeds. As with floods, it may not be the largest (rare and infrequent) nor the smallest (most common) event that moves the greatest amount of material down slopes as valleys develop. Rather, events of moderate magnitude and frequency may play the most important role.

Landslides and related phenomena such as mudflows, earthflows, rockfalls, snow or debris avalanches, and subsidence are natural events that would occur with or without human activity. However, human use and interest has led to both an *increase* in some of these events such as landslides on hillside developments because of oversteeping of slopes and a *reduction* of landslides by means of stabilizing structures or techniques on naturally sensitive slopes likely to experience landslides.

Loss of life and damages from landslides in the United States are substantial. Each year about 25 people are killed by landslides, and this number increases to between 100 and 150 if we include collapses of trenches and other excavations. Total annual cost of damages exceeds $1 billion (1).

Earth materials on slopes may fail and move or deform in several ways, including flowage, sliding, falling, and subsidence (Figure 5.1); a typical downslope movement event (landslide) often involves more than one type of failure (Figure 5.2). Table 5.1 classifies the common types of downslope movements and reflects the terminology used in this discussion and in other chapters. The important variables in classifying downslope movements are type of movement, slope material type, amount of water present, and rate of movement. In general, a landslide is considered to move rapidly if the motion can be discerned with the naked eye; otherwise, it is classified as slow. Actual rates vary from a slow creep at rates of a few millimeters or centimeters per year, to very rapid, at 1.5 meters per day, to extremely rapid, at several tens of meters per second (2;.

SLOPE STABILITY

The nature and extent of slope stability for the various types of downslope movements generally can be ex-

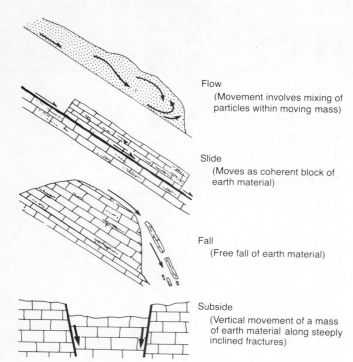

FIGURE 5.1
Common ways that earth materials fail and move in landslides and related phenomena.

Flow (Movement involves mixing of particles within moving mass)

Slide (Moves as coherent block of earth material)

Fall (Free fall of earth material)

Subside (Vertical movement of a mass of earth material along steeply inclined fractures)

plained by examining forces on slopes in terms of the role of these significant, and somewhat interrelated, variables: types of earth materials, slope (topography), climate, vegetation, water, and time.

Forces on Slopes

The causes of many landslides and related types of downslope movement can be examined by studying relations between *driving forces,* those which tend to move earth materials down a slope, and *resisting forces,* those which tend to oppose such movement.

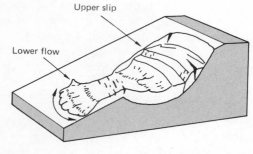

Upper slip

Lower flow

FIGURE 5.2
Block diagram of a common type of landslide consisting of an upper slip motion and a lower flow. (After R. Pestrong, *Slope Stability,* American Geological Institute, 1974.)

TABLE 5.1
Classification of landslides and other downslope movements.

Type of Movement	Materials	
	Rock	Soil
Slides (variable water content and rate of movement)	Slump blocks Translation slide	Slump blocks (Rotational slide) Soil slip (planar)
Falls	Rock fall	Soil fall
slow	Rock creep	Soil creep
Flows	Unconsolidated materials (saturated)	
	Earthflow Mudflow Debris flow	
rapid	Debris avalanche	
Complex	Combinations of slides and flows	

The most common driving force is the downslope component of the weight of the slope material, including anything superimposed on the slope, such as vegetation, fill material, or buildings. The most common resisting force is the shear strength of the slope material acting along potential slip planes. Recall from our discussion of soils in chapter 3 that the shear strength is a function of cohesion and internal friction.

Slope stability is evaluated by computing a safety factor *(SF)* defined as the ratio of the resisting forces to the driving forces. If the safety factor is greater than one, the resisting forces must exceed the driving forces and the slope is considered stable. If, on the other hand, the safety factor is less than one, then the driving forces exceed the resisting forces and a slope failure can be expected. Of course, anything that reduces the resisting forces or increases the driving forces will lower the safety factor and thus increase the chances of a landslide or other type of downslope movement. Driving and resisting forces are not static; rather, they tend to change with time. Thus, depending on changes in local conditions, the safety factor may increase or decrease. For example, consider construction of a road cut in the toe of a slope with a potential slip plane (Figure 5.3). The effect of the road cut in the critical toe of the slope will reduce the driving forces because some of the slope material is removed, and also reduce the resisting forces because the length of the slip plane is reduced, and it is this plane along which the resisting force (shear strength) acts. The overall effect of the road cut will be to lower the safety factor, because the reduction of the driving force is small compared to the reduction of the resisting force. That

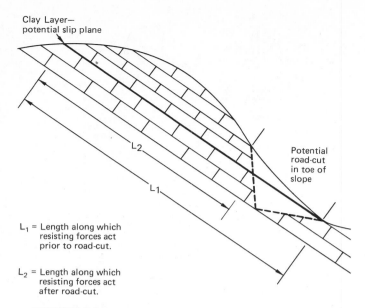

Clay Layer—
potential slip plane

L₂

L₁

Potential
road-cut
in toe of
slope

L_1 = Length along which
resisting forces act
prior to road-cut.

L_2 = Length along which
resisting forces act
after road-cut.

FIGURE 5.3
Illustration of the effects of a road-cut in the toe of a slope
on slope stability. (See text for explanation.)

is, only a small portion of the potential slide mass is
removed, but a relatively large portion of the total
length along which resisting forces act is removed.
Thus, the safety factor must be reduced.

Role of Earth Material Type

The type of downslope movement is partly a function
of slope material type. For landslides, two basic patterns of movement are recognized: *rotational* and
translational (Figure. 5.4). Rotational slides are most
common for soil slopes, but are also associated with
slumping in rock slopes, especially where a permeable
rock such as sandstone overlies a weak rock such as
shale. If, however, a very resistant rock overlies a rock
of very low resistance, then rapid undercutting of the
resistant rock may cause a slab failure (rock fall)
(Figure 5.5).

Translation slides occur along geologic planes of
weakness within rock slopes. Common translation slip
planes include fractures, bedding planes, clay partings,
and foliation planes. In some areas, very shallow slides
parallel to the slope, known as *soil slips,* also occur. For
soil slips, the slip plane is usually above the bedrock
but below the soil within slope material known as *colluvium,* a mixture of weathered rock and other material.

The strength of the slope materials may greatly
influence the magnitude and frequency of landslides
and related events. For example, creep, earthflows,
slumps, and soil slips are much more common on shale
slopes or slopes on weak pyroclastic (volcanic) materials than on slopes on more resistant rock such as well
cemented sandstone, limestone, or granite. In fact,
shales are so notorious for landslide activity that what
is called ''shale terrain'' is characterized by irregular,
hummocky topography produced by a variety of downslope movement processes. For all types of land use,
from agricultural to urban, on shale or other weak rock
slopes, one must carefully consider the possible landslide hazard prior to development.

Role of Slope and Topography

Angle greatly affects the driving forces on slopes. Figure 5.6 illustrates that, as the angle of a potential slip
plane increases, the driving force also increases. Thus,
everything else being equal, landslides should be more
frequent on steep slopes. In general, this is true; for
example, a study of landslides that occurred during two
rainy seasons in California's San Francisco Bay area established that 75 to 85 percent of landslide activity is
closely associated with urban areas on slopes greater
than 15 percent, or 8.5 degrees (Figure 5.7) (3). On a
national scale, the coastal mountains of California and
Oregon, the Rocky Mountains, and the Appalachian
Mountains have the greatest frequency of landslides.
All the types of landslides listed in Table 5.1 occur in
those locations.

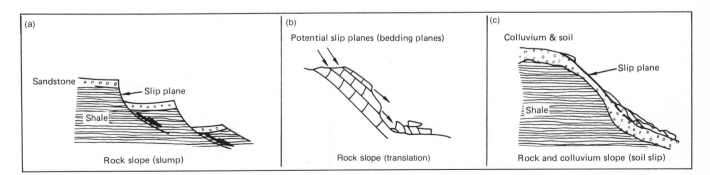

(a)

Sandstone

Slip plane

Shale

Rock slope (slump)

(b)

Potential slip planes (bedding planes)

Rock slope (translation)

(c)

Colluvium & soil

Slip plane

Shale

Rock and colluvium slope (soil slip)

FIGURE 5.4
Some examples of rotational slide (a) and translational slides (b and c).

FIGURE 5.5
Development of a slab slide (type of rock fall). Undercutting of resistant "cap" rock at time (b) causes the development of tension fractures and eventual failure by rock fall at time (c).

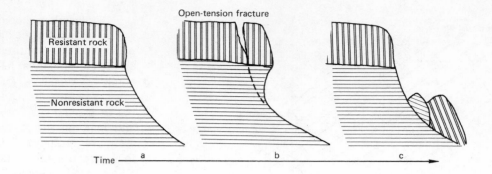

FIGURE 5.6
Effect of slope on the driving force. Notice that, for a hypothetical slide mass, the driving force increases as the slope increases.

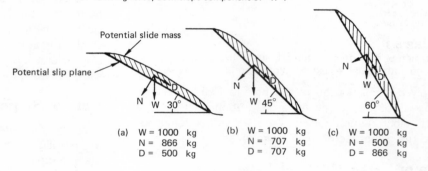

Symbols: W = Weight of slope materials above a possible slip plane.
N = Component of "W" acting into the slope (is a resisting force).
D = Driving force; downslope component of "W".

(a) W = 1000 kg
N = 866 kg
D = 500 kg

(b) W = 1000 kg
N = 707 kg
D = 707 kg

(c) W = 1000 kg
N = 500 kg
D = 866 kg

FIGURE 5.7
Relationship between landslide frequency and urban areas on slopes in the San Francisco Bay area exceeding 15 percent. (After T. H. Nilsen, F. A. Tayler, and R. M. Dean, U.S. Geological Survey Bulletin 1424, 1976.)

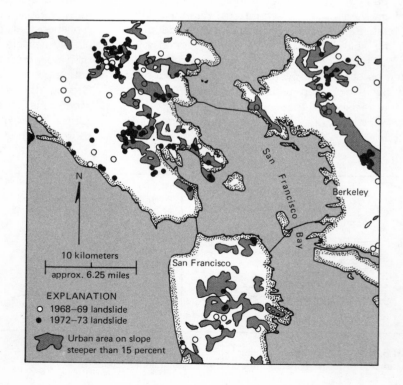

To some extent, the type of landslide activity is also a function of slope and topography. For example, rock falls and debris avalanches are associated with very steep slopes, and shallow soil slips in southern California are common on steep saturated slopes. These soil slips are often transformed downslope into debris flows that are extremely hazardous (Figure 5.8). At the other extreme, earthflows may occur on moderate slopes, and the effect of creep may be observed on slopes of only a few degrees.

Role of Climate and Vegetation

Climate and vegetation can influence the type of landslide or other downslope movement that occurs on a particular slope. Climate is significant in influencing downslope movement because it controls the nature and extent of the precipitation and thus the moisture content of slope materials. For example, both earthflows (which usually occur on slopes) and mudflows (which initially may be confined to channels) involve downslope movement of water-saturated earth materials; but earthflows are common and mudflows relatively rare in humid regions. This results because a good deal of water infiltrates into slopes in the humid areas, facilitating earthflows. Furthermore, humid regions have many perennial streams that continuously transport the materials delivered from slopes out of the drainage basin. Thus, little debris builds up in the

FIGURE 5.8
Home in southern California destroyed by debris flow that originated as a soil slip. This 1969 event claimed two lives. (Photo courtesy of Los Angeles Department of Building and Safety.)

basin, so little material is available for mobilization in a mudflow.

The role of vegetation in landslides and related phenomena is complex because the vegetation in an area is a function of several factors, including climate, soil type, topography, and fire history, each of which also influences what happens on slopes. Vegetation is a significant factor in slope stability for three reasons. First, vegetation provides a cover that cushions the impact of rain falling on slopes, thus facilitating infiltration of water into the slope while retarding grain-by-grain erosion on the surface. Second, vegetation has root systems that tend to provide an apparent cohesion to the slope materials. And third, vegetation adds weight to the slope.

Most problems concerning slope stability and vegetation result from disturbance or removal of vegetation from slopes. In some cases, however, vegetation increases the probability of a landslide, especially for specific types of shallow soil slips on steep slopes. For example, one type of soil slip in southern California coastal areas occurs on steep-cut slopes covered with ice plant (Figure 5.9). During especially wet winter months, the shallow-rooted ice plants take up water, adding considerable weight to steep slopes and increasing the driving forces. The plants also cause an increase in the infiltration of water into the slope, which decreases the resisting forces. When failure occurs, the plants and several centimeters of roots and soil slide to the base of the slope.

Soil slips on natural steep slopes in southern California are a more serious problem (Figure 5.8), and again vegetation, in this case chaparral (dense shrubs or brush), facilitates an increase in water infiltrating into the slope which tends to lower the safety factor. One study concluded that, in some instances, susceptibility to soil-slip may be greater on vegetated slopes than on slopes where vegetation had been recently removed by fire. This should not be interpreted to mean that burning reduces the landslide hazard. Even though soil slips may sometimes be reduced by removal of vegetation, they are not eliminated; in addition, the grain-by-grain erosion caused by rain splash and sheet wash on the surface greatly increases. The eroded sediment tends to fill up the ravines (steep stream valleys) with a meter or more of debris that may be mobilized into mudflows during wet winters (4).

The type of vegetation also affects the frequency and morphology of shallow soil slips in southern California. To compare grass-covered with chaparral-covered slopes, soil slips on grassy slopes are three to five times more frequent, have a lower minimum angle at which failure occurs, and tend to be shorter and wider than those on chaparral-covered slopes (4).

FIGURE 5.9
Shallow soil slips on steep slopes covered with shallow rooted ice plants near Santa Barbara, California.

Fire may facilitate development of hydrophobic soil horizons that retard infiltration of water and establishment of vegetation after burning. The "nonwettability" or hydrophobic properties are produced when waxy organic coatings on soil particles are mobilized by the fire and accumulate in a soil horizon (4). Hydrophobic soils have been recognized in areas with chaparral vegetation as well as forest cover, and the problem is more serious than generally realized because the behavior of water in the near surface environment is critical to the establishment of vegetation and retardation of soil erosion. Unfortunately, our often necessary practice of not allowing periodic natural burns has increased the chances of formation of hydrophobic soils because the less frequent fires have more fuel and burn very hot, increasing the likelihood of hydrophobic properties developing in the soils.

Disturbance or removal of vegetation by logging has also been associated with an increase in landslides. Clearcutting, or removal of all trees, has caused several problems. First, the rate of transpiration (loss of soil water through the trees) is greatly reduced; thus soil moisture increases, tending to reduce the stability of the slopes. Second, in specific instances, infiltration of water into a slope may be increased. This is especially likely if a permeable soil on relatively low-gradient slopes is covered in winter with thick snow pack that slowly melts in the spring; as the slope becomes saturated, resisting forces are lowered. Third, with the exception of redwood trees that regenerate after logging, the roots of cut trees decay with time, reducing their strength and thus the apparent cohesion of the soil. This tends to reduce resisting forces within the slope, helping to explain the increased frequency in landslides several years following timber harvesting (5).

Role of Water

Water is almost always directly or indirectly involved with landslides, and its role is thus particularly important. Water is the universal solvent, and much weathering of rocks, which slowly reduces their shear strength, takes place near the surface of the earth. The weathering contributes to the instability of rocks and is especially significant in areas with limestone, which is very susceptible to chemical weathering. It has been argued that water has a lubricating effect on individual soil grains and potential slide planes. This notion is incorrect; water is not a lubricant (6). On the other hand, the amount of water in a soil is very significant. For example, dry sand will form a cone when dumped, whereas moist sand will stand nearly vertical because surface tension in the water holds the grains together, and with all its pore spaces filled, saturated sand will flow as mud.

The effects of water on slopes and landslides are quite variable. First, saturation of earth materials causes a rise in pore water pressure. In general, as the water pressure in slopes increases, the shear strength (resisting force) of the slope decreases and the weight (driving force) increases; thus, the net effect is to lower the safety factor. This is thought to be a significant factor in the development of soil slips and debris avalanches, as well as other types of landslides. Soil slips generally occur during heavy rainfall when near-surface temporary or *perched* water table conditions may be present. During a rainstorm, the rate of surface infiltration in the unsaturated zone of the soil or colluvium exceeds the rate of deep percolation in the rock below the colluvium, and even though some water moves as seepage parallel to the slope, a temporary (perched) water table develops (Figure 5.10). Failure of the slope

FIGURE 5.10
Idealized diagram showing development of a perched water table in colluvial material during heavy rainfall and relationship to increased instability of the slope. (After R. H. Campbell, U.S. Geological Survey Profesional Paper 851, 1975.)

occurs when resisting forces are reduced sufficiently, that is, when the safety factor becomes less than one. The safety factor will be a minimum when the perched water table rises to the surface, indicating that the potential slide mass is entirely saturated.

A rise in water pressure is present prior to many landslides, and it is safe to say that most landslides are caused by an abnormal increase of the water pressure in the slope-forming materials (6).

Rapid draw-down, when the water level in a reservoir or river is lowered quickly at a rate of at least a meter per day, is a second way that water may affect the stability of slopes. When the water is at a relatively high level, a large amount may enter the banks, producing a bank storage phenomenon. Then, when the water level suddenly drops, the water stored in the banks is left unsupported. This produces an abnormal pore water pressure that reduces the resisting forces; simultaneously, the weight of the stored water increases the driving forces. For this reason, bank failures (slumps) tend to occur along streams after flood water has receded.

A third way that water affects landslides is by contributing to spontaneous liquefaction of clay-rich sediment or *quick clay*. When disturbed, some clays lose their shear strength, behave as a liquid, and flow. The shaking of clay below Anchorage, Alaska, during the 1964 earthquake produced this effect and was extremely destructive (Figures 5.11 and 5.12). Other examples of slides associated with sensitive clays are found in Quebec, Canada, where several large earth

movements have destroyed numerous homes and killed about seventy people in recent years. The slides occur on river valley slopes in initially solid material that is converted into a liquid mud as movement begins (7). These slides are often initiated by river erosion at the toe of the slope and, although they start in a small area, may develop into large events. Since they often involve the reactivation of an older slide, future problems may be avoided by restricting development. Fortunately, older slides, even though masked from ground view by vegetation, are often visible on aerial photographs. The Quebec slides/flows are especially interesting because the liquefaction of clays occurs without earthquake shaking, unlike those in Anchorage.

Seepage of water from artificial sources, such as reservoirs and unlined canals, into adjacent slopes may also affect slope stability. Seepage can lower the cohesion and thus the shear strength by removing cementing materials, as was the case in the St. Francis Dam failure. Seepage may also cause an increase of water pressure in adjacent slopes, causing a reduction in the resisting force.

The ability of water to erode also affects the stability of slopes. Stream or wave erosion may remove and steepen the toe area of a slope, thus reducing the safety factor. This problem is particularly critical if the toe of the slope is an old, inactive landslide that is likely to move again if stability is reduced. Therefore, it is important to recognize old landslides along potential road cuts and other excavations prior to construction, to isolate and correct potential problems.

FIGURE 5.11
Block diagram showing how landslides developed in response to the 1964 Alaskan earthquake in Anchorage. The graben is a down-drop block of earth material. (From W. R. Hansen, U.S. Geological Survey Professional Paper 542A, 1965.)

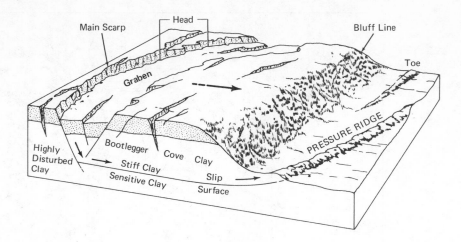

(a)   (b)

FIGURE 5.12
Landslides in Anchorage, Alaska, caused by the 1964 earthquake. The slides are similar to those diagrammed in Figure 5.11: Government Hill School slide (a), aerial view of slide near the hospital on Fourth Avenue (b), and slides that caused the collapse of Fourth Avenue near C Street (c). (Photos [a] and [b] by W. R. Hansen, courtesy of U.S. Geological Survey. Photo [c] courtesy of U.S. Army.)

(c)

Role of Time

The forces on slopes often change with time. For example, both driving and resisting forces may change seasonally as the moisture content or water table position alters. Furthermore, these changes are greater in especially wet years, as reflected by the increased frequency of landslides in wet years. In other slopes, a continuous reduction in resisting forces may occur with time, perhaps due to weathering, which reduces the cohesion in slope materials, or to a regular increase in water pressure from natural or artificial conditions. A slope that is becoming less stable with time may have an increasing rate of creep until failure occurs (Figure 5.13a). A slope's safety factor might also decrease with time because of progressive wetting, which causes disarrangement of the soil particles in the slope, lowering internal friction and thus the strength of the slope materials. Figure 5.13b illustrates this phenomenon as a result of road building and culvert installation that periodically dumps water downslope. This situation emphasizes an important point:

we may design slopes that are stable when constructed, but stability has a way of changing with time. Therefore, slope design should include provisions to minimize processes that might progressively weaken slope materials.

CAUSES OF LANDSLIDES

In many cases, the real causes of landslides—increase in driving force or decrease in resisting force—are masked by immediate causes, such as earthquake shocks, vibrations, or a sudden increase in the amount of moisture in slope material. The distinction between real and immediate causes is very important in court when a landslide case is heard and a definitive statement by an earth scientist concerning the cause of a landslide is expected (8). For example, a translation slide may have as an *immediate* cause heavy rains that saturated the earth material, while the *real* cause is the potential to slide upon long, weak, clay layers. A similar hypothetical example might be given for the slide of

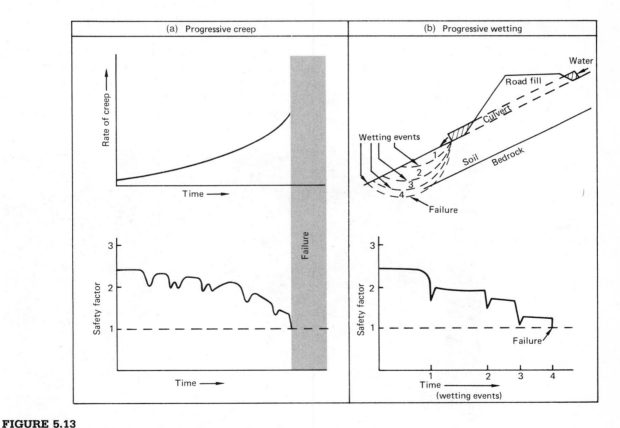

FIGURE 5.13
Idealized diagrams showing the influence of time on the development of a landslide: progressive creep (a) and progressive wetting (b). Progressive creep is symptomatic of a decreasing safety factor, whereas progressive wetting may cause a reduction in the resisting forces and thus produce a lower safety factor. (Diagram [b] after C. S. Yee and D. R. Harr, *Environmental Geology*, vol. 1, p. 374, 1977.)

an artificial slope in a housing development in which the immediate cause is an earthquake shock, but the real cause is a poorly designed slope.

Causes of landslides may also be grouped according to whether they are external or internal. External causes produce an increase in the shear stress (driving force per unit area) at relatively constant shear strength (resisting force per unit area). Examples of external causes include loading of a slope, steepening of a slope by erosion or excavation, and earthquake shocks. Internal causes produce landslides without any recognized external changes, and include processes that reduce the shear strength. Examples include such changes as increase in water pressure or decrease in cohesion of the slope materials. In addition, some causes of landslides are intermediate, having some attributes of both external and internal causes. For example, rapid drawdown involves an increase in the shear stress accompanied by decrease in shear strength. Other intermediate causes include spontaneous liquefaction, and subsurface weathering and erosion (6).

HUMAN USE AND LANDSLIDES

Effects of human use and interest on the magnitude and frequency of landslides very from nearly insignificant to very significant. In cases where our activities have little to do with the magnitude and frequency of landslides, we need to learn all we can about where,

when, and why they occur to avoid hazardous areas and minimize damage. In cases where human use has increased the number and severity of landslides, we need to learn how to recognize, control, and minimize their occurrence wherever possible.

Mixtures of adverse geologic conditions such as weak soil or rock and potential slip plains on steep slopes with torrential rains, heavy snowfall, or seasonably frozen ground (permafrost) will continue to produce landslides, mudflows, and avalanches regardless of human activities. These are natural processes reacting to natural conditions. For example, the night of January 22, 1967, produced a landslide disaster of tremendous magnitude in Brazil. Following a three-hour electrical storm and cloudburst, an area of about 194 square kilometers was devastated by landslides and erosion (floods) that claimed the lives of about 1,700 people. Damage to property and effects on industry were inestimable, as tens of thousands of landslides turned green hills into wastelands and valleys into mud baths (Figure 5.14). All the slides were debris avalanches and mudflows (9). The slopes in disaster areas characteristically have only a thin veneer of residual soil over hard rock, resulting in a contact of relatively easy parting (Figure 5.15). More avalanches cut areas of high vegetation than grass-covered slopes, and many destroyed forests left in their natural state for many years. The combination of high-frequency vibrations of the electrical storm, pressure of the falling water, rising water pressure, groundwater movement,

FIGURE 5.14
Floresta Creek Valley mudflow at the base of the Sierra Das Araras Escarpment in Brazil. Before the mudflow, the valley bottom contained a village and a highway construction camp. Several hundred people lost their lives when the mudflow, which was about 4 meters deep, moved through the area. (Photo by F. O. Jones, courtesy of U.S. Geological Survey.)

FIGURE 5.15
Debris avalanche in thin, residual soils in Brazil. (Photo by F. O. Jones, courtesy of U.S. Geological Survey.)

rock. The average amount of rock and soil debris moved was 2,500 cubic meters, or about 36,000 metric tons (10). Although the loss of life in Virginia was terrible, it was relatively low because of the sparse population. In 1970, inhabitants of Yungay and Ranrahirca, Peru, were not so fortunate when a debris avalanche triggered by an earthquake roared 3,660 meters down Mt. Huascaran at a speed in excess of 300 kilometers per hour, killing about 20,000 persons, depositing many meters of mud and boulders, and leaving only scars where the villages had been (Figure 5.18) (11).

Many landslides have been caused by interactions among adverse geologic conditions, excess moisture, and artificial changes in the landscape and slope material. Examples include the Vaiont Reservoir slide of 1963 in Italy, the Handlova, Czechoslovakia slide in 1960, landslides associated with timber harvesting, and

vegetation, and erosion were responsible for producing the slope failures. The magnitude of this landslide activity, while unique in recorded history, has probably been relatively frequent along the mid-southern coast of Brazil during the geologic evolution of the slopes (9).

The experience in Brazil is similar to a widespread episode of debris avalanches that occurred in August of 1969. Remnants of hurricane Camille, moving eastward from Kentucky and the Appalachian Mountain ridges, mixed with a mass of saturated air to produce thunderstorms of catastrophic proportions. These storms locally produced 71 centimeters of rain in eight hours and triggered a great many debris avalanches in central Virginia (Figures 5.16 and 5.17). The storms claimed 150 lives. The greatest loss of life was the result of flooding, although most people died from broken bones and blunt-force injuries rather than drowning. It is impossible to estimate how many died as a result of the avalanches, but the debris delivered to channels in floods certainly was significant. The avalanches generally followed preexisting depressions, moved a layer of soil and vegetation 0.3 to 1 meter thick, and left a pronounced linear scar of exposed bed-

FIGURE 5.16
Debris avalanche scar in Virginia. (Photo by G. P. Williams, courtesy of U.S. Geological Survey.)

FIGURE 5.17
Debris avalanche deposits of logs, boulders, and other particles along the east branch of
Hat Creek in Virginia. (Photo by G. P. Williams, courtesy of U.S. Geological Survey.)

numerous slides in urban areas such as Rio de Janeiro, Brazil; Los Angeles, California; Hamilton County, Ohio; and Allegheny County, Pennsylvania.

The world's worst dam disaster occurred on October 9, 1963, when approximately 2,600 lives were lost at the Vaiont Dam in Italy. As reported by George Kiersch, the disaster involved the world's highest thin-arch dam (267 meters at the crest), and, strangely, no damage was sustained by the main shell of the dam or the abutments (12). The tragedy was caused by a huge landslide in which more than 238,000,000 cubic meters of rock and other debris moved at speeds of about 95 kilometers per hour down the north face of the mountain above the reservoir, and completely filled it with slide material for 1.8 kilometers along the axis of the valley to heights of nearly 152 meters above the reservoir level (Figure 5.19). The rapid movement created a tremendous updraft of air and propelled rocks and water up the north side of the valley, higher than 250 meters above the reservoir level. The slide, and accompanying blasts of air and water and rock, produced strong earthquakes recorded many miles away. It blew the roof off one man's house well over 250 meters above the reservoir and pelted the man with rocks and debris. The filling of the reservoir produced waves of water over 90 meters high that swept over the abutments of

the dam. More than 1.5 kilometers downstream, the waves were still over 70 meters high, and everything for kilometers downstream was completely destroyed. The entire event, slide and flood, was over in less than seven minutes. The landslide followed a three-year period of monitoring the rate of creep on the slope which varied from less than 1 to as many as 30 centimeters per week, until September, 1963, when it increased to 25 centimeters per day. Finally, on the day before the slide, it was about 100 centimeters per day. Engineers expected a landslide, but one of much smaller magnitude, and they did not realize until October 8, one day before the slide, that a large area was moving as a uniform, unstable mass. Animals grazing on the slope had sensed danger and moved away on October 1.

The slide was caused by a combination of factors. First, adverse geologic conditions, including *weak rocks* and *limestone* with open fractures, sinkholes, and clay partings which were inclined toward the reservoir, produced unstable blocks (Figure 5.20), and *very steep topography* created a strong driving (gravitational) force. Second, water pressure was increased in the valley rocks because of the impounded water in the reservoir. Groundwater migration into bank storage raised the water pressure and reduced the resistance to shearing (resisting force) in the slope. The rate of creep before the slide increased as the water table rose in response to higher reservoir levels. Third, heavy rains from late September until the day of the disaster further increased the weight of the slope materials, raised the water pressure in the rocks, and produced runoff that continued to fill the reservoir even after engineers tried to lower the reservoir level. It was concluded that the immediate cause of the disaster was an increase in the driving force accompanied by great decrease in the resisting force; but the real cause was excess groundwater that raised the water pressure, in turn changing the weight of the slope material and producing a buoyancy effect along plains of weakness in the rock (12).*

The Handlova Slide in Czechoslovakia is an interesting example of an extremely hazardous landslide caused by a mixture of human use and adverse geo-

*G. A. Kiersch, *Civil Engineering* 34 (1964), pp. 32–39.

FIGURE 5.18
Oblique aerial view of the debris avalanche that destroyed Yungay and Ranrahirca, Peru, killing about 20,000 people. (Photo by George Plafker, courtesy of U.S. Geological Survey.)

logic conditions (Figure 5.21). A large coal-burning power plant in Handlova used a soft coal which emitted a large amount of fly ash that prevailing winds carried and deposited to the south. The land was changed by the accumulation of ash to such an extent that some land formerly used for grazing purposes had to be plowed. The plowing allowed rains to infiltrate at a greater rate and disturb an already delicate groundwater situation. After a heavy rainfall in 1960, which further raised the water table, a large landslide involving approximately 20,000,000 cubic meters of earth material developed. The slide moved nearly 152 meters in one month and threatened to severely damage the town. A well-organized program to control the slide and stop the movement by draining the slide material was successful within 60 days. Even so, 150 homes were destroyed (7).

Possible cause and effect relationships between timber harvesting and erosion in northern California, Oregon, and Washington is a controversial topic. Landslides become important in the discussion because there is good reason to believe that landslides, especially shallow debris avalanches and more deeply seated earthflows, are responsible for much of the erosion. In fact, one study in the western Cascade Range of Oregon concluded that shallow slides are the dominant erosion process in the area. Furthermore, whereas timber-harvesting activities (clearcutting and road building), over approximately a twenty-year observation period on geologically stable land, did not greatly increase landslide-related erosion, over the same period of time, logging on weak, unstable slopes did increase landslide erosion by several times relative to forested land (13).

There is no doubt that clearcutting and construction of logging roads can increase erosion by activating old landslides and causing new ones. Figure 5.22 shows before and after photographs of an area in Humboldt County, California, that was logged. One large landslide and two smaller ones are clearly contributing a good deal of sediment to the Mattole River. Although slides A and D were both present before logging, their size increased dramatically following timber-harvesting activities. It might be argued that Figure 5.22 depicts a nightmare from the past, because new regulations on logging practices would prevent this from happening now. To some extent this is true; however, even with stringent regulations, logging on unstable slopes is still likely to influence, initiate, or increase landslide erosion. For example, one of the largest landslides in 25 years of history in the Oregon Coast Ranges occurred in 1975 during logging operations. The slide was a rather fast-moving, deep-seated earth flow that completely blocked the headwater of Drift Creek, producing

FIGURE 5.19

Sketch map of the Vaiont Reservoir showing the 1963 landslide which displaced water that overtopped the dam and caused severe flooding and destruction over large areas downstream. A-A' and B-B' are section lines shown on Figure 5.20 (After Kiersh, *Civil Engineering* 34, 1964.)

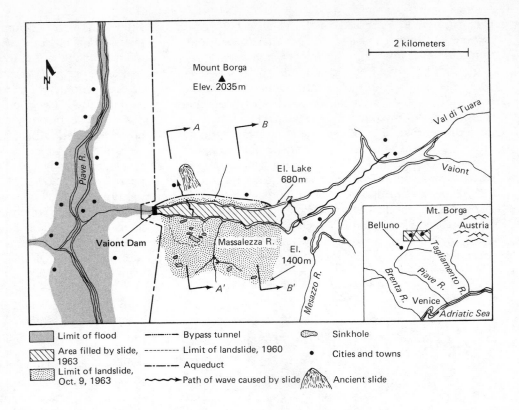

	Limit of flood		Bypass tunnel		Sinkhole
	Area filled by slide, 1963		Limit of landslide, 1960		Cities and towns
	Limit of landslide, Oct. 9, 1963		Aqueduct		
			Path of wave caused by slide		Ancient slide

a small lake (Figure 5.23). Although there is a fair chance the slide would have occurred even in the absence of logging, it is likely that building roads and removing trees reduced the stability of the slope.

The construction of roads in areas to be logged is an especially serious problem because roads may interrupt surface drainage, alter subsurface movement of water, and adversely change the distribution of mass on a slope by cut-and-fill (grading) operations (13). As we learn more about erosional processes in forested areas, we should be able to develop improved management procedures to minimize the adverse effects of timber harvesting. Until then, we will not be out of the woods with respect to landslide erosion problems.

Human use and interest in the landscape are most likely to cause landslides in urban areas where there are high densities of people and supporting structures such as roads, homes, and industries. Examples from Rio de Janeiro, Brazil, and Los Angeles, California, illustrate the situation.

Rio de Janeiro, with a population exceeding 4 million people, may have more slope-stability problems than any other city its size (9). Combinations of steep slopes and fractured rock mantled with surficial deposits contribute to the problem. In earlier times, many such slopes were logged for lumber and fuel and to clear space for agriculture. This early activity was fol-

lowed by landslides associated with heavy rainfall. Recently, lack of room on flat ground has led to increased urban development on slopes. Vegetation cover had been removed, and roads leading to development sites at progressively higher areas are being built. Excavations have cut the toe (lower part) of many slopes and severed the soil mantle at critical points (Figure 5.24). In addition, placing slope fill material below excavation areas has increased the load (driving force) on slopes already unstable before the fill. Furthermore, this area periodically experiences tremendous rainstorms. Thus, it is easy to understand why Rio de Janeiro has a serious problem. In 1966, following heavy rains, numerous landslides occurred. The terrible storms of 1967 that we discussed earlier missed the city; if they had not, the results would have been catastrophic (9).

Los Angeles, and more generally Southern California, has experienced a remarkable frequency of landslides associated with hillside development. For example, in one storm period, landslides, soilslides, and mudflows in the Los Angeles area claimed two lives and forced evacuation of more than 100 homes. Millions of dollars in property damage occurred to structures such as homes, swimming pools, patios, and utilities (Figure 5.25) (14). Landslides in Southern California result from complex physical conditions, including great contrasts in topography, rock and soil types, climate, and vegetation. Interactions between the nat-

FIGURE 5.20
Generalized geologic cross sections through the slide area of the Vaiont River Valley. Location of sections are shown on Figure 5.19. (After Kiersh, *Civil Engineering* 34, 1964.)

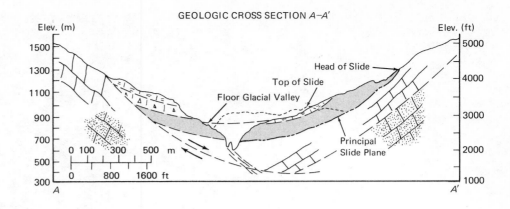

GEOLOGIC CROSS SECTION A–A'

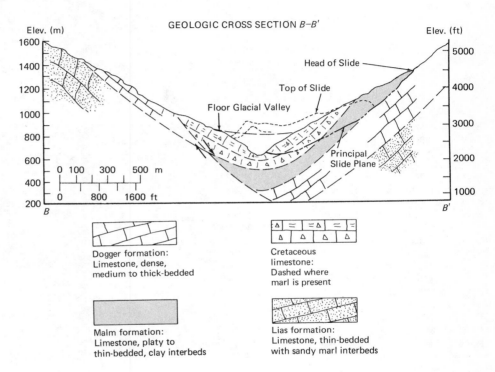

GEOLOGIC CROSS SECTION B–B'

Dogger formation:
Limestone, dense,
medium to thick-bedded

Cretaceous
limestone:
Dashed where
marl is present

Malm formation:
Limestone, platy to
thin-bedded, clay interbeds

Lias formation:
Limestone, thin-bedded
with sandy marl interbeds

ural environment and human activity are complex and notoriously unpredictable. For this reason, the area has the sometimes dubious honor of showing the ever-increasing value of urban geology (14). As a result, more consulting geologists in Southern California are employed in the analysis of slope stability than in any other field, and Los Angeles has led the nation in developing codes concerning grading (artificial excavation and filling) for development.

Even if urban development had never occurred in Southern California, landslides would be fairly common. The bulk of material that moves from valley sides and sea cliffs in Southern California is probably moved by landslides. Figure 5.26 shows one of the larger recent slides, the Portuguese Bend slide of 1956. As computed from geologic mapping, slides affect 60 percent of the

length of sea cliffs, and the retreat of the cliff is probably controlled by landslides (Figures 5.27 and 5.28) (14). Similar estimates for slopes are not available, but the complex geology and terrain features, as well as evidence from old landslide scars and landslide deposits, suggest that slopes historically have been active. But, human activity has tremendously increased the magnitude and especially the frequency of landslides.

The grading process in Southern California has been responsible for many landslides. It took natural processes many thousands, if not millions, of years to produce valleys, ridges, and hills. In this century, our technology has developed the machines to grade them. Leighton writes: "With modern engineering and grading practices and appropriate financial incentive, no hillside appears too rugged for future development"

FIGURE 5.21
Part of the Handlova landslide.
(Photo courtesy of Professor M.
Matula.)

FIGURE 5.22
Mattole River area, Humboldt
County, California. Note the contrast between the area prior to
logging in 1948 (a) and after logging in 1972 (b). Landslides A
and D were present in 1948, but
were much smaller than in 1972;
slide E did not exist in 1948.
Streams B and C were nearly
hidden by vegetation in 1948,
but were open scars in 1972. (After M. E. Huffman, "Geology for
timber harvest planning," *California Geology,* vol. 30, no. 9, pp.
195–201, 1977. Photos courtesy
of California Department of Water
Resources [a] and California Division of Mines Geology [b].)

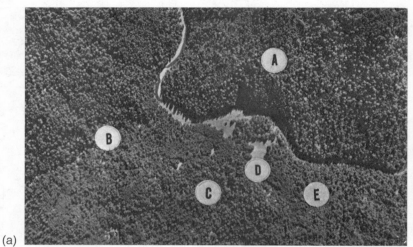

(a)

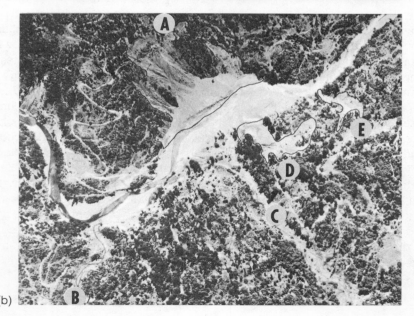

(b)

FIGURE 5.23
Large landslide in the Coast Ranges of Oregon that was associated with timber harvesting activities.

FIGURE 5.24
Landslide that demolished several houses and two apartment buildings in Rio de Janeiro. More than 132 people died. The large slide was evidently facilitated by a smaller landslide caused by a highway cut that overloaded the slope. (Photo by F. O. Jones, courtesy of U.S. Geological Survey.)

FIGURE 5.25
Damage to property in Los Angeles caused by a mudflow. Note the enormous size of material that mudflows can carry. (Photo courtesy of *Los Angeles Times.*)

(14). No earth material can withstand the serious assault of modern technology; therefore, human activity is a geological agent capable of carving the landscape as do glaciers and rivers, but at a tremendously faster pace. We can, for example, convert steep hills almost overnight into a series of flat lots and roads, and such conversions have led to numerous artifically-induced landslides. As shown in Figure 5.29, oversteepened slopes mixed with increased water from sprinkled lawns or septic systems, as well as the additional weight of fill material and a house, make formerly stable slopes unstable. Any project that steepens or saturates a slope, increases its height, or places an extra load on it, may cause a landslide (Figure 5.30) (14).

Landslides on both private and public land in Hamilton County, Ohio, have been a serious problem. The slides occur in glacial deposits (mostly clay, lakebeds, and till) and colluvium and soil developed on shale; the average cost of damage exceeds $5 million per year. In 1974, a major landslide in Cincinnati, Ohio, damaged a highway under construction as well as several private structures. The landslide may be one of the

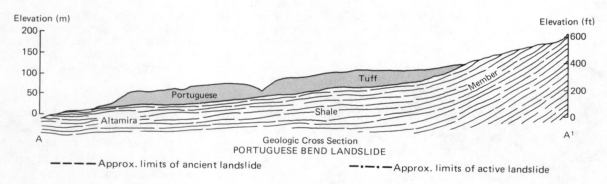

Elevation (m)

Elevation (ft)

Geologic Cross Section
PORTUGUESE BEND LANDSLIDE

- - - - Approx. limits of ancient landslide   - · - · - Approx. limits of active landslide

FIGURE 5.26
Aerial view of Palos Verdes showing the approximate known limits of a large, ancient
landslide and the outline of the active Portuguese Bend landslide. Arrows show the direction of movement. Geologic cross section shows that volcanic tuff (consolidated ash) is
sliding over shale. (Photo and cross section courtesy of Los Angeles County, Department of
County Engineer.)

most expensive in the history of the United States. Estimates in 1979 for permanent repair were $22 million,
but litigation among agencies, private landowners, at-
torneys, and consultants is still in progress (1).

Modification of sensitive slopes associated with
urbanization in Allegheny County, Pennsylvania, is
estimated to be responsible for 90 percent of the landslides, which produce an average of about $2 mil-
lion in damages each year. Most of the landslides are
slow-moving, but one rockfall, a number of years ago
in an adjacent county, crushed a bus and killed 22
passengers. Most of the landslides in Allegheny County
result from construction activity that loads the top
of a slope, cuts into a sensitive location such as the
toe of a slope, or alters water conditions on or in a
slope (15).

FIGURE 5.27
How wave erosion of sea cliffs facilitates landslides. Notice the cracking, an early sign that a slide may be imminent.

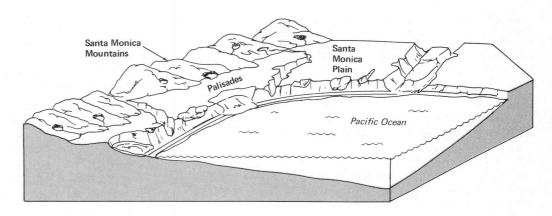

FIGURE 5.28
Block diagram of a portion of the Santa Monica Mountains, Santa Monica Plain, and Pacific Palisades in Southern California. Landslides are common along the sea cliffs, especially in the Pacific Palisades area. (Reprinted, by permission, from F. B. Leighton, "Landslides and Urban Development," *Engineering Geology in Southern California* [Whittier, California: Association of Engineering Geologists], 1966.)

FIGURE 5.29
Development of artificial translation landslides. Stable slopes may be made unstable by removing support from the bedding plane surfaces. The cracks shown in the upper part of the diagram are one of the early signs that a landslide is likely to occur soon. (Reprinted, by permission, from F. B. Leighton, "Landslides and Urban Development," *Engineering Geology in Southern California* [Whittier, California: Association of Engineering Geologists], 1966.)

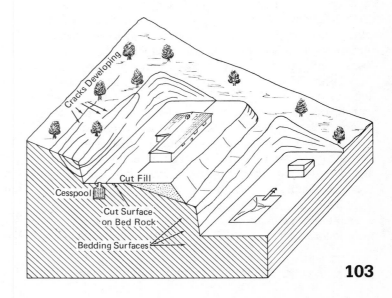

103

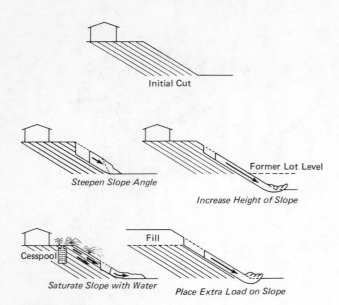

Initial Cut

Steepen Slope Angle

Increase Height of Slope

Former Lot Level

Cesspool

Saturate Slope with Water

Fill

Place Extra Load on Slope

FIGURE 5.30
Four ways in which a stable cut may be made unstable by human activity. (Reprinted, by permission, from F. B. Leighton, ''Landslides and Urban Development,'' *Engineering Geology in Southern California* [Whittier, California: Association of Engineering Geologists], 1966.)

IDENTIFICATION, PREVENTION, AND CORRECTION OF LANDSLIDES

To minimize the landslide hazard, it is necessary to *identify* areas in which landslides are likely to occur, *design* slopes or engineering structures to *prevent* landslides, and *control* and *stop* slides after they have started moving.

Identification of Landslides

Identifying areas with a high potential for landslides is a first step in developing a plan to avoid landslide hazards. Slide tendency can be recognized by examining geologic conditions and identifying previous slides. This information can then be used to produce slope stability maps (Figure 5.31). The utility of these maps in predicting a possible hazard is well illustrated in the before-and-after sequence of photographs of San Clemente, California (Figure 5.32). The house in the upper left corner of the top photograph (a) was mapped as sitting on unstable earth material (see the arrow on Figure 5.31), and, as shown in the lower oblique air photograph (b), a landslide eventually destroyed the home. Of course, we need this type of mapping before homes are constructed, but this example certainly verifies the validity and worth of such mapping.

The individual homeowner, buyer, or builder can evaluate the landslide hazard on hillside property by looking for specific physical evidence that may indicate a potential or real landslide problem. Signs include cracks in buildings or walls around yards, doors, and windows that stick or jam, retaining walls, fences, or posts that are not aligned in a normal way, breakage of underground pipes or other utilities, leaks in swimming pools, tilted trees and utility poles with taut or sagging wires, cracks in the ground, hummocky or steplike ground features, and seeping water from the base or toe of a slope (15). This list was derived for a landslide-prone area in western Pennsylvania, but is applicable to many areas. When applying this information, however, keep in mind that the presence of one or more of the features is not absolute proof that a landslide is likely. For example, cracks in walls may also be caused by expansive soils or creep. Other features, such as hummocky or steplike ground on moderately steep slopes (greater than 15 percent, or 15 meters fall in 100 meters horizontal distance), probably do represent a potential landslide hazard that should be evaluated by a geologist. Furthermore, it is advisable not to limit your inspection only to the property in which you are interested; landslides are often larger than individual lots. Inspect adjacent areas, especially those that are upslope and downslope from your property (15).

After areas with a high landslide risk are identified and mapped, as with our example from San Clemente, the information must be combined with engineering data to implement plans to minimize the landslide hazard.

Motivation to initiate a plan to reduce the landslide hazard through grading codes in the Los Angeles area came in the aftermath of very high losses of property and numerous lives to landsliding in the 1950s and 1960s. During this period, the Portuguese Bend landslide (Figure 5.26) was responsible for damaging or destroying more than 150 homes. This slide is really part of an older, larger slide that was reactivated partly by road-building activities and partly by alteration of a delicate subsurface water situation by urban development. Interestingly, the slide is still moving today, and a few homes are steadily moving toward the ocean. One home has moved about 25 meters in 20 years, is consistently shifting its position, and has to be adjusted every year or so with hydraulic jacks. Other homes have moved up to 50 meters in the same time. The people living in the homes, now about 1 kilometer from the ocean, seem to have adjusted to the slow movement and the ever-changing view. With one exception, no new homes have been constructed since the landslide began to move. The remaining occupants have elected to adjust to the landslide rather than bear the total loss of their properties. Fortunately, the slide has become more stable in recent years. Nevertheless, few geologists would live there now, and in 1978 an

FIGURE 5.31
Relative slope stability map of a portion of the San Clemente area in California. (After Ian Campbell, *Environmental Planning and Geology,* U.S. Department of Housing and Urban Development, 1971.)

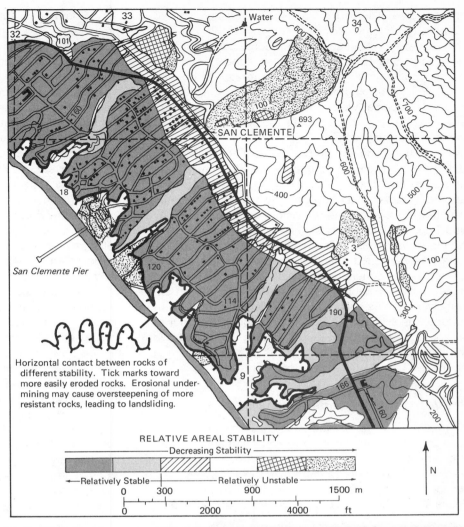

Horizontal contact between rocks of different stability. Tick marks toward more easily eroded rocks. Erosional undermining may cause oversteepening of more resistant rocks, leading to landsliding.

RELATIVE AREAL STABILITY

←——— Decreasing Stability ———→

←—Relatively Stable— | Relatively Unstable——→

(a)

(b)

FIGURE 5.32
Before (a) and after (b) photographs of an area mapped as unstable (see arrow in Figure 5.31) that subsequently failed. (Photos courtesy of George B. Cleveland, California Division of Mines and Geology; Ian Campbell, California Academy of Sciences; and U.S. Geological Survey at Menlo Park.)

adjacent area, also part of an ancient landslide, started to move. This new activity, known as the Abalone Cove slide, has damaged several homes, a beachfront park, and numerous utility lines, streets, and the coastal highway. A vigorous program to control the landslide by pumping out excess groundwater through a series of special wells (thus increasing the resisting forces in the slope by reducing the water pressure) is now being attempted.

Grading codes in the Los Angeles area have been effective in reducing damage to hillside homes from landslides and floods. Table 5.2 illustrates this well; since detailed engineering geology studies have been required, the percentage of hillside homes damaged by landslides and floods has been greatly reduced. Although original costs are greater because of the strict codes, they are more than balanced by the reduction of losses in subsequent wet years. Therefore, we conclude that, even though landslide disasters during extremely wet years will continue to plague us, the application of geologic and engineering information prior to hillside development can help minimize the hazard. We will now discuss specific ways to prevent or control landslides.

Prevention of Landslides

Prevention of large, natural landslides is nearly impossible, but common sense and good engineering practice can do much to minimize the hazard. For example, loading the top of slopes, cutting into sensitive slopes, placing fills on slopes, or changing water conditions on slopes should be avoided or done very cautiously (15).

Common engineering techniques for landslide prevention include provisions for surface and subsurface drainage, removal of unstable slope materials (grading), construction of retaining walls, or some combination of these (2, 11).

Drainage control is usually an effective way to stabilize a slope. The basic idea is to keep water from running across or infiltrating the slope. Surface water may be diverted around the slope by a series of gutters. This practice is common for roadcuts. The amount of water infiltrating a slope may also be controlled by covering the slope with an impermeable layer such as soil cement, asphalt, or even plastic.

Groundwater may be inhibited from entering a slope by excavating a cutoff trench. The trench is filled with gravel or crushed rock and placed in a position to intercept and divert groundwater away from a potentially unstable slope (2).

Grading slopes is another way to increase slope stability. Two common techniques are reducing the gradient of a slope by a single cut-and-fill operation, and benching. In the first case, material from the upper part of a slope is removed and placed near the base. The overall gradient is thus reduced and material is removed from an area where it contributes to the driving force, and placed at the toe of the slope, where it increases the resisting forces. This method is not practical on very steep, high slopes. As an alternative, the slope may be cut into a series of benches or steps. The benches are designed with surface drains to divert runoff. The benches do reduce the slope and, in addition, are good collection sites for falling rock and small slides (2).

TABLE 5.2
Landslide and flood damages to hillside homes in Los Angeles County, California, from before 1952 to 1969. (After Slosson, J. E., and Krohn, J. P., 1977, "Effective Building Codes," *California Geology* 30, no. 6, pp. 136–319. Data from Los Angeles Department of Building and Safety.)

| Construction Dates and Legal Requirements | Number of Homes Built on Hillside Sites | Damaged Homes | | Total Damage | Average Cost Prorated for Total Number of Homes Built |
		Number	Percent of Total (%)		
Pre–1952 No legal requirement for soils engineering or engineering geology studies	10,000	1,040	10	$3,300,000	$300
1952–1963 Soils engineering studies required. Minimum engineering geology studies	27,000	350	1.3	$2,767,000	$100
1963–1969 Extensive engineering geology and soils engineering studies required	11,000	17	0.15	$ 80,000	$ 7

Retaining walls, constructed from concrete cribbing, gabions (stone-filled wire baskets), or piles (long concrete, steel or wooden beams driven into the ground) are designed to provide support at the base of a slope. They should be keyed in well below the base of the slope, backfilled with permeable gravel or crushed rock, and provided with drain holes to reduce the chances of water pressure building up in the slope (Figure 5.33).

A less common method of increasing slope stability involves insertion of heavy bolts (rock bolts) through holes drilled through potentially unstable rocks into stable rocks. This technique was used to secure the slopes at the Glen Canyon Dam on the Colorado River and the Hanson Dam on the Green River in Washington (16).

FIGURE 5.33
Retaining wall (concrete cribbing) used to help stabilize a roadcut.

Landslide Warning Systems

Landslide warning systems do not prevent landslides, but they can provide time to evacuate people and their possessions, and to stop trains or reroute traffic. Surveillance provides the simplest type of warning. Hazardous areas can be visually inspected for apparent changes and small rock falls on roads and other areas can be noted for quick removal. Having people monitor the hazard tends to have advantages of reliability and flexibility, but becomes disadvantageous during adverse weather and in hazardous locations (17). Other warning methods include electrical systems, tilt meters, and geophones that pick up vibrations from moving rocks. Shallow wells can be monitored to signal when slopes contain a dangerous amount of water and, in some regions, monitoring rainfall is useful for detecting when a threshold precipitation has been exceeded and shallow soil slips become much more probable.

Landslide Control

After movement of a slide has begun, the best way to stop it is to attack the process that started the slide. In most cases, the cause of the slide is an increase in water pressure, and in such cases, an effective drainage program must be initiated. This may include surface drains at the head of the slide to keep additional surface water from infiltrating and subsurface drainpipes or wells to remove water and lower the water pressure. Draining will tend to increase the resisting force of the slope material and therefore stabilize the slope.

The tremendous success of drainage is demonstrated by this description from Karl Terzaghi (6). After a high-magnitude rainstorm, movement on a 30-degree slope of deeply weathered metamorphic rock was noted. The surface of sliding was approximately 40 me-

ters below the surface, and the slide area was about 150 meters wide by 300 meters long and involved 400,000 cubic meters of material. The slide was close to a hydroelectric power station, so immediate action was deemed necessary. Field work established that if the water level could be lowered approximately 5 meters, then the increase in resisting force would be sufficient to stabilize the slide. Drainage was accomplished by trenches and horizontal drill holes extending into the water-bearing zones of the rock. After drainage, the movement stopped, and even though the next rainy season brought record rainfall, no new movement was observed (6).

When slope materials are fine-grained and relatively impermeable, drainage is difficult at best and will probably be ineffective. In this situation, common correction methods include reducing the slope angle and constructing artificial barriers such as retaining walls. Unusual methods have also been used; for example, tunnels have been excavated in the slope and hot air circulated to dry out the unstable material. In other cases, unstable soil material has been frozen during construction to stabilize the slope. Drying the material with hot air was apparently successful in the Pacific Palisades area of Southern California, and the freezing technique was used when a slide threatened to impede construction of the Grand Coulee Dam. Freezing was accomplished by using 377 freezing points. The refrigeration system provided 73,000 kilograms of ice per day, and a frozen dam 12 meters high, 6 meters thick, and 30 meters long formed to stabilize the slope material (8).

SNOW AVALANCHE

Approximately 10,000 snow avalanches occur each year in the mountains of the western United States, and about one percent of these cause loss of human life or property damage, killing an average of 7 people and inflicting $300,000 in damage (18). Loss of life is increasing, however, as more people venture into mountain areas for recreation during the winter.

Avalanches may occur in dry or wet snow and are of two general types: *loose-snow avalanches* that occur in cohesionless snow and tend to be relatively small and shallow failures; and *slab avalanches* that may initially vary from about 100 to 10,000 square meters in area and 0.1 to 10 meters in thickness (18). It is the large slab avalanches that are most dangerous, releasing tremendous energy by mobilizing up to a million tons of snow and ice and moving downslope at velocities of 5 to 30 meters per second (18 to 100 km per hour) or more. Horizontal thrust (or impact) from such events tends to vary from 5 to 50 tons per square

meter, but may in extreme cases exceed 100 tons per square meter. By comparison, a thrust of only about 3 tons per square meter is necessary to collapse a frame house, and 100 tons per square meter will move reinforced concrete structures (18).

Avalanches are initiated when a mass of snow and ice on a slope fails because of the overload of a large volume of new snow, or when internal changes in a snow pack produce zones of weakness (low shear strength) along which failure occurs. When conditions are right, even the weight of a single skier can start an avalanche.

Avalanches tend to follow certain paths, chutes, or tracks (see Figure 5.34) that are often well-channeled, but unconfined tracks on open slopes also occur. Avalanche tracks often have several branches near the top that coalesce downslope; thus, it is possible for

FIGURE 5.34

Avalanche chutes or tracks in the Swiss Alps. Left photo shows two chutes, one of which is quite close to the small village; right photo shows a closeup of an avalanche chute devoid of vegetation.

several avalanches to move through the main track in a short period of time as snow packs in upper branches fail. Failure to recognize this possibility has caused loss of several lives when workers clearing debris from a first avalanche are struck by a second (18).

The avalanche hazard can be reduced by avoiding dangerous areas; stabilizing slopes by clearing them with carefully-placed explosions; building structures to divert or retard avalanches; and reforesting avalanche paths, since large avalanches are seldom initiated on densely forested slopes (18).

Avalanches are primarily a threat to skiers on high, steep mountain slopes, but they also threaten mountain resorts, villages, railways, highways, and even sections of some cities. For example, Alaska's capital, Juneau, has a significant avalanche hazard. In the last 100 years, a major avalanche chute above Juneau has released snow and ice six times, events that reached the sea. There have been no damaging avalanches in the last 20 years, however, so an entire subdivision has been constructed across the chute. If another large event occurs, it will destroy about 30 homes, part of a school, and a motel, and eventually roar into the harbor where several hundred boats are docked. It has been estimated that a home in the chute area, with a 40-year life span, has a 96 percent probability of being struck by an avalanche, yet the people who live there have been almost nonchalant about the hazard (19).

SUBSIDENCE

Various interactions among geologic conditions and human activity have also been factors in numerous incidents of subsidence. Most subsidence is caused by withdrawal of fluids from subsurface reservoirs or collapse of surface and near-surface soil and rocks over subterranean voids.

Withdrawal of fluids such as oil with associated gas and water, groundwater, and mixtures of steam and water for geothermal power have caused subsidence (20). In all cases, the general principles are the same. Fluids in earth materials below the earth's surface have a high fluid pressure that tends to support the material above. This is why a large rock at the bottom of a swimming pool seems lighter. Buoyancy is produced by the liquid that tends to lift it. If that support or buoyancy is removed by pumping out the fluid, the support is reduced, and subsidence at the surface may result. The actual subsidence mechanism involves compaction of individual grains of the earth material as the grain-to-grain load increases because of a lowering of fluid pressure. Subsidence of oil fields generally in-

volves considerable reduction of fluid pressure, up to 280 kilograms per square centimeter, at great depth (thousands of meters) over a relatively small area, less than 150 square kilometers. On the other hand, subsidence resulting from withdrawal of groundwater generally involves a relatively low reduction of fluid pressure, often less than 14 kilograms per square centimeter, at relatively shallow depths (less than 600 meters), over a large area, sometimes many hundreds of square kilometers (20).

The Wilmington oil field in the harbor area of Long Beach near Los Angeles, California, is a spectacular incidence of land subsidence that has caused over $100 million in damage to structures, several hundred ruptured oil wells, and localized flooding (Figure 5.35). Subsidence was first noted there in 1940, and by 1974, had increased to 9 meters in the central area (Figure 5.36). Predictions of ultimate subsidence as great as 13.7 meters promoted a joint effort to stop it. A repressuring program was begun in 1958. Tremendous quantities of water were injected into the rocks from which the oil was pumped out, raising the fluid pressure. This procedure actually reversed the subsidence, and by 1963, there had been appreciable rebound amounting to as much as 15 percent of the initial subsidence in some parts of the oil field (20). Figure 5.37 shows the vertical movement at Pier A from 1952 to 1974, where the rebound was significant.

Thousands of square kilometers of the central valley of California have subsided. More than 5,000 square kilometers in the Los Banos-Kettleman City area alone have subsided more than 0.3 meters, and within this area, one 113-kilometer stretch has subsided an average of more than 3 meters, with a maximum of 8.5 meters (Figure 5.38). The cause was overpumping of groundwater from a deep confined aquifer. As the water was mined, the fluid pressure was reduced and the grains were compacted (21). The effect at the surface was subsidence. Similar examples of subsidence caused by overpumping are documented near Phoenix, Las Vegas, and Houston–Galveston. In some instances, the subsidence is accompanied by surface faulting, producing linear scarps and fissures which, in turn, erode to form gullies. In other instances where a great volume of groundwater has been removed, a few centimeters' uplift has actually been recorded. This anomalous uplift is apparently due to elastic rebound when the load of groundwater on near-surface rocks is removed relatively suddenly.

Another example of subsidence caused by overpumping is found in Mexico City. The population in Mexico City increased from fewer than one-half million in 1895 to 1 million in 1920 and 5 million in 1960. As the population grew, so did the demand for ground-

FIGURE 5.35
Flooding at Long Beach, California, caused by subsidence resulting from withdrawal of oil.
(Photo courtesy of City of Long Beach.)

FIGURE 5.36
Subsidence in the Long Beach
Harbor area, California, from
1928 to 1974. (Photo courtesy of
Port of Long Beach.) (2 to 29 ft
= 0.6 to 8.8 m)

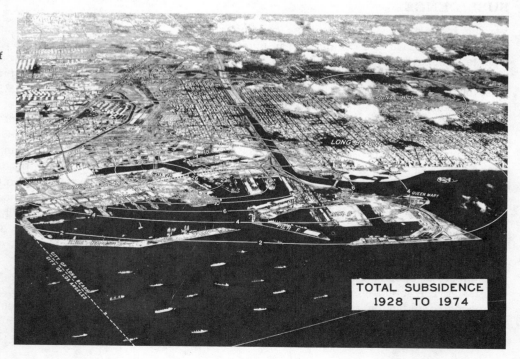

FIGURE 5.37
Vertical movement of bench
mark (point of known elevation)
on Pier A, Long Beach, from
1952 to 1974. Location is just
right of center of Figure 5.36.
(After data, courtesy of Port of
Long Beach.)

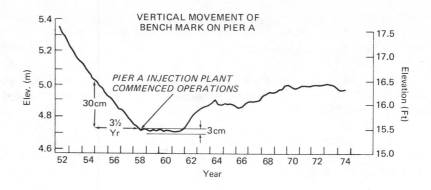

VERTICAL MOVEMENT OF
BENCH MARK ON PIER A

*PIER A INJECTION PLANT
COMMENCED OPERATIONS*

water, which is pumped from sand and gravel aquifers separated by clay and silt deposits. These conditions are continuous at depths of about 60 meters to over 500 meters. The discharge of thousands of private and municipal wells far exceeds the natural recharge of the aquifers (20). Reduction of fluid pressure as the water table lowers has resulted in as much as 7 meters of subsidence, most of which has occurred since 1940. The continuous sinking has produced many problems in drainage and building construction (20). One building, the Palace of Fine Arts, constructed in 1934, has subsided approximately 3 meters, and, reportedly, steps that once went up to the first floor now go down to that floor.

Subsidence is also caused by removal of subterranean earth materials by natural or artificial processes. Rocks such as limestone and dolomite are soluble, and

subterranean voids often form. Lack of support for overlying rock may lead to collapse and the formation of large sinkholes, some of which are over 30 meters across and 15 meters deep. One near Tampa, Florida, collapsed suddenly in 1973, swallowing part of an orange orchard.

What may be the largest sinkhole in the United States formed in 1972 near Montevallo, Alabama (22). A massive hole 120 meters wide and 45 meters deep, named the *December Giant* by the press, developed suddenly when topsoil and subsurface clay collapsed into an underlying limestone cavern (Figure 5.39). The collapse was caused by loss of support to the clay and soil over the cavern. A nearby resident reported hearing a roaring noise accompanied by breaking timber and earth tremors that shook his house. Sinkholes of this type have caused considerable damage to highways, homes, sewage facilities, and other structures. Natural or artifical fluctuations in the water table are probably the trigger mechanism. High water table conditions favor solutional enlarging of the cavern, and the

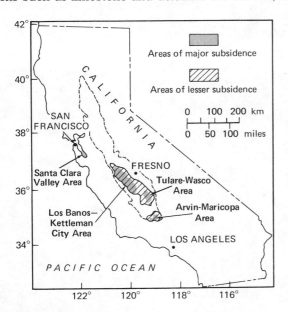

FIGURE 5.38
Principal areas of land subsidence in California resulting from groundwater withdrawal. (After W. B. Bull, Geological Society of America Bulletin 84, 1973. Reprinted by permission.)

FIGURE 5.39
A very large sinkhole known as the *December Giant* near Montevallo, Alabama. (Photo courtesy of Bill Warren and Geological Survey of Alabama.)

buoyancy of the water helps support the overburden. Lowering of the water table eliminates some of the buoyant support and facilitates collapse. This was dramatically illustrated on May 8, 1981, in Winter Park, Florida, when a large sinkhole began developing. The sink grew rapidly for three days, swallowing part of a community swimming pool, parts of two businesses, several automobiles, and a house (Figure 5.40). Damage caused by the sinkhole is in excess of $2 million. Sinkholes form nearly every year in central Florida when the groundwater level is lowest. The Winter Park

sinkhole formed during a drought, when groundwater levels were at a record low. Although exact positions cannot be predicted, their occurrence is greater during droughts; in fact, several smaller sinks developed about the same time as the Winter Park event.

Serious subsidence events have been associated with mining of salt, coal, and other minerals. Salt is often mined by solution methods in which water is injected through wells into salt deposits. The salt dissolves, and water supersaturated in salt is pumped out. The removal of salt leaves a cavity in the rock and

FIGURE 5.40
The Winter Park, Florida, sinkhole of May, 1981. The aerial view (upper) shows the extent of the damage, and the lower photograph is a closer view of it. (Upper photo by George Remaine; lower photo by Barbara Vitaliano; both of *Orlando Sentinel Star*.)

FIGURE 5.41
Large subsidence pit resulting from subsurface solutional mining of salts near Saltville, Virginia.

weakens support for the overlying rock, which may lead to large-scale subsidence. In 1970, one event near Detroit produced a subsidence pit 120 meters across and 90 meters deep. Another near Saltville, Virginia, produced 75 meters of subsidence relatively quickly (Figure 5.41). Two homes went down with the Saltville subsidence. According to local residents, one family moved out the day before the event because a family member had dreamed the mountain was falling. As of 1975, large and still actively growing fractures in the rocks surrounding the subsidence area indicated that the area involved with the subsidence was considerably larger than what actually subsided into the subterranean void below.

Large sinks associated with bedded salt may also occur without solution mining. For example, in June of

(a)

(b)

FIGURE 5.42
Photographs of the Wink Sink. Depth to water in (a) is about 10 meters. Note the road and automobiles for scale. Photograph (b) is a view from the ground and shows some of the oil storage tanks near this sinkhole. (Photographs courtesy of Robert W. Baumgardner, Jr. and the Texas Bureau of Economic Geology.)

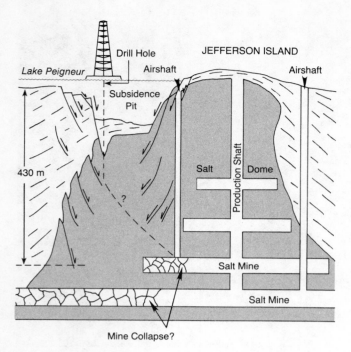

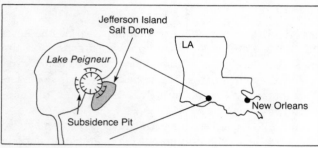

FIGURE 5.43
Idealized diagram showing the Jefferson Island Salt Dome collapse that caused a large subsidence pit to form in the bottom of Lake Peigneur, Louisiana.

1980, a large depression southwest of Kermit, Texas, known as the Wink Sink, developed over a period of about forty-eight hours. At the end of that time, the sinkhole was approximately 110 meters across and 34 meters deep. The Wink Sink and other similar features evidently form by natural processes when ground water, even if it is salty (brine), slowly dissolves caverns in bedded salt which underlie less soluble rock such as sandstone. When caverns reach critical size, the overlying rocks can no longer be supported, and collapse occurs. Because this is a natural process and other caverns undoubtedly exist, future sinks probably will develop in the area without warning (23).

Even though the Wink Sink developed in an oil field, the impact on operations was minimal. Damage consisted of a broken pipeline carrying crude oil to

tanks located approximately 300 meters from the sinkhole (Figure 5.42). In addition, two water disposal lines were broken by the expanding sinkhole, reducing oil production by about 150 barrels of oil per day (23).

A bizarre example of subsidence associated with a salt mine occurred on November 21, 1980, when shallow Lake Peigneur (with an average depth of 1 meter) in southern Louisiana drained following collapse over a salt mine. The collapse occurred after an oil-drilling operation apparently punched a hole into an abandoned mine shaft of a still active multimillion-dollar salt mine (at a depth of about 430 meters) in the Jefferson Island salt dome. As the hole enlarged because of water entering the mine, scouring and dissolving pillars of salt, the roof of the mine collapsed, producing a large subsidence pit (Figure 5.43). The lake drained so fast that ten barges, a tugboat, and an oil drilling barge disappeared in a whirlpool of water into the mine, which in places has tunnels as wide as four-lane freeways and 25 meters high. (Mining is done with the aid of trucks and bulldozers.) The subsidence also claimed more than 25 hectares of Jefferson Island including historic botanical gardens, greenhouses, and a one-half million dollar private home. What is left of the gardens is disrupted by large fractures that roughly step the land down to the new edge of the lake. These fractures are tensional in origin and are commonly found on the margins of large subsidence pits. Future movement may take place on these fractures, enlarging the pit.

Lake Peigneur refilled as water from a canal to the Gulf of Mexico was drawn in, and nine of the barges popped to the surface two days later. There was fear that potentially larger subsidence would take place as pillars of salt holding up the roof of the salt dome dissolved. Fortunately, debris in the way of soil and lake sediment that was pulled into the mine apparently sealed the hole, and it is hoped that further major collapses will not occur.

Approximately 15 million cubic meters entered the salt dome and the mine is a total loss. Fortunately, the 50 miners and 7 people on the oil rig escaped. The previous shallow lake now has a large deep hole in the bottom which undoubtedly will change the aquatic ecology. In a 1983 out-of-court settlement, the salt mine reportedly was compensated $30 million by the oil company involved. The owners of the botanical garden and private home apparently were compensated $13 million by the oil company, drilling company, and salt mine.

The flooding of the mine raises important questions concerning the structural integrity of salt mines. The federal strategic petroleum reserve program is planning to store 75 million barrels of crude oil in an old salt mine of the Welks Island salt dome about 19 kilometers from Jefferson Island. On the other hand, the role of the draining lake in the collapse is very significant, and few salt domes have lakes above them. The Jefferson Island subsidence is thus a very rare type of event.

Full recovery where all the coal is removed from subsurface mines has produced subsidence problems. The Pittsburgh area is a good example. Mining has been going on there for more than 100 years. In early years, companies purchased mining rights permitting removal of the coal with no responsibility for surface damage. The results were not so serious when mining was conducted under farmland, but as recent rapid urbanization has progressed faster than coal can be extracted, problems have resulted. If all the coal is removed, the chance of subsidence and damage to homes is high; however, if about 50 percent of the coal is left, this amount is usually sufficient support (Figure 5.44). The Bituminous Mine Subsidence and Land Conservation Act of 1966 provided for protection of public health, welfare, and safety by regulating coal mining, but this act will cause hundreds of millions of tons of coal to remain in the ground, attesting to the nature of trade-offs when there is conflict in surface and subsurface human use of the land (24).

Subsidence incidents have also been reported over coal mines that have not been worked for more than 50 years (24). On a January morning in 1973, a few residents of Wales were driving when a section of the road suddenly collapsed into a pit 10 meters deep. Their car tottered on the brink while they scrambled to safety. The collapse was over an air shaft of a lost mine. The subsidence disrupted some utility service. Other similar subsidences have happened in the past and are likely to occur in the future.

PERCEPTION OF THE LANDSLIDE HAZARD

The common reaction of homeowners in Southern California is, "It could happen on other hillsides, but never this one" (14). As with flooding, landslide hazard maps will probably not prevent people's moving into hazardous areas, and prospective hillside occupants who are initially unaware of the hazards probably cannot be swayed by technical information. Furthermore, the infrequency of large slides tends to reduce awareness of the hazard, where evidence of past events is not readily visible. Unfortunately, it often takes catastrophic events, such as the recent massive landslide in the Laguna Hills area of California, which claimed numerous expensive homes, to bring the problem to the attention of many people. In the meantime, people in many parts of the Rocky Mountains, Appalachian Mountains, and other areas continue to build homes in areas subject to future landslides.

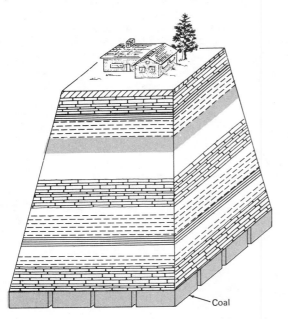
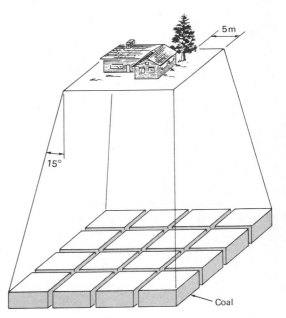

FIGURE 5.44
Idealized diagram showing necessary support for a home over a coal mine. Support based on a rock fracture of 15 degrees on all four sides of the house foundation, with a 5-meter-wide safety area around the foundation. (After A. E. Vandale, *Mine Engineering* 19, 1967).

SUMMARY AND CONCLUSIONS

The most common landforms are slopes, dynamic, evolving systems in which surficial material constantly is moving downslope at rates varying from imperceptible creep to thundering avalanches.

Important aspects of landslides are the type of earth material on the slope, topography, climate, vegetation, water, and time. The cause of most landslides can be determined by examining the relations between the forces that tend to make earth materials slide (*driving* forces) and forces that tend to oppose movement (*resisting* forces). The most common driving force is the weight of the slope materials, and the most common resisting force is the shear strength of the slope materials.

The role of water in landslides is especially significant and nearly always directly or indirectly involved in landslides. Water in streams, lakes, or oceans erodes the toe area of slopes, increasing the driving forces. Excess water increases the weight of the slope materials while raising the water pressure, which in turn decreases the resisting forces in the slope materials. A rise in water pressure occurs before many landslides, and, in fact, most landslides are a result of an abnormal increase in water pressure in the slope-forming materials.

Effects of human use on the magnitude and frequency of landslides vary from insignificant to very significant. Where human activities have little effect on landslides, we need to learn all we can about where, when, and why landslides occur so that we can avoid development in hazardous areas or, if necessary, provide protective measures. In cases in which human use has increased the number and severity of landslides, we need to learn how to recognize, control, and minimize these occurrences.

To minimize landslide hazard, it is necessary to establish identification, prevention, and correction procedures. Monitoring and mapping techniques facilitate identification. Prevention of large natural slides is nearly impossible, but good engineering practices can do much to minimize the hazard when it cannot be avoided. Correction of landslides must be designed to attack the processes that started the slide.

Snow avalanches present a serious hazard on snow-covered, steep slopes. Loss of human life because of avalanches is increasing as more people venture into mountain areas for winter recreation.

Withdrawal of fluids such as oil and water and subsurface mining of salt, coal, and other minerals has led to a subsidence hazard. In the case of fluid withdrawal, the cause of subsidence is a reduction of fluid pressures that tend to support overlying earth materials. In the case of solid material removal, subsidence may result from loss of support for the overlying material.

Perception of the landslide by most people, unless they have prior experience, is negligible. Furthermore, hillside residents, like floodplain occupants, are not easily swayed by technical information.

REFERENCES

1 FLEMING, R. W., and TAYLOR, F. A. 1980. *Estimating the cost of landslide damage in the United States.* U.S. Geological Survey Circular 832.

2 PESTRONG, R. 1974. *Slope stability.* American Geological Institute. New York: McGraw-Hill.

3 NILSEN, T. H.; TAYLOR, F. A.; and DEAN, R. M. 1976. *Natural conditions that control landsliding in the San Francisco Bay Region.* U.S. Geological Survey Bulletin 1424.

4 CAMPBELL, R. H. 1975. *Soil slips, debris flows, and rainstorms in the Santa Monica Mountains and vicinity, southern California.* U.S. Geological Survey Professional Paper 851.

5 BURROUGHS, E. R., JR., and THOMAS, B. R. 1977. *Declining root strength in Douglas fir after felling as a factor in slope stability.* USDA Forest Service Research Paper INT-190.

6   TERZAGHI, K. 1950. *Mechanisms of landslides.* The Geological Society of America: Application of Geology to Engineering Practice, Berken volume: 83-123.

7   LEGGETT, R. F. 1973. *Cities and geology.* New York: McGraw-Hill.

8   KRYNINE, D. P., and JUDD, W. R. 1957. *Principles of engineering geology and geotechnics.* New York: McGraw-Hill.

9   JONES, F. O. 1973. *Landslides of Rio de Janeiro and the Sierra das Araras Escarpment, Brazil.* U.S. Geological Survey Professional Paper 697.

10 WILLIAMS, G. P., and GUY, H. P. 1973. *Erosional and depositional aspects of Hurricane Camille in Virginia, 1969.* U.S. Geololgical Survey Professional Paper 804.

11 OFFICE OF EMERGENCY PREPAREDNESS. 1972. *Disaster preparedness* 1, 3.

12 KIERSCH, G. A. 1964. Vaiont Reservoir disaster. *Civil Engineering* 34:32–39.

13 SWANSON, F. J., and DRYNESS, C. T. 1975. Impact of clear-cutting and road construction on soil erosion by landslides in the Western Cascade Range, Oregon. *Geology* 3, no. 7:393–396.

14 LEIGHTON, F. B. 1966. Landslides and urban development. In *Engineering geology in southern California,* ed. R. Lung and R. Proctor, pp. 149–197, a special publication of the Los Angeles Section of the Association of Engineering Geology.

15 BRIGGS, R. P.; POMEROY, J. S.; and DAVIES, W. E. 1975. *Landsliding in Allegheny County, Pennsylvania.* U.S. Geological Survey Circular 728.

16 MORTON, D. M., and STREITZ, R. 1975. Mass movement. In *Man and his physical environment,* ed. G. D. McKenzie and R. O. Utgard, pp. 61–70. Minneapolis: Burgess.

17 PITEAU, D. R., and PECKOVER, F. L. 1978. Engineering of rock slopes. In *Landslides,* ed. R. Schuster and R. J. Krizek. Transportation Research Board, Special Report 176:192–228.

18 PERLA, R. I., and MARTINELLI, M., Jr. 1976. *Avalanche handbook.* U.S. Department of Agriculture, Forest Service, Agriculture Handbook 489.

19 CUPP, D. 1982. Battling the juggernaut. *National Geographic* 162:290–305.

20 POLAND, J. F., and DAVIS, G. H. 1969. Land subsidence due to withdrawal of fluids. In *Reviews in engineering geology,* ed. D. J. Varnes and G. Kiersch, pp. 187–269, The Geological Society of America.

21 BULL, W. B. 1974. Geologic factors affecting compaction of deposits in a landsubsidence area. *Geological Society of America Bulletin* 84:3783–3802.

22 CORNELL, J., ed. 1974. *It happened last year—earth events—1973.* New York: Macmillan.

23 BAUMGARDNER, R. W., GUSTAVSON, T. C., and HOADLEY, A. D. 1980. Salt blamed for new sink in W. Texas, *Geotimes,* 25, no. 9: 16–17.

24 VANDALE, A. E. 1967. Subsidence: a real or imaginary problem. *Mining Engineering* 19, no. 9: 86–88.

Chapter 6

Earthquakes and Related Phenomena

The earth is a dynamic, evolving system. Its outer layer of lithosphere is broken into several large and numerous small *plates* that move relative to one another. As new lithosphere is produced at oceanic ridge systems and older lithosphere is either consumed at subduction zones or slides past another plate, stress is produced and strain builds up in the rocks. When the stress exceeds the strength of the rocks, the rocks fail, and energy is released in the form of an earthquake. This action can be compared to pushing together and sliding two rough boards past one another. Friction along the boundary slows their motion, but rough edges break off and motion occurs at various places along the boundary. This process is similar to what happens at plate boundaries where one plate slides past another or one overrides another. The rocks undergo strain, and if the stress continues, they eventually break along weak zones called *faults*. This breaking of rocks and resulting movement along faults produces seismic waves that cause the ground to vibrate. Some of the waves travel within the earth (body waves), whereas others travel along the surface (Figure 6.1). Body waves, compressional *(P)* and shear *(S)*, are characterized by relatively high frequencies, 2 to 10 Hertz (cycles per second), whereas the more complex surface waves (Love and Rayleigh waves) have lower frequencies (less than 1 Hertz).

Buildings often have natural frequencies of vibration in the same range as earthquakes. This is unfortunate because a building's shaking is facilitated when the frequency of vibration of earthquake waves is close to the building's natural frequency. Low buildings have higher natural frequencies than tall buildings. As a result, compressional and shear waves, with relatively high frequencies, tend to cause low buildings to vibrate, whereas surface waves with lower frequencies tend to cause tall buildings to vibrate. High frequency waves decay much more quickly with distance from a generating earthquake than do low frequency waves. Thus tall buildings may be damaged at relatively long distances (100 km) from large earthquakes (1), whereas low buildings are more sensitive to shaking when they are near where an earthquake is generated.

The point or area within the earth where the first motion along a fault takes place is called the *focus* of the earthquake. The *epicenter* is generally the point on the surface of the earth directly above the focus, although in some cases, as for an earthquake originating deep in a subduction zone, the epicenter may not be directly over the focus. The location of an earthquake as reported by the news media is the epicenter.

Because most natural earthquakes are initiated near plate boundaries (Figure 3.2), there tend to be linear or curvilinear continuous zones along which most

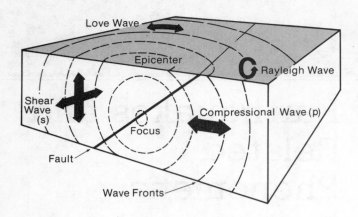

FIGURE 6.1

Idealized diagram showing directions of vibration of body (P and S) and surface waves (Love and Rayleigh) generated by an earthquake associated with the illustrated fault. Also shown are the focus and epicenter of the earthquake event. (See text for further explanation of seismic waves.) Source: W. W. Hays, 1981, *Hazards from Earthquakes.* In *Facing Geologic and Hydrologic Hazards,* ed. W. W. Hays. U.S. Geological Survey Professional Paper 1240-B.

seismic activity will take place (Figure 6.2). Figure 6.3 shows the locations of damaging historical earthquakes in the U.S. Most are in the western part of the country near the boundary between the North American and Pacific plates. However, large earthquakes have occurred far from plate boundaries. For example, during the winter of 1811–1812, a series of strong earthquakes struck the central Mississippi River valley, nearly destroying the town of New Madrid, Missouri, and killing an unknown number of people. The earthquakes, which caused church bells to ring in Boston over 1,600 km away, produced intensive surface deformation over a wide area from Memphis, Tennessee, north to the confluence of the Mississippi and Ohio Rivers; forests were flattened; fractures opened so wide that people had to cut down trees to cross them; the land sank several meters in some areas, causing flooding; and the Mississippi River supposedly reversed its flow during the shaking (2). The earthquakes occurred along a seismically active structure known as the Mississippi Embayment. The embayment (a downwarped trough or rift, at depth) is an area where the crust and litho-

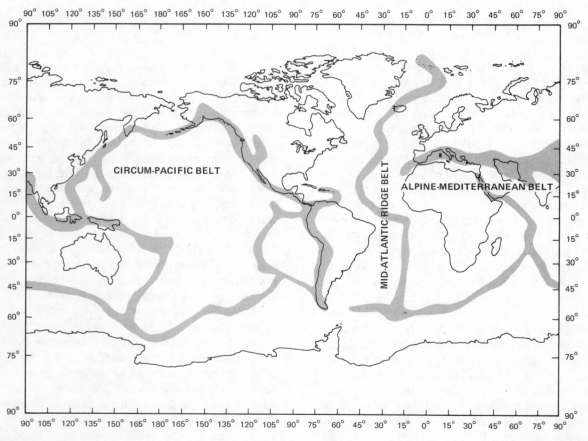

FIGURE 6.2

Map of the world showing the major earthquake belts as shaded areas. (Base map from NOAA.)

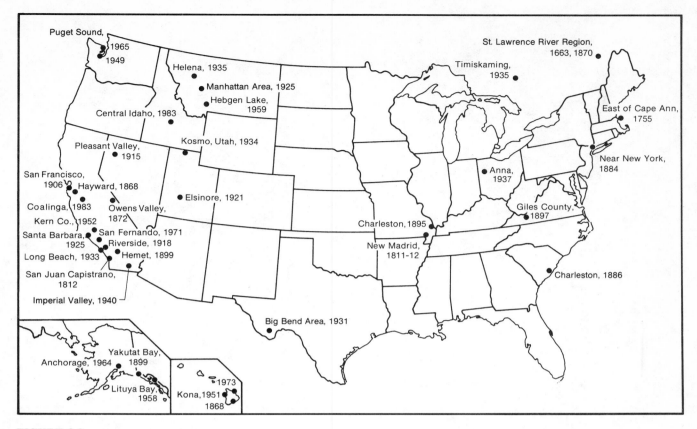

FIGURE 6.3
Locations of historical earthquakes in the United States that have caused damages.
Source: W. W. Hays, 1981, *Hazards from Earthquakes.* In *Facing Geologic and Hydrologic Hazards,* ed. W. W. Hays. U.S. Geological Survey Professional Paper 1240-B.

sphere are relatively weak and breaks repeatedly as stress associated with plate motion (even though plate boundaries are far away) strains the rocks. The recurrence interval for large earthquakes is estimated at 600–700 years; therefore, the inevitability of future damage demands that earthquake hazard be considered in design and construction of critical facilities such as nuclear power plants. Another example of a large intraplate earthquake is the 1886 event that devastated Charleston, South Carolina, claiming 60 lives and causing $23 million in property damage. More recently, in 1983, a large earthquake shook much of central Idaho, producing surface ruptures along a zone 36 km long in this area of relatively low population density.

A great earthquake ranks as one of nature's most catastrophic and devastating events. Earthquakes and their related hazards have destroyed large cities and taken thousands of lives in a matter of seconds. One sixteenth century earthquake in China reportedly claimed 850,000 lives; in 1923, an earthquake near Tokyo claimed 143,000 lives; and in 1976, a catastrophic earthquake in China claimed several hundred thousand

lives. It has been estimated that if a large earthquake were to hit San Francisco again, as it did in 1906, thousands of people might die in that one event. Table 6.1 lists some of the major historical earthquakes that have occurred in the United States.

Seismic risk maps for the United States (Figure 6.4) are based on either relative hazard or probability. Assignment of seismic risk based on past earthquake history and on an arbitrary relative numerical scale (Figure 6.4a) has serious problems because it may assign the same risk to several areas, even though there is not an equal probability of earthquake occurrence. Therefore, a probabilistic approach to seismic risk (Figure 6.4b), while more complicated, is preferable. This map shows contours of maximum values of horizontal ground acceleration caused by seismic shaking that, with a 90 percent probability, are not likely to be exceeded in 50 years. Another way of looking at this is that the darkest area on Figure 6.4b represents the area of greatest seismic hazard, interpreted as an area where the horizontal acceleration from seismic shaking of greater than 4 meters per second per second (40 per-

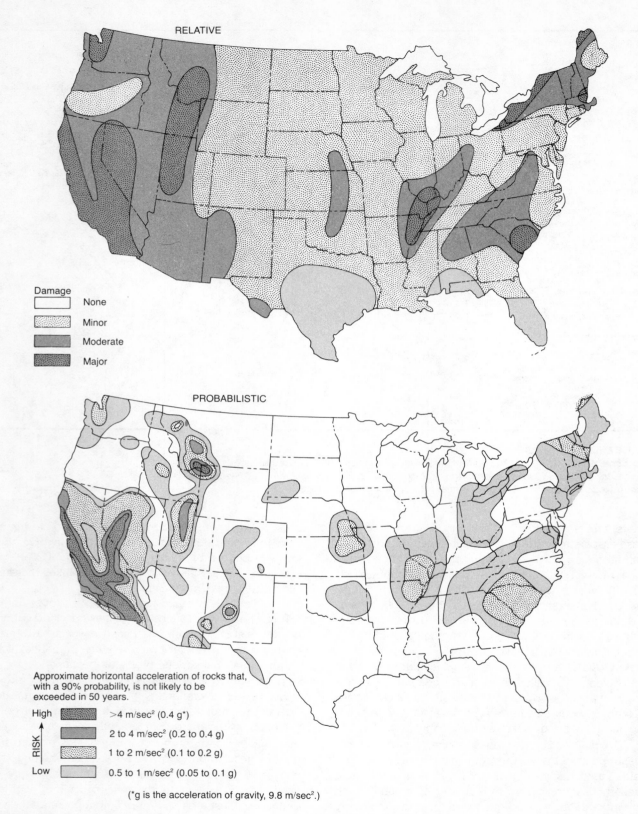

FIGURE 6.4

Two views of seismic hazards. The upper diagram is based on where earthquakes have occurred in the past. The lower diagram is a probabilistic approach (after S. T. Algermissen and D. M. Perkins, 1976, U.S. Geological Survey Open File Report 76–416), now in common use. The probabilistic approach provides more useful information as well as a more realistic evaluation of the earthquake hazard.

TABLE 6.1
Selected major earthquakes
in the United States

Year	Locality	Damage $ million	Lives Lost
1811–12	New Madrid, Missouri	Unknown	
1886	Charleston, South Carolina	23.0	60
1906	San Francisco, California	524.0	700
1925	Santa Barbara, California	8.0	13
1933	Long Beach, California	40.0	115
1940	Imperial Valley, California	6.0	9
1952	Kern County, California	60.0	14
1959	Hebgen Lake, Mont. (damage to timber and roads)	11.0	28
1964	Alaska and U.S. West Coast (includes tsunami damage from earthquake near Anchorage)	500.0	131
1965	Puget Sound, Washington	12.5	7
1971	San Fernando, California	553.0	65
1983	Coalenga, California	31.0	—
1983	Central Idaho	15.0	2

Modified from U.S. Department of Commerce, OEP data, 1971.

cent of the acceleration of gravity) can be expected, with odds of only one in ten to be exceeded in a 50-year period. Thus, when one uses the probabilistic method, the total number of earthquakes as well as other information becomes important in the analysis. Although regional earthquake hazard maps are valuable, considerable more data are necessary to identify hazardous areas more precisely to assist in developing building codes and determining insurance rates.

Like the natural hazards we have already discussed, earthquakes often produce primary effects and secondary effects. Primary effects are caused directly by the phenomenon. They include violent ground motion accompanied by surface rupture and, perhaps, permanent displacement of a meter or so; for example, the 1906 earthquake in San Francisco produced 5 meters of horizontal displacement. These violent motions produce sudden surface accelerations of the ground which can snap and uproot large trees and knock people to the ground. This motion may shear or collapse large buildings, bridges, dams, tunnels, pipelines, and other rigid structures (3). Damage from the 1964 Alaskan earthquake, excluding personal property and income, is shown in Figure 6.5. Extensive damage occurred to transportation systems—railroads, airports, and buildings (Figures 6.6 and 6.7). Fortunately, Alaska has a relatively low population density. What would be the effect of a large earthquake in a highly populated area? A hint was given by the 1971 earthquake that occurred on the fringe of a highly populated area of the San Fernando Valley in California. The earthquake released less than 1 percent of the energy of the Alaskan earthquake, yet it claimed one-half the lives and did more property damage than the Alaskan event of much greater magnitude. Damage to homes and larger build-

ings was extensive in the city of San Fernando (Figures 6.8 and 6.9) and the lower Van Norman Dam above the highly populated valley was severely damaged (Figure 6.10) and on the brink of catastrophic failure, threatening the lives of 80,000 people who evacuated their homes (4, 5). They returned safely to their homes four days later, after the water in the reservoir had been lowered to a safe level.

Secondary effects of earthquakes include a variety of *short-range* events, such as liquefaction, landslides, fires, tsunamis, and floods; and *long-range* effects, including phenomena such as regional subsidence or emergence of land-masses and regional changes in groundwater levels.

Liquefaction is transformation of water-saturated granular material from a solid state to a liquid state as

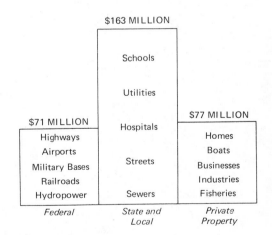

FIGURE 6.5
Earthquake damages in Alaska as estimated by the Office of Emergency Planning. (Data from Hansen and Eckel, U.S. Geological Survey Professional Paper 541, 1966.)

a result of an increase in the pore-water pressure caused by intense shaking. Earthquakes may cause liquefaction of near-surface, water-saturated silts and sands, causing the materials to lose their shear strength and flow. Three types of failure in response to liquefaction are recognized: flow landslides *(earthflows)* may occur on even moderate slopes; laterally spreading landslides may occur on gentle or nearly flat slopes, in which case the ground pulsates with the quaking and is accompanied by cracks, fissures, and differential settling; and quick condition failure may occur, characterized by a complete loss of shear strength (bearing capacity), in which case buildings may tilt and sink into the liquefied sediments while buried tanks may rise buoyantly (5).

Landslides of all varieties, in addition to those associated with liquefaction, are triggered or directly caused by earthquakes. They can be extremely destructive and can cause great loss of life, as demonstrated by the 1970 Peru earthquake. In that earthquake, more than 70,000 people died, of whom 20,000 were killed by a gigantic avalanche. The 1964 Alaskan earthquake produced thousands of landslides and avalanches, some of which devastated Anchorage (7,8).

Damage and even catastrophic destruction have resulted from *fires* indirectly caused by earthquakes. Disruption of electrical power lines and broken gas lines can start fires that are difficult to control because firefighting equpiment may be damaged and essential water mains may be broken. Access to fires is often hampered by blocked and damaged roads. Earthquakes in Japan and the United States have been accompanied by terrible fires. The San Francisco earthquake of 1906 has repeatedly been referred to as the "San Francisco fire," and, in fact, 80 percent of the damage was caused by terrible fires that ravaged the city for several days. The 1923 earthquake in Japan killed 143,000 people, 40 percent of whom died in a fire storm that engulfed an open space where people had gathered in an unsuccessful attempt to reach safety (3).

FIGURE 6.8
Damage to homes in San Fernando, California, caused by the 1971 earthquake. (Photo [a] by R. Castle, courtesy of U.S. Geological Survey. Photo [b] courtesy of Los Angeles City Department of Building and Safety.)

(a)

(b)

Tsunamis, or seismic sea waves (giant waves), are secondary effects generated by coastal or submarine earthquakes, as well as other submarine disturbances such as volcanic activity. Tsunamis are extremely destructive and one of the most serious secondary effects caused by earthquakes. These giant waves claimed the vast majority of lives lost in the 1964 Alaskan earthquake.

Regional changes in land elevations are another secondary effect of some large earthquakes. The Alaskan earthquake of 1964 caused vertical deformation (uplift and subsidence) or more than 250,000 square kilometers (7). The deformation includes two major zones of warping, each about 1,000 kilometers long and more than 210 kilometers wide (Figure 6.11). The seaward zone is one of uplift as great as 10 meters; the landward zone is characterized by subsidence as great as 2.4 meters centered in the vicinity of Portage. Effects of this nature vary from severely disturbed coastal marine life to changes in groundwater levels. In addition, flood damage occurred in some communities that experienced subsidence (Figure 6.12), whereas in areas of uplift, canneries and fishermen's homes were placed above the reach of most high tides. Before the earthquake, these structures and their docks were on the water (Figure 6.13).

FIGURE 6.9
Rescue operations at Veterans Hospital in San Fernando, California, after the 1971 earthquake. (Photo courtesy of Los Angeles City Department of Building and Safety.)

TECTONIC CREEP

Slow, nearly continuous movement along a fault zone not accompanied by felt earthquakes is known as *tectonic creep.* This process can slowly damage roads, sidewalks, building foundations, and other structures (Figure 6.14). It has damaged culverts under the football stadium at the University of California at Berkeley (Figure 6.15). Periodic repairs have been necessary as the cracks develop, and movement of 3.2 centimeters in eleven years was measured (9). Faster rates of tectonic creep have been recorded on the Calaveras fault zone near Hollister, California. There, a winery located on the fault is slowly being pulled apart at about 1 centimeter per year (10). Damage from tectonic creep generally occurs along narrow zones that delineate a fault in which movement is slow and continuous.

MAGNITUDE, INTENSITY, AND FREQUENCY OF EARTHQUAKES

The *magnitude* of an earthquake is a measure of the amount of energy released and is useful in comparing earthquakes quantitatively. It is determined by the amplitude of the largest wave recorded on a *seismograph,* an instrument for recording earthquake waves. The amplitude is measured on a logarithmic scale, so a magnitude of 6, for example, produces a displacement on the seismograph ten times larger than does a magnitude of 5. However, the energy released may be 30 to 60 times greater. Thus, between 27,000 and 216,000

FIGURE 6.10
Severely damaged as a result of the 1971 earthquake, possible failure of the lower Van Norman Dam threatened the lives of thousands of people. (Photo by R. E. Wallace, courtesy of U.S. Geological Survey.)

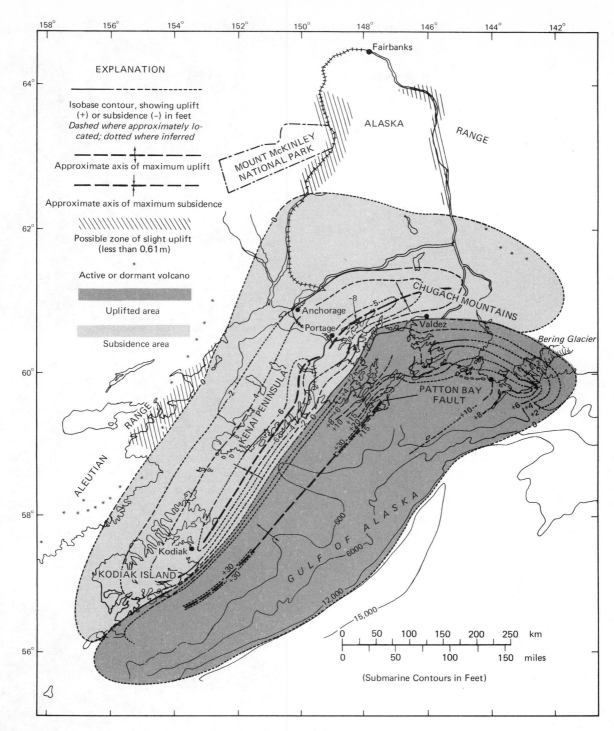

FIGURE 6.11

Map showing the distribution of tectonic uplift and subsidence in south central Alaska caused by the Alaskan earthquake of 1964. (From E. B. Eckel, U.S. Geological Survey Professional Paper 546, 1970.) (2 to 30 ft = 0.6 to 9.2 m)

shocks of magnitude 5 are required to release as much energy as a single earthquake of magnitude 8. There are probably about one million earthquakes a year that can be felt; however, only a few are strong enough to be recorded any considerable distance from their source. It has been estimated that every 100 years

FIGURE 6.12
Flooding of the Portage area of
Alaska caused by regional subsidence resulting from tectonic
land changes accompanying the
1964 earthquake. (Photo courtesy
of U.S. Geological Survey.)

there are about 200 great earthquakes with a magnitude exceeding 7.75.

In practice, the magnitude of an earthquake can
be obtained as illustrated in Figure 6.16. The maximum
amplitude and difference in arrival time of P and S
waves (P waves travel faster than S waves) from a distant earthquake are measured from a seismograph. For
example, Figure 6.16 shows the seismic record of an
earthquake with amplitude of 85 mm and difference in
arrival time of 34 seconds. The line connecting the amplitude and difference in arrival time indicates the
magnitude is 6 and the distance from the epicenter is
about 300 km. Records from several seismographs in the
region are necessary to precisely locate the epicenter.

An earthquake's intensity is based on personal
observations concerning the severity of its shaking. Table 6.2 is the Modified Mercalli Scale (abridged), with
12 divisions of intensity. A particular earthquake has
only one magnitude, but depending on proximity to the
epicenter and local geologic and engineering features,
many levels of intensity can be recorded. Table 6.3
shows approximate relationships among magnitude,
intensity, and average peak horizontal acceleration of
the ground for earthquakes, with the 1971 (magnitude
6.6) San Fernando Valley earthquake as a specific example. Maps produced from questionnaires collected
after an earthquake showing the variability of intensity
(see Table 6.3) are valuable as a crude index of ground

FIGURE 6.13
Canneries and fishing homes
along Orca Inlet in Prince William Sound after the 1964 Alaskan earthquake. The photograph
was taken at a 3-meter tidal
stage, which would have
reached beneath the docks prior
to the earthquake. (Photo courtesy of U.S. Geological Survey.)

shaking and provide a detailed picture of the earthquake, including how and where it was felt. Thus, while the magnitude of an earthquake provides information concerning the amount of energy released, the intensity reflects how people perceive the shaking.

EARTHQUAKES CAUSED BY HUMAN USE OF THE LAND

Various human activities have increased or caused earthquake activity. The damage is regrettable, but the lessons we have learned may help in the future to control or stop large catastrophic earthquakes. Three ways people have caused earthquakes are by loading the earth's crust, disposing of waste into deep wells, and setting off underground nuclear explosions (11).

About 600 local tremors occurred during the first ten years following completion of Hoover Dam, which supplies Lake Mead in Arizona and Nevada. Most of the tremors were very small, but one had a magnitude of about 5 and two had magnitudes of about 4 (11). An earthquake in India which killed 200 people and one in Zambia, both with magnitudes of about 6, were also

TABLE 6.2
Modified Mercalli Intensity Scale (Abridged)

Intensity	Effects
I	Not felt except by a very few under especially favorable circumstances.
II	Felt only by a few persons at rest, especially on upper floors of buildings. Delicately suspended objects may swing.
III	Felt quite noticeably indoors, especially on upper floors of buildings, but many people do not recognize it as an earthquake. Standing motor cars may rock slightly. Vibration like passing of truck. Duration estimated.
IV	During the day felt indoors by many, outdoors by few. At night some awakened. Dishes, windows, doors disturbed; walls make cracking sound. Sensation like heavy truck striking building; standing motor cars rocked noticeably.
V	Felt by nearly everyone; many awakened. Some dishes, windows, etc., broken; a few instances of cracked plaster; unstable objects overturned. Disturbance of trees, poles and other tall objects sometimes noticed. Pendulum clocks may stop.
VI	Felt by all; many frightened and run outdoors. Some heavy furniture moved; a few instances of fallen plaster or damaged chimneys. Damage slight.
VII	Everybody runs outdoors. Damage negligible in buildings of good design and construction; slight to moderate in well-built ordinary structures; considerable in poorly built or badly designed structures; some chimneys broken. Noticed by persons driving motor cars.
VIII	Damage slight in specially designed structures; considerable in ordinary substantial buildings with partial collapse; great in poorly built structures. Panel walls thrown out of frame structures. Fall of chimneys, factory stacks, columns, monuments, walls. Heavy furniture overturned. Sand and mud ejected in small amounts. Changes in well water. Disturbs persons driving motor cars.
IX	Damage considerable in specially designed structures; well designed frame structures thrown out of plumb; great in substantial buildings, with partial collapse. Buildings shifted off foundations. Ground cracked conspicuously. Underground pipes broken.
X	Some well-built wooden structures destroyed; most masonry and frame structures with foundations destroyed; ground badly cracked. Rails bent. Landslides considerable from river banks and steep slopes. Shifted sand and mud. Water splashed (slopped) over banks.
XI	Few, if any (masonry), structures remain standing. Bridges destroyed. Broad fissures in ground. Underground pipe lines completely out of service. Earth slumps and land slips in soft ground. Rails bent greatly.
XII	Damage total. Waves seen on ground surfaces. Lines of sight and level distorted. Objects thrown upward into the air.

From Wood and Neuman, 1931 by U.S. Geological Survey, 1974, Earthquake Information Bulletin, v. 6 no. 5, p. 28.

FIGURE 6.14
Curb displaced by fault creep on the Calaveras Fault, central California. This type of movement is generally very slow and not accompanied by earthquakes noticeable to people. (Photo courtesy of U.S. Geological Survey.)

associated with reservoirs. Evidently, fracture zones (faults) are activated by the increased load of water on the land and by increased water pressure in the rocks below the reservoir.

Events in 1966 dramatically suggested that deep-well disposal of liquid chemical waste activated fracture zones and caused more than 600 earthquakes in the Denver, Colorado, area from April, 1962, to November, 1965 (12). The greatest magnitude was 4.3, large enough to knock bottles off shelves in stores. The chemical waste was a by-product of the Rocky Mountain Arsenal's manufacturing of materials for chemical warfare. The arsenal started production in 1942, but evaporation from earth reservoirs disposed of the waste water until 1961, when groundwater pollution became a problem and engineers decided to pump the waste into a 3,600-meter-deep disposal well. The rock accepting the waste was highly fractured metamorphic rock. The new liquid increased the fluid pressure, facilitating slippage along fractures and precipitating earthquakes. The well-known correlation between rate of injection of waste and frequency of earthquakes is shown in Figure 6.17. When the injection stopped, the earthquakes stopped. Scientists speculate as to whether this principle of starting and stopping earthquakes by varying the fluid pressure in rocks can be used to prevent large earthquakes.

Numerous earthquakes have been triggered by nuclear explosions at the Nevada test site (11). After underground explosion of a 1.1-megaton nuclear device, thousands of aftershocks were recorded. The depth of the shocks ranged from the surface down to 7 kilometers. Magnitudes did not exceed 5.0, which is considerably less than the magnitude of 6.3 from the initial explosion. Analysis of the aftershocks suggested they were triggered primarily by release of natural tec-

FIGURE 6.15
Map showing the location of the University of California Memorial Stadium, Berkeley; the active fault or shear zone within the Hayward fault zone; and the stadium culvert where major cracking has taken place. (After Radbruch, et al., U.S. Geological Survey Circular 525, 1966.)

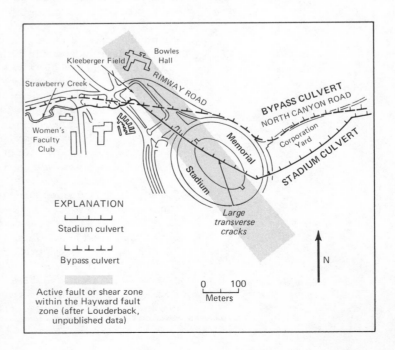

TABLE 6.3

Approximate Relationships Between the Magnitude and Intensity of an Earthquake

Magnitude	Area Felt Over (square kilometers)	Distance Felt (kilometers)	Intensity (maximum expected Modified Mercalli)	Ground Motion: (Average peak horizontal acceleration g = gravity = 9.8 meters per second per second)
3.0–3.9	1,950	25	II–III	Less than 0.15 g
4.0–4.9	7,800	50	IV–V	0.15–0.04g
5.0–5.9	39,000	110	VI–VII	0.06–0.15g
6.0–6.9	130,000	200	VII–VIII	0.15–0.30g
7.0–7.9	520,000	400	IX–X	0.50–0.60g
8.0–8.9	2,080,000	720	XI–XII	Greater than 0.60g

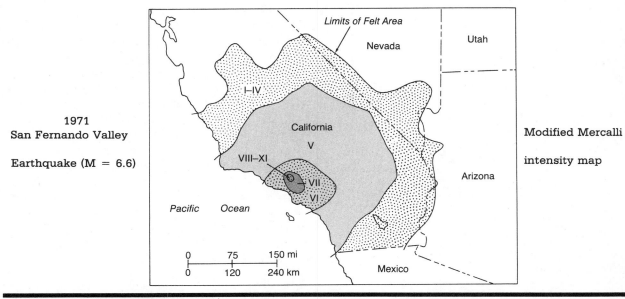

1971 San Fernando Valley Earthquake (M = 6.6)

Modified Mercalli intensity map

Modified after U.S. Geological Survey, *Earthquake Information Bulletin,* Vol. 6, No. 5, 1974, with the 1971 San Fernando Valley earthquake as an example.

tonic stress. Again, scientists wonder if this principle might be used to prevent a large earthquake.

EARTHQUAKE HAZARD REDUCTION

Like landslides and floods, earthquakes are natural processes, and even though human use and interest may affect, to a lesser or greater extent, some aspects of the magnitude and frequency of these processes, they cannot be completely eliminated. Therefore, we must learn to minimize the hazards.

A comprehensive earthquake hazard reduction program must be multifaceted and include recognition of active fault zones and earth materials sensitive to shaking, continued research to better predict and possibly control earthquakes, and development and improvement of possible adjustments to earthquake activity.

Active Fault Zones

A *fault* is a fracture or, more accurately, a zone of fractures along which rocks have been displaced; that is, rocks on one side of the fracture have moved relative to rocks on the other side (Figure 6.18). A *fault zone* is a group of related faults or fault traces that usually form a braided pattern subparallel to the direction of the zone. The fault zone may have considerable width, ranging from a meter or so to several kilometers.

There is no real agreement as to what always constitutes an active fault. Many geologists would label a fault *active* if it can be demonstrated that there has been movement during the last 11,000 years (Holocene time). The Quaternary (last 3 million years) is the most recent period of geologic time; most of our landscape has been produced in that time. Therefore, it seems reasonable to classify any fault that has moved during the Quaternary as *potentially* active (Table 6.4). Faults

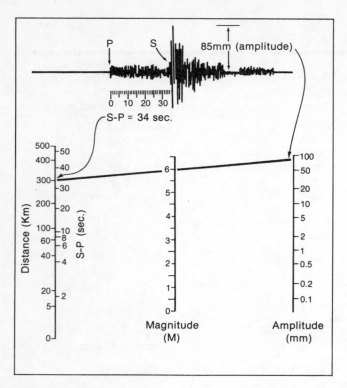

FIGURE 6.16
Idealized diagram showing the procedure for calculating magnitude (M) of an earthquake. For our example, the maximum amplitude (85 mm) is measured from the seismic record; the difference in arrival time between the S and P waves (34 seconds) is also taken from the seismic record; and the approximate magnitude of the earthquake as well as distance from the recording station is obtained by placing a straight line between the amplitude in millimeters and difference in arrival time in seconds, as shown on the diagram. Here, the magnitude is 6 and the distance is approximately 300 km. Modified after B.A. Bolt, 1978, *Earthquakes: A primer.* (San Francisco: W. H. Freeman and Company.)

that have been inactive for the last three million years are generally classified as *inactive,* but, whereas faults accompanying recorded earthquakes certainly are active, it is often difficult to prove the activity of a fault unrelated to such easily measured phenomena.

Prehistoric earthquakes and general fault activity can be estimated by investigating fault-related landform assemblages and soils. Figure 6.19 shows the characteristic landforms found along active strike-slip faults characterized by dominant horizontal motion. Features such as offset streams, sags, linear ridges, and fault scarps may indicate recent faulting, but care must be exercised in interpreting them because some of these forms may have a complex origin.

The study of soils can also be useful in estimating the activity of a fault. For example, soils on opposite sides of a fault may be of similar or quite dissimilar age, thus establishing a relative age for the displace-

ment. This method, in conjunction with other data, was used along parts of the recently discovered Ventura fault in California (Figure 6.20) to show its recent activity.

Prehistoric earthquake activity sometimes can be dated radiometrically (absolutely) if suitable materials can be found. For example, radiocarbon dates from faulted sediments have been obtained from the Coyote Creek fault zone and San Andreas fault zone in southern California.

The Coyote Creek fault is characterized by predominant horizontal movement, but there is a vertical component of motion, and a trench across the fault after a 1968 earthquake revealed three older displacements of the flat-lying sediments that were dated radiometrically (Carbon 14) (Figure 6.21). These data suggest that the recurrence interval for earthquakes the size of the 1968 event is about every 200 years (13).

A recent trench across the right-lateral, strike-slip San Andreas fault at Pallett Creek, California, revealed significant data concerning its earthquake history during the last 1,700 years. During that period, deposition of 4 to 5 meters of sediment, consisting of alternating peat and gravel-sand (flood) deposits, took place. These deposits appear to be displaced by 12 large earthquakes which, by means of radiometric dating, suggest an average recurrence interval of about 145 years. Figure 6.22 shows the deformation of sedimentary units associated with an earthquake that occurred at Pallet Creek about 800 years ago (14).

Seismic Shaking and/or Ground Rupture

Active faults present two basic hazards to people and their structures: *seismic shaking* and/or *surface rupture.* Seismic shaking occurs during sudden movement along a deep portion of a fault, and may or may not be accompanied by surface rupture. On the other hand, rupture of the earth's surface along a fault or fault zone is characterized by fault motion that may be instantaneous or by slow creep, and may or may not be accompanied by an earthquake (15).

In assessing fault hazard, faults are commonly classified according to their recency of movement (see Table 6.4) with the implicit assumption that the younger the earth materials offset by the fault, the more hazardous the fault is likely to be. This assumption faces challenge based on the hypothesis that three types of active faults exist that can be defined in terms of the type of hazard each presents: faults with both ground rupture and seismic shaking hazard; faults with seismic shaking but slight or no ground rupture; and faults with ground rupture but minor if any seismic shaking hazard. Examples of each type of fault in

FIGURE 6.17

Generalized block diagram showing the Rocky Mountain Arsenal well (a) and graph showing the relationship between earthquake frequency and rate of injection of liquid waste for five characteristic time periods (b). (Graph [b] after D. M. Evans, *Geotimes* 10, 1966. Reprinted by permission.)

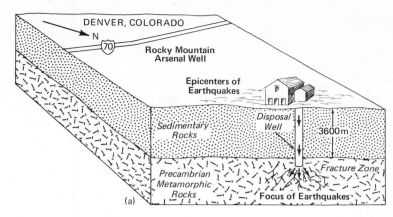

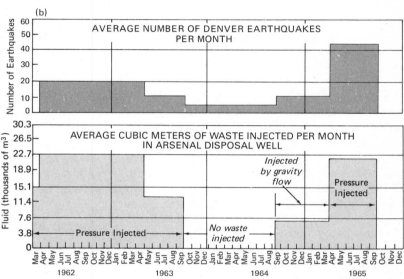

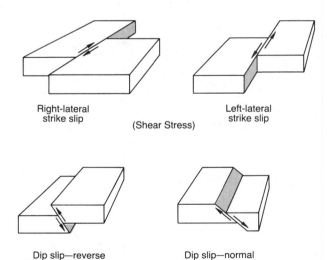

Right-lateral
strike slip

Left-lateral
strike slip

(Shear Stress)

Dip slip—reverse
(Compressive stress)

Dip slip—normal
(Tensional stress)

FIGURE 6.18

Types of fault movement based on the sense of motion relative to the fault and type of stress involved. (From R. L. Wesson, E. J. Helley, K. R. Lajoie, and C. M. Wentworth, U.S. Geological Survey Professional Paper 941A, 1975.)

Southern California are shown in Figure 6.23. It is argued that recognition of the type of fault activity associated with a particular fault is important and should be considered in land-use planning and programs to minimize potential damages from earthquakes or other activities associated with faults or fault zones.

The San Andreas fault in California is an example of a fault with both a seismic shaking and a ground rupture hazard. Historic earthquakes and fault scarps in alluvial fan and other geologically young material demonstrate the potential of the fault to produce large earthquakes and ground rupture.

The Oakridge fault and Newport-Inglewood fault zone (Figure 6.23) both have a potential hazard for seismic shaking, but studies along these faults have suggested that there is little or no evidence for ground rupture.

In several faults near Ventura, California (north of the Red Mountain fault and south of the San Cayetano fault in Figure 6.23), younger alluvial deposits and soils have been displaced or offset by several tens of meters.

FIGURE 6.19
Assemblage of fault-related landforms along a large strike-slip
fault such as the San Andreas
Fault. (After R. L. Wesson, et al.,
U.S. Geological Survey Professional Paper 941 A, 1975.)

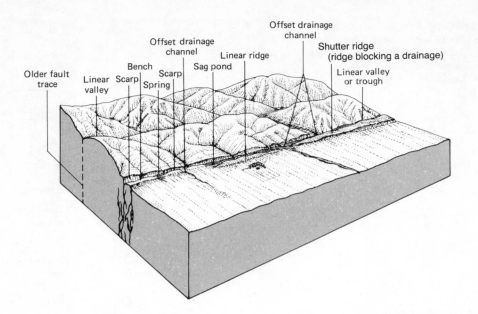

The movement is almost entirely along bedding planes in response to folding of the rocks. These types of faults are known as flexural-slip faults (16), and are clearly different from deep straight faults that are capable of producing both seismic shaking and ground rupture. Figure 6.24 is an idealized drawing comparing deep straight faults with flexural-slip faults. The main point in distinguishing between flexural-slip and other faults is that faulting associated with flexural-slip is usually shallow, does not extend to depths where large earth-quakes are produced, and therefore does not present a high magnitude seismic shaking hazard. Since flexural-

slip faults are often located in areas with high seismic risk, the potential hazard produced by shaking may not be greater or less at a particular site with or without the flexural-slip faulting. However, recognition of the additional surface rupture hazard is important and should be considered in land-use planning.

Earth Materials and Seismic Activity

Surficial earth materials such as mud, alluvium, and bedrock behave differently in response to seismic shak-ing of various frequencies. For example, the intensity

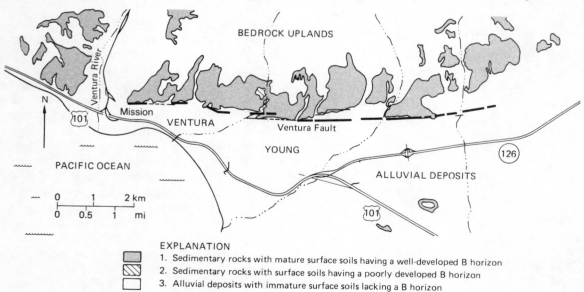

EXPLANATION

1. Sedimentary rocks with mature surface soils having a well-developed B horizon
2. Sedimentary rocks with surface soils having a poorly developed B horizon
3. Alluvial deposits with immature surface soils lacking a B horizon

FIGURE 6.20
Map illustrating the difference in soils on opposite sides of the Ventura Fault in southern
California. (After Wojcicki, et al., U.S. Geological Survey Map MF-781.)

TABLE 6.4
Terminology related to recovery
of fault activity. (After California
State Mining and Geology Board
Classification 1973.)

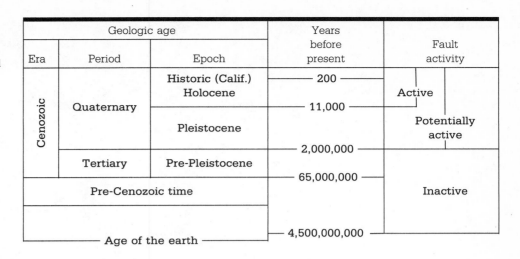

Geologic age			Years before present	Fault activity
Era	Period	Epoch		
Cenozoic	Quaternary	Historic (Calif.) Holocene	— 200 —	Active
			— 11,000 —	
		Pleistocene		Potentially active
			— 2,000,000 —	
	Tertiary	Pre-Pleistocene	— 65,000,000 —	
	Pre-Cenozoic time			Inactive
	Age of the earth		— 4,500,000,000 —	

FIGURE 6.21
Generalized sketch of trench wall
showing vertical displacement of
flat-lying sediments associated
with the Coyote Creek fault, California (a). Older offsets are progressively greater. Graph of vertical deformation against time
(b). Derived from radiocarbon
dates of the faulted sediments.
(After M. M. Clark, A. Grantz,
and R. Meyer, U.S. Geological
Survey Professional Paper 787,
1972.)

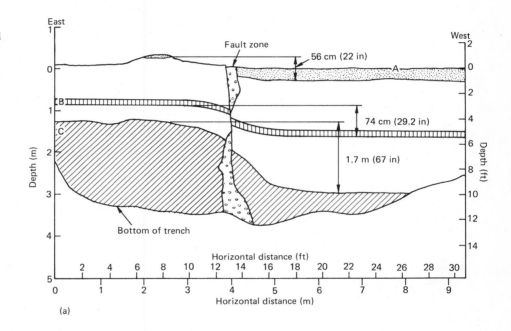

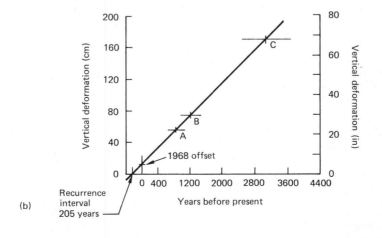

135

FIGURE 6.22
Part of the trench wall crossing the San Andreas fault at Pallett Creek showing a 30-centimeter offset. Notice that the strata above the 22-centimeter scale (central part of photograph) are not offset. The offset sedimentary layers (white beds) and the dark organic layers were produced by an earthquake that occurred about 800 years ago. (Photo courtesy of Kerry Sieh.)

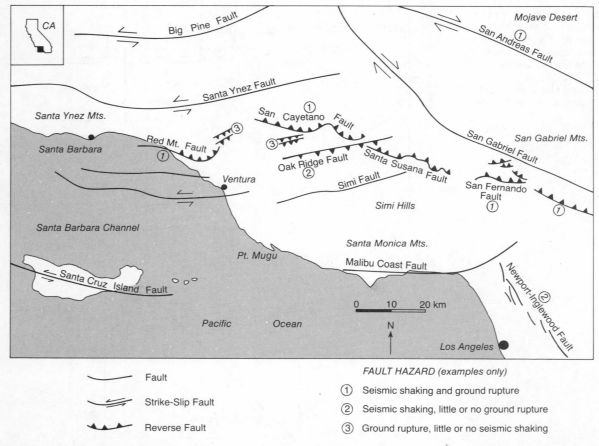

	Fault
	Strike-Slip Fault
	Reverse Fault

FAULT HAZARD (examples only)

① Seismic shaking and ground rupture

② Seismic shaking, little or no ground rupture

③ Ground rupture, little or no seismic shaking

FIGURE 6.23
Major earthquake faults of southern California with specific examples of faults characterized by their type of activity.

of shaking may be much more severe for unconsolidated sediment than for bedrock (Figure 6.25), and, therefore, shaking may be stronger for sites underlain by mud and alluvium than for bedrock sites much closer to the fault (17). As a result, the obvious procedures for minimizing damages associated with shaking are, first, to map areas that are especially sensitive and, second, to make this information available to those responsible for land-use decisions.

As part of earthquake disaster preparedness, it is important to predict the possible geologic effects of a postulated earthquake of particular magnitude. This has been done for the San Francisco Bay region (17). Figure 6.26 shows a generalized geologic cross-section of the south part of the bay through Menlo Park as well as the predicted geologic effects of a 6.5-magnitude earthquake on the San Andreas fault. Notice that, although surface faulting and landsliding is almost en-

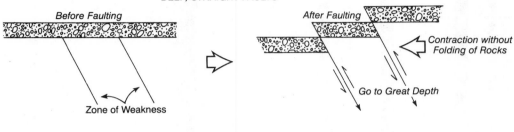

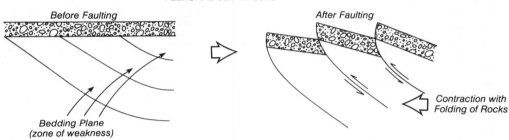

FIGURE 6.24
Idealized drawing comparing deep, straight faults, with both a seismic shaking and ground rupture hazard to shallow, flexural-slip faults with a ground rupture hazard but little seismic shaking hazard (associated with folding of the rocks). (From presentation by Clark and Keller at the Cordilleran Sectional Meeting of the Geological Society of America in Corvallis, Oregon, 1980.)

FIGURE 6.25
Generalized relationship between near-surface earth material and amplification of shaking during a seismic event.

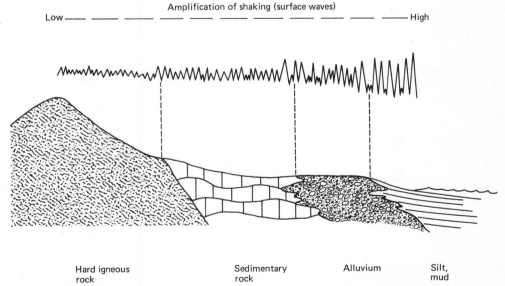

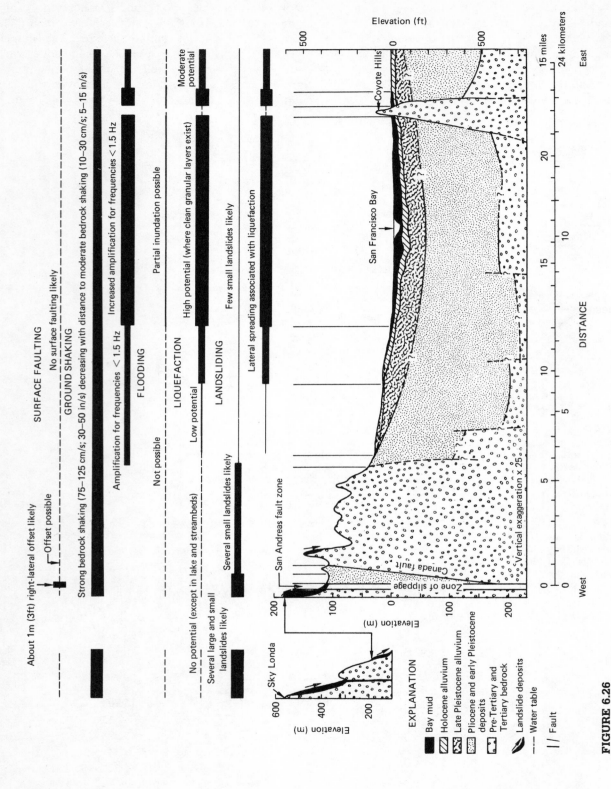

FIGURE 6.26

Predicted geologic effects of a magnitude 6.5 earthquake along the San Andreas fault. (After R. D. Borcherdt, et al., U.S. Geological Survey Professional Paper 941A, 1975.)

138

tirely restricted to the fault zone and strong bedrock shaking decreases with distance from the fault, the bay at 16 kilometers from the fault is very vulnerable to liquefaction and severe shaking because of amplification of surface seismic waves in the bay mud.

EARTHQUAKE PREDICTION

Earthquake prediction is an area of serious research. The Japanese made the first attempts, with some success, using as data the frequency of microearthquakes, repetitive level surveys, water-tube tilt meters, and geomagnetic observations. They found that earthquakes in the area they studied were always accompanied by swarms of microearthquakes that occurred several months before the major shocks. Furthermore, ground tilt correlated strongly with earthquake activity: anomalous tilt occurred shortly before some shocks of a magnitude near 5. Anomalous magnetic fluctuations were also reported (11).

Chinese scientists in 1975 made the first successful prediction of a major earthquake. The Haicheng earthquake of February 4, 1975, had a magnitude of 7.3 and destroyed or damaged about 90 percent of the structures in a city of 90,000 people. The lives of thousands of people undoubtedly were saved by the massive evacuation from unsafe housing just before the earthquake. The short-term prediction was based primarily on a series of foreshocks that began four days prior to the main shock. On February 1 and 2 there were several shocks with a magnitude less than 1. On February 3, less than 24 hours before the main quake, a foreshock of a magnitude of 2.4 occurred, and in the next 17 hours, 8 shocks with magnitude greater than 3 occurred. Then, as suddenly as it began, the foreshock activity became relatively quiet for 6 hours prior to the main earthquake (18,19).

Unfortunately, foreshocks do not always precede large earthquakes, and in 1976, a catastrophic earthquake, one of the deadliest in recorded history, struck near the mining town of T'anshan, China, killing several hundred thousand people. There were no foreshocks!

Optimistic scientists in the United States, Japan, Russia, and China believe that eventually we consistently will be able to make long-range prediction (100,000 years to about 100 years), medium-range prediction (a few years to a few months), and short-range prediction (a few days or hours) for general locations and magnitudes of earthquakes. Earthquake prediction is a complex problem, however, and it will probably be many years before dependable short-range prediction is possible. Such prediction will most likely be based on these factors: previous patterns and frequency of earthquakes, anomalous changes in the stored strain of rocks, magnetic properties of rocks, vertical or horizontal motions (tilting) of rocks, changes in the speed of seismic waves through rocks, seismic gaps along active faults, changes in the electrical resistivity of the earth, changes in the amount of dissolved radioactive gas (radon) in groundwater, and, perhaps, anomalous behavior of animals.

Rates of uplift and subsidence, especially rapid or anomalous change, may be significant in predicting shocks. For example, for more than 10 years before the 1964 earthquake near Niigata, Japan (magnitude 7.5), there was anomalous uplift of the earth's crust (Figure 6.27) (20).

Early in 1976, scientists with the U.S. Geological Survey discovered the so-called "Palmdale Bulge" (Figure 6.28). The uplift, as recognized from regional leveling surveys, apparently developed in the 1960s and early 1970s and is probably continuing today, to a lesser or greater extent. The total uplift is on the order of one-half meter over an area of about 842,000 square kilometers. Parts of the bulge have deflated since it was first recognized; for example, the land near Palmdale rose about 35 centimeters from 1960 to early in 1974, but dropped 17 centimeters from 1974 to 1977.

The main concern with the Palmdale Bulge is that the uplift may represent a long-term precursor to a large earthquake. The area is in the "Big Bend" area of the San Andreas fault where little or no tectonic creep occurs and strain tends to be relieved primarily by infrequent large earthquakes. Some scientists fear the bulge may represent the effect of strain along the fault. On the other hand, events such as the Palmdale Bulge may be a periodic part of tectonic behavior associated with mountain building or other processes that have not been recognized previously. Still other scientists argue that the Palmdale Bulge is just a figment of imagination and results from survey errors. We simply do not know; and this uncertainty is troublesome—perhaps ominous—to the millions of people in southern California.

Changes in speed of primary seismic waves may decrease for months, and then increase to normal just before an earthquake, as shown in Figure 6.29. Such changes may be used to predict earthquakes in the future (21).

Seismic gaps are defined as areas along active fault zones, capable of producing large earthquakes but that have not recently produced a large earthquake.

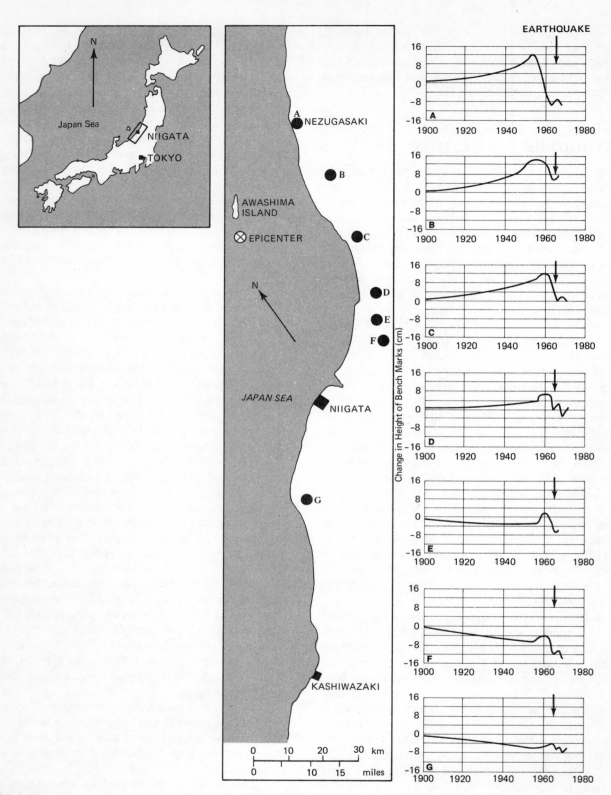

FIGURE 6.27

Anomalous uplift of the crust of the earth was observed for approximately ten years before the 7.5-magnitude earthquake which struck Niigata, Japan, in 1964. The uplift was measured by plotting changes in the elevation of bench marks (points of known elevation) with time. The black dots are the location of the bench marks, and the graphs correspond to these. (From ''Earthquake Prediction,'' Frank Press. Copyright © May 1975 by Scientific American, Inc. All rights reserved.)

FIGURE 6.28
Elevation changes in the area of the Palmdale Bulge from 1959 to 1974. The contour interval is 5 centimeters. (From J. Bennett, Palmdale "Bulge" Update, *California Geology,* vol. 30, no. 8, 1977.) (5 cm = 1.9 in)

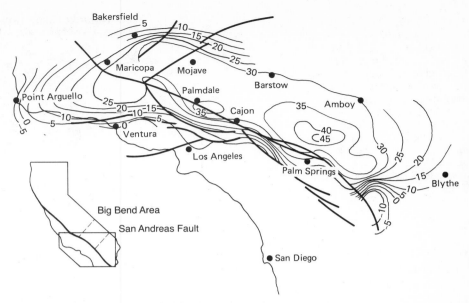

These areas are thought to store tectonic strain and thus are candidates for future large earthquakes (22). The seismic gap principle is generalized by example in Figure 6.30. Rates of movement along parts of the San Andreas fault have been measured before recent earthquakes (11), and four moderate shocks along the San Andreas fault from 1971 to 1973 have tentatively identified such a gap, as generalized in Figure 6.30 (21).

Seismic gaps have been extremely useful in medium-range earthquake prediction. At least eight large earthquakes have been successfully forecast since 1965, including a 1971, $M = 7.2$ event in southeastern Alaska and a 1973, $M = 7.3$ event in Mexico (22).

As earth scientists examine patterns of relative seismicity, two ideas are emerging. First, there are sometimes reductions in small or moderate earthquakes prior to a larger event; and second, small earthquakes may tend to ring an area where a larger event eventually occurs. Such a ring (or doughnut) was noticed to have developed in the 16 months prior to the 1983, $M = 6.2$ earthquake in Coalinga, California. Un-

fortunately, this was not noticed until after the event, and quantitative criteria to identify seismic doughnuts, although the subject of serious research, are not yet refined.

Electrical resistivity is a measure of a material's ability to retard an electrical current. Conductors such as copper and aluminum have a very low resistance, whereas other materials such as quartz (an insulator) have a very high resistance to electric current. In general, the earth is a good conductor, but its resistivity varies with the amount of groundwater present and many other factors. If a current is fed into the earth between two points separated by several kilometers, changes in the voltages will be noted if any change in electrical resistivity takes place in the rocks. Changes before earthquakes have been reported in the United States, Russia, and China (Figure 6.31) (20).

The amount of the radioactive gas radon that is dissolved in the water of deep wells may apparently increase significantly before an earthquake (Figure 6.32). The technique of monitoring radon appears promising

FIGURE 6.29
Earthquake prediction by changes in speed of seismic waves. The speed of the primary wave may decrease for months before an earthquake, then rise to normal just before the earthquake. (After R. E. Wallace, U.S. Geological Survey Circular 701, 1974.)

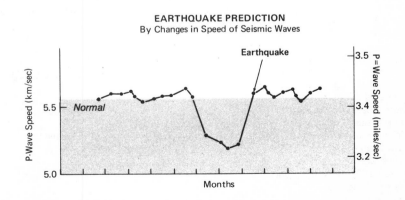

EARTHQUAKE PREDICTION
By Changes in Speed of Seismic Waves

FIGURE 6.30

Earthquake prediction by seismic gaps along active faults. Mapping of microearthquakes has identified slip surfaces along the San Andreas fault related to four moderate earthquakes (of magnitudes 4 to 5). A "gap" may suggest that the next slip and related earthquake will lie between where the other earthquakes occurred. (After R. E. Wallace, U.S. Geological Survey Circular 701, 1974.)

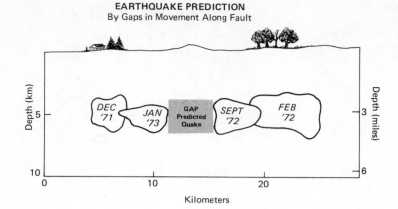

EARTHQUAKE PREDICTION
By Gaps in Movement Along Fault

and is being studied carefully in Russia, China, and the United States.

Anomalous animal behavior is often reported prior to large earthquakes: everything from unusual barking of dogs, chickens that refuse to lay eggs, snakes that crawl out of the ground in winter and freeze, and horses or cattle that run in circles, to rats emerging from the ground and acting as if drugged or dazed. Anomalous behavior was apparently common before the Haicheng earthquake (18). Unusual animal behavior was also observed prior to the 1971 San Fernando earthquake. Unfortunately, the significance and reliability of animal behavior is difficult to evaluate. Nevertheless, there is considerable interest in the topic and research is being conducted.

Toward a Prediction Method

We are a long way from a working, practical methodology to predict earthquakes reliably. On the other hand, a good deal of data is currently being gathered concerning possible precursor phenomena associated with earthquakes. We have recently learned, from many measurements of strain, for example, that the stress pattern along the "Big Bend" area (Figure 6.28) of the San Andreas fault changes. Until the latter part of the 1970s, most measurements suggested the presence of a north-south stress that tended to lock the fault. During the latter part of the 1970s, however, the stress pattern began to change slightly, so that the north-south compression was accompanied by an east-west extension. Other phenomena showed change as well—the amount of radon gas emitted from wells, water-well levels, and tilt meter measurements. This information does not prove there is going to be a large earthquake along the San Andreas fault near Los Angeles in the near future, but neither do the data repudiate this idea. Changes have been observed, some of which are synchronized and may have future significance. The change in the stress field may become particularly important, because an east-west stress may make fault movement easier, and thus make it easier for an earthquake to be produced. Some geologists

FIGURE 6.31

Changes in electrical resistivity of the earth's crust before earthquakes were measured in the USSR, China, and the United States. The data for this graph were obtained for earthquakes monitored in the USSR between 1967 and 1970. The measurements are made by feeding an electric current into the ground and recording voltage changes a few kilometers away. In general, it has been observed that earthquakes are preceded by a decrease in crustal resistivity. (From "Earthquake Prediction," Frank Press. Copyright © May 1975 by Scientific American, Inc. All rights reserved.)

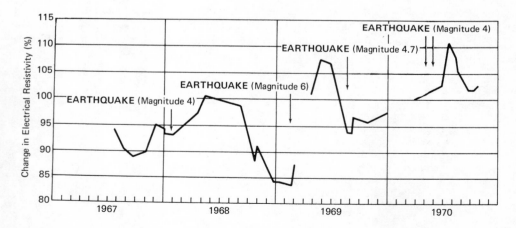

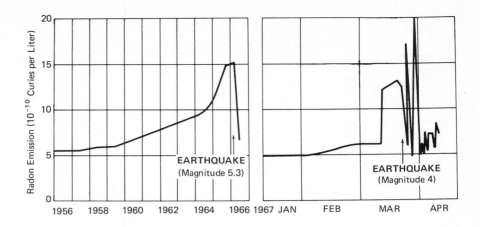

might say that as long as the east-west extension is present, we are in a time period in which a large earthquake along the San Andreas fault in southern California is more likely. Considering that it has been well over 100 years since the last large earthquake along the San Andreas fault in southern California, a large earthquake at any time in the future should not come as a surprise.

Long-term earthquake prediction Although progress on short-range earthquake prediction has not matched expectations, long-range prediction, including hazard evaluation and probabilistic analysis of areas along active faults, has been faster than expected (23). Techniques based on field work have been developed to calculate slip rates on faults and average recurrence intervals of large earthquakes, and geologists are beginning to make statements as to the probability of a large earthquake's occurring on a particular fault segment in the next 20 years. It has even been argued that determining recurrence of earthquakes on faults such as the San Andreas fault in California and the Wasatch fault in Utah is providing more valuable information for earthquake hazard reduction programs through development of appropriate land-use planning, building codes, and engineering design than would short-term earthquake prediction capability (23).

As an example of long-term earthquake prediction, consider California's San Andreas fault. Figure 6.33 shows four main segments of the fault, differentiated by historic and prehistoric activity derived from seismicity, strain measurement, and field observations. The area with the highest probability for a great earthquake ($M = 8$) is the southern segment, where there have been no historic earthquakes and no evidence for large earthquakes in the last several hundred years. The recent geologic record suggests, however, that

large earthquakes should be expected. At one location near Indio, California, several small streams are offset several meters and a late Pleistocene alluvial fan is offset nearly 700 m. The slip rate is estimated to be a few cm/yr for this segment of the fault and the strain rate is about 2 cm/yr across the fault (24). Thus, the southern segment of the San Andreas fault may be a nearly mature seismic gap with as much as 11 m of accumulated strain (deficit fault slip stored in rocks). For these reasons, geologists believe this portion of the fault is a candidate for a large earthquake in the relatively near future. Other general probabilistic statements concerning other sections of the fault are shown in Figure 6.33 (24, 25), but these probabilities are only ball-park estimates and *not* certainties—we may yet be fooled again by the complex behavior of the San Andreas fault. In fact, the April 24, 1984 earthquake ($M = 6.2$) near San Jose, California occurred along a segment of the fault thought to have a low probability of producing such an event.

The Borah Peak earthquake of October 28, 1983, in central Idaho, is another success story for long-range earthquake hazard evaluation. Previous evaluation of the Lost River fault suggested that the fault was active (26). The earthquake, which was about a magnitude 7 event, killed two people and did about $15 million in damage, producing fault scarps up to several meters high and numerous ground cracks along 36 km of the fault. The important fact was that the scarps and faults produced *during* the earthquake were *superimposed* on pre-existing fault scarps, validating the usefulness of carefully mapping scarps produced from prehistoric earthquakes—the ground will break there again!

Even though we are a long way from a working methodology to predict earthquakes, speculating on a possible model is interesting. Assume that we can develop a set of relationships from which an earthquake

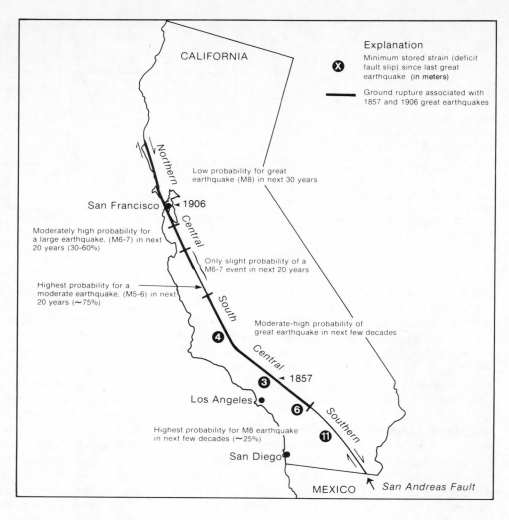

Explanation

X Minimum stored strain (deficit fault slip) since last great earthquake (in meters)

— Ground rupture associated with 1857 and 1906 great earthquakes

CALIFORNIA

Northern

Low probability for great earthquake (M8) in next 30 years

San Francisco — 1906

Central

Moderately high probability for a large earthquake. (M6-7) in next 20 years (30-60%)

Only slight probability of a M6-7 event in next 20 years

Highest probability for a moderate earthquake. (M5-6) in next 20 years (~75%)

South

Moderate-high probability of great earthquake in next few decades

④

Central

③ — 1857

Los Angeles •

⑥

Southern

Highest probability for M8 earthquake in next few decades (~25%)

⑪

San Diego •

MEXICO   ↗ San Andreas Fault

magnitude and date (or time) of occurrence are predicted. The first step is to identify empirical relationships among precursor events (such as foreshock activity, anomalous tilt, radon gas emission) and probable earthquake magnitude. The precursor events may be measured and evaluated for a specific area or by their intensity, and we would expect to find a direct relationship with earthquake magnitude. In other words, after observing many earthquakes, we would empirically derive Figure 6.34a. Then, given a set of precursor observations (point *A* of Figure 6.34a), we could predict the magnitude of a possible earthquake (point *B* of the same graph). We might also assume that there is a relationship between the precursor time interval (time between the start of precursor events and the occurrence of the quake) and the magnitude of an expected earthquake (Figure 6.34). This relationship is also derived empirically from direct observation of known earthquake activity. Now take the predicted earthquake magnitude from Figure 6.34a, point *B,* and use

this in Figure 6.34b to predict the precursor time interval (point *C*). Thus, we now know the magnitude, and, if we know when the precursor events started, we can calculate when the expected earthquake should occur. It sounds simple, but it is pure fiction: none of these relationships have been derived with sufficient precision to use in predicting real earthquakes. Furthermore, the potential for errors of measurement with this kind of work are considerable. Nevertheless, it is valuable to explore one example of how earthquake predictions may eventually come about.

Earthquakes and Critical Facilities

Critical facilities are as those which if damaged or destroyed might cause catastrophic loss of life, property damage, or disruption of society. Examples include large dams, nuclear power plants, and liquid natural gas (LNG) facilities. There are three aspects of the decision process concerning critical facilities and earthquake hazard (27): (1) evaluation of the hazard; (2) evaluation of whether the facility may be designed or modified to accommodate the hazard; and (3) subjective evaluation of an "acceptable risk." The first two factors have a strong scientific component, while risk assessment is a public safety issue since no facility can be rendered absolutely safe.

Consider seismic safety for a nuclear power plant. Evaluation of the earthquake hazard requires estimating *fault capability* and the *maximum credible earthquake* that might be expected along faults in the vicinity of a particular site. A *capable fault,* as defined by the U.S. Nuclear Regulatory Commission, is a fault that has exhibited movement at least once in the last 35,000 years or multiple movements in the last 500,000 years. These are more conservative criteria with a greater safety factor than the definition of an active fault (Table 6.4), reflecting the growing concern for siting nuclear power plants.

Estimation of the maximum credible earthquake can sometimes be accomplished by field work to evaluate effects of historical seismicity and prehistoric activity. Length of surface rupture and displacement per event are both related to earthquake magnitude, and equations are available to estimate magnitude if expected surface rupture length or displacement per event is known or assumed (28). Average recurrence of earthquakes of a particular magnitude can also be estimated if a slip rate (shear rate) is known or assumed (see Figure 6.35).

Unfortunately, the data base concerning fault rupture length and displacement is limited. Therefore, evaluation of the earthquake hazard for many critical facilities will continue to be subjective at best. All geologists can do is gather as much pertinent geologic data as possible to insure that our decisions are based on adequate information. We are presently arguing over what is adequate. Certainly, for a specific fault, we would like to know the length of the fault, style of faulting, total displacement, age of the most recent movement, approximate recurrence interval, and magnitude of the maximum credible earthquake.

In most cases regarding siting of critical facilities, the main problem concerning seismic risk is estimating the activity of a particular fault system. For example, the seismic evaluation for the Auburn Dam site near Sacramento, California, has been controversial. Millions of dollars have been spent and many meters of trenches excavated during the geologic studies, and there is still no agreement as to the hazard associated with the system of possibly active faults in the vicinity of the foundation area for the dam. The general area was considered to have relatively low seismic risk until, in 1975, a magnitude 5.7 earthquake occurred near the Oroville Dam about 80 kilometers away from the Auburn site. This earthquake rekindled concern and initiated a new round of seismic risk evaluation.

Geological evaluation of the Point Conception, California, site for a liquid natural gas (LNG) terminal (1979–1980) has also been controversial as well as time consuming. One of the major points of contention is the existence of numerous small faults that probably are associated with flexural-slip during folding of the rocks. Many large trenches (Figure 6.36) were excavated to study these faults, emphasizing the detailed nature of the geologic investigation required for critical facilities. The flexural-slip faults may not be capable of producing moderate to large earthquakes, but they do present a ground-rupture hazard that must be dealt with in the engineering design for the terminal. Another problem at the LNG site has been determining when these faults were active. Some of the faults cut various materials of differing ages, and obtaining reliable dates has been a problem. The site is actually on a marine platform that may be as old as 100,000 years or more, but subsequent erosion and deposition have taken place so that younger materials are deposited on

FIGURE 6.34
Hypothetical method of how earthquake magnitude and time of event might be predicted. Graph (a) is used to predict the magnitude based on given precursor events, and graph (b) is used to predict the precursor time interval from the earlier predicted magnitude. (See text for full explanation.)

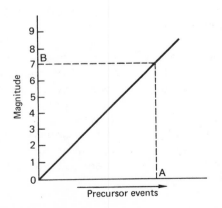

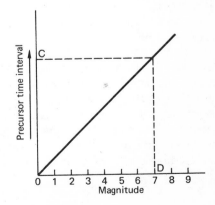

FIGURE 6.35
Relationships between recurrence interval, slip rate, and earthquake magnitude. Source: D. B. Slemmons, 1977, *State-of-the-art for assessing earthquake hazards in the United States*. U.S. Army Corps of Engineers Waterway Experimental Station, Vicksburg, Mississippi.

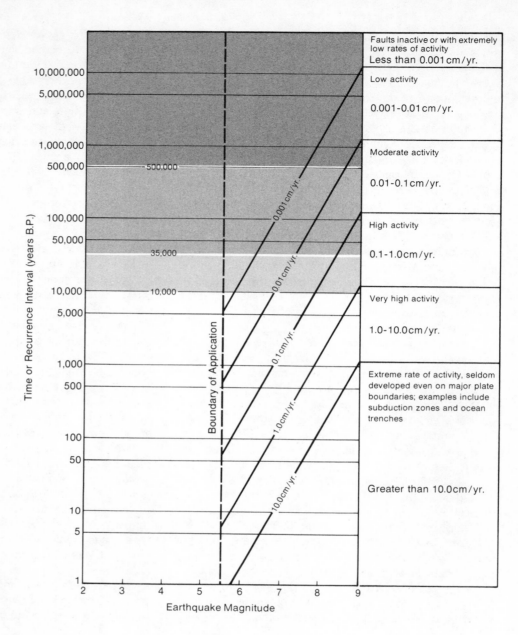

EARTHQUAKE CONTROL

that older surface. All this makes the geological evaluation of the hazard concerning the faults difficult and somewhat speculative.

The problem remains; it is very difficult to absolutely date past earthquake events. Furthermore, the problem will only intensify as more critical facilities are planned in areas with potential seismic risk. Considerable research by many scientists is now being conducted in hopes that we will improve our knowledge of how to identify active faults and estimate the maximum credible earthquake for a particular area.

EARTHQUAKE CONTROL

To aid in possible earthquake control, any system of earthquake prediction must assume that the true nature and causes of earthquakes are fairly well understood. Unfortunately, even though we have a tremendous quantity of empirical observations concerning physical changes in earth materials before, during, and after earthquakes, there is no general agreement on a physical model to explain the observations. Two models are being tested: the *dilatancy-diffusion* theory developed in the United States at Stanford University, and the *dilatancy-instability* theory developed in Russia at the Institute of Physics of the Earth (20).

The American and Russian models of stages preceding an earthquake are shown in Figure 6.37. The first stage of both models is an increase of elastic strain in rocks that causes them to undergo a dilatancy state, which is an inelastic increase in volume that starts after the stress on a rock reaches one-half its breaking strength. During dilatancy, open fractures develop in the rocks, so it is in this stage that the first physical changes take place that might indicate a future earthquake. Here the two models diverge. The American model assumes that the dilatancy and fracturing of the rocks are first associated with a relatively low water content in the dilated rocks (Stage 2, Figure 6.37), which helps to produce lower seismic velocity, more earth movement, higher radon emission, lower electrical resistivity, and fewer minor seismic events. An influx of water then enters the open fractures, causing the pore water pressure to increase (which increases the seismic velocity while further lowering electrical resistivity), weakening the rocks, and facilitating movement along the fractures which is recorded as an earthquake. After the movement and release of stress, the rocks resume many of their original characteristics (20).

In contrast, the Russian model does not rely on an influx of water. Rather, the dilatancy state (Figure 6.37) is accompanied by an avalanche of fractures that releases some stress but produces an unstable situation that eventually causes a large movement along the fractures, recorded as an earthquake (20). There is evidence that, with certain assumptions, both models can roughly fit the known observations. What we need now is more information that can be applied to the problem of predicting and controlling earthquakes.

Scientists realized that earthquakes could be turned on and off by human activity when they noted how effectively the rates of injection of liquid waste into disposal wells controlled earthquake activity in Denver, Colorado. A U.S. Geological Survey project was initiated near Rangely, Colorado, to study the feasibility of actually *starting* earthquakes by raising the fluid pressure in rocks by injecting water into fracture zones, and *stopping* earthquakes by decreasing the fluid pressure by pumping water out of fracture zones. The project demonstrated that earthquake frequency can be controlled by manipulating the fluid pressure (Figure 6.38) (21). It is hoped that results of this research will provide the information necessary to reduce the earthquake hazard by releasing strain in rocks slowly through numerous controlled small shocks rather than

FIGURE 6.36
Large trench excavated as part of the geologic seismic safety study for the proposed liquid natural gas terminal at Point Conception. Note the people and ladders on the walls of the trench for scale. Seismic studies in southern California for critical facilities are taken very seriously and very detailed work is necessary, resulting in large excavations.

FIGURE 6.37

Two models to explain the mechanism responsible for earth-quakes have been proposed. The dilatancy-diffusion model was developed mainly in the United States. The dilatancy-instability model was developed in the USSR. The dashed curves show the expected precursory signals according to the American model. The solid curves show the expected precursory signals according to the Russian model. (From ''Earthquake Prediction,'' Frank Press. Copyright © May 1975 by Scientific American, Inc. All rights reserved.)

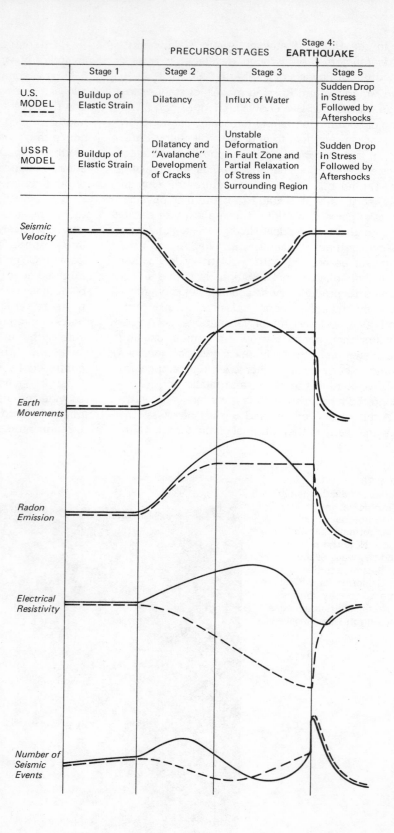

	Stage 1	Stage 2	Stage 3	Stage 4: EARTHQUAKE / Stage 5
U.S. MODEL	Buildup of Elastic Strain	Dilatancy	Influx of Water	Sudden Drop in Stress Followed by Aftershocks
USSR MODEL	Buildup of Elastic Strain	Dilatancy and ''Avalanche'' Development of Cracks	Unstable Deformation in Fault Zone and Partial Relaxation of Stress in Surrounding Region	Sudden Drop in Stress Followed by Aftershocks

PRECURSOR STAGES

Seismic Velocity

Earth Movements

Radon Emission

Electrical Resistivity

Number of Seismic Events

Earthquake Frequency at Rangely Oil Field, Colorado

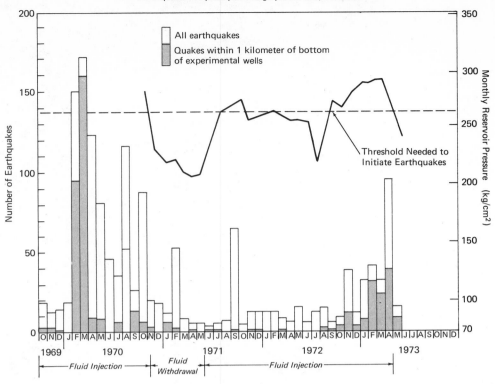

FIGURE 6.38
Earthquake control at the Rangely oil field, Colorado. This is the first example in which earthquakes were intentionally started and stopped by controlling the reservoir pore pressure by controlled pumping. It was found that a particular threshold pressure was needed to initiate earthquakes. (After R. E. Wallace, U.S. Geological Survey Circular 701, 1974.)

having the strain released quickly in a large earthquake. The objective is to drill down into the fault zone (some are accessible to drilling) and then to manipulate the fluid pressure by either pumping out or injecting fluids, initiating a series of small shocks that release the stored strain in the rocks.

Proper use of nuclear explosions also may be effective in earthquake control. We have noted that such explosions probably do release natural tectonic strain. Since it is expected that the longer strain builds up, the larger the eventual earthquake will be, it is prudent to try to reduce that strain. Some scientists believe that properly spaced underground nuclear explosions could be used in conjunction with fluid-injection methods to release strain in rock and thereby limit the severity of earthquakes (29).

ADJUSTMENTS TO EARTHQUAKE ACTIVITY

The mechanism of earthquakes is still poorly understood, and, therefore, such adjustments as warning systems and earthquake prevention are not yet reliable alternatives. There are, however, reliable protective measures we can take. First is *structural protection,* the construction of large buildings able to accommodate at least a moderate shock. Second is *land-use planning*—large structures, schools, hospitals, disaster relief facilities, and communication systems should not be placed on or near active faults or on sensitive earth material such as landfills, unconsolidated clay deposits, or water-saturated silts or sand. These areas are likely to accentuate earthquake waves and, therefore, be

FIGURE 6.39
Damage on different types of
ground caused by earthquake
activity may be very different. At
Varto, Turkey, 12 or 14 buildings
on old stream channel depos collapsed. This material accen-
tuated the earthquake waves.
However, on the bench material
(bedrock), no buildings collapsed.
Information such as this can be
tremendously useful in land-use
planning to avoid high-risk areas.
(From R. E. Wallace, U.S. Geological Survey Circular 701,
1974.)

FIGURE 6.40
Houses should not have been
constructed astride the 1906
break of the San Andreas fault
(white line) or in areas of potential landslides. (Photo by R. E.
Wallace, courtesy of U.S. Geological Survey.)

areas of high damage. Third are insurance and relief measures. Insurance, while not overly expensive in areas with a high earthquake risk, often has a high deductible before it pays benefits, and few people purchase it (30). The United States has federal disaster systems to provide relief and rehabilitation funds, and emergency systems are also provided by such organizations as the Salvation Army and the Red Cross. A fourth alternative is to take no action and pay the consequences. This philosophy is not what we espouse, but it is, in fact, what we often do. At the same time that we are developing ways to predict earthquakes and prepare for relief and rehabilitation, we are increasing the need for prediction and relief by creating new hazards where none existed. When we place any new building or structure on sensitive earth materials or on active faults in areas with an earthquake hazard, we are, in effect, seeking new problems where there were none. We know from studying earthquake damage that building on unconsolidated deposits, such as channel fill, that become unstable (subject to liquefaction) or shake easily (amplify surface waves) during an earthquake is more hazardous than building on bedrock (Figure 6.39), and yet we continue to make these poor land-use choices (Figure 6.40).

To end this discussion on a more positive note, it is encouraging to report that southern California has a cooperative state-federal project to develop a comprehensive preparedness program to respond to a predicted or unexpected catastrophic earthquake. Considering that an $M = 8.3$ earthquake on the San Andreas fault north of Los Angeles may kill thousands of people, cause nearly $20 billion in damages, sever two major aquaducts that serve the region, reduce electrical power to the area by about 40 percent, close all main highways and airports, cause widespread loss of telephone communication, and damage other facilities such as natural gas lines, sewage lines, and fresh water supply lines, it is obviously prudent to develop emergency plans.

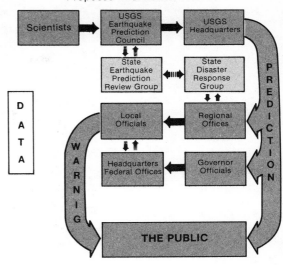

Earthquake Prediction and Warning
Proposed Information Flow

FIGURE 6.41
A federal plan for issuance of earthquake predictions and warning: the flow of information. Source: V. E. McKelvey, 1976, *Earthquake Prediction—Opportunity to Avert Disaster.* U.S. Geological Survey Circular 729.

We hope eventually to be able to predict earthquakes. The federal plan for issuing prediction and warning is shown in Figure 6.41. The general flow is from scientists to a prediction council for verification. A prediction that a damaging earthquake of a specified magnitude will occur (with a probability of certainty) at a particular location over a specific time span may then be issued to state and local officials, who will be responsible for issuing a warning to the public to take defensive action (that has, one hopes, been planned in advance). Potential response to a prediction depends on lead time (Table 6.5), but even a short time—as little as a few days—would be sufficient to mobilize

TABLE 6.5
Potential Response to an earthquake prediction with given lead time.

Lead Time	Buildings	Contents	Lifelines	Special Structures
3 Days	Evacuate previously identified hazards	Remove selected contents	Deploy emergency materials	Shut down reactors, petroleum products pipelines
30 Days	Inspect and identify potential hazards	Selectively harden (brace and strengthen) contents	Shift hospital patients; alter use of facilities	Draw down reservoirs, remove toxic materials
300 Days	Selectively reinforce		Develop response capability	Replace hazardous storage
3,000 Days		Revise building codes and land-use regulations: enforce condemnation and reinforcement		Remove hazardous dams from service

Source: C. C. Thiel, 1976, U.S. Geological Survey Circular 729.

FIGURE 6.42
Example of easements required for building setbacks from active fault traces in Portola Valley, California. Where the location of the fault trace is well known, no new buildings are allowed to be constructed within 15 meters of each side of the fault. Structures with occupancies greater than single family dwellings are required to be 38 meters from the fault. Where the precise location of the fault trace is less well known, more conservative setbacks of 30 meters for single family residences and 53 meters for higher occupancies are required. (After Mader, et al., and U.S. Geological Survey Circular 690, 1972.)

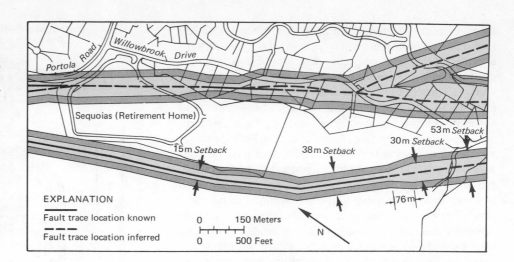

FIGURE 6.43
Oblique aerial photograph showing approximate location of fault traces within a part of the Hayward fault zone, some of which experienced movement during the 1868 earthquake. Present development compatible with a possible earthquake hazard include the undeveloped land, the plant nursery, the cemetery, and the highway. Other such uses might include golf courses, riding stables, drive-in theaters, or other recreational activities. Locating schools, police departments, hospitals, and other public facilities such as these along active fault traces is unwise land-use planning. (Photo by R. E. Wallace, courtesy of U.S. Geological Survey.)

emergency services, shut down reactors, and evacuate particularly hazardous areas.

PERCEPTION OF THE EARTHQUAKE HAZARD

That terra firma is not so firm in places is disconcerting to people who have experienced even a moderate earthquake. The relatively large number of people, especially children, who suffered mental distress following the San Fernando earthquake attests to the emotional and psychological effects. This one experience was sufficient to influence several hundred Los Angeles families formerly from the Midwest to move back to that area.

Regardless of their mental distress after earthquakes, people in areas of potential disaster generally do not really understand the earthquake hazard. In addition, there appears to be a considerable hiatus between what people say and what they do. Near Tokyo, six new earthquake antidisaster centers and a steel mill are being constructed. While the centers are to aid victims of earthquakes, the steel mill is being placed near sea level on landfill that is part of a reclamation project for Tokyo Bay. The site is subject to severe earthquake and tsunami hazard (30). At the same time, across the Pacific Ocean, people in San Francisco continue their casual attitude toward a possible earthquake, which promises to be a tremendous disaster. People in hazardous areas could remove dangerous structures on old buildings that overhang the streets. They could practice better planning for siting buildings and stop filling in bays and other sensitive areas. But they probably will not, because a hazardous event, even a tremendously catastrophic one that comes only once every few generations, is simply not perceived by the general population as a real threat. It is encouraging, however, that some communities are developing ordinances to limit development along active faults (Figure 6.42) and are considering land-use alternatives (Figures 6.43).

SUMMARY AND CONCLUSIONS

Large earthquakes rank among nature's most catastrophic and devastating events. Most earthquakes are located in tectonically active areas where lithospheric plates, on which the continents and ocean basins are superposed, interact.

Primary effects of earthquakes are violent ground motion accompanied by fracturing which may shear or collapse large buildings, bridges, dams, tunnels, and other rigid structures. Secondary effects include *short-range* events, such as fires, landslides, tsunamis, and floods, and *long-range* effects, such as regional subsidence and uplift of landmasses and regional changes in groundwater levels.

Tectonic creep resulting from slow movement along fault zones is a less hazardous process associated with earthquake activity. Nevertheless, it can cause considerable damage to roads, sidewalks, building foundations, and other structures.

The magnitude of an earthquake is a measure of the amount of energy released. It is determined by the amplitude of the largest horizontal trace recorded on a standard seismograph. It has been estimated that there are about one million earthquakes every year, two of which, on the average, have sufficiently high magnitude (greater than 7.75) to cause catastrophic damage.

The intensity of an earthquake is based on the severity of shaking as reported by observers, and varies with proximity to the epicenter and local geologic and engineering features.

Human activity has caused increasing earthquake activity in three ways: first, by loading the earth's crust through construction of large reservoirs; second, by disposing of liquid waste in deep disposal wells, which raises fluid pressures in rocks and facilitates movement along fractures; and third, by setting off underground nuclear explosions. The accidental damage caused by the first two activities is regrettable, but what we learn from all the ways we have caused earthquakes may eventually help us to control or stop large earthquakes.

Reduction of earthquake hazard will be a multifaceted program, including recognition of active faults and earth materials sensitive to shaking, and development of improved ways to predict, control, and adjust to earthquakes.

Prediction and control of earthquakes are now subjects of serious research. Optimistic scientists believe that we will eventually be able to make long-range and short-range predictions of earthquakes. Many believe such predictions may be based on previous patterns and frequency of earthquakes, seismic gaps, anomalous uplift and subsidence, and monitoring of ground tilt, strain in rocks, microearthquake activity, magnetic activity, radon concentration, electrical resistivity changes in earth materials, and anomalous animal behavior. Scientists hope to control earthquakes by manipulating fluid pressure in rocks along faults, thereby causing a series of small earthquakes, rather than let a natural catastrophic event occur. It has also been suggested that nuclear explosions, in conjunction with manipulation of fluid pressure, might be used to release natural tectonic strain slowly before a large earthquake.

To date, long-term earthquake prediction and probabilistic analysis have been much more successful than short-term prediction, and supply important information for land-use planning, developing building codes, and engineering design of critical facilities.

Warning systems and earthquake prevention are not yet reliable alternatives; but communities should develop emergency plans to respond to a predicted or unexpected catastrophic earthquake.

The earthquake hazard in potential disaster areas remains poorly perceived by inhabitants of the areas. Their lack of concern probably results because a process that produces even catastrophic damage may not be perceived as a real threat if it strikes the same area only every few generations.

REFERENCES

1 HAYS, W. W. 1981. Facing geologic and hydrologic hazards. U.S. Geological Survey Professional Paper 1240 B.

2 HAMILTON, R. M., 1980. Quakes along the Mississippi. *Natural History* 89:70–75.

3 OFFICE OF EMERGENCY PREPAREDNESS. 1972. *Disaster preparedness* 1, 3.

4 U.S. GEOLOGICAL SURVEY and the NATIONAL OCEANIC AND ATMOSPHERIC ADMINISTRATION. 1971. *The San Fernando, California, earthquake of February 9, 1971.* U.S. Geological Survey Professional Paper 733.

5 NATIONAL ACADEMY OF SCIENCES and NATIONAL ACADEMY OF ENGINEERING. 1973. The San Fernando earthquake of February 9, 1971: Lessons from a moderate earthquake on the fringe of a densely populated region. In *Focus on environmental geology*, ed. R. W. Tank, pp. 66–76. New York: Oxford University Press.

6 YOUD, T. L.; NICHOLS, D. R.; HELLEY, E. J.; and LAJOIE, K. R. 1975. Liquefaction potential. In *Studies for seismic zonation of the San Francisco Bay region*, ed. R. D. Borcherdt, pp. 68–74. U.S. Geological Survey Professional Paper.

7 HANSEN, W. R. 1965. *The Alaskan earthquake, March 27, 1964: Effects on communities.* U.S. Geological Survey Professional Paper 542A.

8 ECKEL, E. B. 1970. *The Alaskan earthquake, March 27, 1964: Lessons and conclusions.* U.S. Geological Survey Professional Paper 546.

9 RADBRUCH, D. H., et al. 1966. *Tectonic creep in the Hayward fault zone, California.* U.S. Geological Survey Circular 525.

10 STEINBRUGGE, K. V., and ZACHER, E. G. 1960. Creep on the San Andreas fault. In *Focus on environmental geology*, ed. R. W. Tank, pp. 132–137. New York: Oxford University Press.

11 PAKISER, L. C.; EATON, J. P.; HEALY, J. H.; and RALEIGH, C. B. 1969. Earthquake prediction and control. *Science* 166: 1467–1474.

12 EVANS, D. M. 1966. Man-made earthquakes in Denver. *Geotimes* 10: 11–18.

13 CLARK, M. M.; GRANTZ, A.; and MEYER, R. 1972. Holocene activity of the Coyote Creek fault as recorded in the sediments of Lake Cahuilla. In *The Borrego Mountain earthquake of April 9, 1968*, pp. 1112–1130. U.S. Geological Survey Professional Paper 787.

14 SIEH, K. E. 1978. Prehistoric large earthquakes produced by slip on the San Andreas fault at Pallett Creek, California. *Journal of Geophysical Research* 83: 3907–3939.

15 YEATS, R.S., CLARK, M., KELLER, E.A., and ROCKWELL, T. K., (in press) Active fault hazard: Ground rupture vs. seismic shaking. *Geological Society of America Bulletin.*

16 KELLER, E. A., JOHNSON, D.L., CLARK, M. N., and ROCKWELL, T.K. 1980. *Tectonic geomorphology and earthquake hazard, north flank, Central Ventura Basin, California.* Final report, U.S. Geological Survey Contract 14-08-0001-17678.

17 BORCHERDT, R. D., et al. 1975. Predicted geologic effects of a postulated earthquake. In *Studies for seismic zonation of the San Francisco Bay region,* ed. R. D. Borcherdt, pp. 88–95. U.S. Geological Survey Professional Paper 941A.

18 RALEIGH, B., et al. 1977. Prediction of the Haicheng Earthquake. *Transactions of the American Geophysical Union* 58, no. 5: 236–272.

19 SIMONS, R. S. 1977. Earthquake prediction, prevention and San Diego. In *Geologic hazards in San Diego,* ed. A. L. Patrick. San Diego Society of Natural History.

20 PRESS, F. 1975. Earthquake prediction. *Scientific American* 232: 14–23.

21 WALLACE, R. E. 1974. *Goals, strategy, and tasks of the earthquake hazard reduction program.* U.S. Geological Survey Circular 701.

22 RIKITAKR, TSUNEJI. 1983. *Earthquake forecasting and warning.* D. Reidel Publishing Co., London.

23 ALLEN, C. R. 1983. Earthquake prediction. *Geology* 11: 682.

24 RALEIGH, C. B., SYKES, L. R., SIEH, K. E., and ANDERSON, D. L. 1983. Forecasting southern California earthquakes. *California Geology* March: 54–63.

25 SIMON, C. 1983. California's quakes: Narrower odds. *Science News* 124–404.

26 HAIT, M. H. 1978. Holocene faulting, Lost River Range, Idaho. Geological Society of America abstracts with programs 10(5): 217.

27 CLUFF, L. S. 1983. The impact of tectonics on the siting of critical facilities. EOS 64 no. 45: 860.

28 SLEMMONS, D. B. 1977. State-of-the-art for assessing earthquake hazards in the United States (series): Determination of design earthquake magnitude from fault length and maximum displacement data. U.S. Army Engineers Waterways Experimental Station, Vicksburg, Miss.

29 EMILIANI, C., HARRISON, C., and SWANSON, M. 1969. Underground nuclear explosions and the control of earthquakes. *Science* 165:1255–1256.

30 NICHOLAS, T.C., JR. 1974. Global summary of human response to natural hazards: Earthquakes. In *Natural hazards,* ed. G. F. White, pp. 274–284. New York: Oxford University Press.

Chapter 7

Volcanic Activity

Qn a worldwide basis, volcanic activity is a relatively uncommon process that usually affects sparsely populated areas. However, volcanic eruptions can be tremendously destructive, and when such an event occurs near a densely populated area, a catastrophe may be recorded.

Causes of volcanic activity are generally related to plate tectonics, and most active volcanoes are located near plate junctions where *magma* (molten rock) is produced as spreading or sinking lithospheric plates interact with other earth materials. Approximately 80 percent of all active volcanoes are located in the "ring of fire" which circumscribes the Pacific Ocean. This area is essentially the Pacific plate (Table 7.1). In the United States, several volcanoes in Hawaii, about 25 in Alaska, and several in the Cascade Range of the Pacific Northwest can be considered active.

Volcanoes in the Cascade Range of Washington, Oregon, and California continue to pose potential danger to the population and agricultural centers of the Pacific Northwest (Figure 7.1). Table 7.2 summarizes recent activity of major volcanoes in the Cascade Range.

Three common types of volcanoes are the shield and composite volcanoes and the volcanic dome (Figure 7.2). Each type of volcano has a characteristic style

of activity that is primarily a result of the viscosity of the magma, which is related to its silica content.

By far the largest, *shield volcanoes* are characterized by relatively nonexplosive activity, resulting from the low silica content (about 50 percent) of the magma. Shield volcanoes are built up almost entirely from numerous lava flows. They are common in the Hawaiian

TABLE 7.1
Distribution of the world's active volcanoes.

Area	Percentage of Active Volcanoes
Pacific	79
Western Pacific Islands	45
North and South America	17
Indonesian Islands	14
Central Pacific Islands (Hawaii, Samoa)	3
Indian Ocean Islands	1
Atlantic	13
Mediterranean and Asia Minor	4
Other Areas	3
	100

Source: A. Rittman, *Disaster Preparedness*, Office of Emergency Preparedness, 1962.

FIGURE 7.1
Map and plate tectonic setting of
the Cascade Range showing major volcanoes and cities in their
vicinity. See Figure 3.2 for regional tectonic environment.
(Map modified after Crandall and
Waldron, *Disaster Preparedness,*
Office of Emergency Preparedness, 1969.)

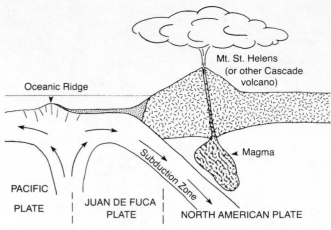

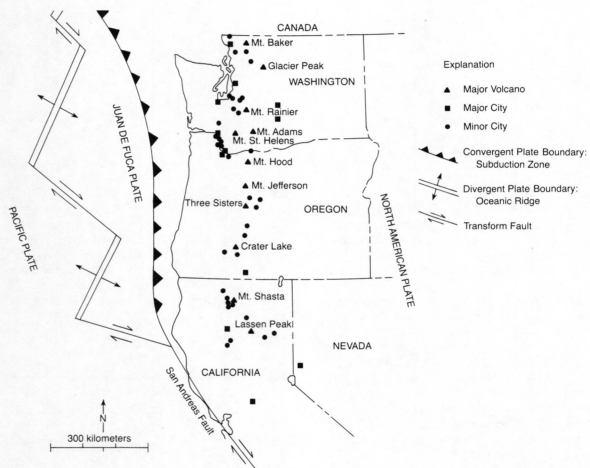

Islands and are also found in some areas of the Pacific Northwest and Iceland.

Composite volcanoes, known for their beautiful cone shape, are associated with a magma of interme-diate silica content (about 60 percent) which is more viscous than the low-silica magma of shield volcanoes. Volcanic activity characteristic of composite volcanoes

is a mixture of explosive activity and lava flows. Ex-amples in the United States include Mt. St. Helens and Mt. Rainier (Figure 7.1), and as the 1980 eruption of Mt. St. Helens demonstrated, these volcanoes are dan-gerous.

Volcanic domes are characterized by viscous magma with a high silica content (about 70 percent).

TABLE 7.2
Summary of recent activity of major volcanoes in the Cascade Range.

	Mt. Baker	Glacier Peak	Mt. Rainier	Mt. St. Helens	Mt. Adams	Mt. Hood	Mt. Jefferson	Three Sisters	Mt. Mazama (Crater Lake)	Newberry Volcano (Newberry Crater)	Mt. Shasta	Glass Mountain area	Lassen Peak— Chaos Crags area
Active in historic time	X		X	X		X					?	?	X
Known products of eruptions, last 12,000 years													
Lava flow	X		X	X	X	X	X	X	X	X	X	X	X
Tephra (airborne rock debris)	X	X	X	X		?		X	X	X	X	X	X
Pyroclastic flow		X	X	X					X		X		X
Mudflow	X	X	X	X	X	X					X		X
Estimated population at risk													
More than 1,000	X		X	X		?			X		X		?
Less than 1,000		?					X	X		X		X	

Source: Crandell and Mullineaux, "Technique and Rationale of Volcanic Hazards." *Environmental Geology* 1 (New York: Springer-Verlag, 1975).

Activity is generally explosive, making volcanic domes very dangerous. Mt. Lassen in northeastern California, which last erupted from 1914 to 1917 (Figure 7.3), is a good example of a volcanic dome. One eruption included a tremendous horizontal blast that destroyed a large area.

A group of volcanic domes known as the Chaos Crags, located just north of Lassen Peak, have a recent history of explosive activity and large rockfall, debris avalanches. Fear of another rockfall or volcanic eruption from the Chaos Crags recently caused closing of the Manzanita Lake campground, lodge, and visitors' center in Lassen National Park.

Although relatively rare, volcanic activity, from a historical standpoint, has destroyed considerable property and taken many thousands of human lives. Table 7.3 lists some of the more catastrophic events. Although this is admittedly a tragic loss of life, it is relatively small compared to the number of deaths caused by floods or earthquakes during roughly the last 2,000 years. Unfortunately, volcanic activity cannot be prevented; therefore, programs to minimize this hazard involve prediction, warning, preparation, and experience gained from past events.

EFFECTS OF VOLCANIC ACTIVITY

Primary effects of volcanic activity include lava flows, ash flows (hot avalanches), lateral blasts, or ash falls. Secondary effects include mudflows, floods, and fires (1). Table 7.4 briefly describes these hazards for the U.S., including predictability, frequency of occurrence, and degree of risk.

Lava Flows

Lava flows are the most familiar product of volcanic activity (Figure 7.4). They result when magma reaches the surface and overflows the crater or volcanic vent along the flanks of the volcano. Lava flows that are low in silica are generally nonexplosive or effusive, while flows with high silica content have eruptive or explosive activity. Volcanoes on islands or continents that have intermediate to high-silica-content magma are generally eruptive, whereas submarine activity involves lava with a low silica content and is usually effusive. Lava flows can be quite fluid and move quite rapidly or relatively viscous and move slowly. Most lava flows are slow enough that people can easily move out of the way as they approach (2).

Several methods, such as bombing, hydraulic chilling, and constructing walls, have been employed to deflect lava flows away from populated or otherwise valuable areas. These methods have had mixed success. They cannot be expected to block large flows, and need further evaluation.

Observation of lava movement in Japan suggests that there are two types of pressure: *hydrostatic pres-*

(a)

(b)

(c)

FIGURE 7.2

(a) Shield volcano has low-silica (approximately 50 percent) magma and relatively nonviolent activity. The largest of all volcanoes on earth, it may measure 200 kilometers in width and is constructed of many lava flows. (b) Composite volcano has magma of intermediate silica content (approximately 60 percent) and activity alternating from explosive to lava flows. The cone is constructed of alternating layers of pyroclastic material (mostly ash) and lava flows. (c) Volcanic dome has high-silica magma (approximately 70 percent). It is the most violent and explosive of all volcanoes. (Photos [a] and [b] courtesy of U.S. Geological Survey. Photo [c] by Mary R. Hill, courtesy of California Division of Mines and Geology.)

sure related to the vertical height of the lava; and *pressure from the momentum of the flow,* which must be considered when building walls to deflect lava. Although even walls of rubble and loosely-laid stone have fortuitously withstood lava flows in Italy and Hawaii, experience suggests that proper planning is necessary for best results (3). The following criteria are suggested by A. C. Mason and H. L. Foster. First, walls about 3 meters high are usually sufficient. Second, the upslope side should be steeper to prevent overriding of the wall by the lava. Third, the wall should be located diagonally to the slope so that the lava will be deflected to

an area where little damage will occur. Fourth, guide channels should be located upslope to facilitate the direction of lava to the proper location. Fifth, walls should be located below likely vents and as far upslope as possible.

Bombing of the lava flows has been attempted to stop their advance. It has proven most effective against lava flows in which fluid lava is confined to a channel by congealing lava on the margin of the flow. The objective is to partly block the channel by bombing, thereby causing the lava to pile up, facilitating an upstream break through which the lava escapes to a less

FIGURE 7.3
Eruption of Lassen Peak in June, 1914. (Photo by B. F. Loomis, courtesy of Loomis Museum Association.)

damaging route. The established procedure is successive bombing at higher and higher points as necessary to control the threat. Bombing has some merit for future research, but it probably provides little protection, is unpredictable, and, in any event, cannot be expected to affect large flows. Poor weather conditions and the abundance of smoke and falling ash can also reduce the effectiveness of bombing (3).

The world's most ambitious program to control lava flows was initiated in 1973 on the Icelandic island of Heimaey, when lava flows with low silica content nearly closed the harbor to the island's main town and thereby threatened continued use of the island as Iceland's main fishing port. The situation prompted immediate action. Experiments on the island of Surtsey in 1967 showed that water could be used to stop the advance of lava, and favorable conditions existed on Heimaey to try it on a large scale. First, the main flows were viscous and slow moving, allowing the necessary time to initiate a control program. Second, transportation by sea and the local road system was adequate to move the necessary pumps, pipes, and heavy equipment. Third, water was readily available. The procedure was to first cool the margin and surface of the flow

with numerous fire hoses fed from a 13-centimeter pipe. Then bulldozers were moved up on the slowly advancing (at less than 1 meter per hour) flow, making a track or road on which the plastic pipe was placed. The pipe did not melt as long as water was flowing in it, and small holes in the pipe also helped cool particularly hot spots along various parts of the flow; at each place, approximately 0.03 to 0.3 cubic meters of water per second was delivered at a distance of about 50 meters in back of the edge of the flow (Figures 7.5 and 7.6). The cooling was done in conjunction with diversion barriers constructed by bulldozers mounding up loosened material in front of the advancing flow. Lava tended to pile up against these barriers. Watering at each location lasted about two weeks, or until the steam stopped coming out of the lava in that particular area. Watering had little effect the first day, but then that part of the flow began to slow down. The program undoubtedly had an important effect on lava flows. It tended to restrict their movement and thus reduced property damage. After the outpouring of lava stopped

TABLE 7.3
Selected historic volcanic events.

Volcano or City	Effect
Vesuvius A.D. 79	Destroyed Pompeii and killed 16,000 people. City was buried by volcanic activity and rediscovered in 1595.
Skaptar Jokull, Iceland, 1783	Killed 10,000 people (many died from famine) and most of the island's livestock. Also killed some crops as far away as Scotland.
Krakatoa, Indonesia, 1883	Tremendous explosion; 36,000 deaths from tsunami.
Mt. Pelee, Martinique, 1902	Ash flow killed 30,000 people in a matter of minutes.
La Soufrière, St. Vincent, 1902	Killed 2,000 people and caused the extinction of the Carib Indians.
Mt. Lamington, Papua, 1951	Killed 6,000 people.
Villarica, Chile, 1963–64	Forced 30,000 people to evacuate their homes.
Mt. Helgafell, Vestmannaeyjar, Iceland, 1973	Forced 5,200 people to evacuate their homes.
Mt. St. Helens, Washington, USA, 1980	Debris avalanche, lateral blast and mudflows killed 68 people, destroyed over 100 homes.

Source: Data partially derived from C. Ollier, *Volcanoes* (Cambridge, Mass.: MIT Press, 1969).

TABLE 7.4
Volcanic hazards for the U.S.A.: Events, effects, predictability, frequency of occurrence, and risk.

	Lava Flows	Ash Flow, Hot Avalanches, Mudflows, and Floods	Ash Fall, Lateral Blast, and Gases
Origin and characteristics	Result from nonexplosive eruptions of molten lava.	Hot avalanches can be caused directly by eruption of fragments of molten or hot solid rock; mudflows and floods commonly result from eruption of hot material onto snow and ice and eruptive displacement of crater lakes. Mudflows also commonly caused by avalanches of unstable rock from volcano.	Produced by explosion or high-speed expulsion of vertical to low-angle columns or lateral blasts of fragments and gas into the air; materials can then be carried great distances by wind. Gases alone may issue nonexplosively from vents. Commonly produced suddenly and move away from vents at speeds of tens of miles per hour.
Location	Flows are erupted slowly and move relatively slowly; usually no faster than a person can walk. Flows are restricted to areas downslope from vents; most reach distances of less than 6 miles. Distribution is controlled by topography. Flows occur repeatedly at central-vent volcanoes, but successive eruptions may affect different flanks. Elsewhere, flows occur at widely scattered sites, mostly within volcanic "fields."	Hot avalanches and mudflows commonly occur suddenly and move rapidly, at tens of miles per hour. Distribution nearly completely controlled by topography. Beyond volcano flanks, effects of these events are confined mostly to floors of valleys and basins that head on volcanoes. Large snow-covered volcanoes and those that erupt explosively are principal sources of these hazards.	Distribution controlled by wind directions and speeds, and all areas toward which wind blows from potentially active volcanoes are susceptible. Zones around volcanoes are defined in terms of whether they have been repeatedly and explosively active in the last 10,000 years.
Size of area affected by single event	Most lava flows cover no more than a few square miles. Relatively large and rare flows probably would cover only hundreds of square miles.	Deposits generally cover a few square miles to a few hundreds of square miles. Mudflows and floods may extend downvalley from volcanoes many tens of miles.	An eruption of "very large" volume could affect tens of thousands of square miles, spread over several states. Even an eruption of "moderate" volume could significantly affect thousands of square miles.
Effects	Land and objects in affected areas subject to burial, and generally they cause total destruction of areas they cover. Those that extend into areas of snow, may melt it and cause potentially dangerous and destructive floods and mudflows. May start fires.	Land and objects subject to burning, burial, dislodgement, impact damage, and inundation by water.	Land and objects near an erupting vent subject to blast effects, burial, and infiltration by abrasive rock particles, accompanied by corrosive gases, into structures and equipment. Blanketing and infiltration effects can reach hundreds of miles downwind. Odor, "haze," and acid effects may reach even farther.
Predictability of location of areas endangered by future eruptions	Relatively predictable near large, central-vent volcanoes. Elsewhere, only general locations predictable.	Relatively predictable, because most originate at central-vent volcanoes	Moderately predictable. Voluminous ash originates mostly at central-vent volcanoes
Frequency, in contiguous United States as a whole	Probably one to several small flows per century that individually cover less than 10 square miles. Flows that cover tens to hundreds of square miles probably occur at an average rate of about one every 1,000 years. (In Hawaii, eruption of many flows per decade would be expected.)		

162

Degree of risk in affected area	To people, low. To property, high.

and are restricted to flanks of volcanoes and valleys leading from them.

Probably one to several events per century caused directly by eruptions.

Probably only about one event per 1,000 years caused directly by eruption at "relatively inactive" volcanoes.

Moderate to high for both people and property near erupting volcano. Risk relatively high to people because of possible sudden origin and high speeds. Risk decreases gradually downvalley and more abruptly with increasing height above valley floor.

volcanoes; its distribution depends mainly on winds. Can be carried in any direction; probability of dispersal in various directions can be judged from wind records.

Probably one to a few eruptions of "small" volume every 100 years. Eruption of "large" volume may occur about once every 1,000 to 5,000 years. Eruption of "very large" volume, probably no more than once every 10,000 years.

Moderate risk to both people and property near erupting volcano; decreases gradually downwind to very low.

Source: D. R. Mullineaux, 1981. U.S. Geological Survey Professional Paper 1240B.

FIGURE 7.4
Lava flows moving slowly through a papaya orchard (a) and a village (b) during the 1959–60 eruption of Kilauea volcano, Hawaii. The lava fountain in (a) is approximately 122 meters high. The white area in the lower right corner of (b) is incandescent lava. (Photos by J. P. Eaton, courtesy of U.S. Geological Survey.)

(a)

(b)

in June, 1973, the harbor was still usable (4). In fact, by fortuitous circumstances, the shape of the harbor was actually improved, since the new rock provides additional protection from the sea.

Pyroclastic Activity: Ash Flow, Fall, and Blast

Pyroclastic activity is characteristic of high-silica magma. It is the eruptive or explosive volcanism in which all types of volcanic debris from ash to very large particles (tephra) are physically blown from a volcanic vent into the atmosphere. Several types of activity are recognized: *volcanic ash eruptions,* in which a tremendous quantity of rock fragments, natural glass fragments, and gas are blown high into the air by explosions from the volcano; *lateral blasts of gas and ash;* and *volcanic ash flows,* which are hot avalanches of ash, rock, and glass fragments that are mixed with gas, are blown out a vent and move very rapidly down the sides of the volcano.

FIGURE 7.5
Volcanic activity threatening homes on the island of Heimaey in 1973. Watering to cool the lava and stop its movement can be seen in the background. (Photo courtesy of Icelandic Air-lines.)

Volcanic ash eruptions can cover hundreds and even thousands of square kilometers with a carpet of volcanic ash. An eruption at Crater Lake about 7,000 years ago blanketed an area of several hundreds of thousands of square kilometers in the northwestern United States with ash. Approximately 2,100 square kilometers were covered with more than 15 centimeters of ash (Figure 7.7). Several volcanoes in the Cascade Range could have similar eruptions in the future (2).

Volcanic ash eruptions create several hazards. First, destruction of vegetation, including crops and trees, may result (Figure 7.8). Second, surface water may be contaminated by sediment, resulting in temporary increase in acidity of the water. The increase in acidity generally lasts only a few hours after the eruption ceases. Third, structural damage to buildings may occur, caused by the increased load on roofs. A depth of one centimeter of ash places an extra 2.5 tons of weight on a roof with a surface area of about 140 square meters. Weight of the ash was a major problem during the 1973 Icelandic eruption. Houses, public buildings, and businesses were literally buried in ash (Figure 7.9). Many buildings collapsed from the excess load, but inhabitants did save many by diligently shoveling the debris from the roofs (4). Fourth, health hazards such as irritation of the respiratory system and eyes are caused by contact with the ash and associated caustic fumes (2).

Volcanic ash flows, which may move as fast as 100 kilometers per hour down the sides of a volcano, can be catastrophic if a populated area is in the path of the flow. It was such a flow of hot, incandescent ash,

FIGURE 7.6
Close-up of an attempt to control the movement of a lava flow with water on the island of Heimaey. (Photo courtesy of Icelandic Air-lines.)

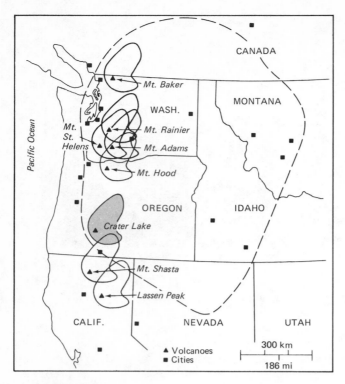

FIGURE 7.7
Map showing area covered by a volcanic ash eruption at Crater Lake about 7,000 years ago. The outer line shows the approximate limit of the ash fall. The shaded pattern shows the area covered by 15 centimeters or more of volcanic ash. This same area is superimposed on other major volcanoes of the Cascade Range. (After Crandell and Waldron, *Disaster Preparedness,* Office of Emergency Preparedness, 1969.)

FIGURE 7.8
Volcanic ash damaged or killed the trees shown here during the 1959–60 Kilauea volcano, Hawaii. (Photo by J. P. Eaton, courtesy of U.S. Geological Survey.)

steam, and other gases that on the morning of May 8, 1902, roared down Mt. Pelee through the West Indies town of St. Pierre, killing 30,000 people. A prisoner in jail was one of the two survivors, and he was severely burned and horribly scarred. Rumor has it that he spent the rest of his life touring circus sideshows as the "Prisoner of St. Pierre." Ash flows are often exploding avalanches that may be as hot as hundreds of degrees Celsius and incinerate everything in their path. They are one of the most lethal aspects of volcanic eruptions. Flows like these have occurred on volcanoes of the Pacific Northwest in the past and can probably be expected in the future. Fortunately, they seldom occur in populated areas.

Another type of pyroclastic flow is *base surge,* which forms when ascending magma comes in contact with water on or near the earth's surface in a violent steam explosion. Such an eruption occurred in 1911 on an island in Lake Taal, Philippines, killing about 1,300 inhabitants on the island and lake shore as a tremendous blast swept across the water. A similar event occurred at the same location in 1965, this time claiming perhaps as many as two hundred lives. Base surge eruptions are commonly associated with small volcanoes with bowl-shaped craters like that of Diamond Head, Hawaii. Many such extinct volcanoes can be found in the Christmas Lake valley region, an ancient lake in south-central Oregon, and in the Tule Lake region of northern California.

Caldera-Forming Eruptions

Giant volcanic craters, or calderas, are produced by very rare but extremely violent eruptions. None have occurred anywhere on earth in the last few thousand years, but at least 10 have occurred in the last 1,000,000 years, three in North America. A large caldera-forming eruption may explosively extrude up to 1,000 cubic kilometers of pyroclastic debris, mostly ash, which is several hundred times that ejected by the May 18 eruption of Mount St. Helens; produce a crater more than 10 km in diameter; and blanket an area of several tens of thousands of square kilometers with ash flow and fall deposits that vary from up to 100 m thick near the crater rim to a meter or so up to 100 km away from the source (5).

The most recent caldera-forming eruptions in North America occurred about 600,000 and 700,000 years ago at Yellowstone National Park in Wyoming and in the Owens Valley, California (Figure 7.10), respectively. The main events in these eruptions can be over very quickly, in a few days to a few weeks, but intermittent, lesser-magnitude volcanic activity can

FIGURE 7.9
Volcanic eruption in 1973, on the island of Heimaey, about 16 kilometers southwest of Iceland (a). This eruption forced the evacuation of 5,200 residents and covered more than half the buildings with ash and lava. House nearly buried in ash (b). (Photos courtesy of Icelandic Airlines.)

(a)

(b)

linger on for a million years. Thus, the Yellowstone event has left us hot springs and geysers, including Old Faithful, while the Owens Valley event (Long Valley caldera) is producing a potential hazard. Actually, both sites are still capable of producing future volcanic activity because hot, partially liquid rock is still present at variable depths beneath the caldera floors. Both calderas are *resurgent calderas* because the floors of the calderas have slowly domed upward since the explosive eruptions that formed them. The problem at Long Valley is that around 1980, the rate of uplift accelerated up to 25 cm in only about two years and was accompanied by earthquake swarms, some of which

were a magnitude of 5 to 6, became harmonic, which is symptomatic of moving magma, and were moving closer to the surface (from an 8-km depth in 1980 to 3.2 km in 1982). Concern over a possible volcanic eruption prompted the U.S. Geological Survey to issue a notice of potential volcanic hazard in 1982. A notice of hazard from continued, severe earthquakes was issued in 1980. Swarms of earthquakes continued into 1983, and the future situation is uncertain (6).

A map and cross-section of the Long Valley caldera and Mammoth Lakes, a popular ski area, is shown in Figure 7.11. The last major volcanic eruption in the caldera since its explosive beginning about 700,000

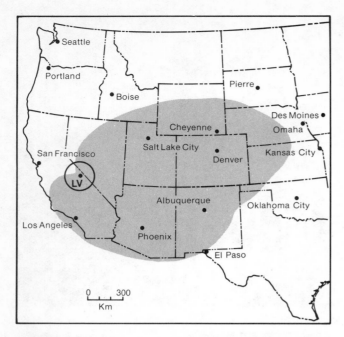

FIGURE 7.10

Area covered by ash from the Long Valley Caldera eruption approximately 700,000 years ago. Circle near Long Valley (LV) has a radius of 120 km and encloses the area subject to at least 1 m of downwind ash accumulation if a similar eruption were to occur again. Also within this circle, hot pyroclastic flows (ash flows) are likely to occur and, in fact, could extend further than shown by the circle. Source: C. D. Miller, D. R. Mullineaux, D. R. Crandell, and R. A. Bailey, 1982. *Potential Hazards from Future Volcanic Eruptions in the Long Valley–Mono Lake Area, East-Central California and Southwest Nevada—A Preliminary Assessment.* U.S. Geological Survey Circular 877.

years ago occurred about 100,000 years ago, but smaller eruptions, including steam explosions and explosive eruptions of rhyolite, occurred as recently as 400 to 500 years ago at the site of the Inyo craters (see Figure 7.11a). The magnitude of potential future eruptions at Long Valley is uncertain, but range from the unlikely very large catastrophic event to more probable smaller eruptions. Geologists now believe that magma is moving up as a shallow intrusion, as shown in Figure 7.11b, but it could reach the surface. Thus, the monitoring of uplift, tilt, earthquakes, and hot springs activity continues. Figure 7.12 shows the potential ash fall and ash flow hazard zones from a possibly large future Long Valley explosive eruption that ejects up to perhaps 10 cubic kilometers of material. Citizens of Mammoth Lakes and other nearby communities are taking the warning seriously, "preparing for the worst and hoping for the best," while pursuing their business in providing recreation to visiting urbanites (6, 7). Given

the potentially catastrophic loss of life and property from a major eruption in this area, great concern is certainly warranted.

Mudflows and Fires

Two serious secondary effects of volcanic activity are *mudflows,* produced when a large volume of loose volcanic ash and other ejecta becomes saturated and unstable and moves suddenly downslope, and *fires.* A small mudflow resulting from the 1915 eruption of Mt. Lassen in California is shown in Figure 7.13. Figure 7.14 shows the effects of mudflows and hot fiery blasts to forested land. What if there had been a town there?

Gigantic mudflows have originated on the flanks of volcanoes in the Pacific Northwest (1). The paths of two mudflows, the Osceola and Electron, that originated on Mt. Rainier are shown in Figure 7.15. Deposits of the Osceola mudflow are 5,000 years old. This mudflow moved over 80 kilometers from the volcano and involved over 1.9 billion cubic meters of debris, equivalent to 13 square kilometers piled to a depth of more than 150 meters. Deposits of the younger, 500-year-old Electron mudflow reached about 56 kilometers from the volcano and involved in excess of 150 million cubic meters of mud. More than 300,000 people now live on the area covered by these old flows, and there is no guarantee that similar flows will not occur again. Figure 7.16 shows the potential risk of mudflows and tephra accumulation for Mt. Rainier. Someone in the valley in view of such a flow would describe it as a wall of mud a few meters high moving at about 30 kilometers per hour, see it at a distance of perhaps 1.6 kilometers, and would need a car headed in the right direction toward high ground to escape being buried alive (2).

Crandell and Waldron emphasize the potential hazard of volcanic mudflows as compared to flood hazards. Floods are usually preceded by heavy rains that cause a gradual rise in water level. People in flood-prone areas generally have time to escape, and when the flood recedes, the water and danger are essentially gone. However, catastrophic mudflows can occur with little or no warning. They are likely to start when the volcano is hidden by clouds of smoke, and after the event, the mud, perhaps a few meters thick, remains. Since mudflows are confined to valleys, there is another possible hazard when the valley is artificially dammed to produce hydroelectric power. A large mudflow could fill a reservoir, pushing the water over the spillway and causing a severe flood downstream. On the other hand, used wisely, reservoirs might be a safety factor for all except the really large mudflows. The water level in reservoirs might be drawn down during an upstream

FIGURE 7.11

Map (a) and idealized diagram (b) illustrating the volcanic hazard near Mammoth Lakes, California. The map (a) shows the location of past volcanic events as well as the area of uplift where magma seems to be moving up, and finally the area where earthquake swarms have occurred near Mammoth Lakes. The geologic cross-section (b) shows a section (northeast-southwest) through the Long Valley Caldera. Shown are inferred geologic relations suggesting the 1980 magma rise that produced the uplift and swarms of earthquakes. Source: R. A. Bailey, 1983. Mammoth Lakes Earthquakes and Ground Uplift: Precursor to Possible Volcanic Activity? In *U.S. Geological Survey Yearbook, Fiscal Year 1982.*

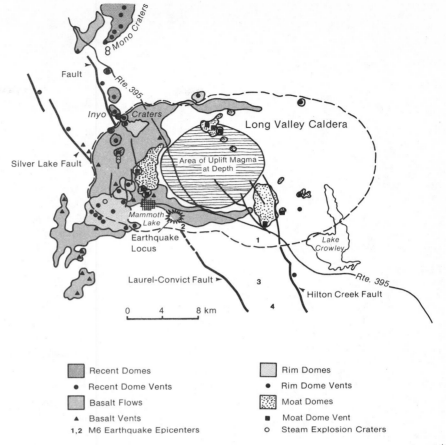

▨	Recent Domes	▨	Rim Domes
●	Recent Dome Vents	●	Rim Dome Vents
▨	Basalt Flows	▨	Moat Domes
▲	Basalt Vents	■	Moat Dome Vent
1,2	M6 Earthquake Epicenters	○	Steam Explosion Craters

(a)

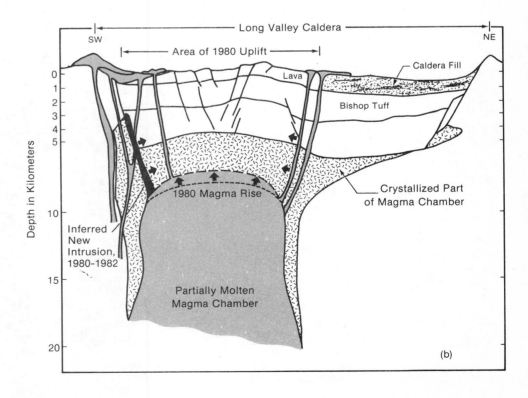

(b)

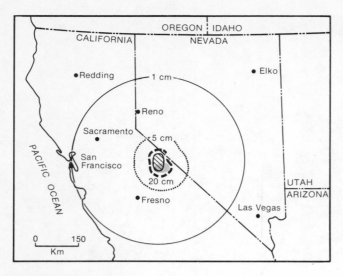

FIGURE 7.12
Potential hazards from a volcanic eruption at the Long Valley Caldera near Mammoth Lakes, California. Stipple patterns and diagonal lines show the hazard from flowage events out to a distance of approximately 20 km from recognized potential vents. Lines surrounding the hazard areas represent potential ash thicknesses at 35 km (dashed line), 85 km (dotted line), and 1 cm at 300 km (solid line). These estimates of potential hazards assume an explosive eruption (moderate volume) approximately 1 cubic km from the vicinity of recently active vents. Source: C. D. Miller, D. R. Mullineaux, D. R. Crandell, and R. A. Bailey, 1982. *Potential Hazards from Future Volcanic Eruptions in the Long Valley-Mono Lake Area, East-Central California and Southwest Nevada—A Preliminary Assessment.* U.S. Geological Survey Circular 877.

volcanic event, and the storage basin behind the dam could be used to contain a possible mudflow. This is not the intended function of the dam, but it is a fortunate safety mechanism (2).

Mt. St. Helens

The May 18, 1980, eruption of Mt. St. Helens in the southwest corner of Washington (Figure 7.1) exemplifies the many types of volcanic events expected from a Cascade volcano. Although the eruption, like many natural events, was unique and complex, making generalizations somewhat difficult, we have learned a great deal from Mt. St. Helens, and the entire story is not yet complete.

Mt. St. Helens awoke in March of 1980, after 120 years of dormancy, with seismic activity and small explosions as groundwater came in contact with hot rock. At 8:32 A.M. on May 18, 1980, a magnitude 5.0 earthquake was registered on the volcano. That earthquake triggered a large landslide/debris avalanche (approximately 2.5 cubic kilometers) that involved the entire northern flank of the mountain on which a bulge had been growing at a rate of about 1.5 meters per day for a period of several weeks. The landslide/debris avalanche shot down the north flank of the mountain, displacing water in nearby Spirit Lake, struck and overrode a ridge 8 km to the north, then made an abrupt turn and moved for a distance of 18 km down the Toutle River. The initial failure of the bulge released internal pressure and Mt. St. Helens erupted with a lateral blast directly from the area previously occupied by the bulge. At nearly the same time, a large vertical cloud rose quickly to an altitude of approximately 19 km (Fig-

FIGURE 7.13
Several small mudflows on the west side of Lassen Peak. Four such mudflows reached Manzita Lake at the base of the mountain in 1915. Some of the dark area at the summit is new lava. (Photo by B. F. Loomis, courtesy of Loomis Museum Association.)

(a)

(b)

FIGURE 7.14
Jessen Meadow with Mt. Lassen in the background. Photograph (a) was taken in August of 1910. Photograph (b) was taken in 1925, from the same viewpoint but after a mudflow and hot blast from the volcano. Notice that all the timber in the 1910 photograph was destroyed. (Photo by B. F. Loomis, courtesy of Loomis Museum Association.)

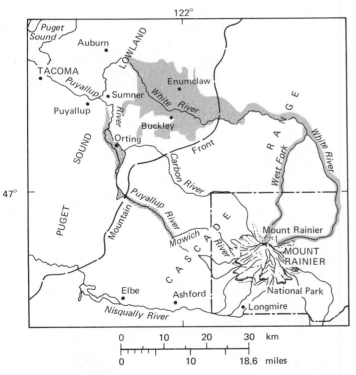

FIGURE 7.15
Map of Mt. Rainier and vicinity showing the extent of the Osceola mudflow in the White River Valley (shaded) and the Electron mudflow (dot pattern) in the Puyallup River Valley. (From Crandell and Mullineaux, U.S. Geological Survey Bulletin 1238, 1967.)

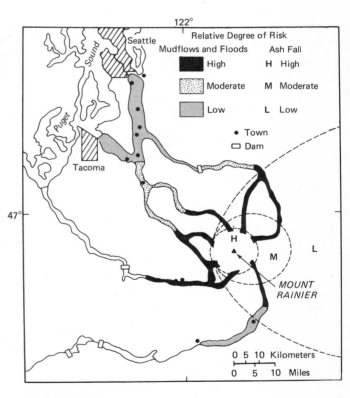

FIGURE 7.16
Sketch map of Mt. Rainier and vicinity showing relative degrees of potential hazards from ash fall and from (tephra), mudflows and floods that might result from an eruption. (From Crandell and Mullineaux, "Techniques and Rationale of Volcanic Hazards," *Environmental Geology*, vol. 1 [New York: Springer-Verlag, 1975].)

171

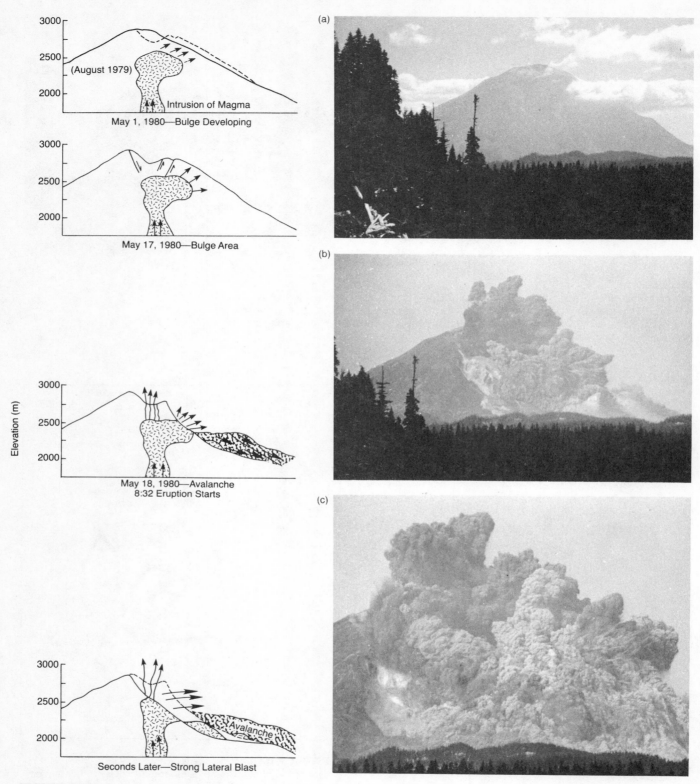

FIGURE 7.17

Idealized diagram and photographs showing the sequence of events for the May 18, 1980 eruption of Mount St. Helens. (Photographs [a], [b], and [c] © 1980 by Keith Ronnholm [the Geophysics Program, Univ. of Washington, Seattle], and photograph [d] by Robert Krimmel, U.S. Geological Survey. Drawings inspired from lecture by James Moore, U.S. Geological Survey.)

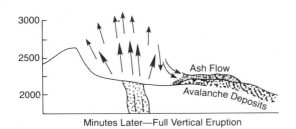

Ash Flow

Avalanche Deposits

Minutes Later—Full Vertical Eruption

(d)

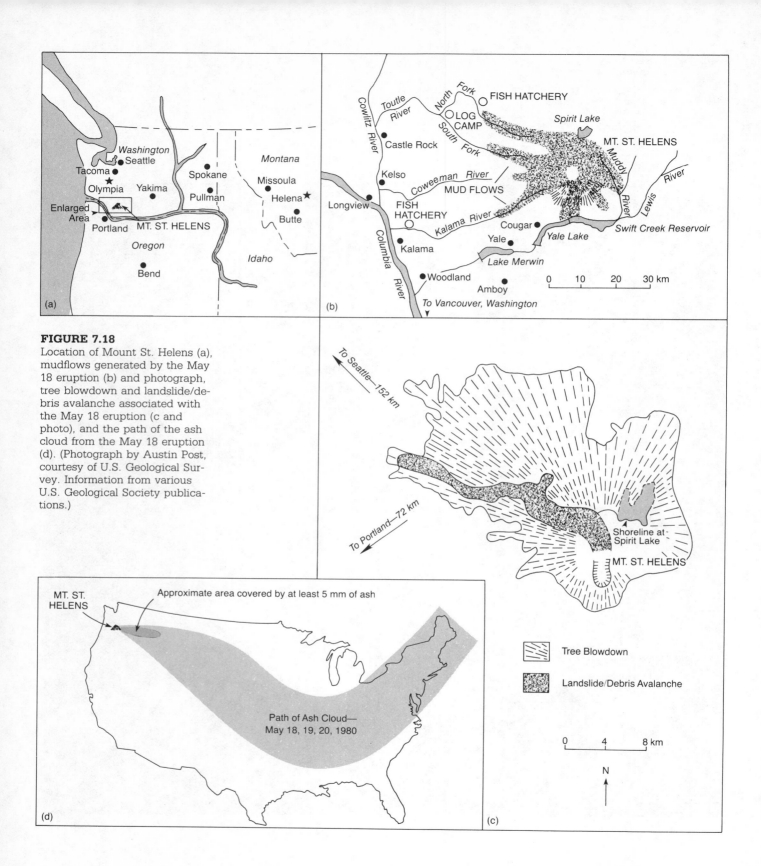

FIGURE 7.18
Location of Mount St. Helens (a), mudflows generated by the May 18 eruption (b) and photograph, tree blowdown and landslide/debris avalanche associated with the May 18 eruption (c and photo), and the path of the ash cloud from the May 18 eruption (d). (Photograph by Austin Post, courtesy of U.S. Geological Survey. Information from various U.S. Geological Society publications.)

174

FIGURE 7.19
Aerial view (a) and ground view (b) of deposits from the May 18, 1980, landslide/debris avalanche associated with the eruption of Mount St. Helens. Notice the hummocky nature of the topography. The person in photograph (b) provides a scale by which the size of some of the large blocks of the debris can be estimated. The total volume of debris is in excess of 2.5 cubic kilometers. (Photographs courtesy of U.S. Geological Survey, Robert Krimmel [a], Harry Glicken [b].)

(a)

(b)

ures 7.17 and 7.18). Eruption of the vertical column continued for more than nine hours, and large volumes of volcanic ash fell on a wide area of Washington, northern Idaho, and western and central Montana (Figure 7.18). The total amount of volcanic ash ejected was several cubic kilometers, and during the nine hours of eruption a number of ash flows swept down the northern slope of the volcano. The entire northern slope of the volcano, the upper part of the north fork of the Tou-

tle River basin, was devastated as forested slopes were transformed into a grey, hummocky landscape consisting of volcanic ash, rocks, blocks of melting glacial ice, narrow gullies, and hot steaming pits (Figure 7.19) (8).

The first of several mudflows, consisting of a mixture of water, volcanic ash, rock, and organic debris (such as logs), occurred minutes after the start of the eruption. Some time later, a young couple fishing on the Toutle River, approximately 36 km downstream

(a)   (b)

FIGURE 7.20
Mount St. Helens before (a) and after (b) the May 18, 1980, eruption. As a result of the eruption, much of the northern side of the volcano was blown away and the altitude of the summit was reduced by approximately 450 meters. (Photographs courtesy of U.S. Geological Survey and Harry Glicken.)

from Spirit Lake, were awakened by a loud rumbling noise from the river which was covered by felled trees. The pair attempted to run to their car, but water from the rising river poured over the road, preventing them from escaping. A mass of mud then crashed through the forest towards the car, and the couple climbed on top of the roof to escape the mud. They were safe only momentarily, as the mud pushed the car over the bank and into the river. They leaped off the roof and fell into the river, which was by now a rolling mass of mud, logs, collapsed train trestles, and other debris. The water also was increasing in temperature. One of them got trapped between logs and disappeared several times beneath the flow but was lucky enough to emerge again. The couple were carried downstream for approximately 1½ km before a family of other campers spotted and rescued them. The flows and accompanying floods raced down the valleys of the north and south forks of the Toutle River at estimated speeds of 29–55 km per hour. Water levels in the river reached at least 4 meters above flood stage and nearly all bridges along the river were destroyed. The hot mud quickly raised the temperature of the Toutle River to as high as 38°C. Mud, logs, and boulders were carried 70 km downstream into the Cowlitz River and eventually 28 km further downstream into the Columbia River. Nearly 40 million cubic meters of material was dumped into the Columbia River, reducing the depth of the shipping channel from a normal 12 m to 4.3 m for a distance of 6 km (8).

When the volcano could be viewed again following the eruption, it was observed that the maximum altitude of the volcano was reduced by about 450 m and the original symmetrical mountain was now a huge, steep-walled amphitheater facing northward (Figure 7.20). The debris avalanche, horizontal blast, pyroclastic flows, and mud flows devastated an area of nearly 400 square kilometers, killing 24 persons and leaving 44 others missing and presumed dead. More than 100 homes were destroyed by the flooding, and approximately 800 million board feet of timber was flattened by the blast (Figure 7.21). The total damage is estimated to be near three billion dollars, but long-range damage to fisheries and other resources is difficult to estimate. The salmon and trout may not return to the rivers for some time, and it is hard to speculate on how much of the downed timber eventually will be salvaged.

PREDICTION OF VOLCANIC ACTIVITY

It is unlikely that we will be able in the near future to predict accurately all volcanic activity, but valuable information is being gathered about phenomena that occur prior to eruptions. One problem is that most prediction techniques require experience with actual eruptions before the mechanism is understood. Thus, we are better able to predict eruptions in the Hawaiian Islands because we have had so much experience there.

Possible methods of predicting volcanic eruptions include geophysical observations of thermal and mag-

(a)

(b)

FIGURE 7.21
The lateral blast from the May 18 eruption of Mount St. Helens downed approximately
800,000,000 board feet of timber. Shown here (a) are some of those trees. The lower photograph (b) is a close-up of one splintered tree trunk where it was severed by the blast.
(Photographs courtesy of U.S. Geological Survey and Harry Glicken.)

netic properties, topographic monitoring of tilting or
swelling of the volcano, monitoring of seismic activity,
and studying the geologic history of a particular volcano or volcanic center (9).

Geophysical monitoring of volcanoes is based on
the fact that, prior to an eruption, a large volume of
magma (liquid rock) moves up into some sort of holding
reservoir beneath the volcano. The hot material
changes the local magnetic and thermal conditions. As
the surrounding rocks heat, the rise in temperature of
the surficial rock may be detected by infrared aerial
photography. Thus, periodic remote sensing of a volcanic chain may detect new hot points that could indicate possible future volcanic activity. This method
was used with some success at Mt. St. Helens prior to
the main eruption on May 18, 1980.

When older volcanic rocks are heated by new
magma, magnetic properties, originally imprinted
when the rocks cooled and crystallized, may change.
These changes might be detailed by ground or aerial
monitoring. Unfortunately, this method has not been
sufficiently researched and tested to be applied to actual prediction of volcanic events (9).

Monitoring of topographic changes and seismic
behavior of volcanoes has been useful in predicting
some volcanic eruptions. The Hawaiian volcanoes, especially Kilauea, have supplied most of the data. The
summit of the volcano actually tilts and swells prior to
an eruption and subsides during the actual outbreak
(Figure 7.22). This movement, in conjunction with

earthquake swarms that reflect the moving subsurface
magma and announce a coming eruption, was used to
predict volcanic activity in the vicinity of the farming
community of Kapoho on the flank of the volcano (Figure 7.23), 45 kilometers from the summit. The inhabitants were evacuated before the activity, which over-
ran, thrust aside, and floated away lava barriers (walls)
and eventually destroyed most of the village (Figure
7.24) (10). Scientists expect that, because of characteristic swelling and earthquake activity before eruptions,
the Hawaiian volcanoes will be predictable. Predicting
eruptions of less active volcanoes, such as in the Cascade Range, is much more difficult, as we learned from
Mt. St. Helens.

Our understanding of volcanic hazards has come
a long way, however. The Mt. St. Helens eruption was
predicted from a long-range standpoint of several decades, and we were fairly successful in making inter-
mediate-range predictions—the increased seismic activity and bulge in the volcano were predicted to lead
to some sort of eruption. Unfortunately, the very short-
range prediction of a few hours for the large explosion
on May 18 was not made. After the May 18 eruption of
Mt. St. Helens, a designated red zone within 30 km of
the volcano and a blue zone within an additional 15 km
to the east and southeast were closed to entry. This
paid off on August 7, when a prediction for an explosive event, based on change in volcanic gas chemistry
and earthquake activity, was made just hours prior to
the eruption. Thus, as the experience with the volcano

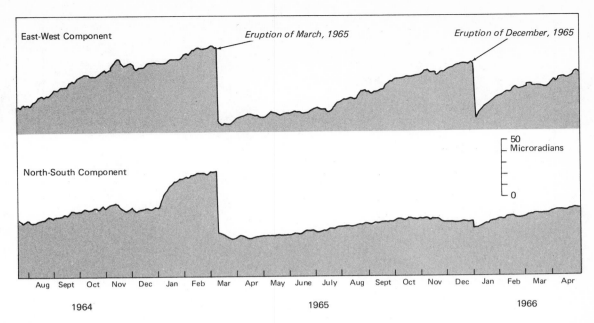

East-West Component

Eruption of March, 1965

Eruption of December, 1965

North-South Component

50
Microradians

0

Aug Sept Oct Nov Dec Jan Feb Mar Apr May June July Aug Sept Oct Nov Dec Jan Feb Mar Apr

1964   1965   1966

FIGURE 7.22
Graph showing east-west component and north-south component of ground tilt recorded
from 1964 to 1966 on Kilauea Volcano, Hawaii. Notice the slow change in ground tilt before eruption and rapid subsidence during eruption. (From R. S. Fiske and R. Y. Koyanagi,
U.S. Geological Survey Professional Paper 607.)

increased, chances for successful prediction increased.
In fact, 13 eruptions, from June 1980 through December 1982, were predicted by monitoring seismicity, deformation of the crater floor, and gas emissions (11).

Cascade volcanoes are complex and do not fit as
easily into a pattern of eruption as do the shield volcanoes of the Hawaiian Islands. Despite this limitation, it
is advantageous to attempt to generalize and make
models for the Cascade volcanoes. As we learn more

from direct observation and experience, the models
should improve our ability to predict eruptions.

Detailed geologic investigations of a volcano or
volcanic center may yield valuable information concerning possible future volcanism. For example, there
is abundant evidence that the Cascade volcanoes of
the Pacific Northwest have a potential for future eruptions, information that was used to predict that Mt. St.
Helens would erupt before the end of the 20th century.

Although each volcano or volcanic center in the
Cascades is somewhat individual with a unique history, we can make some generalizations. Many of the
Cascade volcanoes have developed or evolved through
several stages (Figure 7.25). In the first stage, the main
composite cone is constructed of lava flows and pyroclastic materials with an intermediate silica content
(about 60%), known as *andesite*. The second stage is
characterized by less frequent but divergent activity:
the sporadic development of satellite cones composed
of relatively low-silica volcanic rock (approximately
50%) known as *basalt;* and the development of domes
composed of relatively high-silica (approximately 70%)
volcanic rock known as *rhyolite*. The trend toward volcanic domes may be a dangerous sign, as their eruption can be very explosive (12).

The amount of the materials and the composition
in the evolutionary stages of Cascade volcanoes vary
considerably. While several volcanoes reflect the var-

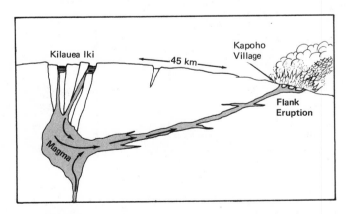

FIGURE 7.23
Idealized diagram showing the 1960 flank eruption on Kilauea which destroyed the village of Kapoho. (After Eaton
and Schmidt, *Atlas of Volcanic Phenomena*, U.S. Geological
Survey.)

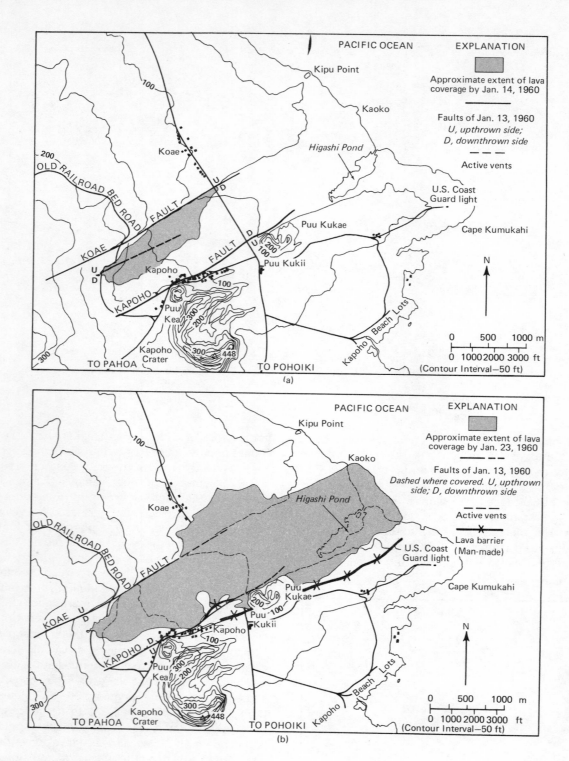

FIGURE 7.24
Volcanic eruption on the flank of Kilauea from January 14 to February 5, 1960. Notice the lava barriers which failed. (After Richter, et al., U.S. Geological Professional Paper 537E, 1970.) (100 ft = 30.5 m)

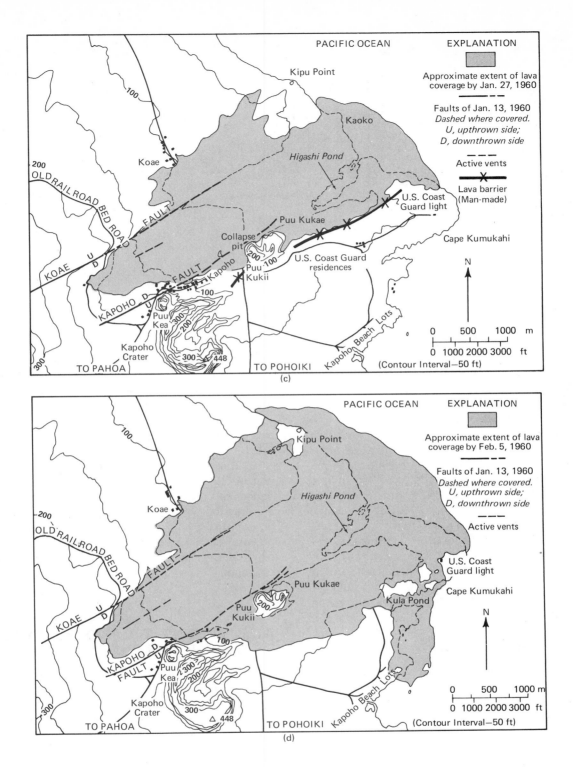

FIGURE 7.24, *continued*

181

FIGURE 7.25

Highly generalized model to illustrate the possible development of a Cascade volcano. Graph (a) shows that the main cone is composed of volcanic rock with a silica content of about 60 percent; the smaller satellite vents and domes have higher and lower silica contents, respectively. Graph (b) shows the development with time. Notice the period of dormancy between the main cone-building stage and the more recent divergent stage. (This figure was developed from ideas presented in Shannon and Wilson, Inc., *Volcanic Hazard Study*, 1976.)

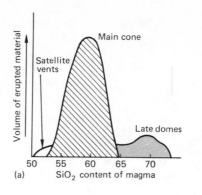

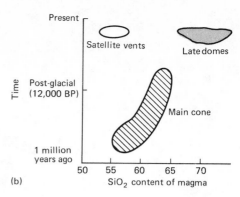

ious stages, no two are exactly alike and no one volcano follows exactly the evolutionary path described in Figure 7.25. Mts. Adams, Baker, and Rainier are still in the main cone-building stage, and all three are composed of a rather uniform andesite (Figure 7.1). Mt. St. Helens may still be in the cone-building stage, as (prior to the May 18, 1980, explosion) the entire upper part of the volcano was constructed during the last 2,500 years, and much of that in the last few hundred years. It is possible that the explosive activity is part of a general cone-building cycle or episode of unknown time length that involves both explosive and constructive eruptions. Thus, in the future, the volcano may again build a cone characteristic of a young composite volcano. On the other hand, the eruptive products, including domes, and their chemical compositions have varied considerably during the last few thousand years at Mt. St. Helens. It is therefore difficult to place the volcano into a general model, emphasizing our earlier point that each volcano may have a unique history.

Other Cascade Volcanoes

Mts. Jefferson and Hood are evidently in the divergent stage, with the development of basalt satellite vents and rhyolite domes. For these two volcanoes, the second stage developed following at least one long period of dormancy (12).

At the end of the second stage, a volcano may become extinct, reenter the first stage, or enter a third stage known as *maturity*. Mount Mazama (Crater Lake in Oregon) is the only known example (Figure 7.1) of this stage. After the second stage, Mazama was dormant for several thousand years, resumed divergent activity, was dormant for another 5,000 years, and, in a spectacular climax, erupted violently about 7,000

years ago, ejecting approximately 60 cubic kilometers of tephra (mostly ash) over an area in excess of 12 million square kilometers. When the eruption was over, the summit of Mount Mazama had disappeared, and the present site of Crater Lake, a caldera nearly 10 kilometers wide and 120 meters deep, was in its place. We conclude that the third, and, presumably, final stage may be very explosive and, thus, very dangerous (12).

The environmental significance of our discussion concerning the evolutionary history of the Cascade volcanoes is that we may be able to estimate the chance of catastrophic eruption of a particular volcano by examining its past activity. The trend toward divergent activity (satellite cones and domes) is a clear danger sign, since this increases the chances of explosive eruptions.

ADJUSTMENT TO AND PERCEPTION OF THE VOLCANIC HAZARD

Apart from the psychological adjustment to losses, the only major human adjustment to volcanic activity is evacuation (13). The small number of adjustments probably reflects the fact that few eruptions occur near populated areas. The experience of about 5,000 Icelandic people on the island of Heimaey offers an example of a hardy people's response to a serious destructive hazard. In January, 1973, the dormant Mt. Helgafell came alive, and subsequent eruptions nearly buried the town of Vestmannaeyjar in ash and lava flows. The harbor, a major fishing port, was nearly blocked. With the exception of about 300 town officials, fire fighters, and police, the people were evacuated. They returned six months later, in July, 1973, to survey the damage and estimate the chances of rebuilding their town and lives. The situation was grim. Ash had drifted up to 4 meters thick and covered much of the island and town. Molten lava was still steaming near the volcano. Their first task was to dig out their homes and shops, salvaging what they could. Then they used the same volcanic debris that had buried their town to pave new roads and an airport to allow materials to be moved in and distributed. They also decided to use the volcano to advantage, and by January, 1974, the first home was heated by lava heat. By 1980, 15 percent of the homes were heated this way, and the system is being expanded to the whole town. The heat source is expected to last at least 15 years. More significantly, two fishmeal factories and three freezing plants were reopened, revitalizing the island's fishing industry (14, and B. Voight, personal communication).

Since eruptions seldom occur near populated areas, little information is available concerning how people perceive the volcanic hazard. One study of perception evaluated volcanic activity in Hawaii; the study makes two suggestions. First, perception of volcanic activity and what people will do in case of an eruption has little to do with proximity to the hazard,

home ownership, knowledge of necessary adjustments, and income level. Second, a person's age and length of residence near the hazard are significant in that person's knowledge of volcanic activity and possible adjustments (13).

Mt. St. Helens provided a valuable example of what kind of human difficulties can be expected during and after a high magnitude physical event that disrupts a large area. The experience should help in devising emergency plans for possible future volcanic eruption or other events such as large earthquakes. For example, the political pressures that arose during the early days of the awakening of Mt. St. Helens are quite interesting. On May 17, just one day before the eruption, a caravan of local cabin owners entered the Spirit Lake area at the base of the north flank of the volcano to inspect their property. They had been requesting admittance for some time, and it was finally granted. A second trip was planned for May 18. If the eruption had not taken place on the 18th, but, say, one week later, one wonders how many people would have returned to the mountain. As we learn more about the magnitude and frequency of natural processes, we are doing a better job of short-range prediction, which should help to save lives in the future.

SUMMARY AND CONCLUSIONS

Volcanic activity is, worldwide, a relatively uncommon process generally related to plate tectonics. Most volcanoes are located at plate junctions where magma is produced as spreading or sinking lithospheric plates interact with other earth material. Regardless of how rare volcanoes are, they have destroyed considerable property and taken many human lives.

Primary effects of volcanic activity include lava flows and pyroclastic activity. Secondary effects include mudflows and fires. All these effects have occurred in the recent history of the Cascade Range of the Pacific Northwest, and there is no reason to believe that further activity will not occur there in the future.

Several methods, including construction of walls, bombing, and hydraulic chilling, have been used in attempts to control lava flows. These methods have had a varying degree of success and require further evaluation.

Pyroclastic activity includes volcanic ash eruptions, which may cover large areas with a carpet of ash; volcanic ash flows, which move as fast as 100 kilometers per hour down the side of a volcano; and lateral blast, which can be very destructive.

Giant caldera-forming eruptions are violent but rare geologic events. Following their explosive beginning, however, they often resurge and may present a volcanic hazard for a million years or longer. Recent uplift and earthquakes at the Long Valley Caldera in California are reminders of the potential hazard.

Mudflows are generated when melting snow and ice mix with volcanic ash. These flows are a serious hazard and can devastate an area many miles from the volcano.

Sufficient monitoring of geophysical properties, topographic changes, seismic activity, and recent geologic history of volcanoes may eventually result in reliable prediction of volcanic activity. Prediction based on ground tilting, earthquake activity, deformation of the crater floor, and gas emissions is being attempted with some success in Hawaii and at

Mt. St. Helens. Applying these techniques to other areas may lead to better prediction of volcanic activity. On a worldwide scale, however, it is unlikely that we will be able to predict accurately all volcanic activity in the near future.

Apart from psychological adjustment to losses, the only major human adjustment to volcanic activity is evacuation.

Perception of the volcanic hazard is apparently a function of one's age and length of residency near the hazard. Direct proximity to a volcano, home ownership, knowledge of necessary adjustments, and income level have little effect on the perception.

REFERENCES

1 OFFICE OF EMERGENCY PREPAREDNESS. 1972. *Disaster preparedness* 1, 3.

2 CRANDELL, D. R., and WALDRON, H. H. 1969. Volcanic hazards in the Cascade Range. In *Geologic hazards and public problems, conference proceedings*, ed. R. Olsen and M. Wallace, pp. 5–18. Office of Emergency Preparedness Region 7.

3 MASON, A. C., and FOSTER, H. L. 1953. Diversion of lava flows at Oshima, Japan. *American Journal of Science* 251:249–58.

4 WILLIAMS, R. S., JR., and MOORE, J. G. 1973. Iceland chills a lava flow. *Geotimes* 18: 14–18.

5 FRANCIS, P. 1983. Giant volcanic calderas. *Scientific American* 248, no. 6:60–70.

6 BAILEY, R. A. 1983. Mammoth Lakes earthquakes and ground uplift: Precursor to possible volcanic activity? *U.S. Geological Survey Yearbook, 1982*:5–13.

7 MILLER, C. D., MULLINEAUX, D. R., CRANDELL, D. R., and BAILEY, R. A. 1982. Potential hazards from future volcanic eruptions in the Long Valley–Mono Lake Area, east central California and southwest Nevada—a preliminary assessment. U.S. Geological Survey Circular 877.

8 HAMMOND, P. E. 1980. Mt. St. Helens blasts 400 meters off its peak. *Geotimes* 25: 14–15.

9 FRANCIS, P. 1976. *Volcanoes.* England: Pelican Books.

10 RICHTER, D. H.; EATON, J. P.; MURATA, K. J.; AULT, W. U.; and KRIVOY, H. L. 1970. *Chronological narrative of the 1959–60 eruption of Kilauea Volcano, Hawaii.* U.S. Geological Survey Professional Paper 537E.

11 SWANSON, D. A., CASADEVALL, T. J., and DZURISIN, D. 1983. Predicting eruptions at Mount St. Helens, June 1980 through December 1982. *Science* 221:1369–1376.

12 WISE, W. S., for SHANNON & WILSON, INC. 1976. *Volcanic hazard study.* Report to Portland General Electric Company. Portland, Oregon.

13 MURTON, BRIAN J., and SHIMABUKURO, SHINZO. 1974. Human response to volcanic hazard in Puna District Hawaii. In *Natural hazards,* ed. G. F. White, pp. 151–59. New York: Oxford University Press.

14 CORNELL, JAMES, ed. 1974. *It happened last year—earth events—1973.* New York: Macmillan.

Chapter 8

Coastal Hazards

Although coastal areas are varied in topography, climate, and vegetation, they are generally dynamic environments. Continental and oceanic processes converge along coasts to produce a landscape that is characteristically capable of rapid change. The impact of hazardous coastal processes is considerable, because many populated areas are located near the coast. This is especially true in the United States where it is expected that most of the population will eventually be concentrated along the nation's 150,000 kilometers of shoreline, including the Great Lakes. Today, the nation's largest cities lie in the coastal zone, and approximately 75 percent of the population lives in coastal states (1).

The most serious coastal hazards are *tropical cyclones*, which claim many lives and cause enormous amounts of property damage every year; tidal floods caused by combination of a high tide with a storm surge; *tsunamis*, or seismic sea waves, which are particularly hazardous to coastal areas of the Pacific Ocean; and *coastal erosion*, which continues to produce considerable property damage that requires human adjustment.

TROPICAL CYCLONES

Tropical cyclones have taken hundreds of thousands of lives in a single storm. A tropical cyclone that struck the northern Bay of Bengal in Bangladesh in November of 1970 produced a 6-meter rise in the sea. Flooding killed approximately 300,000 people, caused $63 million in crop losses, and destroyed 65 percent of the total fishing capacity of the coastal region (2).

Tropical cyclones, known as *typhoons* in most of the Pacific Ocean and *hurricanes* in the Western Hemisphere, cause *damage* and *destruction* from high winds; *river flooding* that results from intense precipitation and usually causes more deaths and destruction than the wind; and *storm surges* (wind-driven oceanic waters) that are the most lethal aspect of tropical cyclones (Figure 8.1). Most deaths in these storms result from drowning (3). Property damage from hurricanes can be staggering. Two hurricanes in the United States in the last few years each caused nearly $1.5 billion in property damage. Despite the increasing population along the Atlantic and Gulf coasts, the loss of lives from hurricanes has decreased significantly because of more effective detection and warning; however, the amount

FIGURE 8.1
Storm surge from Hurricane Camille, 1969. (Photo courtesy of NOAA.)

FIGURE 8.2
Trends of damage and death from hurricanes in the United States from 1915 to 1969. (After *Project Stormfury—1970*, U.S. Department of Commerce, 1970.)

of property damage has greatly increased (Figure 8.2), and there is growing concern that continued population increase accompanied by unsatisfactory evacuation routes, building codes, and refuge sites may contribute to hurricane catastrophes along the Atlantic and Gulf coasts (3).

A secondary effect of tropical cyclones is flash flooding caused by intense rainfall as the storm moves inland. Hurricane Camille, in 1969, first damaged the Louisiana and Mississippi coastlines and, two days later, caused record amounts of precipitation in the mountains of western Virginia, resulting in more than $100 million in damage (3).

The magnitude and frequency of tropical cyclones are not affected by human activity. Most hurricanes form in a belt between 8 degrees north and 15 degrees south of the equator, and the areas most likely to experience cyclones in this zone are those with warm surface-water temperatures. The storms are generated as tropical disturbances and dissipate as they

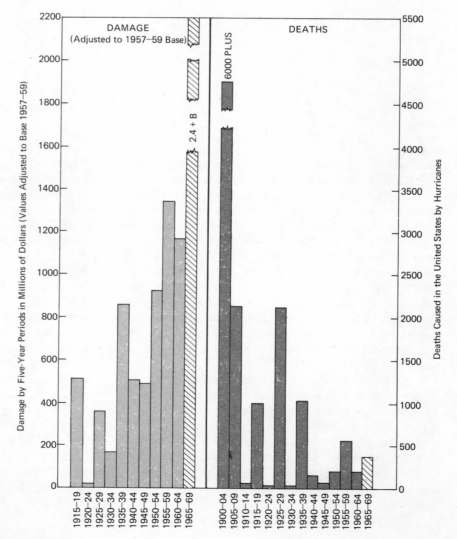

move over the land. Wind speeds in these storms are greater than 100 kilometers per hour, and the winds blow in a large spiral around a relatively calm center called the *eye* of the hurricane. Winds of 100 kilometers per hour or greater are generally recorded over an area about 160 kilometers in diameter, while gale-force winds greater than 60 kilometers per hour are experienced over an area about 640 kilometers in diameter. During an average year, about five hurricanes will develop that might threaten the Atlantic and Gulf coasts. Figure 8.3 shows the storm tracks of and loss of life caused by the several devastating hurricanes that struck the United States from 1964 to 1980 (3).

A fully developed hurricane is an awesome spectacle. Figure 8.4 shows hurricane Allen on Friday, August 8, 1980. The storm, with winds of nearly 300 kilometers per hour and six-meter waves, nearly covered the entire Gulf of Mexico and was one of the largest hurricanes to strike the United States in this century. Approximately 200,000 people heeded the warning and evacuated coastal areas of southern Texas—only a few stubborn people remained. The storm killed more than 100 people before reaching Texas, but fortunately stalled for several hours, losing energy, off the coast of Texas before striking the land in a sparsely populated area. Loss of life in the United States was light (13 people were killed in a helicopter accident while evacuat-

ing offshore oil platforms), but damage to crops from torrential rains that exceeded 25 centimeters was at least several hundred million dollars. On the positive side, the rains ended a month-long drought and thus were welcomed in some areas. In the final analysis, we were lucky with hurricane Allen: the storm stalled offshore and lost energy before striking land; evacuation involved several hundred thousand rather than several million people; and its final path over land was in rural range land. Thus we should not be overconfident and believe that all coastal areas on the Gulf and Atlantic coast are adequately prepared for future large hurricanes.

Tidal Floods

Hurricanes are not the only coastal storms that inflict damage. For example, a storm surge from a lesser tropical disturbance, when combined with a high tide, may produce a "flash" tidal flood. Such an event occurred in Bangor, Maine, on February 2, 1976. The city, located 32 kilometers inland from Penobscot Bay at the confluence of the Kenduskeag Stream and the Penobscot River, was flooded with 3.7 meters of water. The flood resulted when a tidal storm surge, caused by strong south-southeasterly winds up to about 100 kilometers per hour off the coast of New England, moved

FIGURE 8.3
Catastrophic hurricanes affecting the United States from 1964 to 1980. The tracks for Celia and Allen are not shown. (Modified after *Some Devastating North Atlantic Hurricanes of the 20th Century,* U.S. Department of Commerce, 1970. Updated by NOAA.)

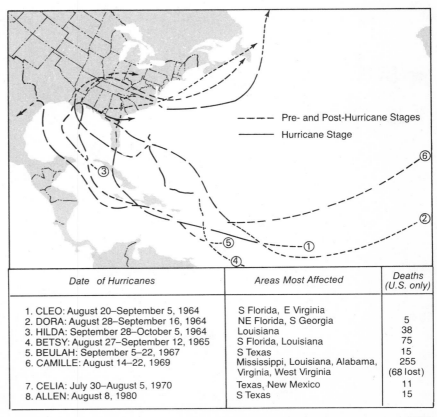

Date of Hurricanes	Areas Most Affected	Deaths (U.S. only)
1. CLEO: August 20–September 5, 1964	S Florida, E Virginia	5
2. DORA: August 28–September 16, 1964	NE Florida, S Georgia	38
3. HILDA: September 28–October 5, 1964	Louisiana	75
4. BETSY: August 27–September 12, 1965	S Florida, Louisiana	15
5. BEULAH: September 5–22, 1967	S Texas	255
6. CAMILLE: August 14–22, 1969	Mississippi, Louisiana, Alabama, Virginia, West Virginia	(68 lost)
7. CELIA: July 30–August 5, 1970	Texas, New Mexico	11
8. ALLEN: August 8, 1980	S Texas	15

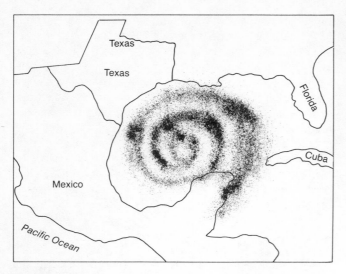

FIGURE 8.4
Hurricane Allen on Friday, August 8, 1980. Idealized drawing (a) shows that the hurricane covered the entire Gulf of Mexico. More detail is shown in the actual satellite photograph (b). (Photograph courtesy of NOAA and Jeff Dozier.)

up the funnel-shaped Penobscot Bay (Figure 8.5). When the surge reached Bangor, flood waters rose very quickly, reaching the maximum depth in less than 15 minutes. Approximately 200 motor vehicles parked in lots along the Kenduskeag Stream were submerged, and when the flood receded about one hour later from

FIGURE 8.5
Bangor, Maine, and the funnel-shaped Penobscot Bay that modified the tidal-storm surge of February 2, 1976. The numbers next to the various locations are the height of the water above that predicted. Notice the dramatic increase in water height from Bucksport to Bangor. (Data from U.S. Geological Survey, Professional Paper 1087, 1979.)

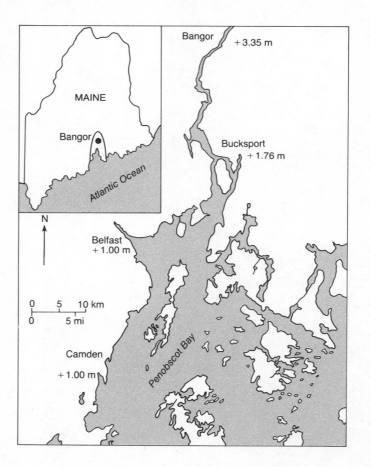

FIGURE 8.6

Map and idealized diagrams showing the concept of the storm surge-tidal barrier recently constructed across the River Thames near London, England. The site of the barrier (a) is just downstream from the main city. A plan view of the river (b) shows the piers and submerged gates across the river where the width is approximately 500 m. The steel gate rests in a concrete sill in the chalk bed of the river (c) and is rotated into place to provide protection from North Sea storm surges as needed.

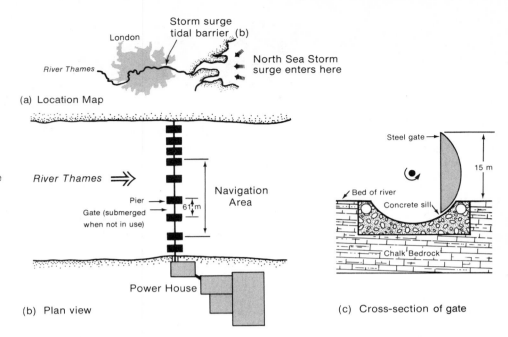

(a) Location Map

(b) Plan view

(c) Cross-section of gate

this first documented tidal flood at Bangor, damages in the downtown area exceeded $2 million (4).

In the first century A.D., the Romans founded the city of Londinium on the banks of the River Thames, and until only recently, the city of London has been at the mercy of tidal floods caused by storm surges. There have been at least seven disastrous floods since the 13th century, two of the more recent in 1928 and 1953. It was not a question of whether London would flood, but *when* the city would be inundated again. Estimates that a great storm surge could do catastrophic damage, perhaps $6 billion worth, threaten the lives of many people, flood buildings, disrupt government, and close the rail system and underground transit, led to construction of the Thames barrier, which was commissioned and tested in 1983. The great barrier cost about $750 million and is constructed of ten huge steel gates separated by nine boat-shaped ribs extending over 500 meters from bank to bank (Figure 8.6). When not in use, the barrier gates rest on concrete sills in the chalk bed of the river and therefore do not present a barrier to navigation. When needed, the gates, powered by electrically driven hydraulics, rotate 90 degrees to stand nearly 15 meters above the bed of the river, and along with downstream embankments and smaller barriers, protect the city of London against the 1,000 year tidal flood (5).

TSUNAMIS

Tsunamis, or seismic sea waves, are often incorrectly called *tidal waves*. They are extremely destructive and

TABLE 8.1

Casualties and damage in the United States caused by tsunamis from 1900 to 1971.

Year	Dead	Injured	Estimated Damage ($000)	Area
1906	—	—	5	Hawaii
1917	—	—	*	American Samoa
1918	—	—	100	Hawaii
1918	40	—	250	Puerto Rico
1922	—	—	50	Hawaii, California, American Samoa
1923	1	—	4,000	Hawaii
1933	—	—	200	Hawaii
1946	173	163	25,000	Hawaii, Alaska, West Coast
1952	—	—	1,200	Midway Island, Hawaii
1957	—	—	4,000	Hawaii, West Coast
1960	61	282	25,500	Hawaii, West Coast, American Samoa
1964	122	200	104,000	Alaska, West Coast, Hawaii
1965	—	—	10	Alaska

*Damage reported but no estimates available.

Source: Eskite, "Analysis of ESSA Activities Related to Tsunami Warning," NOAA.

FIGURE 8.7
Sandy Beach on the Island of
Oahu (a) moments before a tsunami generated by an earthquake in the vicinity of the Aleu-
tian Trench struck the beach.
Note: The arrow on photograph
(a) is the exact location of photograph (b), which shows Sandy
Beach a few minutes later after
the tsunami struck. Notice the
man circled on the lower photograph fleeing from the wave and
the several automobiles, also circled, that were swept off the
highway by the surging wave.
(Photo by Y. Ishii, courtesy of
Honolulu Advertiser.)

(a)

(b)

present a serious natural hazard. Fortunately, they are
relatively rare and are usually confined to the Pacific
basin. The frequency of these events in the United
States is about one every eight years (3). Table 8.1 lists
the casualties and damage in the United States from
tsunamis during the period from 1900 to 1971.

Tsunamis apparently originate when ocean water
is vertically displaced during large earthquakes or other
phenomena. In open water, the waves may travel at
speeds as great as 800 kilometers per hour, and the
distance between successive crests may exceed 100
kilometers. Wave heights in deep water may be less

than one meter, but when the waves enter shallow
coastal waters, they slow to less than 60 kilometers per
hour, and wave heights may increase to more than 15
meters. Figures 8.7 and 8.8 show a tsunami that struck
Hawaii.

Tsunamis claimed most of the lives lost in the
1964 Alaskan earthquake and can cause catastrophic
events thousands of kilometers from where they are
generated, as exemplified by the 1960 seismic waves
that killed 61 people in Hawaii. The waves were
caused by an earthquake in Chile and reached Hawaii
in only 15 hours. Although there is no way to predict

(a)   (b)

FIGURE 8.8
Kuhio Beach on the Island of Oahu (a) a few moments before the tsunami of May, 1960,
struck the beach. (b) The same beach moments after the wave arrived. Notice the ship on
the horizon in both photographs. (Photo by Y. Ishii, courtesy of *Honolulu Advertiser*.)

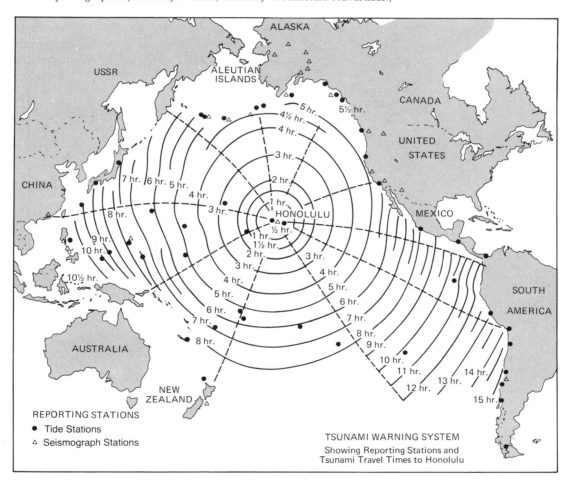

FIGURE 8.9
Tsunami warning system. Map shows reporting stations and tsunami travel times to Honolulu, Hawaii. (From NOAA).

FIGURE 8.10
Tsunami damage to railyards in
Seward, Alaska, caused by the
1964 earthquake. (Photo courtesy
of NOAA.)

tsunamis, it is possible to detect them and warn coastal communities that lie in their path. After an earthquake or another tsunami-generating disturbance has been recognized, the arrival time of a sea wave can be predicted to within plus or minus 1.5 minutes per hour of travel time. This information can be used to produce a tsunami-warning-system map, as shown for Hawaii in Figure 8.9.

Damage caused by tsunamis is most severe at the water's edge, where boats, harbor, buildings, transportation systems, and utilities may be destroyed (Figures 8.10 and 8.11). The waves may also be disastrous to aquatic life and plants in the nearshore environment (3).

COASTAL EROSION

Compared to other natural hazards such as earthquakes, tropical cyclones, or floods, erosion of coasts is

FIGURE 8.11
Tsunami damage to fishing boats
at Kodiak, Alaska, caused by the
1964 earthquake. (Photo courtesy
of NOAA.)

FIGURE 8.12
Breaking waves are powerful agents of erosion. (Photo courtesy of California Division of Beaches and Parks.)

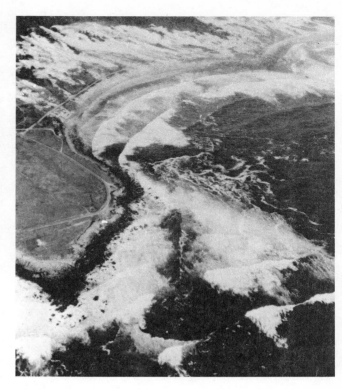

FIGURE 8.13
Oblique aerial photograph showing breaking waves and surf zone. Notice that most of the wave energy is expended on the headland protruding out in the water. This area erodes considerably faster than the bay areas. (Photo courtesy of California Division of Beaches and Parks.)

generally a continuous, predictable process that causes a relatively small amount of damage in a restricted area. Nevertheless, coastal erosion has caused and continues to cause property damage, and large sums of money are spent to control it. As extensive development of coastal areas for vacation and recreational living continues, problems of coastal erosion are certain to become a more serious threat to human use.

Coastal conditions for the East Coast and West Coast are strikingly different. This difference is probably related to plate tectonics. The West Coast is on the leading edge of the North American plate and experiences active uplift and deformation. As a result, the coastline is characterized by rugged sea cliffs and a locally irregular coastline. Such a coast, where rocks project out into the ocean (headlands) is conducive to coastal erosion (Figure 8.12). Upon striking a shore,

waves expend more energy on the protruding areas than on the embayed areas (Figure 8.13). Therefore, the effect of wave erosion is to straighten the shoreline—eroding the headlands and depositing the eroded material in the intervening bays. The East Coast, on the other hand, is on the trailing edge of the North American plate. It is a tectonically quiet coast. For this rea-

FIGURE 8.14
Basic terminology for landforms and wave action in the beach and near-shore environment.

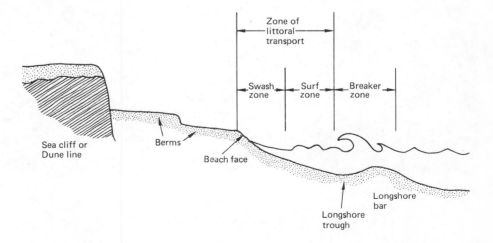

FIGURE 8.15
Block diagram showing beach
drift and longshore drift, which
collectively move sand along the
coast (littoral transport).

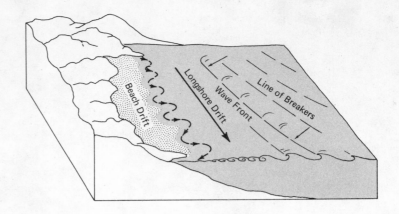

son and others, the major coastline features are chains of barrier islands separated from the mainland by bays, sounds, and lagoons. Exchange of water between lagoons and the open sea takes place through inlets that form when storm waves breach the barrier islands (6).

Beach Form and Process

The sand on beaches is not static. Wave action constantly keeps the sand moving in the surf and swash zones (Figure 8.14), and when waves strike the coast at an angle, the net result is a longshore current and beach drift, which collectively move sand along a coast *(littoral transport)* (Figure 8.15). Along both the East and West Coasts of the United States, the direction of littoral transport (although it can be quite variable) is most often to the south, and transport is on the order of 200,000 to 300,000 cubic meters per year.

Figure 8.14 shows the basic terminology of an idealized near-shore environment. The landward extension of the beach terminates at a natural topographic and morphologic change, such as a seacliff or dune line. The *berms* are flat backshore areas on beaches (where people sunbathe). Berms are thought to form by deposition of sediment as waves recede. The beach face is the sloping portion of the beach below the berm, part of which is exposed by the uprush and backwash of waves *(swash zone)*. It is in the swash

(a)

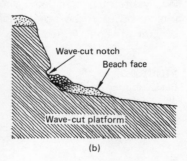

(b)

FIGURE 8.16
Seacliff, beach, and wavecut platform, Santa Barbara, California (a) and idealized cross section (b). (Photo courtesy of Donald Weaver.)

PLATE 9
Winter Park sinkhole that developed in May of 1981. The subsidence caused $2 million in damages and swallowed part of the community swimming pool, parts of two businesses, a house, and several expensive sports cars, trailers, and other vehicles. (Photo by George Remaine of the Orlando Sentinel Star, ©1981.)

PLATE 10
Avalanche protection structure for a Colorado highway. Note the possible older avalanche chute denoted by the zone of young trees in the center part of the photograph. The structure allows avalanches to travel over the highway.

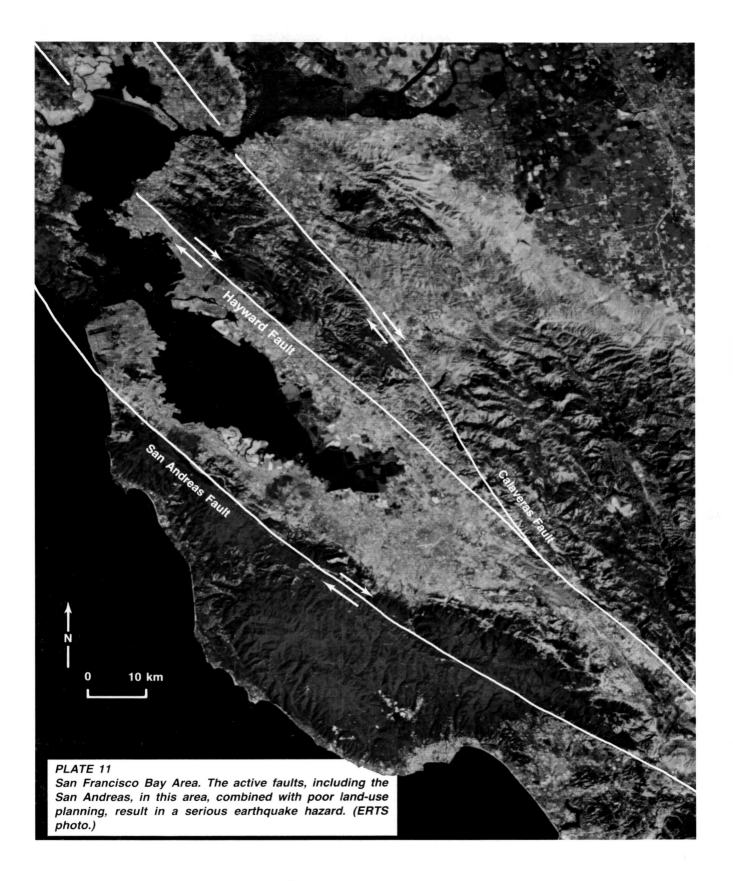

PLATE 11
San Francisco Bay Area. The active faults, including the San Andreas, in this area, combined with poor land-use planning, result in a serious earthquake hazard. (ERTS photo.)

PLATE 12A
Ground rupture along the Lost River fault produced by an earthquake of an approximate magnitude 7 in central Idaho, October 28, 1983.

PLATE 12B
Close-up of ground rupture and fault scarps (2–3 m high) produced by the October 28, 1983 earthquake in central Idaho.   (Photographs courtesy of Paul Link)

PLATE 13
Damage to vegetation in the Toutle River Valley resulting from the 1980 eruption of Mt. St. Helens in Southwestern Washington. Three zones of damage are shown here: nearly total destruction, downed timber, and standing, burned timber. White colored material is volcanic ash. (Photo courtesy of Harry Glicken and the U.S. Geological Survey).

PLATE 14
Coastal housing development, Ventura County, California (top). Recent coastal erosion has moved the mean high-water shoreline as much as 60 m inland in this area, causing damage to or loss of expensive coastal homes. Some houses have been temporarily protected by placing them on pilings (bottom). Coastal erosion in this area is a complex problem related in part to short-term climatic cycles and to the supply of sand that reaches the coastal areas from the Ventura and Santa Clara Rivers. Construction of upstream reservoirs has contributed to the problem by trapping sand that would otherwise nourish the beach environment.

PLATE 15
Large storm waves attacking a beach house in Ventura, California during the winter of 1982–1983 (top). Note large rocks thrown onto the street by wave action (bottom). (Photograph courtesy of C. G. Marshall.)

PLATE 16
Tidal barrier on a coastal river, Hull, England. The ship is going through a large gate (close-up shown in bottom photo) with a rotating barrier that slides down the large pillars to block the channel when tidal flooding is expected.

(a)

(b)

FIGURE 8.17
Erosion of seacliff composed of soft compaction shale near Isla Vista, California. Uncontrolled runoff from storm drains on the seacliff results in serious erosion (a). Controlled runoff released at the base of the seacliff on the beach causes much less erosion (b).

zone that *beach drift*, the up and back movement of beach material along the beach face as idealized in Figure 8.15, occurs. The *surf zone* is that portion of the seashore environment where borelike waves of translation occur after the waves break. It is in the surf zone that *longshore current* occurs. The *breaker zone* is the area where the incoming waves become unstable, peak, and break. The *longshore trough* and *bar* are an elongated depression and adjacent ridge of sand produced by wave action. A particular beach, especially if it is wide and gently sloping, may have a series of long shore bars, troughs, and breaker zones (7).

Erosion Factors

The sand on many West Coast beaches is supplied to the coastal areas by rivers that transport it from areas upstream where it has been produced by weathering of quartz- and feldspar-rich rocks. Our technology has interfered with this material flow of sand from moun-

tains to beach by building dams that effectively trap the sand. As a result, some West Coast beaches are deprived of sediment.

Where a seacliff is present along a coastline, there may be added erosion problems because the seacliff is exposed to both marine and land processes. These processes may work together to erode the cliff at a greater rate than either process could without the other. The problem is further compounded when people interfere with the seacliff environment through inappropriate development.

Figure 8.16 shows a typical southern Californian seacliff environment at low tide. The rocks of the cliff are steeply inclined and folded shale. A thin veneer of sand and coarser material (pebbles and boulders) near the base of the cliff mantles the wave-cut platform. A mantle of sand approximately one meter thick covers the beach during the summer, when long gentle waves construct a wide berm while protecting the seacliff from wave erosion. During the winter, storm waves, which have a high potential to erode beaches, remove the mantle of sand, exposing the base of the seacliff. Thus, it is not surprising that most erosion of the seacliffs in southern California takes place during the winter months.

In addition to wave erosion, processes that attack the seacliff include biological erosion, weathering, rain wash, landslides, and artificially induced erosion (8). Biological processes facilitate and directly cause some erosion of the seacliff; for example, boring mollusks, marine worms, and some sponges can destroy rock. Weathering is significant in weakening the rocks of the seacliff and acts as an aid to erosion: trees on the top of the seacliff may have roots that penetrate the rock and wedge them apart; salt spray may enter small holes and fractures and, as the water evaporates, the salt crystallizes, exerting pressure on the rock that weakens it and can break off small pieces. Rain wash can cause a considerable amount of seacliff erosion; however, the amount of erosion depends upon the nature and extent of the rainfall and the erodability of the rocks that make up the seacliff.

A variety of human activities can induce seacliff erosion. Urbanization, for example, results in increased runoff which—if not controlled, carefully collected, and diverted away from the seacliff—can result in serious erosion. Figure 8.17a shows a drain pipe through which water is simply dumped on the seacliff. The result is active erosion. On the other hand, a short distance away, another drain pipe is routed to the base of the seacliff and the water runs out on the face of the beach (Figure 8.17b), resulting in much less erosion. Watering lawns and gardens on top of a seacliff also adds a good deal of water to the slope. This water tends to migrate toward the base of the seacliff, where it may emerge as small seeps or springs, effectively reducing the stability of the seacliff and facilitating landslides. Figure 8.16 shows a section of seacliff with a park on top. Recreational use of the seacliff is certainly superior to residential use and should be encouraged. The watering of the large lawn, however, has encouraged the continuous flow of several small seeps along the seacliff. Over a period of years, this may lower the resisting forces of the rocks in the cliff and facilitate an increase in the frequency of landslides in an area where they are already all too common.

Building structures such as walls, buildings, swimming pools, and patios may also decrease the stability of the cliff by increasing the driving forces (Figure 8.18). Strict regulation of development in many areas of the coastal zone now forbids most unsafe construction, but we must continue to live with our past mistakes.

The rate of seacliff erosion is variable, and few measurements are available. Along parts of the south-

FIGURE 8.18
Development on the top of a seacliff at Isla Vista, California. (Photo courtesy of Robert Norris.)

ern California coast near Santa Barbara, the rate averages 15 to 30 centimeters per year, depending on the resistance of the rocks and the height of the seacliff (8). These erosion rates are moderate compared to other parts of the world. Along the Norfolk coast of England, for example, erosion rates in some areas are about 2 meters per year. More remarkable, in the North Sea off Germany, one small island with soft erodable sandstone seacliffs had a perimeter of about 200 kilometers in 800 A.D., and by 1900, the perimeter had been reduced to only about 3 kilometers by seacliff retreat. At that time, a concrete sea wall was constructed around the entire island to control the erosion (8).

The main conclusion concerning seacliff erosion and retreat is that it is a natural process that cannot be completely controlled unless large amounts of time and money are invested—and even then there is no guarantee. Therefore, it seems we must learn to live with some erosion. It can be minimized, however, by applying sound conservation practices, such as controlling the water on and in the cliff and not placing homes, walls, large trees, or other loading on top of the cliff.

Sea walls of concrete or rip-rap (large stones) may help retard erosion, but are not always effective because considerable erosion may occur at the top of the cliff above the protective structure. Furthermore, sea walls tend to produce a narrower beach with less sand, particularly if waves are strongly reflected, and, unless carefully designed to complement existing land use, may cause aesthetic degradation. Design and construction of sea walls must thus be tailored to specific sites.

Wave erosion on beaches that do not have a seacliff is also a problem in some West Coast areas. Dams on the coastal rivers decrease the sand supply to the beaches, but that isn't the whole story. During winters of exceptionally large storms and high rainfall, large waves up to five meters high may strike a beach, quickly eroding it. Also, some areas that have wide beaches may erode over a period of several years; for example, one area in Ventura County, California, experienced about 60 meters of beach erosion in the early 1960s, damaging many homes and roads (Figure 8.19).

Beach erosion along the East Coast is a result of tropical cyclones and severe storms; a rise in sea level; and interference of human use and interest with natural shore processes (6). *Tropical cyclones* and *severe storms* can greatly alter a coastline by causing beach erosion and erosion of the foreshore sand dune system and by opening new inlets through the barrier islands. In addition, storm surges cut channels through the foreshore dunes, and eroded sediment is deposited in back of the beach as washover deltas (6).

Recent rise in sea level is eustatic (worldwide) and independent of continental movement, at the rate of about 2–3 millimeters per year. Evidence suggests that the rate of rise has increased during the last 60 years as a result of melting of the polar ice caps and thermal expansion of the upper ocean waters, triggered by global warming, caused by the increased carbon dioxide produced by burning fossil fuels. Sea levels could rise by 700 mm over the next century, insuring that coastal erosion would become an even greater problem than it is today. There is evidence for submergence of the coast in many areas along the East Coast. Along the Outer Banks of North Carolina (Figure 8.20), the Shackleford Banks that were low-lying forests in 1850 are today salt marshes (9).

Human interference with natural shore processes has caused considerable coastal erosion. Most problems have arisen in areas that are highly populated and developed. For example, in some areas, artificial barriers retard the movement of sand, causing beaches to grow in some areas and erode in others, resulting in damage to valuable property.

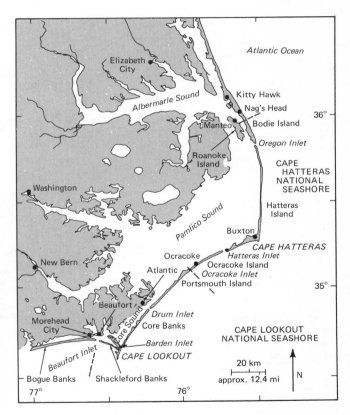

FIGURE 8.20
The Outer Banks of North Carolina showing the locations of Cape Hatteras and Cape Lookout National Seashores. (From Godfrey and Godfrey, *Coastal Geomorphology*, ed. D. R. Coates [Binghamton, New York: Publications in Geomorphology, State University of New York, 1973].)

FIGURE 8.19
Coastal erosion at Oxnard Shores, Ventura County, California, required that this house be placed on a raised foundation. Notice the fireplace that was at one time at ground level. Several homes in this location were destroyed by wave action.

Littoral Cell, Beach Budget, and Wave Climate

The concepts of the littoral cell, beach budget, and wave climate are basic to understanding coastal problems. A *littoral cell* is a segment of coastline that includes an entire cycle of sediment delivery to the coast (usually by rivers), longshore littoral transport, and eventual loss of sediment from the near shore environment (10). Figure 8.21 shows five littoral cells in southern California. Each cell involves the erosion, transport, and deposition of more than 200,000 cubic meters per year of sand. Furthermore, each cell has an annual *budget* of sediment, including sources and losses to the beach environment. Whenever more sediment is transported out of a particular area in a cell than is delivered to that site, erosion results. Negative beach budgets are common, partly because dams on rivers store sediments that would otherwise be delivered to coastal areas, and because coastal structures interfere with longshore transport of sand. The ultimate sink for sand in most of the California littoral cells is transport through submarine canyons to deep sea basins. *Wave climate* is a statistical characterization on an annual basis of wave height, period (time in seconds between arrival of successive waves), and direction, for calculating wave energy at a particular site. Figure 8.22 shows three aspects of wave climate (period, direction, and season of arrival) for the coastal area near Oxnard Shores in Ventura County, California (see also Figure 8.19). The data for Oxnard Shores suggest that the coast there is vulnerable to northwesterly winter waves

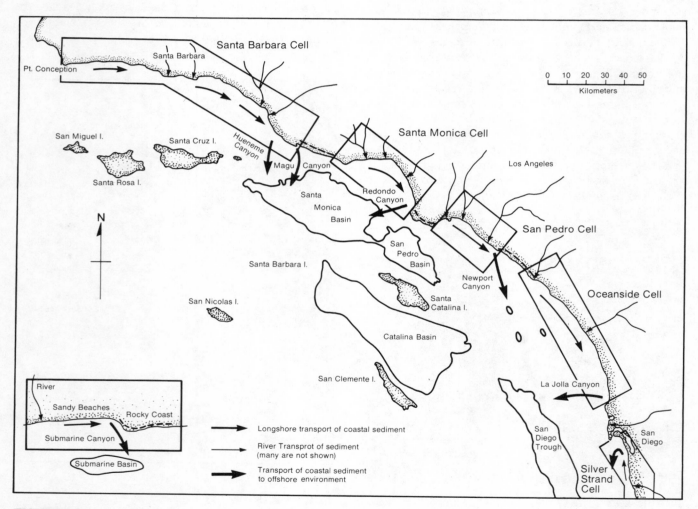

FIGURE 8.21
Littoral cells in southern California. Shown are five of the major cells, four of which deliver sediment transported along the coast to the submarine basin by way of a submarine canyon. (Source: D. L. Inman, 1976, Man's impact on the California coastal zone. State of California, Department of Navigation and Ocean Development.)

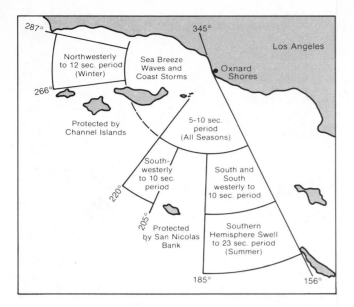

FIGURE 8.22
Part of the wave climate for Oxnard Shores, west of Los Angeles, California. Shown are the directions and periods of dominant waves that strike the area during particular seasons of the year. (Source: D. L. Inman, 1976, Man's impact on the California coastal zone. State of California, Department of Navigation and Ocean Development.)

that move most of the sediment in the Santa Barbara littoral cell (see Figure 8.21).

The beach budget and wave climate provide the basic data necessary for formulating and evaluating beach sediment supply and coastal erosion plans. Without this basic information, comprehensive coastal planning is significantly handicapped. Unfortunately, coastal erosion is usually defended against in a piecemeal manner, community by community, rather than from a littoral cell and beach budget approach that involves planning for a more extensive length of shoreline. As a result, there is little coordination of effort and the problem of beach erosion persists—winter storms in 1983 did over $100 million damage to beach property in California alone. If communities in the same littoral cell worked together, they would be much more successful in fighting coastal erosion.

Examples of Human Activity and Erosion

The Atlantic Coast The Atlantic coast from northern Florida to New York is characterized by barrier islands, many of which have been altered to a lesser or greater extent by human use and interest. Two examples, the barrier island coast of Maryland and the Outer Banks of North Carolina, illustrate the spectrum of human activity.

Demand for the 50 kilometers of Atlantic Ocean-front beach in Maryland is very high, and the limited resource is used seasonally by residents of the Washington, D.C., and Baltimore, Maryland, urban centers (Figure 8.23). Ocean City on Fenwick Island has, since the early 1970s, promoted high rise condominium and hotel development on the waterfront of the narrow island. As a result, the natural frontal dune system has been removed in many locations, resulting in a serious beach erosion problem. Winter storms in 1977 and 1978 removed a good deal of sand, further narrowing the beaches. Attempts by the city to broaden the beaches by moving sand with bulldozers have not been successful, and expensive beach nourishment (bringing in sand from other areas) is under consideration, as loss of the beaches presents a serious economic problem to the resort area (11). More ominous is the almost certain possibility that a future hurricane will cause serious damage to Fenwick Island. The inlet south of Ocean City formed during a hurricane in 1933, and there is no guarantee, despite attempts to stabilize the inlet by coastal engineering, that a new inlet will not form at the present site of Ocean City some time in the future.

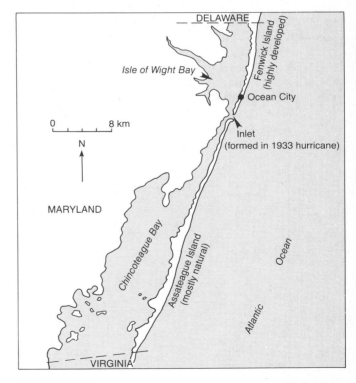

FIGURE 8.23
The barrier island coast of Maryland. Fenwick Island is experiencing rapid urban development and there is concern for potential hurricane damage. What if a new inlet forms (during a hurricane) at the site of Ocean City?

South of Fenwick Island is the nearly natural Assateague Island, where the frontal dune system is preserved and development is prohibited. The island is
used for passive recreation such as sunbathing, swimming, walking, and wildlife observation. Beach erosion
is less on Assateague Island than on Fenwick Island
because the frontal dune system absorbs some of the
wave energy; however, erosion may become a problem
in the future because of the island's close proximity to
Fenwick Island and Ocean City. That is, although people use the two islands for quite different purposes,
they are connected by the same sediment transport
system. Thus, the loss of dunes, the beach nourishment program, and coastal engineering projects on
Fenwick Island may cause less sand to reach the
beaches of Assateague Island, causing erosion. In fact,
the north end of the island is experiencing accelerated
erosion, probably due in part to the installation of engineering structures (jetties) intended to stabilize the
inlet (11). Thus our fundamental principle involving environmental unity holds: we cannot do only one thing,
because everything is connected to everything else.

Another good example of our influence on coastal
processes is found in the Outer Banks of North Carolina
(Figure 8.20). The northern section near Cape Hatteras
(Figure 8.24) has been progressively stabilized and developed (particularly near the beach), whereas the
southern section near Cape Lookout is nearly unaltered
by human activity.

The northern section is characterized by a continuous artificial dune system that was built to control
erosion. The desire to protect roads and buildings behind the dunes led to an attempt at total stabilization.

FIGURE 8.24
The dune line and the beach at Cape Hatteras. (Photo by R.
Anderson, courtesy of National Park Service.)

The idea was to hold everything in place and prevent
the sea from washing over the island. In contrast, the
southern section is characterized by a natural zone of
dunes rather than a line. The zone is frequently broken
by overwash passes that allow water to flood across the
island (Figure 8.25) (9).

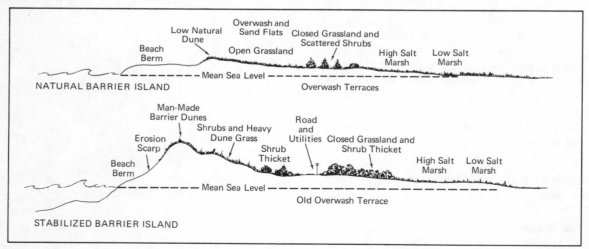

FIGURE 8.25
Cross sections of the Barrier Islands of North Carolina. The upper diagram is typical of the
more natural systems and the lower typical of the artificially stabilized Barrier Islands.
(From Dolan, *Coastal Geomorphology*, ed. D. R. Coates [Binghamton, New York: Publications in Geomorphology, State University of New York, 1973].)

Several studies have shown that building an artificial dune line has resulted in erosion and narrowing of the beach and has necessitated spending millions of dollars to stabilize artificial dunes that tend to erode rapidly (9,12). Even with this effort, the structures and roads behind the dune line will eventually have to be relocated as the shoreline continues to erode and recede. In contrast, where the barrier islands have been left in a natural state, severe storms cause little erosion. As a result, the dunes are not washing away but are moving back by natural processes responsible for the origin and maintenance of the barrier island system. We emphasize, however, that there is considerable controversy as to whether the Core Banks are "typical" barrier islands, since overgrazing with subsequent wind erosion of the frontal dune line may have facilitated the overwash. If this is true, then it is necessary to reevaluate the benefits of frequent overwash.

The Gulf Coast Coastal erosion is also a serious problem along the Gulf of Mexico. One study in the Texas coastal zone suggests that, in specific areas where long-term coastal recession rates can be determined from geologic data, human modification of the coastal zone in the last 100 years has accelerated coastal erosion by 30 to 40 percent over prehistoric rates (13). The human modifications that appear most responsible for the accelerated erosion are construction of coastal engineering structures, subsidence as a result of groundwater withdrawal, and damming of rivers that supply sand to the beaches. Erosion along the Gulf of Mexico in Texas has apparently been going on for several thousand years as a natural process; therefore, it is often difficult to determine the exact extent to which human interference has increased or decreased the erosion. Nevertheless, it appears that the natural rate of landward migration of barrier islands and other coastal beach features in many areas has definitely in-

creased in recent years (historic time). For example, Figure 8.26 shows estimated erosion rates for the Southwestern Matagorda Peninsula in Texas. The recent (1856 to 1956) erosion rates are 34 percent greater than the prehistoric rates. The situation is quite complex, however, because, as the peninsula has retreated nearly two kilometers in the last 1,500 years, the nearby Matagorda Island has grown (accreted) by nearly the same amount. Most of the erosion is in response to high magnitude storms (hurricanes) that produce a net bayward migration of the entire peninsula. The recent situation (post-1965) was further complicated by construction of a ship channel and accompanying jetties to improve navigation that have produced highly variable erosion rates (13).

The Great Lakes Erosion is a periodic problem along the coasts of the Great Lakes, and has been particularly troublesome along the Lake Michigan shoreline. Damage is most severe during prolonged periods of high lake levels that occur following extended periods of above-normal precipitation. The relationship between precipitation and lake level has been documented by the United States Corps of Engineers since 1860. They have shown that the lake level has fluctuated about 1.5 meters during this time. Figure 8.27 shows the level of Lake Michigan from 1900 to 1980. Lake level has been rising in recent years from a low stage in 1964, and in 1973, it again reached the high level of 1952. During a highwater stage there is considerable coastal erosion and many buildings, roads, retaining walls, and other structures are destroyed by wave erosion (Figure 8.28) (14).

During periods of below-average lake level, wide beaches develop that dissipate energy from storm waves and protect the shore. With rising lake-level conditions, however, the beaches become narrow, and storm waves exert considerable energy against coastal

FIGURE 8.26
Estimated rates of coastal erosion for the southwestern Matagorda Peninsula, Texas. (After Wilkinson and McGowen, *Environmental Geology,* vol. 1 [New York: Springer-Verlag, 1977].)

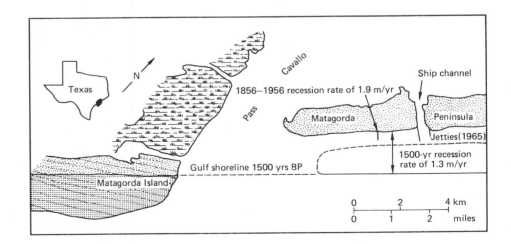

areas. Even a small lake-level rise on a gently sloping shore will inundate a surprisingly wide section of beach (14).

Long-term rates of bluff-top, shoreline recession at many Lake Michigan sites average about 0.4 meters per year (15). Severity of erosion at a particular site depends on factors such as the presence or absence of a frontal dune system (dune-protected bluffs erode at a slower rate); orientation of the coastline (sites exposed to high-energy storm winds erode faster); groundwater seepage (seeps along the base of a coastal bluff cause slope instability, increasing the erosion rate); and existence of protective structures (structures may be locally beneficial, but often accelerate coastal erosion in adjacent areas) (14,15). A few construction options and the relocation alternative are summarized, with respective advantages and disadvantages, in Figure 8.29.

Engineering Structures

Engineering structures in the coastal environment are primarily designed to improve navigation or retard erosion. They include groins, breakwaters, and jetties. Because they tend to interfere with the littoral transport of sediment along the beach, these structures all too often cause undesirable deposition and erosion in their vicinity.

Groins are linear structures placed perpendicular to the shore. They are usually constructed in groups called groin fields (Figure 8.30). The basic idea is that each groin will trap a portion of the sand that is moving in the littoral transport system. A small accumulation of sand will develop updrift of each groin, thus building an irregular but wider beach. The problem is that, while deposition occurs updrift of a groin, erosion tends to occur in the downdrift direction. Thus, a groin or groin field results in a wider, more protected beach in a desired area, but may cause a zone of erosion to develop in the adjacent downcoast shoreline. The erosion results primarily as a groin or groin field becomes filled with trapped sediment. Once a groin is filled, sand is transported around its offshore end to continue its journey along the beach. Therefore, erosion may be minimized by artificially filling each groin; this is known as *beach nourishment,* and requires trucking

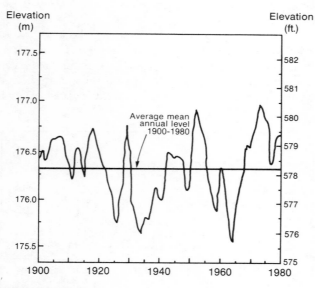

FIGURE 8.27
Lake levels for Lake Michigan from 1900 through 1980. (Source: U.S. Department of Commerce and U.S. Army Corps of Engineers. Modified after Buckler and Winters, 1983.)

FIGURE 8.28
Erosion of a sandy shoreline of Lake Michigan, with damage to residences (upper) and pavement (lower), November, 1972. (Photos courtesy of *Waukegan News-Sun.*)

sand out onto the beach. Thus nourished, the groins will draw less sand from the natural littoral transport system, and the downdrift erosion will be reduced (7).

Even with beach nourishment and other precautions, however, groins may cause undesirable erosion; therefore, their use should be carefully evaluated.

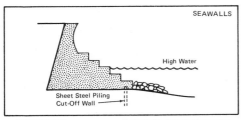

SEAWALLS

High Water

Sheet Steel Piling
Cut-Off Wall

ADVANTAGES

1. Provides protection from wave action and stabilizes the backshore.
2. Low maintenance cost.
3. Readily lends itself to concrete steps to beach.
4. Stabilizes the backshore.

DISADVANTAGES

1. Extremely high first cost.

2. Subject to full wave forces; fail from scour; flanking of foundation.
3. Not easily repaired.
4. Complex design and construction problem. Qualified engineer is essential.
5. Slope design is most important.
6. More subject to catastrophic failure unless positive toe protection is provided.

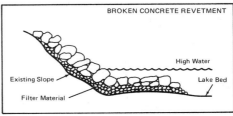

BROKEN CONCRETE REVETMENT

High Water

Existing Slope

Lake Bed

Filter Material

ADVANTAGES

1. Inexpensive.
2. Easy construction.

DISADVANTAGES

1. Large concrete pieces are difficult to obtain.

2. Large pieces required for underlying filter layer because of large void.
3. Extremely unattractive appearance, unless special care is taken in construction.

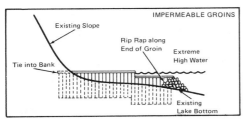

IMPERMEABLE GROINS

Existing Slope

Rip Rap along
End of Groin

Extreme
High Water

Tie into Bank

Existing
Lake Bottom

ADVANTAGES

1. Resulting beach protects upland areas and provides recreational benefit.
2. Moderate first cost and low maintenance cost.

DISADVANTAGES

1. Extremely complex coastal engineering design problem. Qualified coastal engineering services are essential. Groins rarely function as intended.

2. Areas downdrift will probably experience rapid erosion.
3. Unsuitable in areas of low littoral drift.
4. Subject to flanking, must be securely tied into bluff.

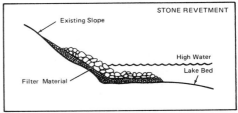

STONE REVETMENT

Existing Slope

High Water

Lake Bed

Filter Material

ADVANTAGES

1. Most effective structure for absorbing wave energy.
2. Flexible — not weakened by light movements.
3. Natural rough surface reduces wave runup.
4. Lends itself to stage construction.
5. Easily repaired — low maintenance cost.
6. The preferred method of protection when rock is readily available at a low cost.

DISADVANTAGES

1. Heavy equipment required for construction.
2. Subject to flanking and moderate scour.
3. Limits access to beach.
4. Moderately high first cost.
5. Difficult construction where access is limited.

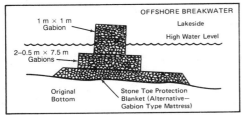

OFFSHORE BREAKWATER

1 m × 1 m
Gabion

Lakeside

High Water Level

2–0.5 m × 7.5 m
Gabions

Original
Bottom

Stone Toe Protection
Blanket (Alternative—
Gabion Type Mattress)

ADVANTAGES

1. Beneficial effect can extend over a considerable length of shoreline.
2. Maintains or enhances recreational value of a beach.
3. Not subject to flanking — can be built in separate reaches.
4. Structure maintenance costs are lower than those of similar

*A gabion is a wire basket filled with rock fragments or coarse gravel.

structures designed for other purposes.

DISADVANTAGES

1. May modify beachline and cause erosion in downdrift areas.
2. Structure is subject to foundation and scour failures. Floating plant and heavy equipment maybe required for construction.

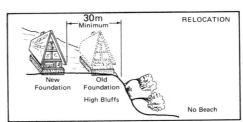

RELOCATION

30m
Minimum

New
Foundation

Old
Foundation

High Bluffs

No Beach

ADVANTAGES

1. It is permanent. In the long run it may be the best method of protection.
2. Adaptable to short reaches of shoreline.
3. Can be accomplished by the individual without coordination through contract with a house mover.

4. No maintenance required.

DISADVANTAGES

1. Special skills and equipment required.
2. Area must be available for relocation of the house.
3. Does not stop erosion.

FIGURE 8.29

Example of several methods of shoreline protection from wave erosion. (From U.S. Army Corps of Engineers.)

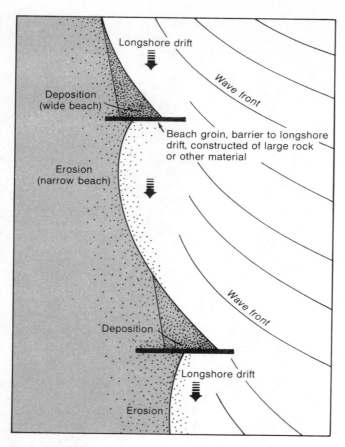

FIGURE 8.30

Idealized diagram showing two beach groins. Deposited sediment builds a wide beach in the updrift direction; in the downdrift direction, the sparsity of sediment for transport can cause erosion to occur. Experiments are underway in California to build expandable or moving groin systems that will interfere with sediment transport only during storms when erosion is a problem. The rest of the year, the expandable groins will not interfere with sediment movement and thus may avert the problems caused by fixed groins.

Breakwaters are designed to intercept waves and provide a protected area (harbor) for boat moorings, and may be attached to the beach or be separated. In either case, they block the natural littoral transport of beach sediment and, thus, locally change the configuration of the coast as new areas of deposition and erosion develop (Figure 8.31). Breakwaters have caused serious erosion problems downdrift of the structures. In addition, they produce a sand trap that continues to accumulate sand in the updrift direction. Eventually the trapped sand may fill or block the entrance to the harbor as a sand spit or bar develops. As a result, a dredging program, or artificial bypass, is often necessary to keep the harbor open and clear of sediment. The sediment that is removed by dredging should be transported and released on the beach downdrift of the

breakwater to rejoin the natural littoral transport system, thus reducing the erosion problem.

Jetties are often constructed in pairs at the mouth of a river (Figure 8.31) or inlet to a lagoon, estuary, or bay (Figure 8.26). They are designed to stabilize the channel, prevent or minimize deposition of sediment in the channel, and generally protect it from large waves (7). Jetties tend to block the littoral transport of beach sediment, thus causing the updrift beach adjacent to the jetty to widen while downdrift beaches erode. The deposition at jetties may eventually fill the channel, making it useless, while downcoast erosion damages coastal development. Mechanically bypassing (dredging) the sediment minimizes but does not eliminate all undesirable deposition and erosion.

There is no way to build a breakwater or jetty along a coast with an active littoral transport system so that it will not interfere with and partially or almost totally block the longshore movement of beach sediment. These structures must therefore be carefully planned and protective measures taken early to eliminate or at least minimize adverse effects. This may include installation of a dredging and artificial sediment-bypass system, beach nourishment program, seawalls, rip-rap, or some combination of these measures (7).

PERCEPTION OF AND ADJUSTMENT TO COASTAL HAZARDS

People adjust to the tropical cyclone hazard either by doing nothing and bearing the loss or by taking some kind of action to modify potential loss. Bearing the loss is probably the most common individual adjustment. Attempts to modify potential loss include *strengthening* the environment with protective structures and land stabilization, and *adapting* behavior by better land-use zoning, evacuation, and warning (16).

Inhabitants in areas with a serious tropical cyclone hazard, such as coastal Bangladesh and the Gulf Coast of the United States, are very much aware of the hazard (17, 18). In Bangladesh, perception of the cyclone hazard was influenced by people's economic and cultural backgrounds. It was found that the people are not particularly concerned about the hazard, and the three socioeconomic classes perceive the cyclones in much the same way; however, the upper class tends to favor earning a living in a less hazardous area.

Although repeatedly devastated by tropical cyclones, coastal areas in Bangladesh do not have an integrated system of private or public adjustment. Although people there are aware of the hazard, the adoption of adjustments is not a function of the frequency of cyclones. Even when shelter is available, people are reluctant to leave their homes. During a

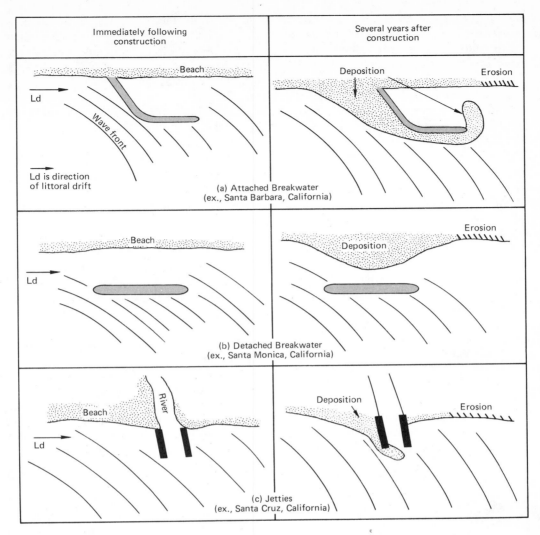

Immediately following construction	Several years after construction

Beach

Ld

Wave front

Ld is direction
of littoral drift

Deposition   Erosion

(a) Attached Breakwater
(ex., Santa Barbara, California)

Beach

Ld

Erosion

Deposition

(b) Detached Breakwater
(ex., Santa Monica, California)

Beach   River

Ld

Deposition   Erosion

(c) Jetties
(ex., Santa Cruz, California)

FIGURE 8.31
Idealized diagrams illustrating the effects of breakwaters and jetties on local patterns of deposition and erosion.

1970 cyclone, approximately 38 percent of the people saved their lives by climbing trees, 5 percent took shelter in a two-story "community center," and 8 percent remained on top of an embankment designed to protect the land against saltwater intrusion. The embankment was incomplete when the storm hit, and was severely eroded and overlapped by storm-surge waves. It is hard to believe that 22 people in a family died only 400 meters from the community center that offered safety (17).

People's attitudes toward hurricane damage and the possibility of preventing or modifying damage are related to their educational level and their hurricane experience. Public awareness programs in areas likely to experience tropical cyclones should thus be matched to the educational level of the target group. Furthermore, since better-educated people tend to be more

positive toward damage prevention, increased attention should be given to the less-educated population to counteract their tendency to lack confidence in damage-prevention measures. Awareness programs also should be designed differently depending on the area's history of hurricane experience (18).

Little information is available concerning the perception of the tsunami hazard. In developed areas with good communications, the established warning system is adequate provided the tsunami is detected far enough out to sea to allow time to warn people in the path of the wave. In undeveloped areas or areas very close to the origin to the seismic sea wave, loss of life can be expected to be greater.

Adjustments to coastal erosion fall into one of several categories: beach nourishment that tends to imitate natural processes; modification of the wave energy (wave climate) through construction of near-shore structures designed to dissipate wave energy; shoreline stabilization through structures such as groins and seawalls; and land use change that attempts to avoid the problem.

The city of Miami Beach, Florida, and the U.S. Army Corps of Engineers began an ambitious beach nourishment program in the mid-1970s to reverse a serious beach erosion problem that had plagued the area since the 1950s and to provide protection from storms. The natural beach had nearly disappeared by the 1950s, and only small pockets of sand could be found associated with various shoreline protection structures, including seawalls and groins. As the beach disappeared, coastal resort areas, including high-rise hotels, became vulnerable to storm erosion (19).

The purpose of the nourishment program for Miami Beach was to produce a positive beach budget, and thus a wide beach, and provide additional protection from storm damage. The project will cost about $62 million over 10 years and involve nourishment of about 160,000 cubic meters of sand per year to replenish erosion losses. By 1980, about 18 million cubic meters of sand had been dredged and pumped from an offshore site onto the beach, producing a 200-meter-wide beach (19). Figure 8.32 shows Miami Beach before and after the nourishment—the change is dramatic. Figure 8.33 shows the cross-section design for the project which includes two components (a wide berm and frontal dune system) to function as a buffer

to wave erosion and storm surge. Although only time will show how well the nourishment protects against a really large storm or hurricane, and although the project is expensive, it has the advantage of providing aesthetic amenities and a degree of protection to valuable resort property. Beach nourishment is certainly preferable to the fragmented erosion control methods that preceded it.

Perception of coastal erosion as a natural hazard depends primarily on an individual's past experience, proximity to the coastline, and the probability of suffering property damage. One study of coastal erosion of seacliffs near Bolinas, California, 24 kilometers north of the entrance to the San Francisco Bay, established that people living very near the coast in an area likely to experience damage in the near future are generally very well informed and see the erosion as a direct and serious threat (20). People living a few hundred meters from a possible hazard, although aware of the hazard, know little about its frequency of occurrence, severity, and predictability. Still further inland, people are aware that coastal erosion exists but have little perception of the hazard.

The Bolinas study is especially interesting because of the controversy that surrounded various alternatives the community considered. Before 1971, coastal-erosion damages and expenditures at Bolinas exceeded $300,000 (Table 8.2). The rate of cliff retreat is approximately 0.5 meters per year under natural conditions and somewhat less when sea walls and other

FIGURE 8.32
Miami Beach before (left) and after (right) beach nourishment. (Photographs courtesy of U.S. Army Corps of Engineers.)

FIGURE 8.33

Idealized diagram showing the cross-section of the Miami Beach nourishment project. The dune and beach berm system provide protection against storm attack. (Source: U.S. Army Corps of Engineers.)

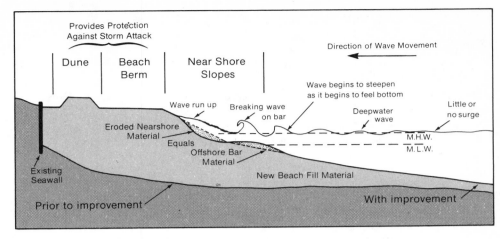

protective measures are used to protect the cliffs. The community considered four alternatives. The first was *zoning* to prevent development in the hazard zone. No structures to slow the erosion of the cliffs would be provided. This, however, would be effective only for undeveloped land, and in several areas, homes and roads are close enough to the retreating cliffs to be damaged eventually regardless of zoning. Second, the community considered *public land acquisition* in the hazard zone. The land would be purchased for open space and recreation. Natural erosion of the cliffs would be allowed, and damage would be restricted to low-cost recreation facilities. Third, *construction of a sea wall* of broken rock at the base of the cliffs to stabilize erosion was considered. This would cost $3 to $4 million, and there would still be up to 20 meters of erosion at the

top of the cliffs unless other protective measures were taken. Fourth, the community considered a *combination of groins, beachfill, and an energy dissipator* (Figure 8.34). The groins and energy dissipator would be constructed of broken rock at an initial cost of $4 to $6 million. This would stabilize the base of the cliff, but erosion on the top would continue until a stable angle of repose was reached, which, as in the case of the sea wall, would be up to 20 meters unless protective measures were taken (20).

The people of Bolinas include weekend and summer residents, retired persons who are permanent residents, long-time agricultural residents, and activist environmentalists. Each, depending on background

TABLE 8.2

Coastal erosion damages and expenditures at Bolinas, California.

	Value ($)[a]
Damages	
Homes (three)	75,000
Real property (the equivalent of 15 lots at $5,000 each)	75,000
Public utilities (roads, pipelines)	50,000
Total	200,000
Expenditures	
Seawall construction	75,000
Cliff drainage and planting	25,000
Rip-rap of road area at top of cliffs	5,000
Other road repairs	10,000
Moving of homes (two)	10,000
Problem studies[b]	10,000
Total	135,000

[a]1971 dollar equivalent.

[b]U.S. Army Corps of Engineers, Bolinas Beach and Erosion Study, $5,000 per year for two years.

Source: *Natural Hazards: Local, National, Global* edited by Gilbert F. White. Copyright © 1974 by Oxford University Press, Inc. Reprinted by permission.

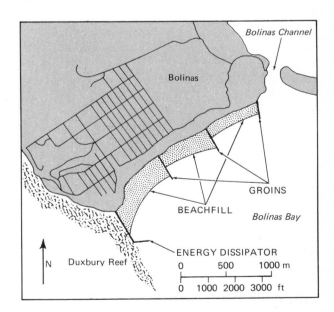

FIGURE 8.34

Proposed adjustments for cliff stabilization at Bolinas, California. (After R. A. Rowntree, Natural Hazards: Local, National, Global, ed. G. F. White [New York: Oxford University Press, 1974].)

and proximity to the hazard, had a different perception of the problem. The town formed a planning group to consider the alternatives and a conservation zone was recommended along with a policy to prevent construction within 50 meters of retreating seacliff. Expensive stabilization plans that would be an inducement to growth and tourism, something the community fears more than cliff erosion, were rejected. The community has, in effect, chosen to accommodate itself to the haz-ard, taking the position that coastal erosion is a natural process rather than a natural hazard. It is likely that the community as a whole will maintain present cliff stabilization measures, but eventually the cliffs will retreat through the first row of houses regardless of interim measures.*

*From *Natural Hazards: Local, National, Global* edited by Gilbert F. White, Copyright © 1974 by Oxford University Press, Inc. Reprinted by permission.

SUMMARY AND CONCLUSIONS

The coastal environment is one of the most dynamic areas on earth, and rapid change is a regular occurrence.

Migration of people to coastal areas is a continuing trend, and approximately 75 percent of the population in the United States now lives in coastal states.

The most serious coastal hazard is tropical cyclones. These violent storms bring high winds, storm surges (wind-driven oceanic waves), and river flooding. They continue to take thousands of lives and cause billions of dollars in property damage.

Tsunamis, or seismic sea waves, are produced by earthquakes or other phenomena that displace oceanic water and generate long waves that travel at speeds as great as 1,000 kilometers per hour. The wave, less than 0.5 meters high in deep water, may grow to a height of 15 meters or more on reaching the coastline. The majority of the lives lost in the 1964 Alaskan earthquake was attributed to tsunamis.

The combination of high tides with a storm surge can cause serious flooding. The threat of such a flood in London, England, precipitated the construction of a storm surge (tidal) barrier across the River Thames.

Although it causes a relatively small amount of damage compared to other natural hazards such as river flooding, earthquakes, and tropical cyclones, coastal erosion is a serious problem along the East, West, Gulf of Mexico, and Great Lakes coasts of the United States.

Human interference with natural coastal processes such as the building of groins, jetties, breakwaters, artificial dunes, and other structures is occasionally successful, but in some cases it has caused considerable coastal erosion. Most problems occur in areas with high population density, but sparsely populated areas along the Outer Banks in North Carolina are also having trouble with coastal erosion.

Perception of and adjustment to coastal hazards depend on factors such as the magnitude and frequency of the hazard, and the individual's educational level, experience, and proximity to the hazard.

With tropical cyclones, the most common individual adjustment is to do nothing and bear the loss. Community adjustments in developed countries generally attempt to modify the environment by building protective structures designed to lessen potential damage, or encourage change in people's behavior by better land-use zoning, evacuation, and warning.

Adjustment to tsunamis is generally to bear the loss. The tsunami warning system is effective in communities that have good communication systems and are far enough from where the wave is generated to allow the necessary time to warn people in its path.

Adjustment to coastal erosion in developed areas is usually the "technological fix": building sea walls, groins, and other structures to stabilize the beach and prevent erosion. These have had mixed success, and often cause additional problems in adjacent areas. More recently, beach nourishment is being used to replenish beach material lost by erosion. One community in California takes the position that coastal erosion of sea cliffs is a natural process rather than a natural hazard, and has rejected engineering solutions to the problem.

REFERENCES

1. COATES, D. R., Ed. 1973. *Coastal geomorphology.* Binghamton, New York: Publications in Geomorphology, State University of New York.

2. WHITE, A. U. 1974. Global summary of human response to natural hazards: Tropical cyclones. In *Natural hazards: Local, national, global,* ed. G. F. White, pp. 255–65. New York: Oxford University Press.

3. OFFICE OF EMERGENCY PREPAREDNESS. 1972. *Disaster preparedness* 1, 2.

4. MORRILL, R. A., CHIN, E. H., and RICHARDSON, W. S. 1979. *Maine coastal storm and flood of February 2, 1976.* U.S. Geological Survey Professional Paper 1087.

5. SHAW, D. P. 1983. A barrier to tame the Thames. *Geographical Magazine* VLV(3): 129–131.

6. EL-ASHRY, M. T. 1971. Causes of recent increased erosion along United States shorelines. *Geological Society of America Bulletin* 82: 2033–38.

7. KOMAR, P. D. 1976. *Beach processes and sedimentation.* New Jersey: Prentice-Hall.

8. NORRIS, R. M. 1977. Erosion of sea cliffs. In *Geologic hazards in San Diego,* eds. P. L. Abbott and J. K. Victoris. San Diego Society of Natural History.

9. GODFREY, P. M., and GODFREY, M. M. 1973. Comparison of ecological and geomorphic interactions between altered and unaltered barrier island systems in North Carolina. In *Coastal geomorphology,* ed. D. R. Coates, pp. 239–58. Binghamton, New York: Publications in Geomorphology, State University of New York.

10. INMAN, D. L. 1976. Man's impact on the California coastal zone. State of California Dept. of Navigation and Ocean Development.

11. U.S. DEPARTMENT OF COMMERCE. 1978. *State of Maryland coastal management program and final environmental impact statement.*

12. DOLAN, R. 1973. Barrier islands: Natural and controlled. In *Coastal geomorphology,* ed. D. R. Coates, pp. 263–78. Binghamton, New York: Publications in Geomorphology, State University of New York.

13. WILKINSON, B. H., and McGOWEN, J. H. 1977. Geologic approaches to the determination of long-term coastal recession rates, Matagordo Peninsula, Texas. *Environmental Geology:* 359–65.

14. LARSEN, J. I. 1973. *Geology for planning in Lake County, Illinois.* Illinois State Geological Survey Circular 481.

15. BUCKLER, W. R., and WINTERS, H. A., 1983. Lake Michigan bluff recession. *Annals of the Association of American Geographers* 73(1): 89–110.

16. BAUMANN, D. D., and SIMS, J. H. 1974. Human response to the hurricane. In *Natural hazards: Local, national, global,* ed. G. F. White, pp. 25–30. New York: Oxford University Press.

17. ISLAM, M. A. 1974. Tropical cyclones: Coastal Bangladesh. In *National hazards: Local, national, global,* ed. G. F. White, pp. 19–25. New York: Oxford University Press.

18. BAKER, E. J. 1974. Attitudes toward hurricane hazards on the Gulf Coast. In *Natural hazards: Local, national, global,* ed. G. F. White, pp. 30–36. New York: Oxford University Press.

19. CARTER, R. W. G., and OXFORD, J. D. 1982. When hurricanes sweep Miami Beach. *Geographical Magazine* VLIV, no. 8: 442–448.

20. ROWNTREE, R. A. 1974. Coastal erosion: The meaning of a natural hazard in the cultural and ecological context. In *Natural hazards: Local, national, global,* ed. G. F. White, pp. 70–79. New York: Oxford University Press.

THREE

Human Interaction With The Environment

I n this latter part of the twentieth century, Americans are a metropolitan people. Today, nearly 70 percent of the American people live in or near metropolitan areas that have populations greater than 50,000. It is expected that by the year 2000, 85 percent of Americans will be urban residents. Therefore, metropolitan population growth is now a basic feature of the transformation from agrarian to industrial life in the United States.

The evolution of urban areas has been from farm to small town, to city, to large metropolitan areas, to what is now referred to as the "urban region." An *urban region* is an area with at least one million people, in which there is a zone of metropolitan areas with intervening counties never far from a city[1]. In 1920, there were 10 urban regions that contained approximately one-third of the total population of the country. By 1970, there were 16 urban regions with three-fourths of the population, and by the year 2000, there will probably be at least 23 urban regions with five-sixths of the nation's people. If these projections are correct, 54 percent of all Americans will be living in the two largest urban regions—41 percent in the metropolitan belt extending from the Atlantic seaboard and westward past Chicago, and 13 percent in the California region between San Francisco and San Diego.

It is projected that by the year 2000, urban regions will occupy one-sixth of the total land area of the continental United States, and the environmental impact of this urbanization will be staggering. Urbanization renders a large portion of the soils within the urbanized area impervious, modifies rivers, and otherwise changes the previous landscape to suit human use. This process need not totally destroy the land, however. If growth is properly planned and directed, and if regular review and reevaluation of the goals and objectives are conducted, urban areas should be able to maintain at least a rough balance between acceptable environmental degradation and constructive use of the environment.

The stages of conversion from a rural to an urban area are rural, early urban, middle urban, and late urban[2]. During the *rural* stage, the landscape may consist of its virgin state and/or modest to intensive cultivation. Characteristically, there may be occasional farm buildings and dwellings which draw their water supply from wells, springs, streams, or ponds. Sewage is disposed of through outhouses, cesspools, or, perhaps, septic tanks, and garbage may be fed to domestic animals. The *early urban* stage is characterized by city-type homes built on small to large plots of land in a semiorderly to orderly manner. Schools, churches, and shopping facilities are interspersed. The water supply is generally drawn from individual wells. Rubbish is buried or burned and sewage is disposed of in septic tanks or cesspools. The middle urban stage is characterized by large-scale housing developments. There are more schools and shopping centers and some industrial sites. During this stage, systems to provide centralized water systems and sewers to dispose of waste waters may be built. Solid wastes may be collected by trucks. This is the first stage in which environmental degradation from urbanization is easily recognizable. The *late urban* stage is characterized by a large number of homes, apartments, commercial and industrial buildings, sidewalks, and parking lots. A large part of the land surface is made impervious or nearly so. Sanitary waste water systems and storm sewers attempt to remove human and industrial wastes and storm water runoff. In this stage, environmental degradation is obvious to many and may, or has, become a serious threat.

Chapters 9 through 11 consider the relations between people and our physical environment. Because these relations are often most stressed in urban areas, special attention is given to the problems stemming from human activities and inactivities that create our special urban landscape. Several relationships between hydrology and human use are reviewed in Chapter 9. Especially important here are shallow aquifer contamination, sediment pollution, and stream channelization.

Chapter 10 is concerned with the treatment and disposal of wastes (particularly sanitary landfills), ocean dumping, radioactive wastes, septic systems, and hazardous chemical wastes. Recent advancements involving the relations between health and the fields of geology and geography are discussed in Chapter 11. This subject promises to become more significant as our understanding of the relations among earth processes, earth materials, and health problems increases.

[1] COMMISSION ON POPULATION GROWTH AND THE AMERICAN FUTURE. 1972. *Population and the American future.* Washington, D.C.: U.S. Government Printing Office.

[2] SAVINI, J., and KAMMERER, J. C. *Urban growth and the water regimen.* U.S. Geological Survey Water Supply Paper 1591A.

Chapter 9

Hydrology and Human Use

The more obvious and well-known detrimental aspects of human use of the surface water, groundwater, and atmospheric water in the hydrologic cycle include fouled rivers, persistent "suds" on rivers or from wells, and unusual odors from surface water or groundwater. Problems with water supplies and waste waters were among the first environmental problems to be recognized and attacked on a scientific basis. Traditionally, water and waste waters have been considered and treated as separate entities. Although convenient and understandable, such segmentation of complex subjects can compound problems, obscure solutions, and confuse the general public. It has been only within the last 50 years that the interrelationships among and between surface water, groundwater, and atmospheric water have been determined in a reasonable, acceptable manner, but all too frequently, they are still considered as separate entities. Consequently, many local, state, and federal agencies split jurisdictional responsibilities along artificial lines. Fortunately, during the last decade, there has been a distinct trend to bridge these jurisdictional splits.

The fields of engineering, geology, biology, chemistry, and meteorology have each progressed through philosophical ponderings and debates on certain aspects of the "nature of things" in the founding stage of their particular science. Only within this century have the various disciplines joined together to investigate and pursue quantification of elusive segments of the hydrologic cycle.

Water is one of the most abundant of the important renewable resources on earth. The hydrosphere or total world water is approximately 1.4 billion cubic kilometers, sufficient, if the earth were smooth, to cover it to a uniform depth of several kilometers. In other words, on a global scale, water abundance is no problem; water is a heterogenous resource that can be found in either liquid, solid, or gaseous form at a number of locations at or near the surface of the earth. Depending upon the specific location of water, the residence time may vary from a few days to many thousands of years (Table 9.1). More than 99 percent of the total of water on earth is unavailable or unsuitable for beneficial human use because of its salinity (sea water) or location (ice caps and glaciers). Thus, the amount of water for which all the people on earth compete is much less than 1 percent of the total. In spite of this, compared to other resources, water is used in tremendous quantities. In recent years the total amount of water used each year on earth is approximately 1,000 times the world's total production of minerals, including petroleum, coal, metal ores, and non-

213

TABLE 9.1
The world's water supply (selected examples); data from U.S. Geological Survey source.

Location	Surface Area (km)^2	Water Volume (km)^3	Percentage of Total Water	Water: Estimated Average Residence Time
Oceans	361,000,000	1,230,000,000	97.2	Thousands of years
Atmosphere	510,000,000	12,700	0.001	9 days
Rivers and Streams	—	1,200	0.0001	2 weeks
Groundwater: shallow, to depth of 0.8 km	130,000,000	4,000,000	0.31	Hundreds to many thousands of years
Lakes (fresh water)	855,000	123,000	0.009	Tens of years
Ice caps and glaciers	28,200,000	28,600,000	2.15	Up to tens of thousands of years and longer

metals. This adds up to nearly 800 metric tons per person per year on a worldwide basis for all water use excluding hydropower (1). Because of its great abundance, water is generally a very inexpensive resource. On the other hand, because the quantity and quality of water that is available at any particular time is highly variable, statistical statements concerning the cost of water on a global basis are not particularly useful. Shortages of water have led and continue to lead to serious economic disruption and suffering, apparently with increasing frequency because of drought or other reasons (2). Therefore, a more detailed discussion of hydrology and water availability, use, pollution, and management is certainly worthwhile. In addition, we will discuss problems associated with land use—sediment pollution, off-road vehicles, and the channelization controversy.

SURFACE WATER HYDROLOGY

Surface water is the portion of precipitation that dislodges soil and rock particles upon impact with the land surface (Figure 9.1), is rejected from infiltrating, runs overland in rivulets or streams to rivers, physically and chemically picks up minerals, and erodes the

FIGURE 9.1
Raindrops strike the earth with enough force to tear apart unprotected soil and separate soil particles from one another. They wash the soil and generally move soil particles downslope. (Photos courtesy of U.S. Department of Agriculture.)

earth. These waters function within the hydrologic and rock cycles simultaneously. They drain the land, but not uniformly or consistently. Variations reflect differences in the amounts of precipitation, evaporation, transpiration, and other climatic, soil, vegetation, and geologic conditions. The runoff values include groundwaters and bank storage waters released to the rivers.

The amount of surface-water runoff from a watershed or a drainage basin and the amount of sediment carried vary significantly. Both of these characteristics of surface waters are composite results of the interactions between the rock and hydrologic cycles. The many and varied effects are produced by the diverse climatic, geologic, physiographic, and land-use characteristics of the particular drainage basin, together with variations of these factors with time. For instance, even the most casual observer is aware of the difference in the amount of sediment carried by a stream in flood stage compared to what appears to be carried at low-flow conditions.

The principal geologic factors affecting surface-water runoff and sedimentation include rock and soil type, mineralogy, degree of weathering, and structural characteristics of the soil and rock. Fine-grained, dense, clay soil and exposed correlative rock types with few fractures generally have little infiltration capacity; hence, the runoff of precipitation falling upon such materials is comparatively rapid. Conversely, sandy and gravelly soils, well-fractured rocks, and soluble rocks such as those in karst areas generally have high infiltration capacities and absorb a larger amount of precipitation. This condition retards overland runoff to steams until the soils become saturated or the rate of precipitation exceeds the infiltration capacity of the soils.

The structural characteristics of the soil and rock affect the runoff and sedimentation characteristics in several ways. Structure often controls the direction of flow, gradient, and drainage pattern of surface waters and groundwaters. Structural control of surface waters is common in areas where the bedrock is fractured or folded, providing variations in resistance to erosion. Similar control of the direction of movement and the point of discharge of groundwaters is possible, as illustrated in Figure 9.2. There, a portion of the infiltrated waters (recharge to the groundwaters) is diverted from one surface watershed (drainage basin) to an entirely different stream basin. This figure also explains why groundwater basins do not automatically coincide with surface watersheds as is frequently assumed.

Physiographic factors that affect runoff and sediment transport include climate, shape of the drainage basin, relief and slope characteristics, the basin's orientation to prevailing storms, and drainage pattern (spatial arrangement of the natural stream channels).

Climatic factors that influence runoff and stream sedimentation include the type of precipitation that occurs, the intensity with which precipitation occurs, the duration of precipitation with respect to the total annual climatic variation, and the types of storms (whether cyclonic or thunderstorm).

The *shape of a drainage basin* is greatly affected by the geologic conditions, and a principal effect of basin shape upon runoff and sedimentation is its role in governing the rate at which water is supplied to the main stream. Basins that are long and narrow and have

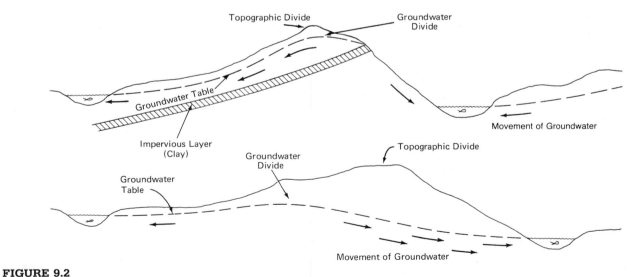

FIGURE 9.2
Idealized diagrams showing that the topographic divide may not necessarily coincide with the groundwater divide. Hence, a portion of the infiltrated waters is diverted from one surface watershed to an entirely different stream basin.

short tributaries receive the flow (especially stream flows) from the tributaries much more rapidly than basins that have long, sinuous tributaries.

The factors of *relief* and *slope* are interrelated; the greater the relief, the more likely the stream is to have a steep gradient and a high percentage of steep, sloping land. Relief and slope are also important because they affect the rate of overland flow, the velocity of water in a stream, the rate at which water infiltrates the soil or rock, and the rate at which surplus soil and groundwater enter a stream.

Orientation of the stream basin to prevailing storms influences the rate of flow, the peak flow, the duration of surface runoff, and the amount of transpiration and evaporation losses. The latter is a factor because of the influence of basin orientation on the amount of heat received from the sun and exposure to prevailing winds.

The *drainage net* is important in its effect on the efficiency of the drainage system. It is controlled by soil, rock, and topographic features. The characteristic shape of a hydrograph from a drainage basin varies with the type of drainage net. For instance, in a well-drained, efficient drainage basin, surface runoff from a rainstorm concentrates quickly to a flood peak. Discharge from such an efficient stream is commonly described as *flashy*. Conversely, in a stream that has relatively small variation between high and low flows, storm waters tend to run off more slowly, and the stream is therefore considered sluggish and less efficient. Figure 9.3 illustrates the difference in the shape of hydrographs from similar storms for an efficient drainage basin and a less efficient one.

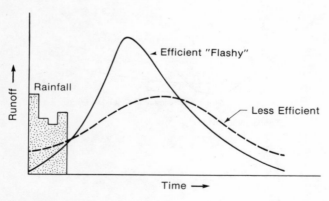

FIGURE 9.3
Idealized diagram illustrating the difference between an efficient drainage basin that removes surface runoff quickly and a less efficient one that takes a longer time to remove the water. Many natural stream basins have less efficient hydrographs, whereas artificial channels tend to be more efficient but may damage the environment.

Figure 9.4 shows the water resource regions of the U.S., and Table 9.2 shows the variations in the natural sediment yield for relatively small river basins. The amount of sediment carried by rivers as a part of their work within the rock cycle varies with geologic, climatic, topographic, physical, vegetative, and other conditions. Hence, some rivers are consistently and noticeably different in their clarity and appearance, as can be inferred from Table 9.3. Although this table reflects varying degrees of human influence, it demonstrates the sizable variations of sediment load per unit area in various parts of the world. For instance, on the average, the Lo River of China carries nearly 200 times more suspended load than does the Nile River of Egypt. In the United States, the Mississippi is not as "muddy" as the Missouri and the Colorado rivers.

The general relationship between size of drainage basin and sediment load suggests that, as basin size increases, the sediment yield per unit area decreases (Table 9.4). This relationship results from the increase in probability of sediment storage and deposition with increased basin size, the fact that smaller basins tend to be steeper (which increases the energy available for erosion), and the decreased probability of total basin coverage by a single storm event with increased basin size.

TABLE 9.2
Estimated ranges in sediment yields from drainage areas of 260 square kilometers or less.

Region	Estimated Sediment Yield (metric tons/km^2/yr)*		
	High	Low	Average
North Atlantic	4,240	110	880
South Atlantic-Gulf	6,480	350	2,800
Great Lakes	2,800	40	350
Ohio	7,391	560	2,780
Tennessee	5,460	1,610	2,450
Upper Mississippi	13,660	40	2,800
Lower Mississippi	28,760	5,460	18,220
Souris-Red-Rainy	1,650	40	175
Missouri	23,470	40	5,250
Arkansas-White-Red	25,760	910	7,710
Texas-Gulf	8,140	320	6,310
Rio Grande	11,700	530	4,550
Upper Colorado	11,700	530	6,310
Lower Colorado	5,670	530	2,100
Great Basin	6,240	350	1,400
Columbia-North Pacific	3,850	120	1,400
California	19,510	280	4,550

*The range in high to low values reflects different years with different discharges and ability to erode and transport sediment.

Source: *The Nation's Water Resources*, Water Resources Council, 1968.

FIGURE 9.4
Water resources regions of the United States. (From Water Resources Council, *The Nation's Water Resources,* 1968.)

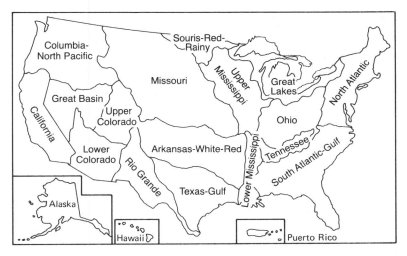

TABLE 9.3
Some major rivers of the world ranked by sediment yield per unit area.

River	Drainage Basin (10^3 km²)	Sediment Load per Year (tons/km²)
Amazon	5,776	63
Mississippi	3,222	97
Nile	2,978	37
Yangtze	1,942	257
Missouri	1,370	159
Indus	969	449
Ganges	956	1,518
Mekong	795	214
Yellow	673	2,804
Brahmaputra	666	1,090
Colorado	637	212
Irrawaddy	430	695
Red	119	1,092
Kosi	62	2,774
Ching	57	7,158
Lo	26	7,308

Source: Data from Holman, 1968.

GROUNDWATER HYDROLOGY

Groundwater is the portion of precipitation that penetrates the soil, moves through the zone of aeration and capillary fringe, and finally reaches the water table and becomes part of the groundwater system. Other sources of groundwater include water that infiltrates from surface waters, including lakes and rivers, storm water retention or recharge ponds, and waste-water treatment systems such as cesspools and septic tank drain fields.

The movement of water to the water table and through aquifers is an integral part of both the rock cycle and the hydrologic cycle. For example, the water may dissolve minerals from material it moves through and deposit them elsewhere as cementing material, producing sedimentary rocks. The water may also transport sediment and/or microorganisms, or it may seemingly do none, some, or all of these chores. What actually occurs varies with the chemical and physical characteristics of the water as it infiltrates through the biological environments and the soil-horizon environments above the water table and moves through the groundwater system below the water table.

Movement of groundwater in the hydrologic cycle may render the water more or less suitable for human use. Natural processes may even render the water toxic; for instance, water may become toxic through the dissolving of sufficient amounts of one or more elements or minerals, such as arsenic. Most often, groundwaters dissolve a mixture of minerals and some gases which, at worst, are nuisances to some human uses. Iron as ferrous hydroxide, calcium carbonate (which creates so-called hardness), and hydrogen sulphide (which creates a "rotten egg" odor) are examples.

Vast amounts of potable (drinkable) groundwaters are available, although not uniformly distributed, and are not always as readily available for human use as the concentrated population in an area might desire. Three problems concerning how people perceive the availability of groundwater and our understanding of this resource are apparent. First, people tend to assume that water is available when, where, and in the amounts they want. We turn on a faucet and expect water—it is someone else's responsibility to see that we have it. Second, since groundwater is out of sight, it is out of mind and/or mysterious. Third, groundwater is not as easily measured quantitatively as surface water. Therefore, precise quantitative values of groundwater reserves are not available, and we rely on estimates of the probable reserves. These estimates are

TABLE 9.4

Arithmetic average of sediment-production rates for various groups of drainage areas in the United States. These data illustrate that as the size of a drainage basin (watershed) increases, the sediment production per unit area decreases.

Watershed-Size Range (km²)	Number of Measurements	Average Annual Sediment-Production Rate (m³/km²)
Under 25	650	1,810.3
25–250	205	762.2
250–2,500	123	481.2
Over 2,500	118	238.2

Source: From *Handbook of Applied Hydrology*, by Ven Te Chow. Copyright 1964 by McGraw-Hill. Used with permission of McGraw-Hill Book Company.

similar to those of petroleum or ore reserves, but groundwater reserves are dynamic, renewable, and constantly fluctuating. That is, some groundwater is in storage; some discharges to our surface waters, providing water to maintain the dry-season flow of perennial streams and springs; and some is lost in evapotranspiration. How much groundwater is in a particular location at any one time defies accurate measurement with available technology and data.

The long residence time for groundwaters (Table 9.1) reflects the deep, insulated type of storage aquifers provide. Not all groundwater takes hundreds of years to rejoin the other, more rapidly moving parts of the hydrologic cycle, but most of it is well below the influence of transpiration by plants and evaporation into the atmosphere. Where it is not that deep, it is most susceptible to evapotranspiration, discharge to streams, and use and/or abuse by humans. The latter is of increasing concern because of the potential long-term damage to this resource and the increasing need to use it as water needs increase.

WATER SUPPLY*

The water supply at any particular point on the land surface of the earth is dependent upon a number of factors involving the hydrologic cycle. Of particular importance are the rates of precipitation, evaporation, stream flow, and subsurface flow. In addition, people's various uses of water may also be significant. One concept that is useful to understanding water supply is the *water budget*. On a continental scale, the water budget for the coterminous United States is shown in Figure 9.5. The amount of water in vapor form that passes

---

*Much of the discussion concerning water supply and use is summarized from U.S. Water Resources Council, 1978, *The Nation's Water Resources, 1975–2000*, v. 1.

over the United States every day is approximately 152,000 million cubic meters (40,000 billion gallons per day [bgd]) and of this, approximately 10 percent falls as precipitation in the form of rain, snow, hail, or sleet. Approximately two-thirds of the precipitation evaporates quickly or is transpired by vegetation. The remaining one-third, or about 5,510 million cubic meters per day (1,450 bgd), enters the surface or groundwater storage systems; flows to the oceans or across the nation's boundaries; is consumed; or evaporates from reservoirs. Unfortunately, only a portion of this water can be developed for intensive uses because of natural variations in precipitation that cause either floods or droughts. Thus, only about 2,565 million cubic meters per day (675 bgd) is considered to be available 95 percent of the time (2).

On a regional scale, it is critical to consider annual precipitation and runoff patterns to develop regional water budgets. Figures 9.6 and 9.7 illustrate the variability for the coterminous United States in terms of precipitation and runoff. Potential problems with water supply can be predicted in areas where average precipitation and runoff are relatively low, generally in the southwestern and Great Plains regions of the United States as well as some of the intermontane valleys in the Rocky Mountain area. The theoretical upper limit of surface water supplies is often taken to be the mean annual runoff, assuming it could be successfully stored. Unfortunately, this is not possible, because of evaporative losses from large reservoirs and the limited number of suitable sites for reservoirs. As a result, shortages in water supply are bound to occur in areas with low precipitation and runoff, and strong conservation practices will be necessary to insure an adequate supply (2).

Because annual variations in stream flow may be substantial, even areas characterized by high precipitation and runoff may suffer periodically from droughts. For example, dry years in 1961 and 1966 in the northeastern United States and in the western United States in 1976 and 1977 produced serious water shortages. Fortunately, in the more humid eastern United States, stream flow tends to vary less than in other regions, and thus the threat of drought is correspondingly less (2).

Groundwater is an important source in many parts of the United States, and nearly half the country's population uses groundwater as a primary source for drinking water. Fortunately, the total amount of groundwater available in the United States is enormous, accounting for approximately 20 percent of all water withdrawn for consumption. Within the coterminous United States, the amount of total groundwater within 0.8 km of the land surface is estimated to be between 125,000 and 224,000 km³. To put this figure in

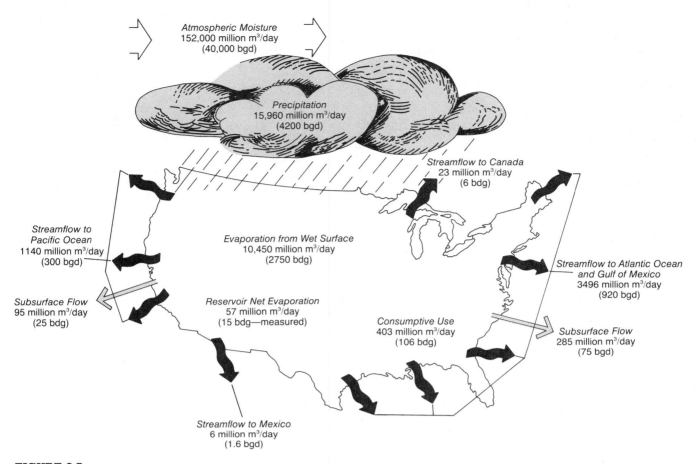

Atmospheric Moisture
152,000 million m³/day
(40,000 bgd)

Precipitation
15,960 million m³/day
(4200 bgd)

Streamflow to Canada
23 million m³/day
(6 bdg)

Streamflow to
Pacific Ocean
1140 million m³/day
(300 bgd)

Evaporation from Wet Surface
10,450 million m³/day
(2750 bdg)

Streamflow to Atlantic Ocean
and Gulf of Mexico
3496 million m³/day
(920 bgd)

Subsurface Flow
95 million m³/day
(25 bdg)

Reservoir Net Evaporation
57 million m³/day
(15 bdg—measured)

Consumptive Use
403 million m³/day
(106 bgd)

Subsurface Flow
285 million m³/day
(75 bgd)

Streamflow to Mexico
6 million m³/day
(1.6 bgd)

FIGURE 9.5
Water budget for the coterminous United States. (From U.S. Water Resources Council, 1978, *The Nation's Water Resources, 1975–2000.*)

perspective, the lower estimate is about equal to the total discharge of the Mississippi River during the last 200 years. Unfortunately, because of the cost of pumping and exploration, much less than the total quantity of groundwater is considered to be an available supply. Figure 9.8 shows the major aquifers in the coterminous United States capable of yielding more than 0.2 m³ per minute (50 gallons per minute) (2).

Protecting groundwater resources is an environmental problem of particular public concern because so many people derive their domestic water supplies from groundwater. The residency time for groundwater aquifers is often measured in hundreds to thousands of years; therefore, once aquifers are damaged by pollutants, it may be difficult or impossible to reclaim them for continued use. Aquifers are also very important because approximately 30 percent of the stream flow in the United States is supplied by groundwater that emerges as springs or other seepages along the stream channel. This phenomenon is known as *base flow* and is responsible for the low flow or dry season flow of

most perennial streams. Therefore, maintaining high quality groundwater is important in maintaining good quality stream flow.

In many parts of the country, groundwater is mined or overdrafted because withdrawal from wells exceeds natural inflow. Groundwater overdraft is a serious problem in the Texas-Oklahoma-High Plains area, California, Arizona, Nevada, New Mexico, and isolated areas of Louisiana, Mississippi, Arkansas, and the south Atlantic-Gulf region. In the Texas-Oklahoma-High Plains area alone, the overdraft amount is approximately equal to the natural flow of the Colorado River (2). In order that groundwater resources be managed rather than simply used, plans should be devised to withdraw groundwater within limits of known fluctuations or natural replenishment rates. However, groundwater in substantial amounts may need to be withdrawn during a time of drought and can then be replenished by artificial recharge or other means during wetter periods. Careful management is desired so that the overall resource is not adversely stressed.

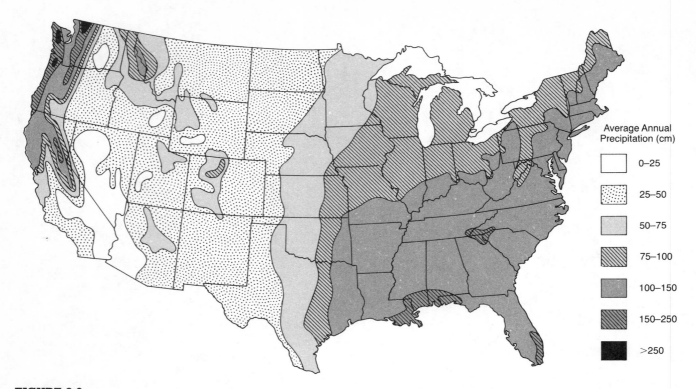

FIGURE 9.6
Average annual precipitation for the United States. (From U.S. Water Resources Council, 1978, *The Nation's Water Resources, 1975–2000.*)

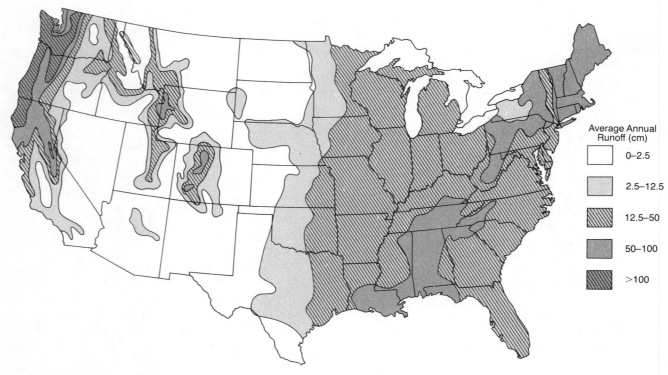

FIGURE 9.7
Average annual runoff for the United States. (From U.S. Water Resources Council, 1978, *The Nation's Water Resources, 1975–2000.*)

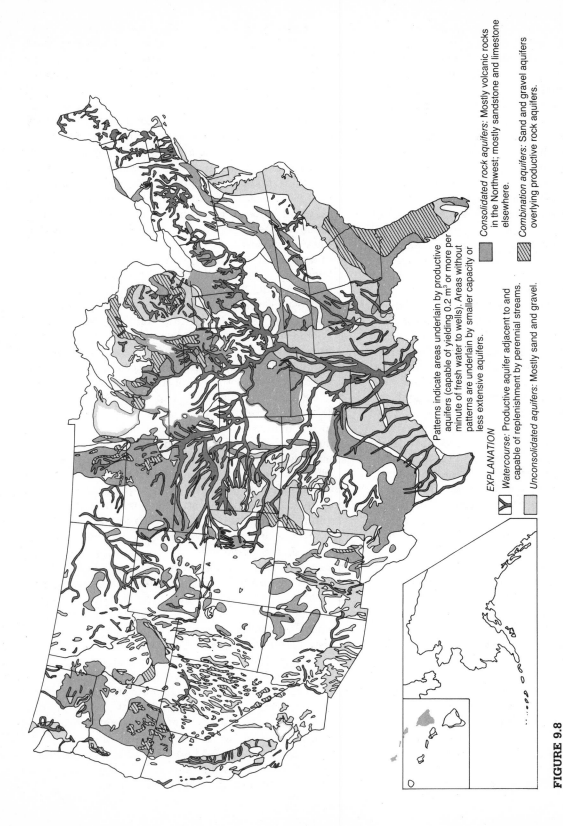

Patterns indicate areas underlain by productive
aquifers (capable of yielding 0.2 m³ or more per
minute of fresh water to wells). Areas without
patterns are underlain by smaller capacity or
less extensive aquifers.

EXPLANATION

Watercourse: Productive aquifer adjacent to and
capable of replenishment by perennial streams.

Unconsolidated aquifers: Mostly sand and gravel.

Consolidated rock aquifers: Mostly volcanic rocks
in the Northwest; mostly sandstone and limestone
elsewhere.

Combination aquifers: Sand and gravel aquifers
overlying productive rock aquifers.

FIGURE 9.8

Major aquifers in the United States. (From U.S. Water Resources Council, 1978, *The Nation's Water Resources, 1975–2000*.)

221

WATER USE

When discussing water use, we must distinguish between *instream* and *offstream* uses.

Instream water uses refer to the water required by a stream or river to provide for navigation, hydroelectric power generation, fish and wildlife, or recreational uses. There is presently controversy surrounding what flow must be maintained to insure that the stream environment is not degraded to the extent that instream uses are damaged. Controversies result because fish and wildlife require certain discharges for maximum biological productivity, whereas generating hydroelectric power requires more fluctuations in discharges to match power needs; navigational needs require still other discharges, and the discharge necessary to move the sediment load in a river may require yet another pattern of flow. Figure 9.9 shows some of these conflicting demands in an idealized diagram. A major problem concerns how much water may be removed from a stream or river and transported to another location without damaging the stream system. This is a particular problem in the Pacific Northwest, where anadromous fish (steelhead and salmon) are on the decline, partly as result of loss of habitat and damage associated with timber harvesting. At the same time, demands are being made on the stream to withdraw water for reservoir systems to feed the cities of southern California. California is somewhat unique in that two-thirds of the runoff in the state occurs north of San Francisco where there is a surplus of water, whereas two-thirds of the water use occurs south of San Francisco where there is a deficit. In recent years, the California Water Project and the Central Valley Project have been involved with the movement of tremendous amounts of water from north to south. The problems of instream uses, however, are not unique to California, as many large cities in the country must seek water from farther and farther away. New York City, for example, has imported water from nearby areas for more than 100 years. We can expect that in the coming years, as more and more water is needed for cities and agricultural purposes, intensive argument and debate will center on instream water use. A fruitful area of research is thus open to those who wish to evaluate precisely what flows are necessary to maintain the natural river system.

Offstream uses are those we often refer to when we speak of water use, and mean withdrawal of a certain amount of water from its source, either groundwater aquifer or surface water. Water that is consumed (consumptive use) refers to that part of the water resource that is withdrawn and subsequently evaporated, transpired, incorporated into agricultural products, or consumed by people and livestock. Consumptive use, therefore, refers to water that does not return to the stream or groundwater resource immediately after use (2).

Today, offstream water use (withdrawals) in the United States amounts to more than 1,700 million m^3 per day (450 bgd), or considerably more than one-half the average flow of the Mississippi River, which drains approximately 40 percent of the coterminous United States. Water consumption amounts to approximately 22 percent of that withdrawn, or 379 million m^3 per day (100 bgd). Figure 9.10 shows present sources and disposition of offstream water withdrawals, amount of withdrawals by the major users (self-supplied industry, agriculture, public supply, and rural), and consumptive use of fresh water. Figure 9.11 shows recent trends in water withdrawals, consumptive use, and population. Examining the information in these figures reveals that industry (self-supplied, including thermoelectric power) and agriculture (irrigation) withdraw 58 and 34 percent respectively, while public supply (domestic, public, commercial, and industrial uses) and rural (domestic and livestock uses) withdraw much less (3). In terms of consumption, agriculture (irrigation) consumes about 80 percent of the total fresh water used (3). Trends in withdrawal and consumption (Figure 9.11) suggest that self-supplied industry, especially power plants, has been responsible for much of the recent growth in water resource utilization. It is expected that by the year 2000, withdrawals will decrease while water consumption will increase slightly. We can hope that projected reductions in offstream withdrawals will be realized from conservation efforts, improved technology, and recycling of water. Increases from the present level of consumption will most likely reflect population growth and greater demand by industries, manufacturing, agriculture, and stream electric generation (2).

Examining water withdrawals and consumption for the United States has an important bearing on future water conservation, which strives to obtain the greatest use from existing supplies. For example, irrigation stands out as a prime candidate for conservation, because it involves large volumes of water and because agriculture accounts for by far the greatest consumptive use. Potential for saving in withdrawals is estimated at between 20 percent and 30 percent if

- Improved systems to deliver water are used, involving such aspects as lined and covered canals, computer monitoring and scheduling of releases, and a more integrated use of surface and ground waters
- Better on-farm water management practices such as night irrigation, improved irrigation systems (sprinklers, drip irrigation), and better land preparation with respect to application of water

FIGURE 9.9
Idealized diagram showing the variability in stream discharge best suited for some instream water uses. The line with a great number of fluctuations is that for hydroelectric power generation, indicating the flux in water as related to power demand and production.

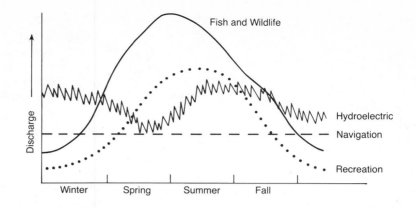

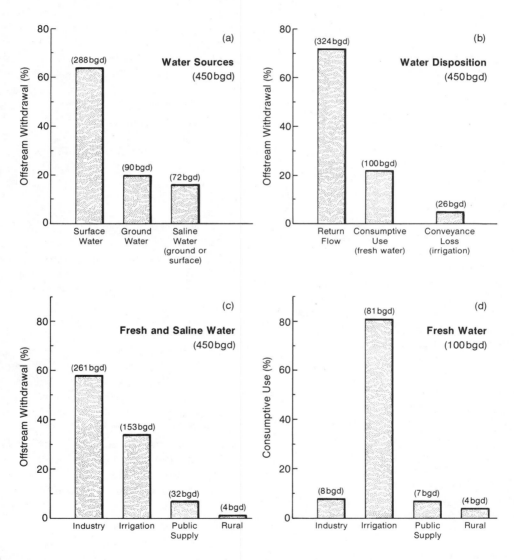

FIGURE 9.10
Water use in the United States for 1980: (a) sources; (b) disposition; (c) offstream withdrawals; and (d) consumptive use; "bgd" = billions of gallons per day. (Data from W. B. Solley, E. B. Chase, and W. B. Mann, IV, 1983, U.S. Geological Survey Circular 1001.)

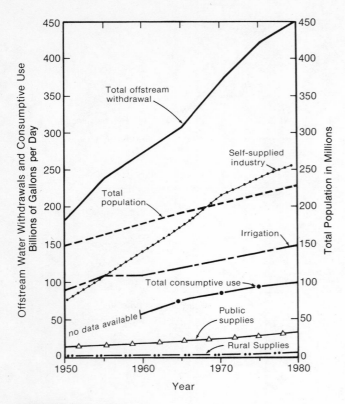

FIGURE 9.11
Offstream water withdrawals and consumptive use, 1950–
1980. (After W. B. Solley, E. B. Chase, and W. B. Mann, IV,
1983, U.S. Geological Survey Circular 1001.)

Domestic use of water accounts for only about 6 percent of the total national withdrawals, but nevertheless
this may be substantially reduced at a relatively small
cost with such practices as more efficient fixtures, reductions in outside water use, and watering at night.
Practicing water conservation in the area of steam
electric generation could also result in as much as a
25–40 percent reduction in withdrawal by means of different types of cooling towers that use less or no water.
Manufacturing and industry might reduce water withdrawals by increasing in-plant treatment and recycling
of water or developing new equipment and processes
that require less water (2). The field of water conservation is changing rapidly and it is expected that a
number of innovations will be forthcoming to reduce
the total withdrawals of water for various purposes. On
the other hand, consumption will continue to increase
as population increases and as there is increased demand from agriculture and industry (2).

SURFACE WATER POLLUTION

Water pollution refers specifically to degradation of water quality as measured by chemical, physical, or bio-

logical criteria. Degradation of water is generally in
terms of the water's intended use, departure from
norm, effects on public health, or ecological impacts.

From a public health or ecologic view, a pollutant
is any substance, biological or chemical, of which an
identifiable excess is known to be deleterious to other
desirable living organisms. Within this framework, excessive amounts of the heavy metals such as lead and
mercury, certain radioactive isotopes, nitrogen, phos-
phorous, sodium, and other useful, even necessary, elements, as well as certain pathogenic bacteria and viruses, are all pollutants. In some instances, a material
may be considered a pollutant to a particular segment
of the world's population although it is not harmful to
other segments. For instance, excessive nitrogen in the
proper compound is deleterious to infants, but much
less so, if at all, to adults. Similarly, excessive sodium,
as in salt, is not generally harmful, but can be to people
on diets that restrict salt intake for medical reasons.

Water pollutants are generally categorized as
being emitted from either a point or non-point source.
Point sources are discrete and confined, such as pipes
that empty into streams or rivers from industrial or municipal sites. In general, point-source pollutants from
industries are controlled through on-site treatment or
disposal and are regulated by permit. Municipal point
sources are also regulated by permit. Most point
sources in urban areas result from combined sewer
systems in older cities in the northeastern and Great
Lakes areas of the United States. These sewer systems
combine storm water flow with municipal waste, and,
during times of intense precipitation, urban storm runoff may exceed the capacity of the sewer, causing it to
back up and overflow, delivering pollutants to nearby
surface waters.

Non-point sources are generally diffused and intermittent, influenced by factors such as land use, climate, hydrology, topography, and geology. Non-point
sources are difficult to control and come from either
urban or rural areas. Common urban non-point sources
include urban runoff that contains all sorts of pollutants, from heavy metals to chemicals and sediment.
Rural sources of non-point pollution are generally associated with agricultural, mining, or forestry practices.

Many types of materials may pollute surface (or
ground) water. Our discussion will focus on a few selected examples. Sediment pollution, a major problem,
will be discussed along with land use and soil erosion.

Selected Water Pollutants
(Excluding Sediment)

Oxygen-demanding wastes (BOD) are produced by a
variety of sources, and many are associated with agri-

cultural practices that release organic material into surface waters. The amount of oxygen required for the oxidative decomposition of material in water measured in milligrams per liter of oxygen consumed over five days at 20°C is defined as the biochemical oxygen demand (BOD). Approximately 33 percent of all BOD in streams results from agricultural activities, but urban areas, particularly with combined sewer systems, also add considerable BOD to streams. The threshold for water pollution alert identified by the Council on Environmental Quality for BOD is a dissolved oxygen content of less than 5 mg per liter of water. Dissolved oxygen content is used because as the BOD increases, the dissolved oxygen decreases. Figure 9.12 is an idealized diagram illustrating the effect of BOD on dissolved oxygen content for a stream in which the BOD has resulted from sewage input. We can recognize three zones: the pollution zone with a high BOD and reduced dissolved oxygen content of the water; an active decomposition zone in which organisms are reducing the biological oxygen demand; and, finally, a recovery zone in which the dissolved oxygen increases and the BOD is much reduced. Many streams have the capability to degrade organic waste after it enters the stream. Problems result when the stream is overloaded with biological oxygen-demanding waste that overpowers the natural cleansing function of the stream.

Pathogens and fecal coliform bacteria are examples of other biological pollutants. Pathogens may consist of water-borne diseases such as cholera or typhoid. The fecal coliform bacteria are one measure of biological pollution, and the threshold for pollution used by

the Council on Environmental Quality is 200 cells of fecal coliform bacteria per 100 milliliters of water.

Nutrients are associated with water pollution, particularly with respect to cultural eutrophication (overfertilization of a water body), which produces algae blooms in ponds, lakes, and slow-moving streams and rivers. The primary pollutant is phosphorus, which is released from agricultural sources such as farm fields and feed lots or urban sewage treatment plants (Figure 9.13). Most of the phosphorus (66 percent) released to streams comes from agricultural sources (2).

Synthetic organic and inorganic compounds are a serious source of water pollution. Many are classified as hazardous chemical wastes and form as a by-product of producing chemicals for a variety of purposes. (See Chapter 10 for further details on hazardous chemical wastes.)

Oil discharged into surface water, usually the ocean, has been associated with major pollution problems. Several large oil spills from submarine oil drilling operations have made news headlines in recent years, such as the 1969 oil spill in the Santa Barbara Channel, California, and the 1979 Yucatán Peninsula spill in Mexico, the world's largest to date, which spewed out about three million barrels of oil before being capped in 1980. Both spills were caused by oil wells blowing out; both caused damage to beaches and marine life when the oil drifted ashore. Favorable winds averted total disaster on Texas beaches and inland wetlands, since the oil touched the shore only briefly. Santa Barbara was not so lucky. Oil is also released into the environment from large ships; and oil tankers have been shipwrecked, releasing all or part of their cargo of oil. More recently, during the Iran-Iraq war, an oil slick originating from a bombed coastal oil refinery reportedly threatened a Saudi desalinization plant.

Heavy metals can be dangerous pollutants. Mercury, zinc, and cadmium are often deposited in the bottom of stream channels along with sediment. If deposited on flood plains, the heavy metals may become incorporated in plants, food crops, and animals.

Radioactive materials in water can be dangerous pollutants. Of particular concern is the possible long-term exposure of humans to low dosages of radioactivity. There are two schools of thought on radioactivity: one holds that there is a threshold beyond which damage results; the second argues that any exposure to radioactivity above background radiation may be dangerous. This particular issue is unresolved to date.

Thermal pollution refers to hot water emissions, primarily from industry and power plants. The discharged water often is several degrees warmer than the surrounding water and thus can hold less oxygen. Warmer water may also facilitate the growth of unde-

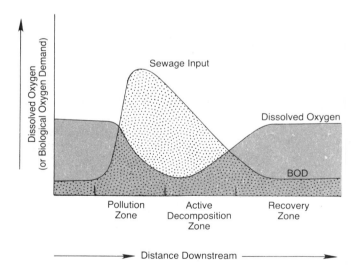

FIGURE 9.12
Idealized diagram showing the relationship between dissolved oxygen and biochemical oxygen demand (BOD) downstream after input of sewage into a stream.

FIGURE 9.13
Relationship between land use and average nitrogen and phosphorus concentration in streams (milligrams per liter). (From Council on Environmental Quality, 1978, Annual Report.)

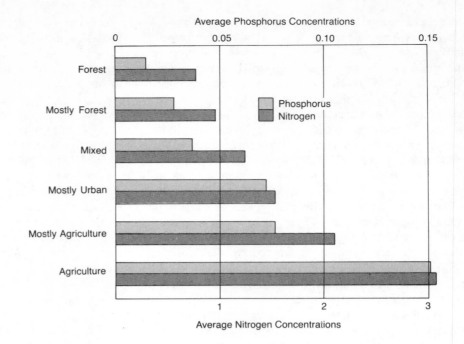

Average Phosphorus Concentrations

Average Nitrogen Concentrations

sirable forms of life, including different types of water plants and fish. On the other hand, during the winter, the warm water may attract and hold certain fish species desired by fishermen.

This list of pollutants is certainly not complete. Others that could be discussed are beyond the scope of this book. Table 9.5 lists some of the thresholds used by the Council on Environmental Quality in the analysis of water pollution. These six items may be considered indicators of water quality.

TABLE 9.5
Thresholds used in Council on Environmental Quality Analysis of National Water Quality.

Indicator	Abbreviation	Threshold Level
Fecal Coliform Bacteria	FC	200 cells/100 ml[a]
Dissolved Oxygen	DO	5. mg/l[b]
Total Phosphorus	Tp	0.1 mg/l[c]
Total Mercury	Hg	2. μg/l[d]
Total Lead	Pb	50. μg/l[d]
Biochemical Oxygen Demand	BOD$_3$	5. mg/l[e]

l = liter; ml = milliliter; mg = milligram; μg = microgram.

[a] Criteria level for "bathing waters" from EPA "Redbook."

[b] Criteria level for "good fish populations" from EPA "Redbook."

[c] Value discussed for "prevention of plant nuisances in streams or other flowing waters not discharging directly to lakes or impoundments" in EPA "Redbook."

[d] Criteria level for "domestic water supply (health)" from EPA "Redbook." Criteria level for preservation of aquatic life is much lower.

[e] Value chosen by CEQ.

Source: Council on Environmental Quality, 1979 Annual Report.

Since the 1960s there has been a serious attempt in the United States to reduce water pollution and thereby increase water quality. The basic assumption is that people have a desire for safe water to drink, to swim in, and to use for agricultural and industrial purposes. At one time, water quality near major urban centers was considerably worse than it is today, and in at least one instance, an American river was inadvertently set on fire. In recent years there have been a number of encouraging success stories, perhaps the best known of which concerns the Detroit River. In the 1950s and early 1960s, the Detroit River was considered a dead river, after becoming an open dump for sewage, chemicals, garbage, and all matter of urban trash. Tons of phosphorus were discharged into the river each day and a film of oil up to 0.5 cm thick was often present. Aquatic life was seriously damaged and thousands of ducks and fish were killed. Although no one would say today that the Detroit River is a clean stream, the improvement resulting from industrial and municipal pollution control has been considerable. Oil and grease emissions were reduced by 82 percent between 1963 and 1975, and now the shoreline is usually clean. Phosphorus and sewage discharges have also been greatly reduced. Fish once again can live in the Detroit River. Other success stories include New Hampshire's Pemigewasset River, North Carolina's French Broad River, and the Savannah River in the southeastern United States. These examples are evidence that water pollution abatement is providing positive results (4).

The Hudson River assessment and cleanup of PCBs, polychlorinated biphenyls that have a chemical structure similar to DDT and Dioxin, is another good example of people's determination to clean up our rivers. PCBs were used mainly in electrical capacitors and transformers; discharge of the chemicals from two outfalls on the Hudson River began about 1950 and terminated in 1977. Approximately 295,000 kg of PCBs are believed to be present in Hudson River sediments. Concentrations in the sediment are as high as 1000 parts per million near the outfalls, compared to fewer than 10 parts per million several hundred km downstream at New York City (5). The U.S. Food and Drug Administration permits fewer than 2.5 parts per million PCBs in dairy products, and the New York state limit for drinking water is 0.1 part per billion. PCBs are carcinogenic and known to cause disturbances of the human liver, nervous system, blood, and immune response system. They are also nearly indestructible in the natural environment and become concentrated in the higher parts of the food chain—thus the concern! Water samples in the 240 km tidal reach of the Hudson River have yielded average PCB concentrations ranging from 0.1 to 0.4 part per billion, but PCBs are concentrated to much higher levels in some fish. As a result, fishermen on the lower Hudson have suffered a significant economic impact from the contamination, because nearly all commercial fishing was banned, and sport fishing was greatly reduced (5,6).

The cleanup plan for the Hudson River, which will cost nearly $30 million, involves removing and treating contaminated river sediment. Dredging in "hot spots" where the concentration of PCBs is greater than 50 parts per million should be completed in the mid 1980s. Dredging will greatly reduce the time necessary for the river to clean itself by the natural process of sediment transport to the ocean.

GROUNDWATER POLLUTION

Approximately one-half of the people in the United States today depend upon groundwater as their source of drinking water. We have long believed that groundwater is, in general, pure and safe to drink, so it is disconcerting to learn that groundwater may, in fact, be quite easily polluted by any one of several sources (Table 9.6), and that the pollutants may be difficult to recognize even when they are toxic or potentially deadly. The hazard associated with a particular groundwater pollutant depends upon the volume of pollutant discharged; the concentration or toxicity of the pollutant; the concentration of the pollutant in the aquifer; persistence of the pollutant in the environment; and degree of exposure to people or other organisms (7).

Numerous examples of chemical and bacterial pollution of soil water and aquifers have been documented in recent years. Probably the best known example comes from Niagara Falls, New York, and the "Love Canal," where burial of chemical wastes has caused serious water pollution and health problems.

There are significant differences in the physical, geologic, and biologic environments associated with groundwater pollution compared with those of surface-water pollution. In the latter, hydraulics of flow and availability of oxygen and sunlight (which may tend to degrade pollutants), along with the rapidity with which processes of dilution and dispersion of pollutants occur,

TABLE 9.6
Classification of sources of groundwater pollution and/or contamination.

Wastes		Nonwastes
Sources designed to discharge waste to land and/or groundwater	Sources that may accidentally discharge waste to land and groundwater	Sources that may discharge a contaminant (not a waste) to land and ground matter
Spray irrigation	Surface impoundments	Buried product storage tanks and pipelines
Septic systems, cesspools, etc.	Landfills	Accidental spills
Land disposal of sludge	Animal feedlots	Highway deicing salt stockpiles
Infiltration or percolation basins	Acid water from mines	Ore stockpiles
Disposal wells	Mine spoil piles and tailings	Application of highway salt
Brine injection wells	Hazardous chemicals waste disposal sites	Agricultural activities

are markedly different than for groundwater, where the opportunity for bacterial degradation of pollutants is generally confined to the soil or a meter or so below the surface. Furthermore, the channels through which groundwater moves are often very small and variable, as shown schematically in Figure 9.14. For these reasons, it is obvious that the rate of movement is much reduced (except, perhaps, in large solution channels within limestones), and the opportunity for dispersion and dilution is limited. In addition, groundwater is usually devoid of oxygen, which is helpful in killing aerobic types of microorganisms but may provide a ''happy home'' for anaerobic varieties.

Like surface water, groundwater flows in response to the force of gravity. The preference for horizontal groundwater motion is related to layering characteristics of consolidated and unconsolidated sediments. For fractured, folded, and faulted rocks, however, structural characteristics can be the dominant factors that control the direction of flow.

The ability of most soils and rocks physically to filter out solids, including pollution solids, is well recognized. However, this ability varies with different sizes, shapes, and arrangements of filtering particles, as evidenced in the use of selected sands and other materials in water filtration plants. Also known, but perhaps not so generally, is the ability of clays and other selected minerals to capture and exchange some elements and compounds when they are dissociated in solutions as positively or negatively charged elements or compounds. Such exchanges, along with sorption and precipitation processes, are important in the capture of pollutants. These processes have definable units of capacity and are reversible. They also can be overlooked easily in designing facilities to correct pollution problems by relying on soils and rocks of the geologic environment for treatment; this oversight can result in possible groundwater pollution. This is especially significant in land application of waste waters.

Much conjecture, debate, and research have gone on about groundwater pollution. The principal concerns of groundwater pollution are the introduction of chemical elements, compounds, and microorganisms that do not occur naturally in aquifers that can be or are used for drinking water. Because of the degradation of the aquifer, and also because of difficulty in detection, the long-term residency, and the difficulty and expense of aquifer recovery, strong arguments can be made that no wastes or possible pollutants should be allowed to enter any part of the groundwater system. This is an impossible dream. Rather, the answer lies in knowing more about how natural processes treat wastes to insure that when soil and rocks are not capable of treating, storing, or recycling wastes, we can develop processes to make the pollutants treatable, storable, or recyclable.

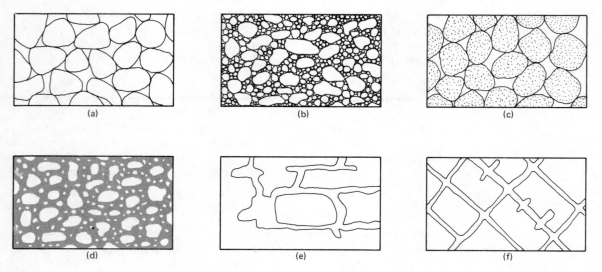

FIGURE 9.14
Idealized diagram showing types of soil and rock openings and their relationships to porosity. Well-sorted sediment with high porosity (a). Poorly sorted sediment with low porosity (b). Well-sorted sediment in which the particles themselves are also porous, resulting in a higher total porosity (c). Well-sorted sediment in which the openings between grains are partially filled with cementing material, reducing the porosity (d). Rock in which porosity is increased because of solutional openings (e). Rock that is porous because of open fractures (f). (From Meinzer, U.S. Geological Survey Water Supply Paper 489, 1923.)

Studies and case histories of groundwater pollution generally show that microorganisms seldom survive for distances greater than 30 meters because either they are filtered out or are killed by the hostile environment (8). They appear to have survived in those geologic environments where porosity and permeability have high values, such as gravel, coarse sand, or fracture and cavern systems in limestone (9,10). The spectrum of possible problems includes contamination by microorganisms from sewage disposal (11,12), chromium pollution from industrial processes (13), contamination from oil-field brines and related industrial wastes (14), and pollution caused by disposal of pharmaceutical wastes (15).

Aquifer pollution is not solely the result of disposal of wastes on the land surface or in the ground. Overpumping or mining of groundwater such that inferior waters migrate from adjacent aquifers or the sea also cause contamination problems. Hence, human use of public or private water supplies can accidentally result in aquifer pollution. Problems of this nature are best shown by examples of intrusion of salt water into freshwater supplies in coastal areas of New York, Florida, and California, although they are not confined to those areas.

Figure 9.15 illustrates the general principle of salt water intrusion. The groundwater table generally is inclined toward the ocean and a wedge of salt water inclined toward the land is present. Thus, with no confining layers, salt water near the coast may be encountered at depth. Because fresh water is slightly less dense than salt water (1.000 compared to 1.025 grams per cubic centimeter), a column of fresh water 41 centimeters high is needed to balance 40 centimeters of salt water. A more general relationship is that the depth of salt water below sea level is 40 times the height (H in Figure 9.15) of the water table above sea level. If wells are drilled, then a cone of depression develops in the fresh water table which may allow intrusion of salt water as the interface between fresh and salt water rises in response to the loss of mass of fresh water (see Figure 9.15).

Long Island, New York, is a good example of an area where groundwater problems are of major concern. Two counties on the island, with a population of several million people, are entirely dependent on groundwater for their water supply. In particular, Nassau County has been carefully studied (Figure 9.16) (16). Two major problems associated with groundwater in Nassau County are intrusion of salt water and shallow aquifer contamination. Figure 9.16 shows the general movement of groundwater under natural conditions for Nassau County. Salty groundwater is restricted from inland migration by the large wedge of fresh water moving beneath the island.

Groundwater resources of Nassau County are in two main aquifers. The upper aquifer is composed of young glacial deposits which yield large amounts of water at depths less than 30 meters. Below the glacial deposits are older marine sedimentary rocks consisting

FIGURE 9.15
Idealized drawings showing how salt water intrusion might occur. The upper drawing shows the groundwater system near the coast under natural conditions and the lower shows a well with both a cone of depression and a cone of ascension. If pumping is intensive, the cone of ascension may be drawn upward, delivering salt water to the well. The "H" and "$40\,H$" represent the height of the fresh water table above sea level and the depth of salt water below sea level respectively.

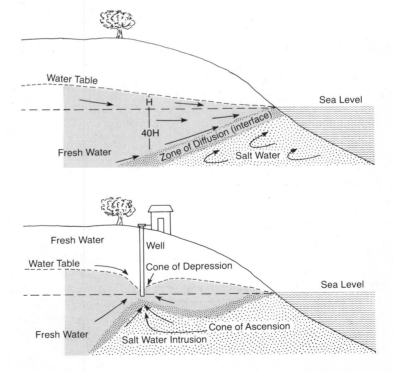

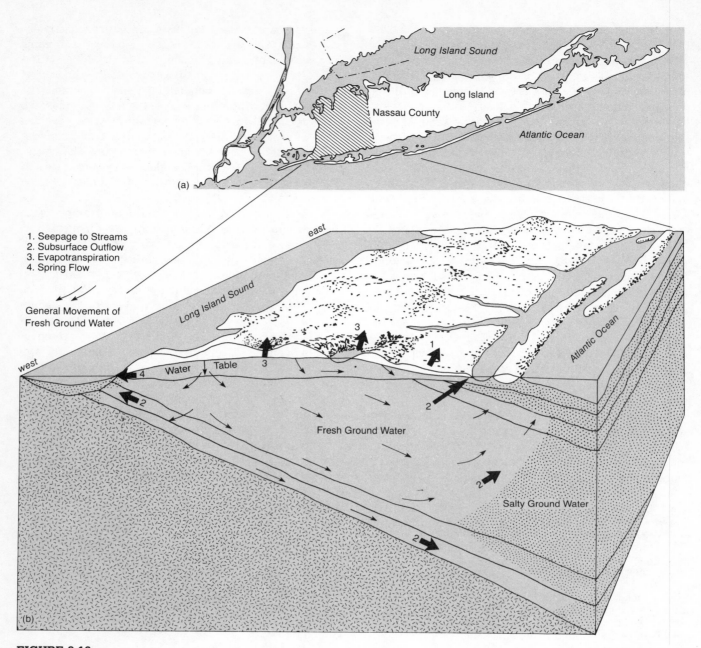

FIGURE 9.16
Idealized drawing showing the general movement of fresh groundwater for Nassau County,
Long Island, New York. (From U.S. Geological Survey Professional Paper 950, 1978.)

of interbedded sands, clays, and silts. Most of the fresh water in Nassau County is pumped from sandy beds at depths below 30 meters from this aquifer. Most of the water-bearing sands are confined by overlying clay and silt beds of low permeability and thus the water is under artesian pressure, which causes the water in wells to rise to within 15 meters of the surface (see Figure 3.21). In terms of the total volume of water and number of people who use it, the groundwater resource for Long Island is one of the world's largest (16).

Despite the huge quantities of water in the groundwater system of Nassau County, intensive pumping in recent years has caused water levels to decline as much as 15 meters in some areas. As groundwater is removed near coastal areas, the subsurface outflow to the ocean decreases, allowing salt water to migrate inland. Salt water intrusion in the deep aquifer has occurred in Nassau County. The mechanism of salt instrusion is more complex than that ideally shown in Figure 9.15. As fresh water from sandy beds in the deep aquifer are intensely pumped, salt water is drawn inland as a series of narrow wedges. Although the problem of salt water intrusion is not yet widespread, the salt water front is being carefully monitored as part of a comprehensive management program.

The most serious groundwater problem on Long Island is shallow aquifer pollution associated with urbanization. Identified sources of pollution in Nassau County include urban runoff, household sewage from cesspools and septic tanks, salt used to de-ice highways, and industrial and solid waste. Pollutants from these sources are introduced into the near-surface water supply and then migrate downward, especially in areas of intense pumping and declining groundwater levels. Figure 9.17 shows the extent of high concentration of dissolved nitrate in deep groundwater zones. The greatest concentrations are located beneath densely populated urban zones where water levels have dramatically declined, and nitrates from sources such as cesspools, septic tanks, and fertilizers are routinely introduced into the hydrologic environment.

Drilling deep wells, such as those for scientific and petroleum exploration, can also cause, and probably has accidentally and unwittingly caused, degradation of freshwater aquifers. This degradation occurs because of the unrealized hazards during the early phases of petroleum and other well drilling (especially for salt) and the lack of information about the total depths at which potable waters are known to occur in many parts of the United States and throughout the world.

Correction of aquifer contamination is not impossible, but it often seems so because of the complex nature of some wastes and the time, expense, and difficulties in locating, monitoring, and confining pollutants. Efforts are being made to improve the methods. Table 9.7 summarizes 118 cases of groundwater contamination from various parts of the United States, listing the type of contaminant, the method of detection, and the remedial action taken. The two most common remedial measures are pumping with subsequent treatment and direct removal of the contaminant. Each of these two methods was used in about 25 percent of the cases.

WATER MANAGEMENT

Luna Leopold suggests that we need a new philosophy of water management—one based on geologic, geographic, and climatic factors, as well as the traditional economic, social, and political factors. He argues that management of water resources cannot be successful

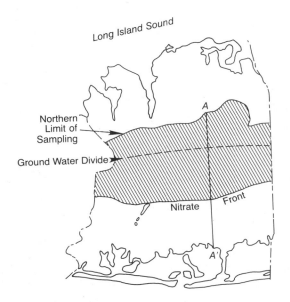

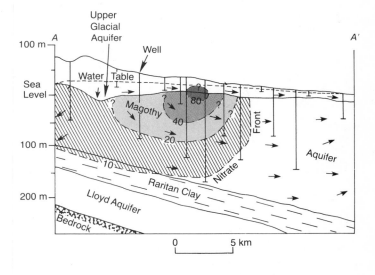

FIGURE 9.17
Extent of high concentration of dissolved nitrate in groundwater zone, Nassau County, Long Island, New York. The greatest concentrations are located beneath densely populated urban zones where water levels have dramatically declined and nitrates are more abundant due to urban waste disposal and horticulture practices. Contours shown in milligrams per liter of dissolved nitrate. (From U.S. Geological Survey Professional Paper 950, 1978.)

TABLE 9.7
Summary of case histories of
groundwater contamination.

Contaminant	Total	Cases affecting or threatening water supplies	Cases causing or threatening fire or explosion
Industrial wastes	40	26	2
Landfill leachate	32	7	0
Petroleum products	18	10	8
Chlorides (road salt and oil field brine)	11	9	0
Organic wastes	11	9	0
Pesticides	3	2	0
Radioactive wastes	1	0	0
Mine wastes	2	1	0
Totals	118	64	10

Means of Detection

Well contamination	58	Not mentioned	4
Investigation	32	Fumes in basement	1
Stream contamination	9	Fumes in ground	1
Spill on ground	5	Fumes in sewer line	1
Leak discovered	4	Animal deaths	1

Total 116

Remedial Action

Direct		Indirect	
Ground water pumped and treated	27	Extent of groundwater contamination determined:	44
Contaminated soil removed	8	Leading to remedial action	25
Trench installed	8	No further action	19
Artificial recharge employed	4	New water supply provided	16
Nutrients added	2	Action being considered	9
Source of contamination eliminated	26	Monitoring begun	2
Surface water collected and treated	7	Damages awarded	4
Landfill site closed	9	Charcoal filters installed	2
Site regraded	3	None mentioned	12
Total	94	Total	89

(From Lindor, D. E. and Cartwright, K., Illinois State Geological Survey, *Environmental Geology Notes* no. 81, 1977.)

as long as it is naively perceived primarily from an economic and political standpoint; however, this is how water use is approached. The term *water use* is appropriate because we seldom really *manage* water (17). The essence of Leopold's water management philosophy is summarized as follows:

Surface water and groundwater are both subject to natural flux with time. In wet years, there is plenty of surface water, and near-surface groundwater resources are replenished. During these years, we hope that our flood control structures, bridges, and storm drains will withstand the excess water. Each of these structures is designed to withstand a particular flow (for example, the twenty-year flood), which, if exceeded, may cause damage or flooding.

All in all, we are much better prepared to handle floods than deficiencies of water. During dry years, which must be expected even though they can not be

exactly predicted, we should practice specific strategies to minimize hardships. For example, some subsurface waters in various locations in the western United States are either too deep to be extracted economically or have marginal water quality. These waters may be isolated from the present hydrologic cycle and, therefore, not subject to natural recharge. Such water might be used occasionally when the need is great. However, advance planning to drill the wells and connect them to existing water lines is necessary if they are to be ready when the need arises. Another possible emergency plan may involve the treatment of waste water. Reuse of water on a regular basis might be too expensive, but advance planning to reuse treated water during emergencies might be a wise decision (17).

When dealing with groundwater that is naturally replenished in wet years, we should develop plans to use surface water when available and not be afraid to use groundwater in dry years; that is, the groundwater might be pumped out at a rate exceeding the replenishment rate in dry years. During wet years, natural recharge and artificial recharge (pumping excess surface water into the ground) will replenish the groundwater resources. This water management plan recognizes that excesses and deficiencies in water are natural, and we can plan for them.

Finally, there exists in most natural systems a harmony controlled by physical laws. Thus, water in rivers and groundwater have expected variations in flow that are dependent on geologic, geographic, and climatic factors which should be considered in managing water resources. Although we have the engineering capabilities to dam and channelize all rivers, to mine irreplaceable groundwater, and to move surface water thousands of kilometers, these technological activities should not be judged solely on the basis of their economic and political values. Rather, we should have a reverence for rivers and other natural systems to insure their and our survival (17).

LAND USE, HYDROLOGY, AND SOIL EROSION

The principal human influences affecting the pattern, amount, and intensity of surface-water runoff, erosion, and sedimentation are the varied land uses and manipulations of our surface water, such as artificial drainage and construction of detention and recharge ponds and dams. Such works are undertaken for several reasons and are justified most frequently as "improvements to the environment" or "for the betterment of civilization." Figure 9.18 summarizes estimated and observed variations in sediment yield under various historical changes in land use, and Figure 9.19 illustrates sediment production for various degrees of soil exposure as a function of drainage area and land use.

Figures 9.18 and 9.19 suggest that the effects of land-use change on the drainage basin and its streams may be quite dramatic. Streams in naturally forested or wooded areas are assumed to be stable, that is, without excessive erosion or deposition. For example, a land-use change that converts forested land for agricultural purposes generally increases the runoff and sediment yield (erosion) of the land. As a result, the streams become muddy and may not be able to transport all the sediment delivered to the channels. Therefore, the channels will aggrade (partially fill with sediment), possibly increasing the magnitude and frequency of flooding.

Urbanization

The change from agricultural, forested, or rural land to highly urbanized land is even more dramatic. First, during the construction phase, there is a tremendous increase in sediment production, which may be accompanied by a moderate increase in runoff. The response of streams in the area is complex and may include both channel erosion (widening) and aggradation, resulting in wide, shallow channels. The combination of increased runoff and shallow channel increases the flood hazard. Following the construction phase, the land is mostly covered with buildings, parking lots, and streets, and thus, the sediment yield drops to a low level. However, the runoff increases further because of the large impervious areas and use of storm sewers, again increasing the magnitude and frequency of flooding. The streams respond to the lower sediment yield and higher runoff by eroding (deepening) their channels.

Off-Road Vehicles

Urbanization is not the only land-use change that causes increased soil erosion and hydrologic changes.

FIGURE 9.18
Graph showing the effect of land-use change on sediment yield in the Piedmont region of the eastern U.S. before the beginning of extensive farming and continuing on through a period of construction and urbanization. (After M. Gordon Wolman, *Geografiska Annaler* 49A, 1966.) (200 tons/sq km = 500 tons/sq mi)

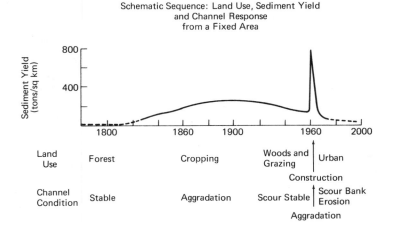

FIGURE 9.19

Relationship between sediment
yield and drainage area for various land uses. Notice that for a
given drainage area (say 26 km²)
the sediment discharge or yield
increases by several hundred
times as land use changes from
wooded to severely exposed. (After Putnam, U.S. Geological Survey Open File Report, 1972.)

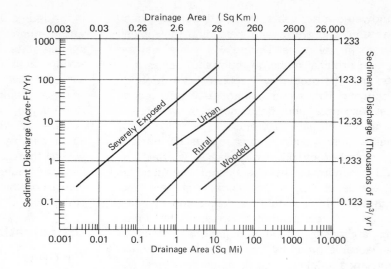

In recent years, the popularity and numbers of off-road
vehicles have increased enormously, and demand for
recreational areas to pursue this interest has led to serious environmental problems and conflicts among various users of public lands.

Widespread use of off-road vehicles (ORVs) is
causing significant impacts on the environment. There
are now more than 12 million ORVs, many of which are
invading the deserts, coastal dunes, and forested
mountains of many parts of the United States. The major areas of concern are soil erosion, changes in hydrology, and damage to plants and animals. The problem
is not small. A single motorcycle need travel only 7.9
kilometers to have an impact on 1,000 square meters,
and a four-wheel-drive vehicle has an impact over the
same area by traveling only 2.4 kilometers. On some
desert areas, the tracks produce scars that may remain
part of the landscape for hundreds of years (18,19).

ORVs cause direct mechanical erosion and facilitate wind and water erosion of materials loosened by
their passing. Runoff from ORV sites is as much as
eight times greater than for adjacent unused areas, and
sediment yields are comparable to those found on construction sites in urbanizing areas (19). Figure 9.20
shows an ORV site in the Mojave Desert. Motorcycles
and dune buggies nearly destroyed the vegetative
cover on the sandy soil, thus contributing to a 1973
dust storm that was visible on satellite imagery. Figure
9.21 shows eroded ORV trails in the San Gabriel Mountains of California. The accumulation of eroded sediment in the dry pond at the base of the hill is approximately 15 by 50 meters. In areas of intensive use, ORVs
are the dominant agents of erosion, turning vegetated
hills into eroding wasteland.

Hydrologic changes from ORV activity are primarily the result of near-surface soil compaction that
reduces the ability of the soil to absorb water. Further-

more, what water is in the soil becomes more tightly
held and thus less available to plants and animals.
Compaction of soils can be measured easily with a
penetrometer; the more compacted soils have a shal-

FIGURE 9.20
South end of the Shadow Mountains, Mojave Desert. Dune
buggies and motorcycles have destroyed the vegetative
cover, allowing the sandy soil to be removed easily by wind
and running water. This area contributed sediment to a dust
storm, recognized on satellite imagery. (Photo by H. G. Wilshire, courtesy of U.S. Geological Survey.)

lower depth of penetration under a given load. Figure 9.22 compares depth of penetration of a standard penetrometer for both in-tracks and out-of-tracks areas in the Mojave Desert. Interestingly, the tank tracks produced 15 to 35 years ago are still visible and the compacted soils have not recovered. Figure 9.23 shows bulk density and soil moisture variations with depth for in- and out-of-motorcycle trails near San Francisco, California. Only one example of each is illustrated, but numerous other sets of data all suggest that the near-surface bulk density is increased and moisture content decreased by ORV use (19). Compaction also changes the variability of the soil temperature (Figure 9.24). This is especially apparent near the surface, where daytime soil temperatures become hotter and nighttime soil temperatures lower.

Animals are killed or displaced and vegetation damaged or destroyed by intensive ORV activity. The damage is a result of a combination of soil erosion, compaction, temperature change, and moisture content change (19).

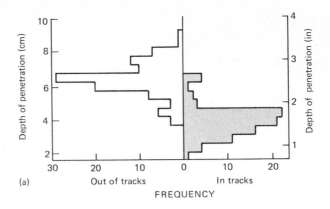

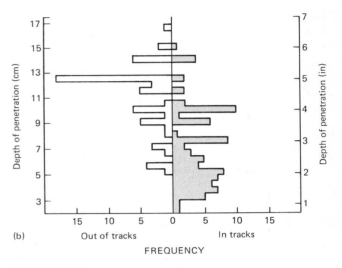

FIGURE 9.22
Histograms showing the frequency of penetrance in and out of off-road vehicle tracks in the Mojave Desert. Penetrance is inversely proportional to the soil compaction and directly proportional to soil strength. Graph (a), based on 178 measurements, is for tank tracks 15 to 35 years old. Graph (b), based on 146 measurements, is for recent motorcycle tracks. (Courtesy of H. G. Wilshire, U.S. Geological Survey.)

FIGURE 9.21
Soil erosion along off-road vehicle trails at Tejon, San Gabriel Mountains, California. Some of the eroded materials have filled in the dry pond (15 × 50 meters) at the base of the slope. (Photo by H. G. Wilshire, courtesy of U.S. Geological Survey.)

Management of ORVs is a difficult task. In 1972, the President issued an Executive Order requiring agencies in charge of managing public lands to adopt specific policies for ORV traffic. The object was to promote safety and minimize conflicts among the various uses of public lands (19).

There is little doubt that some land must continue to be set aside for ORV use. The problems are how much land should be involved, and how to minimize environmental damage. Sites should be chosen in closed basins with minimal soil and vegetation variation. The possible effects of airborne removal (by wind) must be evaluated carefully, as must the sacrifice of nonrenewable cultural, biological, and geological resources (18). A major problem remains: intensive ORV

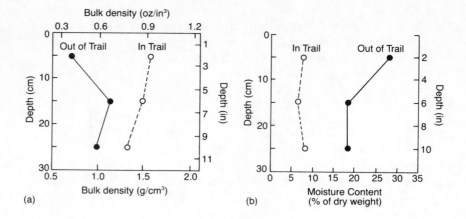

FIGURE 9.23
Relations between bulk density (a) and moisture content (b) with depth for in and out of off-road vehicle trails. Data from an ORV site near San Francisco, California. (Courtesy of H. G. Wilshire, U.S. Geological Survey.)

use is incompatible with nearly all other land use, and it is very difficult to restrict damages to a specific site. Material removed by mechanical, water, and wind erosion will always have an impact on other areas and activities (18,19).

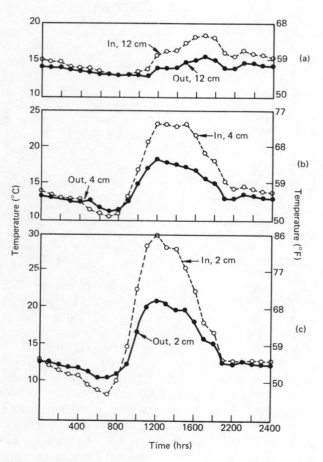

FIGURE 9.24
Diurnal temperature variations at three depths in and out of off-road vehicle trails near Rio Vista, California. (Courtesy of H. G. Wilshire, U.S. Geological Survey.)

An alternative to off-road vehicle use on public lands is legal off-road vehicle recreational activities on private lands. In Los Angeles County, for example, a site known as Indian Dunes (Figure 9.25) has been developed along the Santa Clara River valley. The site has more than 35 kilometers of trails, an international motocross obstacle course, an oval dirt track, hill climbs for motorcycles, and a sand racing basin for motorcycles. The site is managed somewhat as a downhill ski resort is managed, in that the runs are manicured, watered to keep down dust, and generally controlled. Theoretically, the erosion of slopes is minimized, since they are constantly being regraded following intensive ORV use. Indian Dunes might be criticized because it causes environmental degradation, but activity is limited to a particular area and steps are taken to try to minimize danger while providing legal opportunity for people to practice their chosen recreational activity. Certainly sites such as this are preferable to opening many hundreds or thousands of square kilometers to uncontrolled and unsupervised off-road vehicle activity.

SEDIMENT POLLUTION

Sediment is probably our greatest pollutant. In many areas, it chokes the streams, fills in lakes, reservoirs, ponds, canals, drainage ditches, and harbors, buries vegetation, and generally creates a nuisance that is difficult to remove. It is truly a resource out of place in that it depletes a land resource (soil) at its site of origin (Figure 9.26), reduces the quality of the water resource it enters, and may deposit sterile materials on productive croplands or other useful land (Figures 9.27 and 9.28) (20).

Most natural pollutional sediments consist of rock and mineral fragments, ranging in size from sand particles less than 2 millimeters in diameter to silt, clay, or very fine colloidal particles. Examples of the variations in sediment size found in reservoirs in different

FIGURE 9.25
Indian Dunes near Los Angeles, California, a managed off-road vehicle site.

climatic areas are shown in Table 9.8. Rates of storage depletion caused by sediment filling in ponds and reservoirs are shown in Table 9.9. These data suggest that small reservoirs will fill up with sediment in a few decades, while large reservoirs will take hundreds of years to fill.

Some pollutional sediments are debris resulting from the disposal of industrial, manufacturing, and public wastes. Such sediments include trash directly or indirectly discharged into surface waters. (Litter is not confined to roadsides and parks.) Most of these sedi-

ments are very fine-grained and difficult to distinguish from the naturally occurring sediments unless they contain unusual minerals or other particular characteristics.

The principal sources of artificially induced pollutional sediments are disruption of the land surface for construction, farming, deforestation, and channelization works. In short, a great deal of sediment pollution is the result of human use of the environment and civilization's continued change of plans and direction. It cannot be eliminated, but only ameliorated.

FIGURE 9.26
Severe soil erosion and loss of a valuable resource. (Photo by F. M. Roadman, courtesy of U.S. Department of Agriculture.)

FIGURE 9.27
Sediment pollution can damage productivity of farmland. (Photo by R. B. Brunstead, courtesy of U.S. Department of Agriculture.)

FIGURE 9.28
Sediment pollution can also spoil the visual amenities of the urban environment. (Photo by W. B. King, courtesy of U.S. Department of Agriculture.)

TABLE 9.8
Average-grain-size distribution of reservoir deposits. In general, semiarid and arid regions tend to produce more sand and less clay-size sediment than the more humid regions, reflecting the importance of chemical weathering in the latter areas.

Climatic Area	Drainage Area (km²)	Average-Grain-Size Components		
		Clay (%)	Silt (%)	Sand (%)
Humid				
W. Frankfort Reservoir, Ill.	10	19	78	3
Lancaster Reservoir, S.C.	23	28	56	16
Lake Bracken, Ill.	23	52	42	1
Lake Lee, N.C.	132	36	57	7
High Point Reservoir, N.C.	163	45	35	20
Lake Marinuka, Wis.	360	30	64	6
Lake of the Ozarks, Mo.	36,400	46	50	4
Subhumid				
Ardmore Club Lake, Okla.	10	40	24	36
Moran Reservoir, Kans.	13	52	44	4
Mission Lake, Kans.	21	27	17	56
Lake Merritt, Tex.	31	56	41	3
Lake Claremore, Okla.	145	51	46	3
Lake Brownwood, Tex.	4,014	70	28	2
Great Salt Plains Reservoir, Okla.	8,206	51	41	8
Semiarid				
Wellfleet Reservoir, Nebr.	39	9	14	77
Sheridan Reservoir, Kans.	1,134	85	15	
Tongue Reservoir, Wyo.	4,524	35	46	18
Canton Reservoir, Okla.	32,455	41	39	20
Arid				
Conchos Reservoir, N.M.	19,110	22	52	26
Lake Mead	436,800	28	25	47

Source: From *Handbook of Applied Hydrology,* by Ven Te Chow. Copyright 1964 by McGraw-Hill. Used with permission of McGraw-Hill Book Company.

TABLE 9.9
Summary of reservoir capacity and storage depletion of the nation's reservoirs.

Reservoir Capacity (acre-ft)[a]	Number of Reservoirs	Total Initial Storage Capacity (acre-ft)[a]	Total Storage Depletion		Individual Reservoir Storage Depletion		Average Period of Record yr
			(acre-ft)[a]	%	Average %/yr	Median %/yr	
0–10	161	685	180	26.3	3.41	2.20	11.0
10–100	228	8,199	1,711	20.9	3.17	1.32	14.7
100–1,000	251	97,044	16,224	16.7	1.02	.61	23.6
1000–10,000	155	488,374	51,096	10.5	.78	.50	20.5
10,000–100,000	99	4,213,330	368,786	8.8	.45	.26	21.4
100,00–1,000,000	56	18,269,832	634,247	3.5	.26	.13	16.9
Over 1,000,000	18	38,161,556	1,338,222	3.5	.16	.10	17.1
Total or average	968	61,239,020	2,410,466	3.9	1.77	.72	18.2[b]

Source: F. E. Dendy, "Sedimentation in the Nation's Reservoirs." *Journal of Soil and Water Conservation* 23,1968.

[a] 1 acre-ft = approximately 1234 m³.

[b] The capacity-weighted period of record for all reservoirs was 16.1 years.

FIGURE 9.29
Example of a sediment-and ero-sion-control plan for a commercial development. (Courtesy of Braxton Williams, Soil Conservation Service.) (384 to 400 ft = 117 to 122 m)

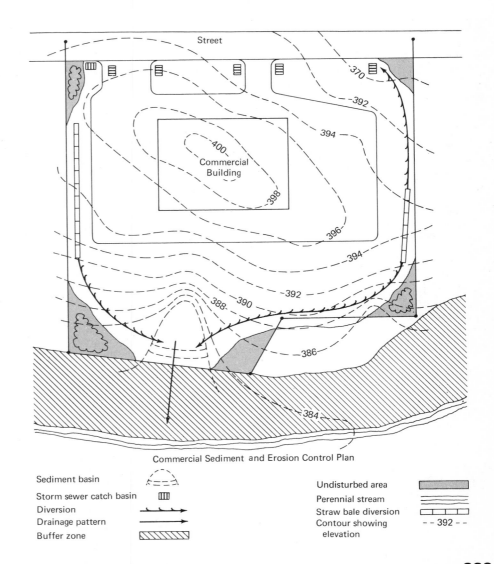

Commercial Sediment and Erosion Control Plan

Sediment basin		Undisturbed area	
Storm sewer catch basin		Perennial stream	
Diversion		Straw bale diversion	
Drainage pattern		Contour showing elevation	- - 392 - -
Buffer zone			

In this century in the United States, emphasis on soil and water conservation has grown considerably. The first programs emphasized stabilization of areas of excessive wind and water erosion, as well as flood control and irrigation water for arid and semiarid land. As the programs progressed, adverse environmental effects from some of the work, such as accelerated stream erosion and rapid reservoir siltation, were discovered. Simultaneously with implementation of new conservation programs, additional needs produced demands for broader and expanded use of completed improvements for recreation, water supply, and navigation, which necessitated research and development to find solutions to new or rediscovered problems. Design changes were developed and applied to correct the erosion and sedimentation problems. These changes included contour plowing, changes in farming practices, and construction of small dams and ponds to hold runoff or trap sediments.

Currently, the same basic problems—erosion control and sediment pollution—occupy a significant portion of the public's attention. Sources of the sediment have expanded, however, to include those from land development, highways, mining, and production of new, unusual chemical compounds and products, all in the interest of a better life. Sediment pollution affects rivers, streams, the Great Lakes, and even the oceans, and the problem promises to be with us indefinitely.

The solution to sediment pollution requires sound conservation practices, particularly in urbanizing areas where tremendous quantities of sediment are produced during construction. Figure 9.29 shows a typical sediment and erosion control plan for a commercial development. The plan calls for diversions to collect runoff and a sediment basin to collect sediment and keep it on the site, thus preventing stream pollution. A generalized cross section and plan (map) view of a sediment control basin are shown in Figures 9.30 and 9.31.

A study in Maryland demonstrates that sediment control measures can reduce sediment pollution in an urbanizing area (21). The suspended sediment transported by the northwest branch of the Anacostia River near Colesville, Maryland, with a drainage area of 54.6 square kilometers, was measured over a ten-year period from 1962 to 1972. During that time, urbanization (construction) within the basin involved about 3 percent of the area each year. The total urban land area in the basin was about 20 percent at the end of the ten-year study.

Sediment pollution was a problem because the soils are highly susceptible to erosion and there is sufficient precipitation to insure their erosion when not protected by a vegetative cover. Most of the sediment is transported during spring and summer rainstorms (21). A sediment-control program was initiated between 1965 and 1971, and the estimated sediment yield was reduced by about 35 percent. The basic sediment control principles were to tailor development to the natural topography, expose a minimum amount of land, provide protection for exposed soil, minimize surface runoff from critical areas, and trap eroded sediment on the construction site. Specific measures included scheduled grading to minimize the time of soil exposure, mulch protection and temporary vegetation to protect exposed soils, sediment diversion berms, stabilized waterways (channels), and sediment basins. This Maryland study concluded that even further sediment control can be achieved if major grading is scheduled during periods of low erosion potential and if better sediment traps are designed to control runoff during storms (21).

CHANNELIZATION

For more than 200 years, Americans have lived and worked on floodplains. Advantages and enticement to do so stem from the abundance of rich, alluvial soil found on the floodplain, conveniences such as abundant water supply and ease of waste disposal, and

FIGURE 9.30
Cross section of a sediment basin. Storm water carrying sediment runs into the sediment basin where the sediment settles out and the water filters through loose gravel and into a pipe outlet. Accumulated sediment is periodically removed mechanically. (After *Erosion and Sediment Control,* Soil Conservation Service, 1974.)

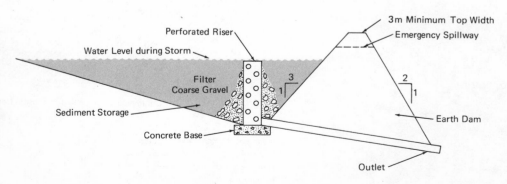

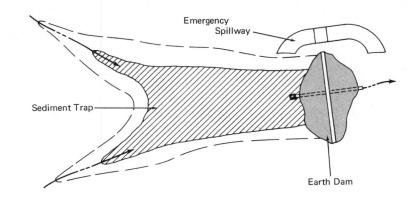

FIGURE 9.31
Plan view of a sediment basin. Sediment enters the basin from several different streams and by overland flow to be trapped behind the dam. At times of high flow, water may bypass the dam by way of the emergency spillway. Sediment basins must be cleaned out periodically to remain efficient. (From *Erosion and Sediment Control,* Soil Conservation Service, 1974.)

Emergency
Spillway

Sediment Trap

Earth Dam

proximity to communications, transportation, and commerce that developed along the rivers. Of course, building houses, industry, public buildings, and farms on the floodplain invites disaster, but floodplain residents have refused to recognize the natural floodway of the river for what it is: part of the natural river system. As a result, flood control and drainage of wetlands became prime concerns. It is not too great an oversimplification to infer that, as the pioneers moved West, they had a rather set procedure for modifying the land: first, clear the land by cutting and burning the trees, then modify the natural drainage. From that history came two parallel trends: an accelerating program to control floods matched by an even greater growth of flood damages (22).

Historically, the response to flooding has been to attempt to control the water by constructing dams, modify the stream by constructing levees, or even rebuild the entire stream to more efficiently remove the water from the land. Every new project has the effect of luring more people to the floodplain in the false hope that the flood hazard is no longer significant. However, we have yet to build a dam or channel capable of controlling the heaviest rainwaters, and when the water finally exceeds the capacity of the structure, flooding may be extensive (22).

Channelization of streams consists of straightening, deepening, widening, clearing, or lining existing stream channels. Basically, it is an engineering technique, with the objectives of controlling floods, draining wetlands, controlling erosion, and improving navigation (22). Of the four objectives, flood control and drainage improvement are the two most often cited in channel improvement projects.

Thousands of kilometers of streams in the United States have been modified, and thousands of more kilometers of channelization are being planned or constructed. Federal agencies alone have carried out several thousand kilometers of channel modification in the

last twenty years (23). Considering that there are approximately 5.6 million kilometers of streams in the United States, and the average yearly rate of channel modification is only a very small fraction of one percent of the total, the impact may appear small. Nevertheless, nearly 1,600 kilometers of channel per year is a significant amount of environmental disruption and, therefore, merits evaluation.

Channelization is not necessarily a bad practice; however, inadequate consideration is being given to its adverse environmental effects. In fact, the picture emerging is that little is known about these effects, and that little is being done to evaluate them (22).

Adverse Effects of Channelization

Opponents of channelizing natural streams emphasize that the practice is completely antithetical to the production of fish and wetland wildlife, and that, furthermore, the stream suffers from extensive aesthetic degradation. The argument is as follows: Drainage of wetlands adversely affects plants and animals by eliminating habitats necessary for the survival of certain species. Cutting trees along the stream eliminates shading and cover for fish, while exposing the stream to the sun, which results in damage to plant life and heat-sensitive aquatic organisms. Cutting of bottomland (floodplain) hardwood trees eliminates the habitats of many animals and birds and also facilitates erosion and siltation of the stream. Straightening and modifying the streambed destroys the diversity of flow patterns, changes peak flow, and destroys feeding and breeding areas for aquatic life. Finally, the conversion of wetlands with a meandering stream to a straight, open ditch seriously degrades the aesthetic value of a natural area (22). Figure 9.32 summarizes some of the differences between natural streams and those modified by people.

It is commonly believed that channelization increases the flood hazard downstream from the modified channel. Although in many cases this is true, it is far from the entire story. One study concluded that, on the contrary, increases of normal and peak flows from channelization are not particularly significant, and in some cases the flood peaks are actually reduced (23). This results partly because the contribution of runoff from the modified channels tends to be a small fraction of the total basin runoff; peak runoff from channelized streams may not coincide with basinwide runoff, and thus the quicker flow from a modified stream may pass prior to the natural flood flow from the entire basin, thereby reducing the normal aggregated peak flow; and many streams that are modified have gradients so low that no amount of straightening could significantly increase the downstream flow. However, we emphasize that channel modification, especially straightening, can increase flooding directly downstream from the project, and the problem may be compounded if there are a number of projects in the same basin.

Examples of channel work projects that have adversely affected the environment are well known. The Willow River in Iowa and Blackwater River in Missouri illustrate some of the possible adverse impacts on streams that are channelized.

FIGURE 9.32
Comparison of a natural stream with a channelized stream. (Modified after Corning, 1975. *Virginia Wildlife*, February.)

NATURAL STREAM

Suitable water temperatures:
  adequate shading;
  good cover for fish life;
  minimal temperature variation;
  abundant leaf material input.

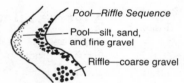

Pool—Riffle Sequence
—Pool—silt, sand, and fine gravel
Riffle—coarse gravel

Sorted gravels provide diversified habitats for many steam organisms.

CHANNELIZED STREAM

Increased water temperatures:
  no shading; no cover for fish life;
  rapid daily and seasonal temperature fluctuations; reduced leaf material input.

Mostly Riffle

Unsorted gravels;
  reduction in habitats; few organisms.

POOL ENVIRONMENT

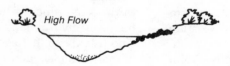

High Flow

Diverse water velocities:
  high in pools, lower in riffles. Resting areas abundant beneath banks, behind large rocks, etc.

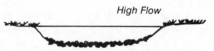

High Flow

May have stream velocity higher than some aquatic life can stand.
Few or no resting places.

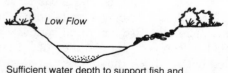

Low Flow

Sufficient water depth to support fish and other aquatic life during dry season.

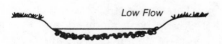

Low Flow

Insufficient depth of flow during dry season to support fish and other aquatic life. Few if any pools (all riffle).

The channelization of the Willow River in Iowa has been extensively studied (24,25). As early as 1851, the river was known for frequent flooding that caused extensive damage to crops. As a result, the valley was generally considered too risky for agriculture. At that time, the river was confined at low flow to a meandering channel about 1 meter below the floodplain. To alleviate flooding, channelization in three stages from 1906 to 1920 converted about 45 kilometers of the Willow River into a straight drainage ditch. This had the effect of producing a much shorter channel with a steeper gradient. Changes in the size and shape of the drainage ditch in the Willow River valley since construction ended in 1920 show that there has been progressive widening and deepening of the channel (Figure 9.33). In 1958, the channel was characteristically U-shaped with nearly vertical walls (Figure 9.34). In 1920, the ditch in this vicinity was only about 5 meters deep and less than 15 meters wide. In 1958, it was over 11 meters deep and 30 meters wide, as shown by the cross sections in Figure 9.33, and the excessive bank erosion is not unique along the ditch. The history is one of continuous enlargement since the project was completed.

The channelization designed for the Willow River was inadequate and resulted in excessive channel erosion. The drainage ditch response is a classic example of adverse channel change caused by artificial reshaping of a natural channel (25).

Channelization of the Blackwater River in Missouri has also resulted in environmental degradation, including channel enlargement, reduced biological productivity, and increase in downstream flooding (26). The natural meandering channel was dredged and shortened in 1910, nearly doubling the natural gradient. This change evidently initiated a cycle of channel erosion that is still in progress, and the channel has shown no tendency to resume meandering (26). Maximum increase in channel cross-sectional area exceeds one thousand percent, and, as a result, many bridges have collapsed and have had to be replaced. One particular bridge over the Blackwater collapsed because of bank erosion in 1930 and was replaced by a new bridge 27 meters wide. This bridge had to be replaced in 1942 and again in 1947. The 1947 bridge was 70 meters wide, but it too collapsed from bank erosion. Since channelization, erosion has progressed at an average rate of 1 meter in width and 0.16 meter in depth per year (26).

Biological productivity in the Blackwater River was reduced because of channelization. This reduction resulted because the channel was excavated in easily erodible shale, sandstone, and limestone, producing a smooth streambed that does not easily support bottom fauna after it has been disturbed. As a result, the channelized reaches contain far fewer fish than the natural reaches (26).

The downstream end of the channelization project of the Blackwater terminates where the rock type changes to a more resistant limestone, making dredging financially unfeasible. Since the channelized portion has a larger channel, it can carry more water without flooding than the natural, downstream channel. As a result, when heavy rains fall, runoff carried by the upstream channelization section is discharged into the unchannelized downstream section; and because the channel cross section is much less there, frequent flooding results (26).

The Blackwater River situation is a good example of possible trade-offs and unforeseen degradations caused by channelization. The upper reaches benefited from the use of more floodplain land; but this benefit must be weighed against loss of farmland to erosion, cost of bridge repair, loss of biological life in the stream, and downstream flooding. A reasonable trade-off might have been to straighten the channel less, still providing a measure of flood protection without causing such rapid environmental degradation. The question is, how much straightening can be done before inducing unacceptable damage to the riverine system?

FIGURE 9.33
Longitudinal (long) and cross-channel profiles of Willow Drainage Ditch during specific years from 1919 to 1960. [After R. U. Ruhe, "Stream Regimen and Man's Manipulation." In *Environmental Geomorphology*, ed. D. R. Coates (State University of New York, 1971) p. 1. (After R. B. Daniels, 1960.) Reprinted by permission.)] (30 m = 100 ft; 10 km = 6.2 mi)

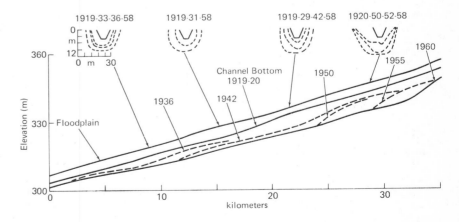

FIGURE 9.34
Willow Drainage Ditch near the mouth of Thompson Creek. The ditch here is about 11 meters deep and 30 meters wide. Note incipient development of meanders at low flow. (Photo courtesy of R. B. Daniels.)

Benefits of Channelization

Not all channelization causes serious environmental degradation, and in some cases, drainage projects can even be beneficial. Benefits are probably best observed in urban areas subject to flooding and in rural areas where previous land use has caused drainage problems. In addition, other examples can be cited to show where channel modification has improved navigation or reduced flooding and has not caused environmental disruption.

Many streams in urban areas scarcely resemble natural channels. The process of constructing roads, utilities, and buildings with the associated sediment production is sufficient to disrupt most small streams. A channel restoration project—to clean urban waste from the channel, allowing the stream to flow freely, and to protect the banks by not removing existing trees and, where necessary, planting additional trees and other vegetation—is often needed. The channel should be made as attractive as possible by allowing the stream to meander and, where possible, by providing for variable, low-water flow conditions—fast and shallow (riffle) alternating with slow and deep (pools). Where lateral bank erosion must be absolutely controlled, the outsides of bends may be defended with large stones known as riprap (Figure 9.35).

An attempt to improve the aesthetic appeal of an open cut ditch—Whiskey Creek near Grand Rapids, Michigan—emphasizes that engineers are beginning to consider channel restoration seriously. The idea was to make the open drain appear as a country stream meandering through substantial residential and commercial properties. Several of the channel's meander bends are shown during construction in Figure 9.36. Although it may be argued that open cut drains are unlikely to be considered pleasing sights under any circumstances, certainly the Whiskey Creek project is something out of the ordinary and is an improvement over a straight ditch designed only for flood control.

Another example of channel restoration is found in the coastal plain of North Carolina where many streams have been channelized. As part of a mosquito-abatement program in Onslow County, the state is attempting to drain some areas and thus destroy the insect's habitat. Examination of some of the streams before improvement reveals that they are nearly blocked by debris and logs. Evidence such as buried, hand-cut

FIGURE 9.35
Idealized diagram showing channel-restoration design criteria for urban streams using a variable cross-channel profile to induce scour and deposition at desired locations. (Modified after Keller and Hoffman, 1977, *Journal of Soil and Water Conservation*, vol. 32, no. 5.)

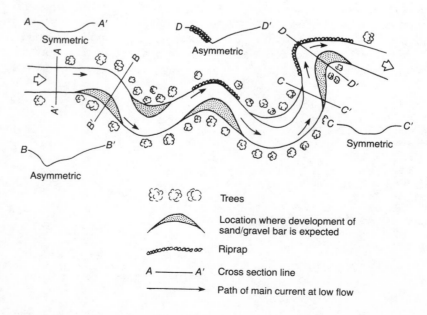

FIGURE 9.36
Channelization of Whiskey Creek near Grand Rapids, Michigan. This is an attempt to improve the aesthetic appeal of a stream project by constructing a series of meandering bends. (Photo courtesy of Williams and Works.)

cyprus logs and handmade cyprus shingles along the stream bank suggests that streams that were once free-flowing have been choked with debris left from logging practices years ago. The channel restoration involves excavating the old prelogging channel (Figure 9.37). Engineers on the project say the old channel can be followed by the high organic soils that form along the channel bank. The resulting restoration is a meandering stream that is allowed to flow freely again (Figure 9.38). The material dredged in restoring the channel is left on the bank and quickly supports new vegetation. Small cyprus seedlings are present only a few months after the work has been completed. Although this particular practice of channel work needs further study and evaluation, it seems to be a step in the right direction. It also appears to be popular with landowners who request that the work be done rather than having to be convinced that it should be, as is often the case with drainage work.

River restoration of the Kissimmee River in Florida may become the most ambitious restoration project ever attempted in the United States. Channelization of the river began in 1962; after nine years of construction at a cost of $24 million, the meandering river had been turned into a 83-kilometer-long straight ditch. Now, at what will probably exceed the original cost of channelization, the river may be returned to its original sinuous path. Work will begin on a 16 km experimental stretch of river, at a cost of about $1 million; if successful, restoration may be undertaken in other sections of the

river. The restoration is being seriously considered because the channelization failed to provide the expected flood protection, degraded valuable wildlife habitat, contributed to water quality problems associated with land drainage, and caused aesthetic degradation.

Channel modification on the Missouri River is an interesting example of river training in which the

FIGURE 9.37
Channel restoration on the coastal plain of North Carolina. The dragline operator is instructed to excavate the old prelogging channel as much as possible.

FIGURE 9.38
Channel restoration on the coastal plain of North Carolina. The channel produced is a meandering stream in which the vegetation soon becomes reestablished. This photograph was taken only a few days after the channel had been re-formed.

power of running water is used to maintain the desired channel. The project involves engineering improvements to produce a series of relatively narrow, gentle meanders rather than the previously existing succession of wide, straight reaches with a steep gradient and abrupt changes in flow direction. That is, before modification, the channel was semibraided in places and contained numerous islands and bars that formed as a result of deposition of large amounts of sediment. This was especially true downstream from the Platte River where it enters the Missouri River (25).

The engineering project that was started in the 1930s to induce the Missouri River to narrow and meander rather than braid involves construction of permeable pile dikes (Figure 9.39) that reduce the stream's velocity, causing deposition of sediment in desired locations. The water passing around the dikes, on the other hand, is constricted, causing an increase in velocity and scour in desired areas. As erosion and deposition progress, the dikes silt in, and natural growth of willows is established. A section of the river showing the improvements and results from 1930 to 1940 is shown in Figure 9.40. This project is one case where human interference has improved the area by providing additional land for agriculture and facilitating navigation (25). However, the dikes have caused a reduction in aquatic habitat by filling in sloughs and marshland downstream from the dikes (Figure 9.41). As a result, the Corps of Engineers have designed a new type of dike, with a notch at the top, that will allow more water to pass through and, it is hoped, will min-

imize the loss of valuable wildlife and fish habitat. Furthermore, there is concern that the river training has reduced the total channel capacity and is thus facilitating more frequent flooding.

The Future of Channelization

Although considerable controversy and justifiable anxieties surround channelization, it is expected that

FIGURE 9.39
Upstream view of the Treville Bend of the Missouri River. The permeable pile dikes on the inside of the bend effectively converge the flow of water toward the outside, where the faster water scours the channel. Thus, the power of running water is used to maintain the channel. (Photo courtesy of U.S. Army Corps of Engineers, Omaha District.)

FIGURE 9.40

Otoe Bend. The diagram on the left shows the channel as it was in 1930 with the 1940 design channel superimposed on it. The diagram on the right shows the systems of permeable dikes used to train the river. Notice that Schemmel Island has developed precisely where deposition was planned. [After R. V. Ruhe, "Stream Regimen and Man's Manipulation." In *Environmental Geomorphology,* ed. D. R. Coates (State University of New York, 1971) p. 16. Reprinted by permission.]

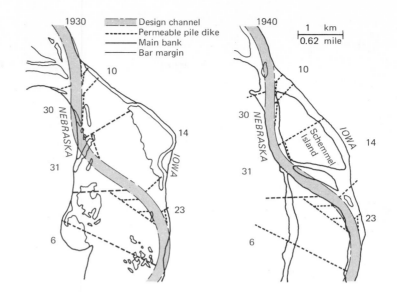

channel modification will remain a necessity as long as land-use changes, such as urbanization or conversion of wetlands to farmland, continue. Therefore, we must strive to design channels that reduce adverse effects. Effective channelization can be accomplished if project objectives of flood control or drainage improvement are carefully tailored to specific needs.

If the primary objective is drainage improvement in areas where natural flooding is not a hazard, then there is no need to convert a meandering stream into a straight ditch. Rather, design might involve cleaning the channel and maintaining a sinuous stream. In addition, for gravel-bed streams with a low gradient, a series of relatively deep areas (pools) and shallow areas (riffles) might be constructed by changing cross-sectional shape, asymmetrically to form pools and symmetrically to form riffles, as found in natural streams

(Figure 9.42). Where this is possible, the resulting channel would tend to duplicate nature rather than be alien to it. In addition, cutting trees along the channel bank would be minimized and new growth would be encouraged. This plan would tend to minimize adverse effects by producing a channel that is more biologically productive and aesthetically pleasing.

Channel design for flood control is more complicated because natural channels are maintained by flows with a recurrence interval of one to two years, whereas flood-control projects may need to carry the 25- or even 100-year flood. A 100-year channel cannot be expected to be maintained by a 2-year flood. Such a channel would probably be braided and choked with migrating sandbars. Therefore, it is suggested that a pilot channel be constructed where possible (Figure 9.43). A meandering channel designed to be main-

FIGURE 9.41

Upstream view of Browers Bend on the Missouri River. Notice the Iowa Public Service power plant on the right bank. A new design of permeable pile dikes is now under experimentation. It will allow water to flow through notches in the dikes, thus causing less filling-in of aquatic habitat (marsh and slough), shown on the left bank behind the dikes. The dikes shown here are of the conventional type, and filling of the marsh is clearly reducing wildlife and aquatic habitat. (Photo courtesy of U.S. Army Corps of Engineers, Omaha District.)

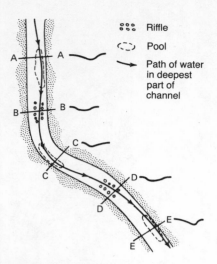

FIGURE 9.42
Idealized diagram showing pools and riffles and alternating asymmetric and symmetric cross-channel profiles. (From E. A. Keller, ''Channelization: A Search for a Better Way,'' *Geology Five,* The Geological Society of America, 1975. Reprinted by permission.)

tained by the 2-year flood is superimposed on the larger flood-control channel. Addition of pools and riffles in the pilot channel, when possible, would help provide fish habitat and better low flow conditions. The large channel might be vegetated, and the untrained observer might not easily recognize it. Obviously this

plan will not work in urbanizing areas where sediment production is high and property is not available to be purchased for the larger channel. However, if sediment is reduced by good conservation practice and the property is available to be purchased, the urban area will benefit from a more aesthetically pleasing and more useful stream.

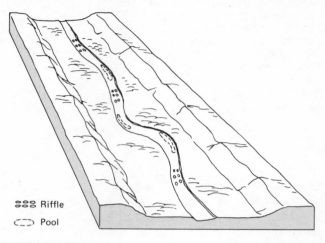

FIGURE 9.43
Pilot channel (channel within a channel) suggested for some flood-control projects. Pools and riffles are constructed in the meandering channel within the larger flood-control channels. (From E. A. Keller, ''Channelization: A Search for a Better Way,'' *Geology Five,* The Geological Society of America, 1975. Reprinted by permission.)

SUMMARY AND CONCLUSIONS

Obvious and well-known detrimental aspects of human use of surface water, groundwater, and atmospheric water include pollution of rivers and groundwaters. We are beginning to understand that solutions to many hydrologic problems require integrating all aspects of the water cycle.

Water is one of the most abundant and important renewable resources on earth. However, most water (more than 99 percent) is unavailable or unsuitable for beneficial human use because of its salinity or location. The supply and use pattern of water on earth at any particular point on the land's surface depends upon a number of factors involving interactions between the hydrologic cycle and rock cycles. Of particular importance in evaluating the water source and use patterns for a region is to develop a water budget and define the natural variability and availability of water.

During the next several decades, we expect that the total water withdrawn from streams and groundwater will decrease slightly, but that consumptive use will increase as a result of greater demands from a growing population and industry.

A major conflict concerning water withdrawn from streams is to define the instream needs for adequately maintaining fish and wildlife, navigation, and other competitive uses for the water.

Water pollution specifically refers to degradation of water quality as measured by physical, chemical, or biological criteria. Degradation of the water is generally in terms of the intended use for the water, departure from norm, effects on public health, or ecological impacts. Water pollutants are often defined as being either from point or non-point sources. The major pollutant types are oxygen-demanding waste (BOD), pathogens and fecal coliform bacteria, nutrients, synthetic organic and inorganic compounds, oil, heavy metals, radioactive materials, and thermal. Since the 1960s there has been a serious attempt to improve water quality in the United States, and the program seems to be successful; however, water quality in many areas is still substandard.

The movement of water down to the water table and through aquifers is an integral part of the rock and hydrolic cycles. In moving through an aquifer, groundwater may improve in quality, but it may also be rendered unsuitable for human use by natural or artificial contaminants.

In the case of groundwater pollution, the physical, biologic, and geologic environments are considerably different from those of surface water. The ability of many soils and rocks to degrade pollutants physically or otherwise is well known, but less generally known is the ability of clays and other earth materials to capture and exchange certain elements and compounds.

Pollution of an aquifer may result from disposal of wastes on the land surface or in the ground. It can also result from overpumping of groundwater in coastal areas, which leads to intrusion of salt water into freshwater aquifers.

Water resource management is in need of a new philosophy that considers geologic, geographic, and climatic factors and employs creative alternatives.

Principal human influences affecting runoff and sediment production include the varied land uses (especially urban) and construction projects to control floods, such as channel improvement and reservoirs.

Sediment pollution, natural or artificial, is certainly one of the most significant pollution problems. A great deal of sediment pollution is a result of human use, particularly construction, agriculture, and urbanization. Although the problem cannot be completely eliminated, it can be minimized.

Channelization is the straightening, deepening, widening, cleaning, or lining of existing streams. The most commonly cited objectives of channel modification are flood control and drainage improvement. Although many cases of environmental degradation resulting from channelization can be cited, it need not necessarily be a bad practice. In fact, in several cases, channel modification has clearly been a beneficial practice. Nevertheless, the present state of channel modification technology is not sufficient for us to be reasonably sure that environmental degradation will not result. Therefore, until new design criteria compatible with natural stream processes are developed and tested, channelization should be considered only as a solution to a particular flood or drainage problem where the impact and alternatives are carefully studied to minimize environmental disruption.

REFERENCES

1. COUNCIL ON ENVIRONMENTAL QUALITY. 1980. *The global 2000 report to the President,* The Technical Report, v. 2, 764pp.

2. U.S. WATER RESOURCES COUNCIL. 1978. *The nation's water resources 1975–2000,* v. 1, A summary, 86pp.

3. SOLLEY, W. B., CHASE, E. B., and MANN, W. B. IV. 1983. *Estimated use of water in the United States in 1980.* U.S. Geological Survey Circular 1001.

4. COUNCIL ON ENVIRONMENTAL QUALITY. 1979. *Environmental quality: The tenth annual report of the Council on Environmental Quality.*

5. ANONYMOUS. 1982. *U.S. Geological Survey Activities, Fiscal Year 1981.* U.S. Geological Survey Circular 875:90–93.

6. GEISER, K. and WANECK, G., 1983. PCB's and Warren County. *Science for the People* 15:13–17.

7. PYE, V. I. and PATRICK, R. 1983. Groundwater contamination in the United States. *Science* 221:713–718.

8. MALLMAN, W. L. and MACK, W. N. 1961. *A review of biological contamination of ground water.* Proceedings of Symposium on Ground Water Contamination, Taft Sanitary Engineering Center, Cincinnati, Ohio.

9. LeGRAND, H. E. 1968. Environmental framework of ground water contamination. *Journal of Ground Water* 6:14–18.

10. DEUTSCH, M. 1965. Natural controls involved in shallow aquifer contamination. *Journal of Ground Water* 3:37–40.

11. BOGAN, R. H. 1961. *Problems arising from ground water contamination by sewage lagoons at Tieton, Washington.* Public Health Service Technical Report W61–5.

12. VOGT, J. E. 1961. *Infectious hepatitis outbreak in Posen, Michigan.* Public Health Service Technical Report W61–5.

13. DEUTSCH, M. 1961. *Incidents of chromium contamination of ground water in Michigan.* Public Health Service Technical Report W61–5.

14. PETTYJOHN, W. A. 1971. Water pollution by oil-field brines and related industrial wastes in Ohio. *Ohio Journal of Science* 71:257–69.

15. BURT, E. M. 1972. The use, abuse and recovery of a glacial aquifer. *Journal of Ground Water* 10:65–72.

16. FOXWORTHY, G. L. 1978. Nassau County, Long Island, New York—Water problems in humid country. In *Nature to be Commanded,* ed. G. D. Robinson and A. M. Spieker. U.S. Geological Survey Paper 950:55–68.

17. LEOPOLD, L. B., 1977. A reverence for rivers. *Geology* 5:429–30.

18. WILSHIRE, H. G. and NAKATA, J. K. 1976. Off-road vehicle effects on California's Mojave Desert. *California Geology* 29:123–32.

19. WILSHIRE, H. G. et al. 1977. *Impacts and management of off-road vehicles.* Geological Society of America. Report to the Committee on Environment and Public Policy.

20. ROBINSON, A. R. 1973. Sediment, our greatest pollutant? In *Focus on environmental geology,* ed. R. W. Tank, pp. 186–92. New York: Oxford University Press.

21. YORKE, T. H. 1975. Effects of sediment control on sediment transport in the northwest branch Anacostia River Basin, Montgomery County, Maryland. U.S. Geological Survey, *Journal of Research* 3:487–94.

22. HOUSE REPORT No. 93–530. 1973. *Stream channelization: What federally financed draglines and bulldozers do to our nation's streams.* Washington, D.C.: U.S. Government Printing Office.

23. ARTHUR D. LITTLE, INC. 1972. *Channel modifications: An environmental, economic and financial assessment.* Report to the Council on Environmental Quality.

24. DANIELS, R. B. 1960. Entrenchment of the Willow drainage ditch, Harrison County, Iowa. *American Journal of Science* 225:161–76.

25. RUHE, R. V. 1971. Stream regimen and man's manipulation. In *Environmental geomorphology,* ed. D. R. Coates, pp. 9–23. Binghamton, New York: Publications in Geomorphology, State University of New York.

26. EMERSON, J. W. 1971. Channelization: A case study. *Science* 173:325–26.

Chapter 10

Waste Disposal

One view of waste disposal might consider the ultimate technology to be a system capable of accepting an unlimited amount of waste and safely containing it forever outside the human sphere of life. In the first century of the Industrial Revolution, the volume of waste produced was relatively small, and the concept of "dilute and disperse" was adequate. Factories were located near rivers because the water provided easy transport of materials by boat, ease of communication, sufficient water for processing and cooling, and easy disposal of waste into the river. With few factories and sparse population, "dilute and disperse" seemed to remove the waste from the environment (1).

As industrial and urban areas expanded, the concept of "dilute and disperse" became inadequate, and new concepts such as "concentrate and contain" became popular. It became apparent, however, that containment was and is not always achieved. Containers, natural or artificial, may leak or break and allow waste to escape. Thus, another new concept developed—that of converting wastes to useful material, in which case they are no longer wastes but resources. But even with today's technology, there are large volumes of waste that cannot be economically converted or that are essentially indestructible; hence, we still have waste-disposal problems (1).

Disposal or treatment of liquid and solid wastes by state and municipal agencies costs billions of dollars every year (Figure 10.1). It is one of the government's most expensive environmental problems, accounting for about 55 percent of total environmental expenditures (2).

All societies produce waste, but industrialization and urbanization have caused an ever-increasing affluence and have greatly compounded the problem of waste management. Although tremendous quantities of liquid and solid waste from municipal, industrial, and agricultural sources are being collected and recycled, treated, or disposed of, new and innovative programs remain a necessity. The popular notion is to consider all so-called wastes as essentially "resources out of place." Although we will probably never be able to recycle all wastes, it seems apparent that increasing costs of raw materials, energy, transportation, and land will make it financially feasible to recycle more resources as well as to reuse the land where wastes that have not been recycled are buried, creating new land resources for development.

For discussion purposes, it is advantageous to break the management, treatment, and disposal of waste into several categories: solid-waste disposal, hazardous chemical waste management, radioactive

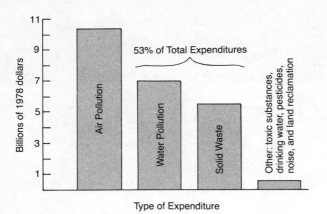

FIGURE 10.1
U.S. pollution abatement and environmental quality expenditures for 1978. (Data from Council on Environmental Quality, Annual Report, 1979.)

waste management, ocean dumping, septic-tank sewage disposal, and waste-water treatment. Although some of the categories overlap—for example, ocean dumping of solid waste—each has sufficiently different aspects to warrant separate discussion.

SOLID-WASTE DISPOSAL

Disposal of solid waste is primarily an urban problem. In the United States alone, urban areas produce about 640 million kilograms of solid waste each day. That amount of waste is sufficient to cover more than 1.6 square kilometers of land every day to a depth of 3 meters (3). Figure 10.2 summarizes major sources and types of solid waste (refuse), and Table 10.1 lists their compositions. It is no surprise that paper is by far the most abundant solid waste; however, we emphasize that this is only an average content, and considerable variation can be expected as factors such as land use, economic base, industrial activity, climate, and season of the year vary.

The common methods of solid-waste disposal summarized from a U.S. Geological Survey report include: on-site disposal, composting, incineration, ocean dumps, and sanitary landfills (3). Of these, the geologic factor is most significant in planning and siting a sanitary landfill. Without careful consideration of the soils, rocks, and hydrogeology, a landfill program may not function properly.

On-Site Disposal

By far the most common on-site disposal method in urban areas is the mechanical grinding of kitchen food waste. Garbage disposal devices are installed in the waste-water pipe system from a kitchen sink, and the garbage is ground and flushed into the sewer system. This effectively reduces the amount of handling and quickly removes food waste, but final disposal is transferred to the sewage treatment plant where solids such as sewage sludge still must be disposed of (3).

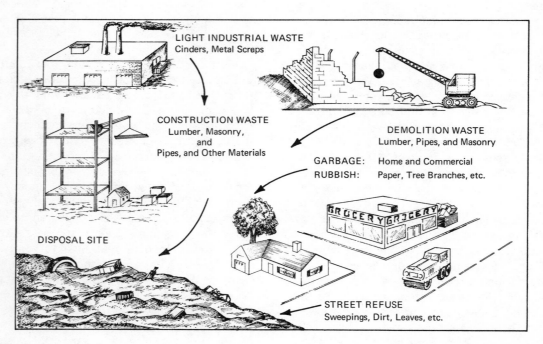

FIGURE 10.2
Types of materials or refuse commonly transported to a disposal site.

TABLE 10.1
Generalized composition of solid waste likely to end up at a disposal site.

Type of Waste	Average (%)
Paper products	43.8
Food wastes	18.2
Metals	9.1
Glass and ceramics	9.0
Garden wastes	7.9
Rock, dirt, and ash	3.7
Plastics, rubber, and leather	3.1
Textiles	2.7
Wood	2.5
Total	100.0

Another method of on-site disposal is incineration. This method is common in institutions and apartment houses (3). It requires constant attention and periodic maintenance to insure proper operation. In addition, the ash and other residue must be removed eventually and transported to a final disposal site.

Composting

Composting is a biochemical process in which organic materials decompose to a humuslike material. It is rapid, partial decomposition of moist, solid, organic waste by aerobic organisms. The process is generally carried out in the controlled environment of mechanical digesters (3). Although composting is not common in the United States, it is popular in Europe and Asia where intense farming creates a demand for the compost (3). A major drawback of composting is the necessity of separating out the organic material from the other waste, so it is probably economically advantageous only when organic material is collected separately from other waste (4).

Incineration

Incineration is the reduction of combustible waste to inert residue by burning at high temperatures (900°–1,000°C). These temperatures are sufficient to consume all combustible material, leaving only ash and noncombustibles. Incineration effectively reduces the volume of waste that must be disposed of by 75 to 95 percent (3).

Advantages of incineration at the municipal level are twofold. First, it can effectively convert a large volume of combustible waste to a much smaller volume of ash to be disposed of at a landfill; and second, combustible waste can be used to supplement other fuels in generating electrical power. Disadvantages are high capital outlay, high maintenance cost, additional handling to remove materials that do not burn, and air pollution. Because of the disadvantages, it seems unlikely that incineration will become a common way for urban areas to dispose of waste (4). Nevertheless, about 600 central incinerator plants operating in the United States burn about 135,000 metric tons of waste per day (3).

Open Dumps

Open dumps (Figure 10.3) are the oldest and most common way of disposing of solid waste, and although thousands have been closed in recent years, many are still being used. In many cases, they are located wherever land is available, without regard to safety, health hazard, and aesthetic degradation. The waste is often piled as high as equipment allows. In some instances, the refuse is ignited and allowed to burn; in others, the refuse is periodically leveled and compacted (3).

Although well known, the potential for water pollution caused by indiscriminate dumping has not been studied thoroughly enough to determine the long-range effects of dumping or the extent of area of the pollution. One interesting example was reported from Surrey County, England, where household wastes were dumped into gravel pits, polluting the groundwater (3). Refuse was dumped directly into 7-meter-deep pits with about 4 meters of water in the bottom. The maximum rate of dumping over a six-year period (1954 to 1960) was about 90,000 metric tons per year. Observation of water quality after the dumping was terminated revealed that chloride concentrations at the dump were 800 milligrams per liter, which is over three times the maximum concentration of 250 milligrams per liter con-

FIGURE 10.3
Open dump. (Photo by W. J. Schneider, courtesy of U.S. Geological Survey.)

sidered safe for a public water supply. Adjacent gravel pits in the down-gradient direction revealed chloride concentrations of 290 milligrams per liter, and pits as far as 800 meters away had concentrations of 70 milligrams per liter (3). In addition, bacterial pollution was also detected within 800 meters of the disposal site.

As a general rule, open dumps tend to create a nuisance by being unsightly, breeding pests, creating a health hazard, polluting the air, and sometimes polluting groundwater and surface water. Fortunately, the day of open dumps is giving way to the better planned and managed sanitary landfills that will eventually take their place.

Sanitary Landfill

The sanitary landfill, as defined by the American Society of Civil Engineering, is a method of solid-waste disposal that functions without creating a nuisance or hazard to public health or safety. Engineering principles are used to confine waste to the smallest practical area, reduce it to the smallest practical volume, and cover it with a layer of compacted soil at the end of each day of operation, or more frequently, if necessary. Covering the waste with compacted soils makes the landfill "sanitary." The compacted layer effectively denies continued access to the waste by insects, rodents, and other animals. It also isolates the refuse from the air, thus minimizing the amount of surface water entering into and gas escaping from the wastes (Figure 10.4) (4).

The sanitary landfill as we know it today emerged in the late 1930s, and by 1960, more than 1,400 cities disposed of their solid waste by this method. Two types are used: *area landfill* on relatively flat sites and *depression landfill* in natural or artificial gullies or pits.

The depths of landfills vary from about 2 to 13 meters. Normally, refuse is deposited, compacted, and covered by a minimum of 15 centimeters of compacted soil at the end of each day. A 1-to-4 cover ratio is common, that is, 30 centimeters of soil for every 120 centimeters of compacted waste. The finishing cover is at least 50 centimeters of compacted soil designed to minimize infiltration of surface water (3). Compaction and subsidence can be expected for years following completion of a landfill. Therefore development on the site after completion that cannot accommodate potential subsidence should be avoided.

A significant possible hazard from a sanitary landfill is groundwater or surface-water pollution. If waste buried in a landfill comes into contact with water percolating down from the surface water or groundwater moving laterally through the refuse, obnoxious, mineralized liquid capable of transporting bacterial pollutants called *leachate* is produced (5). For example, two landfills dating from the 1930s and '40s in Long Island, New York, have produced leachate plumes in a groundwater resource that are several hundred meters wide and have migrated several kilometers from the disposal site.

The nature and concentration of leachate produced at a disposal site depends on the composition of the waste, the length of time the infiltrated water is in contact with the refuse, and the amount of water that infiltrates or moves through the waste (3). Table 10.2 shows a generalized analysis of landfill leachates at three stages of stabilization compared to raw sewage and slaughter-house waste. Note that landfill leachate has a much higher concentration of pollutants. Fortunately, however, the amount of leachate produced from urban waste disposal is much less than the amount of raw sewage produced.

FIGURE 10.4
Sanitary landfill near Charlotte, North Carolina. (Photo courtesy of R. R. Phelps.)

TABLE 10.2
Composition of liquid wastes: landfill leachate, raw sewage, and slaughterhouse waste in ppm (parts per million) or mgll (milligrams per liter).

Constituents	Landfill Leachate			Raw Sewage[a]	Slaughter-house[b] Wastes
	Less than 2 yr Old	6 yr Old	17 yr Old		
BOD[c]	54,610	14,080	225	104	3,700
COD[d]	39,680	8,000	40	246	8,620
Total solids	19,144	6,794	1,198		2,690
Chloride	1,697	1,330	135		320
Sodium	900	810	74		
Iron	5,500	6.3	0.6	2.6	
Sulfate	680	2	2		370
Hardness	7,830	2,200	540		66
Misc. heavy metals	15.8	1.6	5.4	1.3	

[a]Data provided by the Metropolitan Sanitary District of Greater Chicago.
[b]Data from the files of the Illinois Department of Public Health.
[c]Biochemical oxygen demand.
[d]Chemical oxygen demand.
Source: G. M. Hughes and K. Cartwright, "Scientific and Administrative Criteria for Shallow Water Disposal," *Civil Engineering* 42 (New York: American Society of Civil Engineers, 1972).

Another possible hazard from landfills is uncontrolled production and escape of methane gas, generated as organic wastes decompose. For example, gas generated in an Ohio landfill migrated several hundred meters through a sandy soil to a housing area, where one home exploded and several others had to be evacuated. Properly managed, though, methane gas is a resource.

Factors that control the feasibility of sanitary landfills include topographic relief, location of the groundwater table, amount of precipitation, type of soil and rock, and the location of the disposal zone in the surface-water and groundwater flow system. The best sites are those in which natural conditions insure reasonable safety in disposal of solid waste; conditions may be safe because of climatic, hydrologic, or geologic conditions, or combinations of these (6).

The best sites are in arid regions (Figure 10.5). Disposal conditions are relatively safe there because in a dry environment, regardless of whether the burial material is permeable or impermeable, little leachate is produced. On the other hand, in a humid environment, some leachate will always be produced, so an acceptable level of leachate production must be established to determine the most favorable sites. The acceptable level varies according to local water use, local regulations, and the ability of the natural hydrologic system to disperse, dilute, and otherwise degrade the leachate to a harmless state.

The most desirable site in a humid climate is shown in Figure 10.6. There the waste is buried above the water table in relatively impermeable clay and silt soils. Any leachate produced will remain in the vicinity of the site and be degraded by natural filtering action and base exchange of some ions between the clay and the leachate. This would also hold for high water table

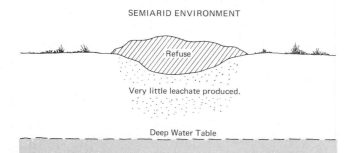

FIGURE 10.5
Sanitary landfill site in a semiarid environment. (After Bergstrom, Environmental Geology Notes No. 20, Illinois State Geological Survey, 1968.)

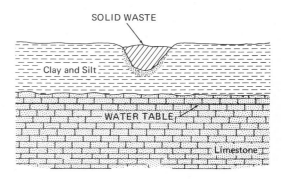

FIGURE 10.6
Most desirable landfill site in a humid environment. Waste is buried above the water table in a relatively impermeable environment. (After W. J. Schneider, U.S. Geological Survey Circular 601F, 1970.)

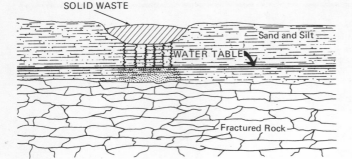

FIGURE 10.7
Waste-disposal site where the refuse is buried above the water table over a fractured rock aquifer. Potential for serious pollution is low because leachate is partially degraded by natural filtering as it moves down to the water table. (After W. J. Schneider, U.S. Geological Survey Circular 601F, 1970.)

conditions, as is often the case in humid areas, provided that the impermeable material is present. Ideally, a minimum of 10 meters of relatively impermeable material is desired between the base of the landfill and the shallowest aquifer. If less than 10 meters of relatively impermeable material is present, then the nature of the bedrock must be considered (7).

If the refuse is buried above the water table over a fractured-rock aquifer, as shown in Figure 10.7, the potential for serious pollution is low because the leachate is partly degraded by natural filtering as it moves down to the water table and the dispersion of contaminates is confined to the fracture zones (3). If the water table were higher or the cover material thinner and permeable, then widespread groundwater pollution of the fractured-rock aquifer might result.

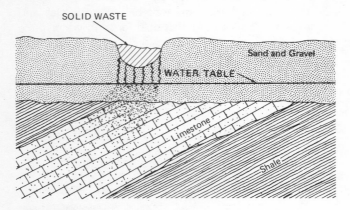

FIGURE 10.8
Solid-waste-disposal site where the waste is buried above the water table in permeable material in which leachate can migrate down to fractured bedrock (limestone). The potential for groundwater pollution may be high because of the many open and connected fractures in the rock. (After W. J. Schneider, U.S. Geological Survey Circular 601F, 1970.)

If the geologic environment of a landfill site is characterized by an inclined limestone-rock aquifer, overlain by permeable sand and gravel soils as shown in Figure 10.8, considerable contamination of the groundwater might result. This happens because the leachate quickly moves through the permeable soil and enters the limestone, where open fractures or cavities may transport the pollutants with little degradation other than dispersion and dilution. Of course, if the inclined rock is all shale, which is impermeable, little pollution will result (3).

In summary, the following guidelines should be followed in site selection of sanitary landfills (7). First, limestone or highly fractured rock quarries and most sand and gravel pits make poor landfill sites because these earth materials are good aquifers. Second, swampy areas, unless properly drained to prevent disposal into standing water, make poor sites. Third, clay pits, if kept dry, may provide satisfactory sites. Fourth, flat upland areas are favorable sites if an adequate layer of impermeable material such as clay is present above any aquifer. Fifth, floodplains that are likely to be periodically inundated by surface water should not be considered acceptable sites for refuse disposal. Sixth, any permeable material with a high water table is probably an unfavorable site. Last, in rough topography, the best sites are near the heads of gullies where surface water is at a minimum. While these general criteria are useful, they do not preclude a hydrogeological investigation including drilling to obtain samples, permeability testing, and other tests to predict the movement of leachate from the buried refuse (4).

Once a site is chosen for a sanitary landfill, monitoring of the movement of groundwater should begin before filling commences. After the operation starts, continued monitoring of the movement of leachate and gases should continue as long as there is any possibility of pollution. This is particularly important after the site is completely filled and the permanent cover material is in place, because a certain amount of settlement always occurs after a landfill is completed, and if small depressions form, then surface water may collect, infiltrate, and produce leachate. Monitoring and proper maintenance of an abandoned landfill reduces its pollution potential (4).

Hazardous waste pollutants from a solid-waste-disposal site can enter the environment by as many as six paths (Figure 10.9). First, compounds volatilized in the soil and fill after placement, such as methane, ammonia, hydrogen sulfide, and nitrogen gases, may enter the atmosphere. Second, heavy metals, such as lead, chromium, and iron, are retained in the soil. Third, soluble material, such as chloride, nitrate, and sulfate, readily pass through the fill and soil to the groundwater

FIGURE 10.9
Idealized diagram showing several ways that hazardous waste pollutants from a solid-waste-disposal site may enter the environment.

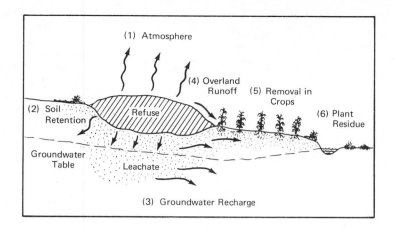

system. Fourth, overland runoff may also pick up leachate and transport it into the surface-water network. Fifth, some crops and cover plants growing in the disposal area may selectively take up heavy metals and other toxic materials to be passed up the food chain as people and animals eat them. Sixth, if the plant residue left in the field contains toxic substances, it will return these materials to the environment through soil-forming and runoff processes (8).

A thorough monitoring program will consider all six possible paths by which pollutants enter the environment. Generally, atmospheric pollution by gas from landfills is not a problem, and this trend will probably continue as more and more food wastes are ground and flushed into the sewage system. Many landfills have no surface runoff, so it is not necessary to monitor surface water; however, thorough monitoring is required if overland runoff does occur. Monitoring of soil and plants should include periodic chemical analysis at prescribed sampling locations. A minimum of nine sample sites is recommended. It is advisable to collect plant samples at harvest time or, if plants are left in the field, after they become dormant. Soil sample sites should be analyzed twice a year: in early summer and late fall. The need for monitoring groundwater quality through observation wells is minimal if the landfill is in relatively impermeable soil overlying dense, permeable rock. In this case, leachate and groundwater movement may be less than 30 centimeters per year. However, if permeable, water-bearing zones exist in the soil or bedrock, then observation wells are necessary to sample the water quality periodically and monitor the movement of the leachate (8). It is also important to monitor the unsaturated zone above the groundwater table to provide early warning of potential pollution problems before groundwater resources are contaminated and remedial methods become expensive and difficult. This can be done by installing suction lysimeters (devices to collect soil water samples) at various

depths below the surface. Samples should be collected at least four times per year.

HAZARDOUS CHEMICAL WASTES

The creation of new chemical compounds has proliferated tremendously in recent years. In the United States alone, approximately 1,000 new chemicals are marketed each year and a total of about 50,000 chemicals are currently on the market. Although many of the chemicals have been beneficial, approximately 35,000 of the chemicals used in the United States are classified as either definitely or potentially hazardous to human health (see Table 10.3). The United States currently generates about 57 billion metric tons of hazardous chemical waste per year, and of that, only about 10 percent has been safely handled in the recent past. Until recently, as much as half the total volume of wastes was being dumped indiscriminately (9).

In the United States, of the 32,000 to 50,000 uncontrolled waste disposal sites, probably 1,200 to 2,000 contain sufficient waste to pose a serious threat to public health and the environment. For this reason, many scientists believe that management of hazardous chemical materials may be the most serious environmental problem ever to face the United States.

Uncontrolled dumping of chemical waste can pollute the soil and groundwater resources in several ways (Figure 10.10). First, if chemical waste is stored in barrels either on the surface of the ground or buried at a disposal site, the barrels eventually corrode and leak, with a potential to pollute surface, soil, and groundwater. Second, liquid chemical waste may be dumped in an unlined "lagoon" where the contaminated water may then percolate through the soil and rock, and eventually to the groundwater table; and third, liquid chemical waste is sometimes illegally dumped in deserted fields or even along dirt roads.

TABLE 10.3
Examples of products we use and potentially hazardous waste they generate.

Products we use	Potentially hazardous waste
Plastics	Organic chlorine compounds
Pesticides	Organic chlorine compounds, organic phosphate compounds
Medicines	Organic solvents and residues, heavy metals (mercury and zinc, for example)
Paints	Heavy metals, pigments, solvents, organic residues
Oil, gasoline, and other petroleum products	Oil, phenols and other organic compounds, heavy metals, ammonia salts, acids, caustics
Metals	Heavy metals, fluorides, cyanides, acid and alkaline cleaners, solvents, pigments, abrasives, plating salts, oils, phenols
Leather	Heavy metals, organic solvents
Textiles	Heavy metals, dyes, organic chlorine compounds, solvents

Source: U.S. Environmental Protection Agency. SW-826, 1980.

Hazardous landfills and other sites for disposal of chemical waste are not merely a potential problem. Examples from Nevada, Kentucky, New Jersey, and New York illustrate that they are already affecting the lives of a growing number of people. As recently as a few years ago, waste generated from producing pesticides and organic solvents at Henderson, Nevada, was probably contaminating Lake Mead, the water supply for Las Vegas; and at a location near Louisville, Kentucky, known as the "Valley of the Drums," thousands of leaking drums of waste chemicals were stored on the surface or buried, oozing toxic materials.

Near Elizabeth, New Jersey, are the remains of about 50,000 charred drums, stacked four-high in places, next to a brick and steel building once owned by a now-bankrupt chemical corporation. The drums and other containers had been left to corrode for nearly ten years and many had been either improperly labeled or have been burned so that the nature of the chemicals could not be determined simply from outside markings. Leaking barrels allowed unknown waste to seep into an adjacent stream that flows eventually into

the Hudson River. Clean-up efforts are extremely difficult because there is little precedent for what to do with unknown dangerous chemicals. Identification of some of the materials at the site showed that there were two containers of nitroglycerine, numerous barrels of biological agents, cylinders of phosgene and pyrophoric gasses (which are extremely volatile and ignite when exposed to the air), as well as a variety of heavy metals, pesticides, and solvents, many of which are extremely dangerous and potentially explosive. It took months of work with a large crew of workers to remove most of the material from the New Jersey site. Unfortunately, however, it is difficult to know if all the wastes have been removed; additional material may be buried or sitting at other locations that are difficult to find (10).

The best known example of problems associated with chemical waste disposal comes from the town of Niagara, New York, and the "Love Canal," which is really a ditch that was originally intended as a canal to connect major markets with industrial sites in 1892. The plan for the canal never materialized, and the ditch known as Love Canal remained until the mid-1920s, when chemical dumping began. For 30 years, at least 80 different chemicals were buried in the ditch. Finally, in 1953, the dump site was covered with about 1 meter of soil. In that same year, the company that had been dumping the waste donated the land to the city of Niagara Falls for $1, and eventually an elementary school and more than 200 homes were built in and around the site (Figure 10.11). For a number of years, no one suspected that a time bomb was ticking beneath the site as the disposal drums slowly began corroding and waste leaked into the soil and near-surface water. In 1976, following heavy rains in the mid-1970s and heavy snowfall during the winter of 1976-77, Love Canal began to gain national attention because of the serious pollution problem that was developing. Trees and gardens were dying; a swimming pool popped up from its foundation, floating in a bath of chemicals; dogs that

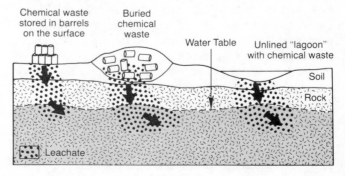

FIGURE 10.10
Ways that uncontrolled dumping of chemical waste may pollute soil and/or groundwater.

FIGURE 10.11
Former location of the Love
Canal extending from lower right
to upper left and directly beneath
the school yard in the center of
the photograph. The white,
patchy areas indicate sections
where vegetation will not grow
in the canal area. (Photograph by
John George and courtesy of the
New York State Department of
Environmental Conservation.)

sniffed in the landfill developed sores that would not
heal; and children who played in the area could only
watch and wonder as the rubber on their tennis shoes
and tires disintegrated. In some locations, puddles of
toxic, noxious substances were oozing to the surface.
A study of the site revealed the old dumpsite and identified a number of substances present that were suspected of being carcinogens, including benzene,
dioxin, dichlorethylene, and chloroform. Although officials readily admitted that little was known about the
impact of these chemicals and others at the site, there
was grave concern for the people living in the area.
The concern has proved warranted, because of the al-
leged higher-than-average rates of miscarriages, blood
and liver abnormalities, birth defects, and chromosome
damage. However, a study by the New York authorities
suggests that no chemically-caused health effects have
been absolutely established (11,12,13).

The cost to clean up the Love Canal site and re-
locate residents may eventually cost $100 million. In
addition, approximately $3 million per year may be
needed to monitor the area in the future. One of the
methods for controlling leachate at Love Canal was to
cover the dump site with a thick layer of compacted

clay, topped with soil and grass. The purpose of the
clay is to inhibit infiltration of surface water. Lateral
movement of water will be inhibited from entering the
site by construction of impervious walls, and subsurface seepage from the site is treated (11,12,13).

The real tragedy of Love Canal is that it is not an
isolated incident. There are probably many hidden
Love Canals across the country—time bombs waiting
to explode. It is important to identify and treat these
sites. To this end, Congress passed the Environmental
Response, Compensation and Liability Act in 1980, establishing a $1.6 billion revolving fund (popularly called
the "superfund") to clean up several hundred of the
worst known abandoned hazardous chemical waste
disposal sites around the country. Although the program has experienced significant management problems and is way behind schedule, a small number of
sites have been treated. Unfortunately the $1.6 billion
is not enough to pay for decontamination of all the tar-
geted sites—that would cost many times more. Thus
the problem of abandoned disposal sites is likely to per-
sist for some time to come.

Safe management of hazardous chemical wastes
includes several options: on-site processing to recover

byproducts with commercial value, microbial breakdown, and disposal by incineration, secure landfill, or deep-well injection. A number of technological advancements have been made in the field of toxic waste management, and as land disposal becomes more and more expensive, the recent trend toward on-site treatment is likely to continue. On-site treatment will not eliminate all hazardous chemical wastes, however, so disposal will remain necessary. Of the available disposal techniques, the secure landfill and deep-well injection both involve the geologic environment, so we will discuss them in detail.

Secure Landfill

The basic idea of the secure landfill is to confine waste to a particular location, control the leachate that drains from the waste, and collect and treat the leachate. Figure 10.12 demonstrates these procedures. A dike and liner (made of clay or other impervious material such as plastic) confine the waste, and a system of internal drains concentrates the leachate in a collection basin, from which it is pumped out and transported to a waste-water treatment plant. The function of the impervious liner is to insure that the leachate does not contaminate soil, and in particular, groundwater resources. However, this type of waste disposal procedure must have a well designed system to monitor both the unsaturated zone above the groundwater table and the groundwater. Sampling the unsaturated zone is difficult but necessary to detect leachate *before* it reaches the groundwater, where correcting and/or cleaning up pollutants is more difficult and costly.

It has recently been argued that there is no such thing as a really secure landfill, implying that all landfills will leak to some extent. Unfortunately this is probably true; impervious liners can fail and drains can become clogged, causing overflow. But landfills that are carefully sited and engineered can minimize problems. The preferable sites have good natural barriers or buffers to migration of leachate, such as thick clay deposits, arid climate, and deep water table. It is thus feasible to site secure landfills that function with a minimum of problems, and it is necessary if we are going to safely dispose of the large volumes of hazardous chemical wastes that are currently generated. Disposal, however, should be limited to the specific chemicals that are compatible and suitable for land disposal.

Deep-Well Disposal

In the practice of waste disposal by injection into deep wells, the term *deep* refers to rock, not soil, which is below and completely isolated from all freshwater aquifers, thereby assuring that injection of waste will not contaminate or pollute existing or potential water supplies. This generally means that the waste is injected into a permeable rock layer several thousand meters below the surface in geologic basins confined above by relatively impervious, fracture-resistant rock, such as shale or salt deposits (1).

Deep-well injection of oil-field brine (salt water) has been important in the control of water pollution in oil fields for many years, and huge quantities of liquid waste (brine) pumped up with oil have been injected back into the rock. Today, more than four billion liters per day are pumped into subsurface rocks (14). In recent years, the technique has been used more commonly for permanent storage of industrial waste deep underground, and by mid-1973, there were nearly 300 deep-well disposal units pumping industrial waste underground (14). A typical well is about 700 meters deep, and wastes are pumped into a 60-meter-thick zone at a rate of about 400 liters per minute (15).

FIGURE 10.12
Idealized diagram of the secure landfill for hazardous chemical waste. The impervious liner and systems of drains are an integral part of the system to insure that leachate does not escape from the disposal site. Monitoring in the unsaturated zone is important and involves periodic collection of soil water with a suction device.

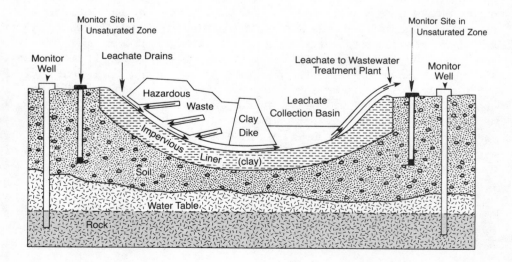

Deep-well disposal of industrial wastes should not be viewed as a quick and easy solution to industrial waste problems (16). Even where geologic conditions are favorable for deep-well disposal, there are natural restrictions because of the limited number of suitable sites, and because there is limited space for disposal of waste within these sites. Possible injection zones in porous rock are usually already filled with natural fluids, usually brackish or brine water. Therefore, to pump in waste, some of the natural fluid must be displaced by compression (even very slight amounts of compression of the natural fluids in a large volume of permeable rock can provide considerable storage space) and by slight expansion of the reservoir rock as the waste is being injected (15).

The most favorable sites from a geologic consideration are synclinal basins and coastal plains, because they generally contain thick sequences of sedimentary rocks bearing salt water. Figure 10.13 shows geologic features that are significant in evaluating sites for deep waste-injection wells and locating some existing injection systems. The most favorable basins are tectonically stable and range from tens to hundreds of kilo-

meters in width. Since natural flow of fluids is slow, less than about 1 meter per year, deep-well disposal should be restricted vertically and much less restricted horizontally (1).

Problems with deep-well disposal   Several problems associated with disposal of liquid waste in deep wells have been reported (15,16). Perhaps the best known are the earthquakes between 1962 and 1965, which we discussed earlier, caused by injection of waste from the Rocky Mountain Arsenal near Denver, Colorado. The injection zone was fractured gneiss at a depth of 3.6 kilometers, and the increased fluid pressure evidently initiated movement along the fractures. This is not a unique case; similarly initiated earthquakes have been reported in oil fields in western Colorado, Texas, and Utah (15). A similar activation of faults in southern California caused by injection of fluids into the Inglewood oil field for secondary recovery is thought to have contributed to the failure of the Baldwin Hills Reservoir. In a different sort of incident, in 1968, a disposal well on the shore of Lake Erie in Pennsylvania "blew out," releasing several million liters

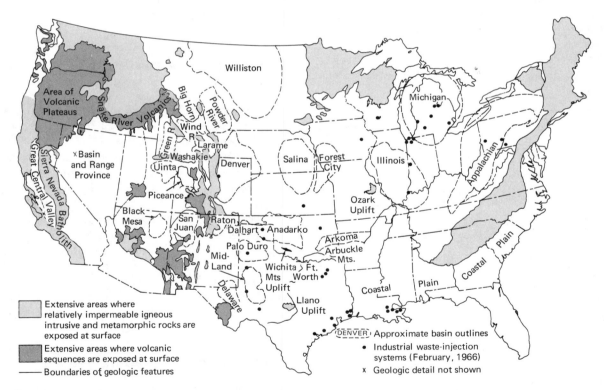

FIGURE 10.13
Geologic features significant in deep-well disposal. Notice that most of the industrial-waste injection systems (wells) are located in geologic basins or on coastal plains. Both of these environments are characterized by thick sequences of sedimentary rocks. (After D. L. Warner, American Association of Petroleum Geologists Memoir 10, 1968.)

of harmful chemicals from a paper plant into the lake (16). Regardless of these incidents, most oil-field brine and industrial disposal wells are functioning as designed and are not degrading the near-surface environment.

Feasibility and general site considerations for disposal wells The feasibility of deep-well injection as the best solution to a disposal problem depends on four factors: the geologic and engineering suitability of the proposed site, the volume and the physical and chemical properties of the waste, economics, and legal considerations (17).

Geologic consideration for disposal wells are twofold (17). First, the injection zone must have sufficient porosity, thickness, permeability and areal extent to insure safe injection. Sandstone and fractured limestone are the commonly used reservoir rock (16). Second, the injection zone must be below the level of fresh-water circulation and confined by relatively impermeable rock such as shale or salt, as shown in Figure 10.14.

We achieve optimal use of limited underground storage space if deep-well injection is used only when more satisfactory methods of waste disposal are not available and when the volume of injected wastes is minimized by good waste management (17). Optimal use includes thorough evaluation of the physical and chemical properties of the waste to insure that it will not adversely affect the ability of the rock reservoir to accept it. Adverse effects can be minimized in some instances by preinjection treatment of the waste to en-hance compatibility with the reservoir rock and the natural fluids present there. It is also advisable to take advantage of natural buffers; for example, if the waste is acidic, use of limestone as a reservoir rock may be advantageous because the acid waste tends to increase the reservoir's permeability by chemically attacking and enlarging natural fractures, thus allowing more waste to be injected. Care must be taken, however, because certain acids, such as sulfuric acid, may react to plug the porosity of limestone.

Construction and operating costs, particularly those related to geologic factors, often determine the feasibility of deep-well injection as the preferred method of waste treatment. Important geologic factors pertaining to construction costs are the depth of the well and the ease with which drilling proceeds. Operating costs depend on the permeability, porosity, and thickness of the injection zone, and the fluid pressure in the reservoir. All of these are important in determining the rate at which the reservoir will accept liquid waste (17).

Consideration of legal aspects of deep-well disposal suggests that existing laws, regulations, and policies are probably adequate. However, regulatory agencies do need more funds and trained personnel to regulate successfully the growing number of disposal wells (14).

Monitoring disposal wells An essential part of any disposal system is monitoring. It is important to know exactly where the wastes are going, how stable they are, and how fast they are migrating. It is especially important in deep-well disposal that involves toxic or otherwise hazardous materials.

Effective monitoring requires that the geology be precisely defined and mapped before initiating the disposal program. It is especially important to locate all freshwater-bearing zones and old or abandoned oil or gas wells that might allow the waste to migrate up to freshwater aquifers or to the surface (Figure 10.15). A system of deep observation-wells drilled into the disposal reservoir in the vicinity of the well can monitor the movement of waste, and shallow observation-wells drilled into freshwater zones can monitor the water quality to quickly identify any upward migration of the waste.

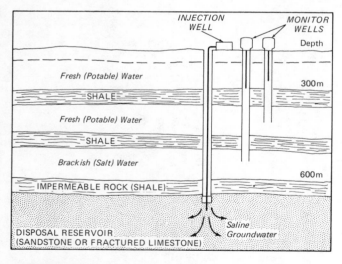

FIGURE 10.14
Idealized diagram showing a deep-well injection system. The disposal reservoir is a sandstone or fractured limestone capped by impermeable rock and isolated from all fresh water. Monitor wells are a safety precaution to insure that there is no undesirable migration of the liquid waste into either the formation above or fresh water. (100 m = 327 ft)

Radioactive Waste Management

Radioactive wastes are by-products that must be expected as electricity is produced from nuclear reactors, or weapons are produced from plutonium. Considering waste disposal procedures, radioactive waste may be grouped into two general categories: low-level waste

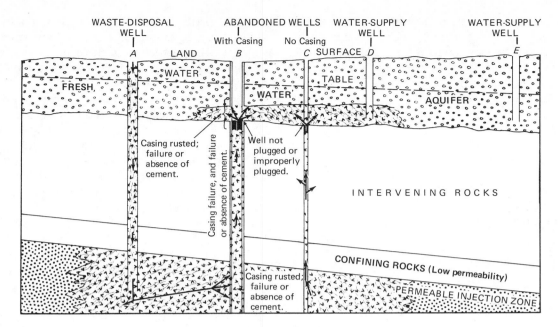

FIGURE 10.15
Idealized diagram showing how liquid waste might enter a freshwater aquifer through abandoned wells. This diagram illustrates why the location of all abandoned wells should be known, and it emphasizes the necessity of monitoring wells. (After Irwin and Morton, U.S. Geological Survey Circular 630, 1965.)

and high-level waste. In addition, the tailings from uranium mines and mills must also be considered hazardous (Figure 10.16). The existence of more than 20 million tons of abandoned tailings in the western United States that will produce radiation for at least 100,000 years emphasizes the problem.

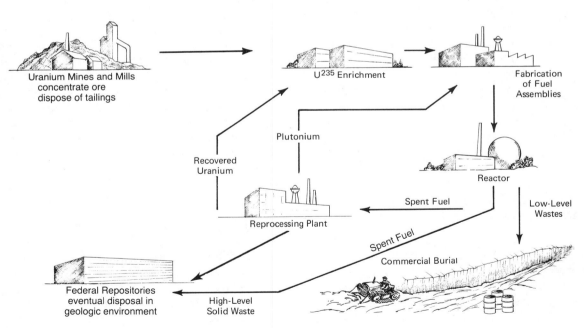

FIGURE 10.16
Idealized diagram showing the nuclear fuel cycle. The U.S.A. does not now reprocess spent fuel. Disposal of tailings, which because of their large volume can be more toxic than high-level wastes, has been treated casually.

Low-level waste contains low enough concentrations or quantities of radioactivity that it does not present a significant environmental hazard if properly handled. Included here are a wide variety of items such as residuals or solutions from chemical processing; solid or liquid plant waste, sludges, and acids; and slightly contaminated equipment, tools, plastic, glass, wood, fabric, and other materials (18).

The philosophy for management of low-level waste is "dilute and disperse." Experience suggests that low-level radioactive waste can be buried safely in carefully controlled and monitored near-surface burial areas in which the hydrologic and geologic conditions severely limit the migration of radioactivity (18).

High-level wastes are produced when fuel assemblages in nuclear reactors become contaminated with large quantities of fission products. This spent fuel must be removed periodically and reprocessed or disposed of. Because fuel assemblies will probably not be reprocessed in the near future in the United States, present waste management problems involve removal, transport, storage, and eventual disposal of spent fuel assemblies (19).

Hazardous radioactive materials produced from nuclear reactors include fission products such as krypton-85 (half-life of 10 years), strontium-90 (half-life of 28 years), and cesium-137 (half-life of 30 years). The half-life is the time required for the material's radioactivity to be reduced to one-half its original level. Generally at least 10 half-lives (and preferably more) are minimal before a material is no longer considered a health hazard. Therefore, a mixture of the fission products we have mentioned will require hundreds of years of confinement from the biosphere. Reactors also produce a small amount of plutonium-239 (half-life of 24,000 years), which is an artificial element. Because plutonium and the fission products must be isolated from the biological environment for exceedingly long periods of time (a quarter of a million years or longer), their permanent disposal is a geologic problem.

Disposal of High-Level Waste— A Geologic Perspective

High-level waste is extremely toxic, and a sense of urgency surrounds its disposal, since the total volume of spent fuel assemblies slowly accumulates. But there is also conservative optimism that the waste disposal problem will be solved.

It has been projected that without a disposal program, by the year 2000, several hundred thousand spent fuel elements from commercial reactors will be in storage awaiting disposal or eventual reprocessing to recover plutonium and unfissioned uranium. With reprocessing, resulting solid high-level waste would only occupy several thousand cubic meters—a very small volume that would not even cover a football field to a depth of one meter (19).

Military activities to produce plutonium for weapons also generates high-level waste. At present, several hundred thousand cubic meters of liquid and solid high-level waste are being stored at U.S. Department of Energy repositories at Hanford, Washington, Savannah River, Georgia, and Idaho Falls, Idaho.

Serious problems with radioactive waste have occurred with liquid waste buried in underground tanks. Sixteen leaks involving 1,330 cubic meters were located at Hanford from 1958 to 1973. An incident in 1973 involved a leak of 437 cubic meters of low-temperature waste. Since then, improvements such as stronger, double-shelled storage tanks, reduction of the volume of liquid waste stored through the solidification program, and increased reserve capacity have been made, and it is hoped these improvements will reduce the chance of future incidents (20).

Storage of high-level waste is at best a temporary solution that allows the federal government to meet its commitments for accepting waste. Regardless of how safe any "storage" program is, it requires continuous surveillance and periodic repair or replacement of tanks or vaults. It is therefore desirable to develop more permanent "disposal" methods in which retrievability may be possible but not absolutely necessary.

There is fair agreement that the geologic environment can provide the most certain safe containment of high-level radioactive waste. Because "disposal" of high-level waste is a necessity, the federal government is actively pursuing and developing possible alternative methods. Although the possibilities of disposal into polar ice caps and sediment in deep ocean basins are being explored, stable bedrock offers the most promise. A comprehensive geologic disposal development program should have a number of objectives, such as these:

■ To identify sites that meet broad geologic criteria of tectonic stability and slow movement of groundwater with long flow paths to the surface
■ To conduct intensive subsurface exploration of possible sites to positively determine geologic and hydrologic characteristics
■ To predict future behavior of potential sites
■ To evaluate risk associated with various predictions
■ To make a political decision as to whether the risks are acceptable to society, although this is not a geologic responsibility (19).

The Nuclear Waste Policy Act of 1982 initiated a comprehensive federal/state, high-level nuclear waste disposal program. The Department of Energy is responsible for investigating several potential sites in three rock types, including a salt dome at Richton, Mississippi; bedded salt in the Paradox Basin, Utah, and Deaf Smith County, Texas; basalt at Hanford, Washington; and tuff (welded volcanic ash) at the Nevada test site (21). The Act originally called for the President to recommend a site by 1987, but now it appears that it will be extended to at least 1990. The Act also establishes the right of the host state to object to the site, but an objection can be overridden by agreement of both Houses of Congress.

The geologic environment that received early attention for permanent disposal of high-level waste was bedded salt. There are several advantages to disposing of nuclear waste in salt. First, salt is relatively dry and impervious to water. Second, fractures that might develop in salt tend to be self-healing. Third, salt permits dissipation of larger quantities of heat than is possible in other rock types. Fourth, salt is approximately equal to concrete in its ability to shield harmful radiation. Fifth, salt has a high compressive strength when dry and is generally located in areas of low earthquake activity. Sixth, salt deposits are relatively abundant, and using some for waste disposal will cause a negligible resource loss. Many scientists have studied the feasibility of salt as a disposal for high-level wastes and have generally concluded that salt should effectively isolate the waste from the biosphere for hundreds of thousands of years (20). On the other hand, salt has recently been criticized because it has a relatively low ability to absorb radioactive nuclides from the waste in an insoluble form. Furthermore, pockets of brine (very salty water) have been found in some salt deposits, which would certainly create serious problems and jeopardize waste retrieval if a repository were accidentally flooded. The presence of brine or other groundwater also decreases the strength of salt. Finally, it is suspected that bedded salt may slowly flow, and even a cm of movement could be hazardous (21). On the

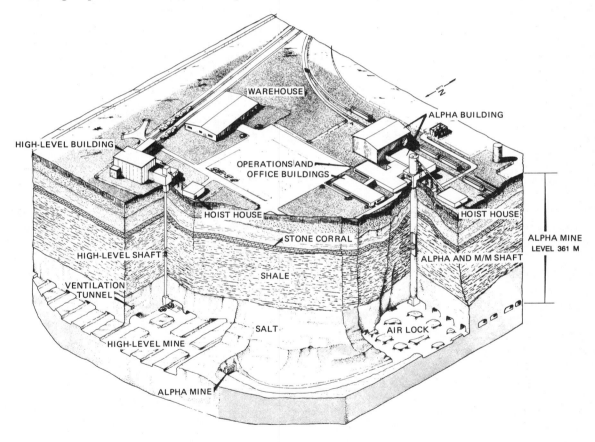

FIGURE 10.17
Schematic diagram showing how underground bedded salt may be used as a disposal site for solidified, high-level radioactive wastes. (Courtesy of ERDA.)

positive side again, careful site evaluation could probably identify areas where brine might cause problems, waste could be wrapped in materials that readily adsorb harmful nuclides from the waste, and areas of potential salt flowage might be predicted.

The federal government has considered disposal of radioactive waste in salts since 1955, and, during 1966 and 1967, initiated the Project Salt Vault in which high-level solid-waste disposal procedures were simulated in abandoned salt mines in Kansas. The project was considered a success, and in 1970, it was announced that the abandoned Carey Salt Mine near Lyons, Kansas, was to be used as a demonstration site for waste disposal and, if successful, the site would eventually become the National Radioactive Waste Repository. The Lyons site project was abandoned because of scientific, public, and state concern over two factors: geologic uncertainty—several thousand abandoned mine shafts and drill holes in the area might allow the waste to migrate up the shafts or drill holes, and land-use conflicts—present land use of partly urbanized to farming and grazing land is in conflict with the proposed disposal plan (20). Bedded salt in the Paradox Basin, Utah, is one of the sites currently under evaluation for waste disposal, and Figure 10.17 is an idealized diagram of a possible disposal operation.

The potential of tuff at the Nevada test site and basalt at Hanford, Washington, for disposal of high-level radioactive wastes is not being evaluated because these rocks are geologically favorable, but because disposal would be on an existing nuclear reservation, minimizing potential social and political opposition. Some argue that the Washington site is a poor choice because of the proximity of the Columbia River and problems of predicting groundwater movement through basalt. The tuff is a better prospect because of the present arid climate, low rate of groundwater recharge, little chance of surface water contamination, and its location far from population centers (21,22).

Salt domes in Mississippi and Texas are a poor choice for high-level nuclear waste disposal sites because, in addition to having the same potential problems of bedded salt, they are not nearly as stable as bedded salt, having moved thousands of meters during implacement of the dome. Furthermore, salt domes tend to be located in a highly populated part of the country with a wet climate (21).

We can see from this discussion that disposal of radioactive wastes in the geologic environment is not a simple task. No rock or location is likely to be absolutely satisfactory, so we will have to choose the lesser evil. Furthermore, considering the long time that high-level wastes remain dangerous and the many variables in the disposal process, it is likely that an accident will eventually happen (21). It is thus critical to select a site where accidental contamination will not produce a potentially catastrophic problem.

Current plans for respositories for high-level waste call for retrievability for 10 to as long as 50 years. This presents problems because it may necessitate a shallower disposal than would otherwise be necessary. It has been suggested that the waste could be disposed of at shallow depths (up to several hundred meters) either in thick, unsaturated zones or in shallow, carefully engineered and monitored tunnels in the arid western United States. Certainly these options also deserve careful consideration (20,23).

A major problem with disposal of high-level radioactive waste remains: how credible are long-range (thousands to a few million years) geologic predictions (19)? There is no easy answer to this question, since geologic processes vary over both time and space. Climates change over long periods of time, as do areas of erosion, deposition, and groundwater activity. For example, large earthquakes hundreds or even thousands of kilometers away from a site may permanently change groundwater levels. The seismic record for most of the United States extends back in time only for a few hundred years, and, therefore, estimates of future activity are tenuous at best. The bottom line is that geologists can suggest sites that have been relatively stable in the geologic past, but they cannot absolutely guarantee future stability. Therefore, decision makers, not geologists, need to evaluate the uncertainty of prediction in light of pressing political, economic, and social concerns (19). These problems do not mean that the geologic environment is not suitable for safe containment of high-level radioactive waste, but that we must take care to insure that the best possible decisions are made on this critical and controversial issue.

OCEAN DUMPING*

The oceans of the world, 363 million square kilometers of water, cover more than 70 percent of the earth and play a part in maintaining the world's environment by providing the water necessary to maintain the hydrologic cycle, contributing to the maintenance of the oxygen–carbon-dioxide balance in the atmosphere, and affecting global climate. In addition, the oceans are valuable to humans because they provide such necessities as foods and minerals.

---

*The section on ocean dumping is summarized from *Ocean Dumping: A National Policy* (Washington, D.C.: Council on Environmental Quality, 1970), pp. 1–25.

It seems reasonable that such an important resource as the ocean would receive preferential treatment, yet in 1968 alone, more than 43 million metric tons of waste at 246 sites were dumped off the coasts of the United States. Federal legislation and regulation by the Environmental Protection Agency in 1972 reduced the number of sites to 118, but ocean dumping continues to degrade the oceanic environment (2). Furthermore, if population growth in coastal regions continues as expected, increased amounts of waste that may end up in ocean dump sites will inevitably result.

The types of wastes that have been dumped in the oceans off the United States include: *dredge spoils*—solid materials such as sand, silt, clay, rock, and pollutants deposited from industrial and municipal discharges that are removed from the bottom of water bodies to improve navigation; *industrial waste*—acids, refinery wastes, paper mill wastes, pesticide wastes, and assorted liquid wastes; *sewage sludge*—solid material that remains after municipal waste water treatment; *construction and demolition debris*—cinder block, plaster, excavation dirt, stone, tile, etc.; *solid waste*—refuse, garbage or trash; *explosives;* and *radioactive waste* (24).

The 1972 Marine Protection Research and Sanctuaries Act prohibits ocean dumping of radiological, chemical, and biological warfare agents and any high-level radioactive waste. It also provides for regulation of all other waste disposal in the oceans off the United States by the Environmental Protection Agency or, in the case of dredge spoil, the U.S. Army Corps of Engineers. In addition to materials prohibited by law from being dumped, the Environmental Protection Agency has prohibited material whose effect on marine ecosystems cannot be determined; persistent inert materials that float or remain suspended, unless they are processed to insure that they sink and remain on the bottom; and material containing more than trace concentrations of mercury and mercury compounds, cadmium and cadmium compounds, organohalogen compounds (organic compounds of chlorine, fluorine, and iodine,

etc.), and compounds that may form from such substances in the oceanic environment; as well as crude oil, fuel oil, heavy diesel oil, lubricating oils, and hydraulic fluids. The Environmental Protection Agency also requires special permits and strictly regulated dumping of arsenic, beryllium, chromium, and lead; low-level nuclear waste; organo-silicon compounds; organic and inorganic processing wastes, including cyanide, fluoride, and chlorine; oxygen-demanding waste; organic chemicals; and petrochemicals. A major objective of the act is to prevent ocean dumping of wastes that, before recent federal laws were enacted, were discharged into rivers and the atmosphere (2).

Ocean Dumping and Pollution Problems

Ocean dumping contributes to the larger problem of ocean pollution, which has seriously damaged the marine environment and caused a human health hazard in some areas. Shellfish have been found to contain pathogens such as polio virus and hepatitis, and at least 20 percent of the nation's commercial shellfish beds have been closed because of pollution. Beaches and bays have been closed to recreational uses. Lifeless zones in the marine environment have been created. Heavy kills of fish and other organisms have occurred, and profound changes in marine ecosystems have taken place (24).

Major impacts of marine pollution on oceanic life include: killing or retarding of growth, vitality, and reproductivity of marine organisms by toxic pollutants; reduction of dissolved oxygen necessary for marine life because of increased oxygen demand from organic decomposition of wastes; biostimulation by nutrient-rich waste, causing excessive blooms of algae in shallow waters of estuaries, bays, and parts of the continental shelf, resulting in depletion of oxygen and subsequent killing of algae that may wash up and pollute coastal areas; and habitat change caused by waste-disposal practices that subtly or drastically change entire marine ecosystems (24).

TABLE 10.4
Concentrations of heavy metals in dredge spoils.

Metal	Concentrations in Dredge Spoils (ppm)	Natural Concentrations in Seawater (ppm)	Concentrations Toxic to Marine Life (ppm)
Cadmium	130	0.08	0.01–10.0
Chromium	150	0.00005	1.0
Lead	310	0.00003	0.1
Nickel	610	0.0054	0.1

Source: *Ocean Dumping: A National Policy,* Council on Environmental Quality, 1970.

TABLE 10.5

Concentrations in parts per million (ppm) of heavy metals in sewage sludge.

Metal	Concentrations in Sewage Sludge (ppm)			Natural Concentrations in Seawater (ppm)	Concentrations Toxic to Marine Life (ppm)
	Min.	Avg.	Max.		
Copper	315	643	1,980	0.003	0.1
Zinc	1,350	2,459	3,700	0.01	10.0
Manganese	30	262	790	0.002	—

Source: *Ocean Dumping: A National Policy,* Council on Environmental Quality, 1970.

Major impacts on people and society caused by marine pollution include: production of a public health hazard caused by marine organisms transmitting disease to people; loss of visual and other amenities as beaches and harbors become polluted by solid waste, oil, and other materials; and economic loss. In 1969 alone, loss of shellfish from pollution in the United States was $63 million. In addition, a great deal of money is spent cleaning up solid waste, liquid waste, and other pollutants in the coastal areas (24).

Alternatives to Ocean Dumping

The ultimate solution to the problem of ocean dumping is development of economically feasible and environmentally safe alternatives.

Disposal of dredge spoils represents the vast majority of all ocean dumping. Dredge spoils are removed primarily to improve navigation and are usually redeposited only a few kilometers away. Approximately one-third are seriously polluted with heavy metals such as cadmium, chromium, lead, and nickel (Table 10.4), as well as other industrial, municipal, and agricultural wastes. As a result, disposal in the marine environment can be a significant source of pollution (24).

The long-range alternative is to phase out ocean disposal of polluted dredge spoils, but this is not possible at present because of the great volume involved. Until land-base disposal can handle the volume, interim techniques should be developed, such as determining which spoils are polluted before disposal and then hauling the polluted spoils further from the dredging site to a safe disposal site. The main disadvantage here is the greater cost of longer hauls (24).

Environmental problems from disposal of sewage sludge at sea are important in terms of volume, toxic and possibly pathogenetic materials, and possible effects on marine life. Table 10.5 shows minimum, average, and maximum concentrations of three heavy metals found in sewage sludge, compared to normal concentrations in seawater and what concentrations are toxic to marine life. These data suggest that sewage sludge, like polluted dredge spoils, may be ex-

tremely harmful to marine life. Based on a conservative estimate of 54 grams of sludge generated per person per day, by the year 2000, more than 1.8 million metric tons of sludge will be generated in the coastal regions, an increase of 50 percent in thirty years. Most sewage sludge is currently disposed of on land or incinerated, but in some areas, tremendous quantities are disposed of by ocean dumping. In 1968, approximately 180,000 metric tons of sludge were disposed of at sea, compared to about 2.7 million metric tons on land (24). The city of Los Angeles has had an ongoing battle with the Environmental Protection Agency over continued ocean dumping. Los Angeles argues that in its case, ocean disposal may be safer than and preferable to on-land disposal. To resolve the disagreement, studies are going on to determine the effects of sludge disposal in a submarine canyon head at Santa Monica Bay.

Alternatives to ocean dumping include various methods of land disposal. Unfortunately, land disposal is considerably more expensive than near-shore marine disposal. However, if the sludge is digested (treated to control odors and pathogens) or barged long distances from shore, the cost becomes comparable, and land disposal may even be cheaper (24).

Perhaps the best long-range alternative to ocean dumping is the use of digested sludge for land reclamation, especially strip-mined land reclamation. Many strip mines in other areas are in great need of reclamation, and the nutrient value and organic material in sludge can improve texture, composition, and nutrient levels of low productivity soils. Unfortunately, the largest supplies of sludge are often far from the sites where they might be used.

SEPTIC-TANK SEWAGE DISPOSAL

Population movement in the United States continues to be from rural to urban, or urbanizing, areas. Often construction of an adequate sewage system has not kept pace with growth. As a result, the individual septic tank (soils absorption system) that has been used for years in both rural and urban areas continues to be an important method of sewage disposal. All land, how-

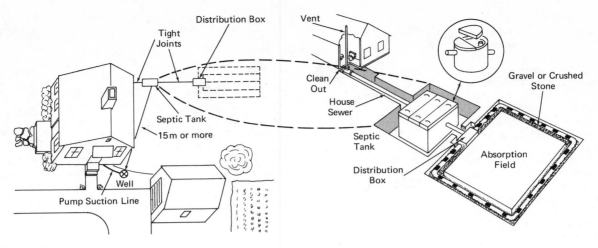

FIGURE 10.18
Generalized diagram showing septic-tank sewage disposal system (right) and location of the absorption field with respect to the house and well. (After Indiana State Board of Health.)

ever, is not suitable for installation of a septic-tank disposal system, and evaluation of individual sites is necessary and often required by law before a permit can be issued to construct a subsurface sewage disposal system. As mentioned earlier, the most satisfactory method of sewage disposal is the use of municipal sewers and sewage treatment facilities.

The basic parts of a septic-tank disposal system are shown in Figure 10.18. The sewer line from the house leads to an underground septic tank in the yard. The tank is designed to separate solids from the liquid, digest and store organic matter through a period of detention, and allow the clarified liquids to discharge into the seepage bed, or absorption field, which is a system

FIGURE 10.19
Graph used to determine the size of absorption field needed for a septic system for a private residence. (After Agriculture Information Bulletin 349, Soil Conservation Service, 1971.)

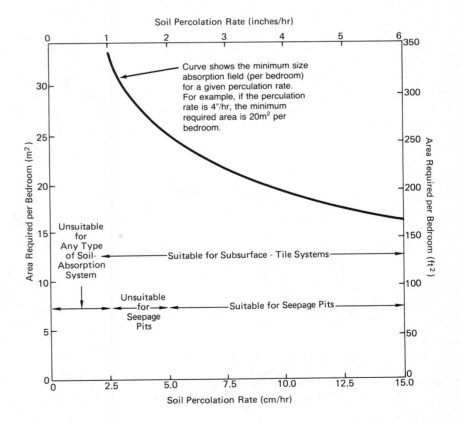

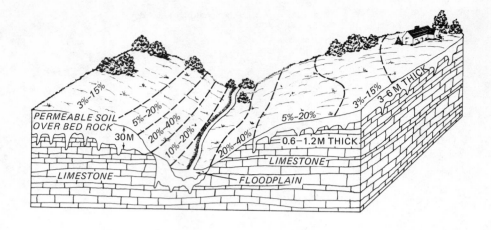

FIGURE 10.20
Block diagram illustrating some of the limitations of a septic system on limestone with a variable cover. The deep soil over the limestone at the left has moderate limitations, and construction may be difficult on the steeper slopes. The soil that is 3 to 6 meters thick on the extreme right may not be suitable for a septic system if the water supply is to come from a well. The shallow soil (0.6 to 1.2 meters thick) has severe limitations, and if a septic system is placed there, the stream at the bottom of the hill may become polluted. (After Agriculture Information Bulletin 349, Soil Conservation Service, 1971.)

of piping through which treated sewage seeps into the surrounding soil.

Geologic factors that affect the suitability of a septic tank disposal system include type of soil, depth to the water table, depth to bedrock, and topography. These variables are generally listed with soil descriptions associated with a soil survey of a county or other area. Soil surveys are published by the Soil Conservation Service and, when available, are extremely valuable in interpreting possible land use, such as suitability for a septic system. Since the reliability of a soils map for predicting limitations of soils is limited to an area larger than a few thousand square meters, and because soil types can change in a few meters, it is often desirable to have an on-site evaluation by a soil scientist. On some sites, it is also helpful to determine the rate at which water moves through the soil, which is best done by a percolation test. The test also is use-

ful in calculating the size of the absorption field needed (Figure 10.19). Soils with a percolation rate of less than 2.5 centimeters per hour are always undesirable for a septic-tank disposal system. Other criteria are these: the maximum seasonal elevation of the groundwater table should be at least 5 meters below the bottom of the absorption field; bedrock should be at a depth greater than 1.2 meters below the absorption field; and the surface slope generally should be less than 15 percent, or a 15-meter drop per 100-meter horizontal distance (25,26). Figures 10.20 and 10.21 demonstrate some of these limitations.

Sewage absorption fields may fail for several reasons. The most common cause is poor soil drainage (Figure 10.22), which allows the effluent to rise to the surface in wet weather. Common reasons for poor drainage are the existence of clay or compacted soils with low permeability, high water table, impermeable

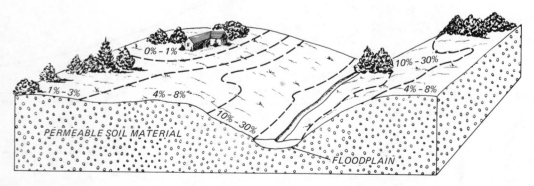

FIGURE 10.21
Block diagram illustrating the effect slope has on usage of a septic system. Septic systems work very well in deep, permeable soils. However, layout and construction problems may be encountered on slopes of more than 15°, and floodplains are not suitable for absorption fields. (From Agriculture Information Bulletin 349, Soil Conservation Service, 1971.)

FIGURE 10.22
Effluent from a septic-tank sewage disposal system rising to the surface in a backyard. The soil is poorly drained, and septic systems in such soils will not function well during wet weather. (Photo by W. B. Parker, courtesy of U.S. Department of Agriculture.)

rock near the surface, and frequent flooding. Another factor that causes failure is steep topography; however, if the absorption field is laid on the slope contours, a steeper slope may be accommodated (Figure 10.23) if there are no horizontal, impervious clay layers in the soil near the surface. Such layers allow the effluent to flow above them until it is discharged on the surface and runs unfiltered down the slope (25).

WASTE-WATER TREATMENT

Waste-water treatment and water pollution control are two of the greatest environmental expenses for which industry and municipalities pay. In 1979, the bill was about $12 billion. Figure 10.24 shows projected changes in expenses for waste-water treatment and water pollution control from 1979 to 1988. The greatest increase is projected for municipal waste-water treatment.

The main purpose of waste-water treatment is to reduce the amount of suspended solids, bacteria, and oxygen-demanding materials in waste water. In addition, new techniques are being developed to remove whatever harmful dissolved inorganic materials may be present. The new techniques, however, are more likely to reinforce rather than replace present technology (27).

Existing waste-water treatment generally falls into two broad classes (Figure 10.25): *primary treatment*, involving removal of grit, screening, grinding, flocculation, and sedimentation; and *secondary treatment*, involving controlled biological assimilation and degradation processes that occur in nature by microorganisms. The name *tertiary treatment* is sometimes used for chemical purification or other additional treatment measures designed to further purify the water (27).

The most troublesome aspect of waste-water treatment is the handling and disposal of sludge produced in the treatment process. The amount of sludge produced is conservatively estimated at about 54 to 112 grams per person per day, and its disposal accounts for 25 to 50 percent of the capital and operating

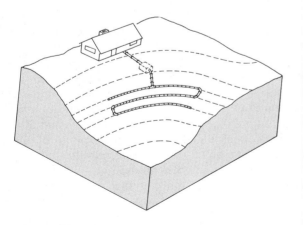

FIGURE 10.23
Block diagram illustrating how a septic field may be constructed on a steeper slope. The tile lines are laid out on the contour. This type of configuration is necessary on most sloping fields or in fields where there may be a change in soil type. (From Agriculture Information Bulletin 349, Soil Conservation Service, 1971.)

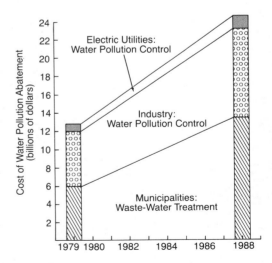

FIGURE 10.24
Cost of water pollution abatement for U.S. municipalities, industry, and electrical utilities from 1979 projected to 1988. (Data from Council on Environmental Quality, 1980.)

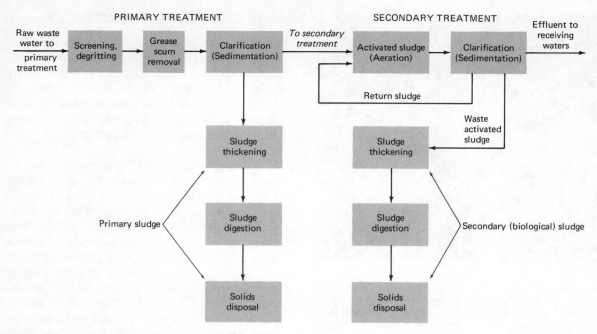

Raw waste water to primary treatment → Screening, degritting → Grease scum removal → Clarification (Sedimentation) → *To secondary treatment* → Activated sludge (Aeration) → Clarification (Sedimentation) → Effluent to receiving waters

Return sludge

Waste activated sludge

Sludge thickening

Sludge thickening

Primary sludge — Sludge digestion

Sludge digestion — Secondary (biological) sludge

Solids disposal

Solids disposal

FIGURE 10.25

Flow chart showing various procedures in primary and secondary treatment of waste water. (Reprinted from *Cleaning Our Environment—The Chemical Basis for Action,* a report by the Subcommittee on Environment Improvement, Committee on Chemistry and Public Affairs, American Chemical Society, 1969, p. 107. Reprinted by permission of the copyright owner.)

cost of a treatment plant. An example from the Metropolitan Sanitary District of Greater Chicago emphasizes the problem. Waste-water treatment there in the late 1960s, serving 5.5 million people and an industrial equivalent of 3 million more, processed about 5.7 million cubic meters of waste water per day, producing more than 900 tons of dry sludge daily. Disposal costs of this sludge were about $14.5 million, or about 46 percent of the total operating and maintenance costs (27).

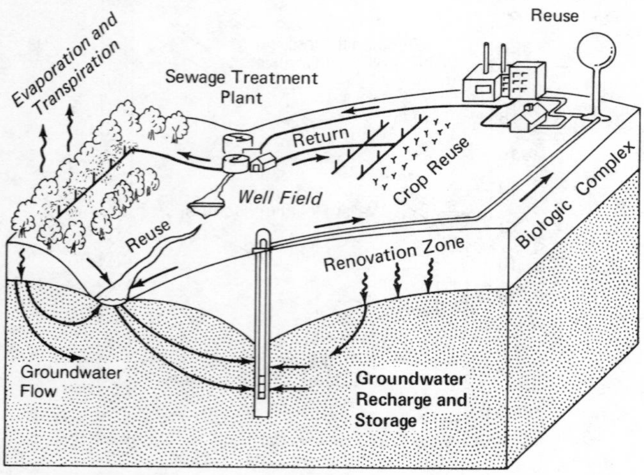

Sludge handling and disposal have four main objectives: (1) to convert the organic matter to a relatively stable form; (2) to reduce the volume of sludge by removing liquid; (3) to destroy or control harmful organisms; and (4) to produce by-products whose use or sale reduces the cost of processing the sludge (27). Final disposal of sludge is accomplished by incineration, burying it in a landfill, using it for soil reclamation, or dumping it in the ocean. From an environmental standpoint, the best use of sludge is to improve soil texture and fertility in areas disturbed by activity such as strip mining and poor soil conservation.

Although it is unlikely that all the tremendous quantities of sludge from large metropolitan areas can ever be used for beneficial purposes, many smaller towns and many industries, instutitions, and agricultural activities can take advantage of municipal and animal wastes by converting them into resources. This process of locally recycling liquid waste is called the "Waste Water Renovation and Conservation Cycle," shown schematically in Figure 10.26. The major pro-

FIGURE 10.26

Block diagram showing the waste-water renovation and conservation cycle. (From Parizek and Myers, American Water Resources Administration, 1968.)

FIGURE 10.27

(a) Two effluent-application areas, one near Muskegon, the other near Whitehall, Michigan, will eliminate the need to discharge inadequately treated industrial and municipal wastes to surface waters. These two subsystems will replace four municipal treatment facilities [total flow, 11 mgd (41,500 m³/d)] and eliminate direct discharges of five industrial plants [combined flow of 19 mgd (71,800 m³/d)]. The effluent-on-land system will handle the 1992 requirements of the county, with a population of 170,000 generating a flow of 43.4 mgd (165,000 m³/d), including an industrial flow of 24 mgd (90,600 m³/d). Irrigation land consists of 6,000 acres (24.3 million m²) surrounding the treatment cells and storage basins. (b) Raw sewage from the sewers of Muskegon is transported via a pressure main to this Muskegon-Mona Lake reclamation site. There, the sewage flows through three aerated treatment lagoons (1, 2, and 3) and is then diverted either into the large storage basin or into the settling lagoon (4) before going into the outlet lagoon (5). Effluent leaving the outlet lagoon is chlorinated and then pumped through a piping system that carries it to 55 spray-irrigation rigs. The circles symbolize the radius swept out by the spray rigs. After the effluent trickles down through the soil, it is collected by a network of underground pipes and pumped to surface waters. A series of berms around the lowland edges of the irrigation circles will contain storm-water runoff. These waters will pond within the circles until handled by the drainage system. [Reprinted, by permission, from E. I. Chaiken, S. Poloncsik, and C. D. Wilson, *Civil Engineering* 42 (New York: American Society of Civil Engineers, 1973.)]

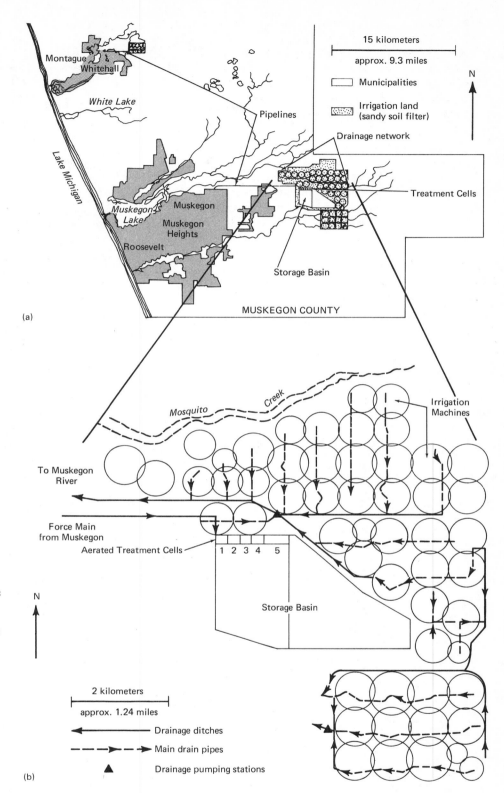

cesses in the cycle are: return of the treated waste water to crops by sprinkler or other irrigation system; renovation, involving natural purification by slow percolation of waste water through soil to eventually recharge the groundwater resource with clean water; and reuse of the water by pumping it out of the ground for municipal, industrial, institutional, or agricultural purposes (28). Of course, all aspects of the cycle are not equally applicable to a particular waste-water problem. Considerable variability between waste recycling from cattle feed lots and recycling from industrial or municipal sites is obvious. But the general principle of recycling is valid, and the processes are similar in theory.

When recycling waste water, the return and renovation processes are crucial, and soil and rock type, topography, climate, and vegetation play significant roles. Particularly important here are the ability of the soil to safely assimilate waste, the ability of the selected vegetation to use the nutrients, and knowledge of how much waste water can be applied (29).

Waste water is recycled on a large scale near Muskegon and Whitehall, Michigan (Figure 10.27). Raw sewage from homes and industries is transported by sewers to the treatment plant where it receives primary and secondary treatment. The waste waters are then chlorinated and pumped into a piping network which transports the effluent to a series of spray irrigation rigs. After the waste water trickles down through the soil, it is collected in a network of tile drains and transported to the Muskegon River for final disposal. This last step is an indirect tertiary treatment, using the natural environment as a filter. The Michigan project is controversial, and there is concern for possible pollution of surface water and groundwater, as well as prob-

TABLE 10.6

Removal Efficiency, Muskegon land treatment system (1976–78).

Wastewater Resource	3-Year Average Concentration (mg/liter)		Percent Removal
	Influent	Effluent	
BOD$_1$	300	3–4	99
Suspended Solids	290	7	98
Phosphorus	3	0.08	97
Nitrogen	9	2.5	72
Potassium	10.5	3	71

Source: Y. A. Demirjian and Council on Environmental Quality, Annual Report, 1979.

lems associated with an elevated groundwater table. However, it is a possible alternative for direct (at a treatment plant), tertiary treatment, and experience gained from this project should be valuable in evaluating other possible sites for recycling waste water. Table 10.6 summarizes three years of water treatment performance in Michigan. The results are encouraging; most of the potential pollutants as well as heavy metals, color, and viruses are removed.

An interesting example that served as a model project for subsequent recycling of waste water comes from Santee, California, near San Diego. Even though the Santee project was eventually terminated, partly because of lack of customers for the reclaimed water (there is an abundant alternative water supply), it was innovative. After regular treatment of the sewage, the waste water was stored for a month in a holding pond (Lake 1 in Figure 10.28) to allow further oxidation.

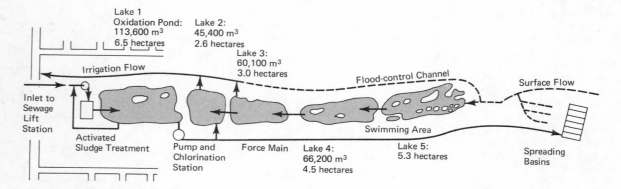

FIGURE 10.28

Sketch map showing a water renovation cycle at Santee, California. After leaving the treatment plant, water is taken to spreading basins where eventually it goes into the surface-water system and through a series of lakes to be used eventually for irrigation and other purposes. Although this project was terminated in the mid 1970s, it has served as a model for other water renovation projects. (Reprinted by permission from *POWER*, June 1966.)

Then, after chlorination, the water was pumped to a spreading area, where it trickled into the ground and flowed through a sand-gravel soil (an indirect tertiary treatment). The water surfaced again and entered a string of lakes, overflowing to form a stream used to irrigate a golf course. The geology of the Santee site in Sycamore Canyon is particularly suitable for lakes because a layer of impermeable clay underlies (at a depth of 3 to 5 m) surficial sands and gravels. The slope of the canyon and the existence of the sands and gravels overlying clay allows for lateral movement and easy collection of the waste water. The clay also seals the lakes, decreasing the possibility that waste water will contaminate other water resources. Microbiological investigation suggested that the lake water was safe, and, in addition to providing wildlife habitat, the area was accepted by 12,500 residents and other people who knew where the water came from and used the lake area for a variety of recreational activities (30, 31).

A final example of an innovative system which uses naturally occurring earth materials advantageously to purify water for public consumption is reported from a Michigan community on the shore of Lake Michigan. The city of Ludington, with a summer population of 11,000, uses sands and gravels below the lake bottom to prefilter and thus treat the lake water for municipal use. The procedure is shown schematically in Figure 10.29. A system of lateral intakes is buried in sand and gravel 4 to 5 meters below the lake bottom, where the water depth is at least 5 meters. The water is pumped out for municipal use, and in some cases the only additional treatment is chlorination.

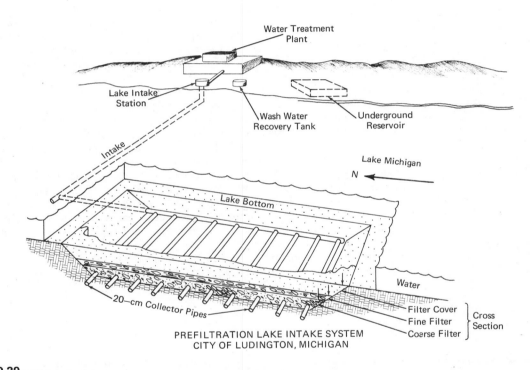

FIGURE 10.29
Idealized diagram showing how the water supply for Ludington, Michigan, is derived from the waters of Lake Michigan, using the natural, sandy bottom of the lake as a filter. (Courtesy of Williams and Works.)

SUMMARY AND CONCLUSIONS

The history of waste-disposal practices since the Industrial Revolution has moved from the practice of "dilution and dispersion" to a new concept of "concentrate, contain, and recycle."

A fast-growing disposal practice for solid waste is the sanitary landfill, but other methods, such as on-site disposal, composting, incineration, and open dumping, are still used. Of these methods, the geologic factor is most significant in siting sanitary landfills.

Hazardous chemical waste management may be the most serious environmental problem in the United States. Hundreds or even thousands of uncontrolled disposal sites may be time bombs that will eventually cause serious public health problems. Realistically, we know we are going to continue to produce hazardous chemical wastes. Therefore, it is imperative that carefully engineered secure landfills and other safe disposal methods be developed and used.

Disposal of toxic or otherwise hazardous waste by deep-well injection into an appropriate geologic environment can, in some instances, be a wise decision. Such disposal should not be considered a quick and easy solution to industrial-waste problems, however, and the method should be used only when more appropriate alternatives are not available. Furthermore, careful, continuous monitoring is necessary to insure safety when wastes are disposed of by deep-well injection.

Radioactive-waste management presents a serious and ever-increasing problem. High-level wastes that remain hazardous for thousands of years and are currently being stored will have to be permanently disposed of eventually. A likely method is disposal in a carefully and continuously monitored appropriate geologic environment, possibly bedded salt. Apparently, low-level waste can be buried safely and carefully controlled and monitored at near-surface sites.

Ocean dumping can be a significant source of marine pollution, and efforts to control indiscriminate dumping are in effect. Alternatives to ocean dumping of materials such as polluted dredge spoils, sewage sludge, and other potentially hazardous materials are being developed, but in many cases they are not yet practical or economically feasible.

Continued urban expansion will insure that the septic-tank sewage disposal system remains an important and common method of waste disposal. However, suitable soil and topography are necessary to reduce the chance of the system's failing, which would result in pollution of the environment and possible production of a public health problem.

The costs of waste-water treatment are the highest environmental costs facing state and local governments, and this trend is expected to continue. One of the largest expenses of operating and maintaining a waste-water treatment facility is that of the handling and disposal of sewage sludge.

An exciting possiblity for some municipal, industrial, and agricultural waste water is the concept of a "Waste Water Renovation and Conservation Cycle." The basic idea is to convert treated waste water into a resource for local recycling.

REFERENCES

1. GALLEY, J. E. 1968. Economic and industrial potential of geologic basins and reservoir strata. In *Subsurface disposal in geologic basins: A study of reservoir strata*, ed. J. E. Galley, pp. 1–19. American Association of Petroleum Geologists, Memoir 10.

2. COUNCIL ON ENVIRONMENTAL QUALITY. 1973. *Environmental quality—1973*. Washington, D.C.: U.S. Government Printing Office.

3. SCHNEIDER, W. J. 1970. *Hydraulic implications of solid-waste disposal*. U.S. Geological Survey Circular 601F.

4. TURK, L. J. 1970. Disposal of solid wastes—acceptable practice or geological nightmare? In *Environmental geology*, pp. 1–42. Washington, D.C.: American Geological Institute Short Course, American Geological Institute.

5. HUGHES, G. M. 1972. *Hydrologic considerations in the siting and design of landfills*. Environmental Geology Notes, no. 51. Illinois State Geological Survey.

6. BERGSTROM, R.E. 1968. *Disposal of wastes: Scientific and administrative considerations.* Environmental Geology Notes, no. 20. Illinois State Geological Survey.

7. CARTWRIGHT, K., and SHERMAN, F. B. 1969. *Evaluating sanitary landfill sites in Illinois.* Environmental Geology Notes, no. 27. Illinois State Geological Survey.

8. WALKER, W. H. 1973. Monitoring toxic chemical pollution from land disposal sites in humid regions. *Ground Water* 12:213–18.

9. ENVIRONMENTAL PROTECTION AGENCY. 1980. *Everybody's problem: Hazardous waste.* SW-826.

10. MAGNUSON, E. 1980. The poisoning of America. *Time* 116:12: 58–69.

11. ABELSON, P. H. 1983. Waste management. *Science* 220:1003.

12. ELLIOTT, J. 1980. Lessons from Love Canal. *Journal of the American Medical Association* 240: 2033–2034, 2040.

13. KUFS, C. and TWEDELL, D. 1980. Cleaning up hazardous landfills. *Geotimes* 25: 18–19.

14. McKENZIE, G. D., and PETTYJOHN, W. A. 1975. Subsurface waste management. In *Man and his physical environment,* ed. G. D. McKenzie and R. O. Utgard, pp. 150–56. Minneapolis: Burgess Publishing.

15. PIPER, A. M. 1970. *Disposal of liquid wastes by injection underground: Neither myth nor millennium.* U.S. Geological Survey Circular 631.

16. COMMITTEE OF GEOLOGICAL SCIENCES. 1972. *The earth and human affairs.* San Francisco: Canfield Press.

17. WARNER, D. L. 1968. Subsurface disposal of liquid industrial wastes by deep-well injection. In *Subsurface disposal in geologic basins: A study of reservoir strata,* ed. J. E. Galley, pp. 11–20. American Association of Petroleum Geologists, Memoir 10.

18. OFFICE OF INDUSTRY RELATIONS. 1974. *The nuclear industry, 1974.* Washington, D.C.: U.S. Government Printing Office.

19. BREDEHOEFT, J. D., ENGLAND, A. W., STEWART, D. B., TRASK, J. J., and WINOGRAD, I. J. 1978. *Geologic disposal of high-level radioactive wastes—Earth science perspectives.* U.S. Geological Survey Circular 779.

20. MICKLIN, P. P. 1974. Environmental hazards of nuclear wastes. *Science and Public Affairs* 30: 36–42.

21. HUNT, C. B. 1983. How safe are nuclear waste sites? *Geotimes* 28, no. 7: 21–22.

22. HEIKEN, G. 1979. Pyroclastic flow deposits. *American Scientist* 67: 564–571.

23. HAMMON, R. P. 1979. Nuclear wastes and public acceptance. *American Scientist:* 146–150.

24. COUNCIL ON ENVIRONMENTAL QUALITY. 1970. *Ocean dumping: A national policy.* Washington, D.C.: U.S. Government Printing Office.

25. SOIL CONSERVATION SERVICE. 1971. *Soils and septic tanks.* Agriculture Information Bulletin 349. Washington, D.C.: U.S. Government Printing Office.

26. U.S. DEPARTMENT OF HEALTH, EDUCATION, AND WELFARE. 1967. *Manual of septic tank practice.* Rockville, Maryland: Bureau of Community Management.

27. AMERICAN CHEMICAL SOCIETY. 1969. *Clean our environment: The chemical basis for action.* Washington, D.C.: U.S. Government Printing Office.

28. PARIZEK, R. R., and MYERS, E. A. 1968. Recharge of ground water from renovated sewage effluent by spray irrigation. *Proceedings of the Fourth American Water Resources Conference,* pp. 425–443.

29. FOGG, C. E. 1972. Waste management-nationally. *Soil Conservation,* May 1972.

30. POWER. 1966. *Water: A special report.* New York.

31. McCAULEY, D. 1977. Water reclamation and re-use in six cities. In *Water re-use and the city,* ed. R. E. Kasperson and J. X. Kasperson, pp. 49–73. Hanover, New Hampshire: The University Press of New England.

Chapter 11

The Geologic Aspects of Environmental Health

The geologic cycles—tectonic, rock, geochemical, and hydrologic—are responsible on a global scale for the geographic pattern of continents, ocean basins, and climates. On a smaller scale, the cycles produce the spatial arrangement of all landforms, rocks, minerals, soils, groundwater, and surface water. As a member of the biological community, the human race has carved a niche with other members of the community—a niche highly dependent on complex interrelations among the biosphere, atmosphere, hydrosphere, and lithosphere. We are only beginning to inquire into and gain a basic understanding of the total range of factors that affect our health and well-being. As we continue our exploration of the geologic cycle—from the scale of minute quantities of elements in soil, rocks, and water to regional patterns of climate, geology, and topography—we are making startling discoveries about how these factors might influence the incidence of certain diseases and death rates. In the United States alone, the death rate varies considerably from one area to another. Generally, the highest death rates are near the East Coast, with the rates about twice those in the Great Plains and some mountain areas (1).

Disease is defined as an imbalance resulting from a poor adjustment between an individual and the environment (2). It seldom has a one-cause, one-effect re-

lationship. The geologist's contribution is to help isolate aspects of the geologic environment that may influence the incidence of disease. This tremendously complex task requires sound scientific inquiry coupled with interdisciplinary research with physicians and other scientists. Although the picture is now rather vague, the possible rewards of the emerging field of medical geology are exciting and may eventually play a significant role in environmental health.

DETERMINING HEALTH FACTORS

To study the geologic aspects of environmental health, one must consider cultural and climatic factors associated with patterns of disease and death rates. This procedure helps to isolate the geologic influence. Cultural aspects of a society reflect the sum total of concepts and techniques that a group of people have developed to survive in their environment; these aspects influence the occurrence of disease by linking or separating disease and people. The nature and extent of these links depend on such factors as local customs and degree of industrialization. Primitive societies that live directly off the land and water are plagued by a different variety of health problems than urban society. Industrial societies have nearly eliminated diseases such as cholera, typhoid, hookworm, and dysentery, but are more likely to suffer from lung cancer and other diseases related to air, soil, and water pollution.

The high incidence of stomach cancer in Japanese people is an example of a relationship between culture and disease (3). The Japanese prefer rice that is polished and powdered; unfortunately, the powder contains asbestos as an impurity. Asbestos, a fibrous mineral, is either a true carcinogen (cancer-causing material) or takes a passive role as a carrier of trace-metal carcinogens.

Incidence of lead poisoning also suggests cultural, political, and economic relations with patterns of disease. Effects of lead poisoning can include anemia, mental retardation, and palsy. Lead is found in some moonshine whiskey and has resulted in lead poisoning among adults and even unborn or nursing infants whose mothers drank it (4). It has even been suggested that one of the reasons for the fall of the Roman Empire was widespread lead poisoning. Some estimates are that the Romans produced about 55,000 metric tons of lead per year for 400 years. Lead was used in pots in which grape juice was processed into a syrup for a preservative and sweetener of wine, in cups from which the Romans drank wine, and in cosmetics and medicines. The ruling class also had water piped into their homes through lead pipes. Historians argue that gradual lead poisoning among the upper class resulted in their eventual demise through widespread stillbirths, deformities, and brain damage. This hypothesis has gained support because of the high lead content found in the bones of ancient Romans (4).

Climatic factors such as temperature, humidity, and amount of precipitation are sometimes intimately related to disease patterns. Serious health hazards exist in the tropics where two of the worst climatically controlled diseases, schistosomiasis and malaria, are found. Schistosomiasis, called *snail fever,* is a significant cause of death among children and saps the energy of millions of people in the world. Its effects have tremendous socioeconomic consequences, and some researchers consider it the world's most important disease (2).

Malaria and schistosomiasis, among other diseases, are clearly related to climatic factors because the vectors necessary to carry the disease, that is, mosquitoes and snails, are climatically controlled. Other diseases, such as Burkett's tumor—a debilitating, but seldom fatal, cancerous disease of the lymph system—is associated in Africa with three climatic conditions: annual rainfall greater than 51 centimeters, elevations lower than 1,500 meters, and mean annual temperature of at least 15° Celsius (3). Figure 11.1

shows the climatic conditions and the occurrence of Burkett's tumor in Africa. We emphasize that the relationship between climate and the disease may not be a cause-effect relationship; in fact, some workers believe that Burkett's tumor is caused by a virus. Nevertheless, the association of the disease with particular climatic conditions is evident.

Assumed relations between culture and climate and the incidence of disease must be viewed with some skepticism because there is seldom a simple answer to environmental health problems. For example, if schistosomiasis were controlled only climatically, then all areas with an appropriate climate, such as much of the Amazon River Basin, would have the disease. Fortunately, this is not the case, and in some instances the reason is geologic. The conditions for the disease in the Amazon River Basin are nearly optimal, yet it occurs only in two very limited areas, primarily because there is not a sufficient amount of calcium in the water to support the snail, the intermediate host. In other areas, the acidity of the water in the presence of copper and other heavy metals may be responsible for the absence of the necessary snails in an otherwise suitable environment for schistosomiasis (1).

From this discussion, we can see some of the complex relations between disease patterns and envi-

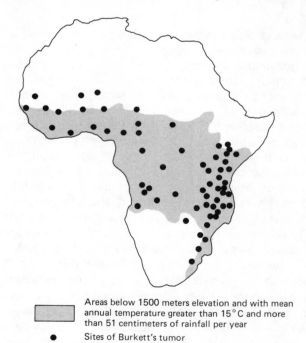

Areas below 1500 meters elevation and with mean annual temperature greater than 15°C and more than 51 centimeters of rainfall per year

● Sites of Burkett's tumor

FIGURE 11.1
Map of Africa showing areas below 1,500 meters in elevation with mean annual temperature greater than 15°C and more than 51 centimeters of rainfall, superimposed on sites where Burkett's tumor has been identified. (From Armed Forces Institute of Pathology.)

ronment. With this in mind, let us now focus on the geologic aspects of health problems by considering geologic factors of environmental health, occurrence and effects of trace elements on health, and the significance of the geologic environment to the incidence of heart disease and cancer, the leading causes of death in the United States.

SOME GEOLOGIC FACTORS OF ENVIRONMENTAL HEALTH

The soil that we cultivate to provide nutrients for subsistence, the rock on which we build our homes and industries, the water we drink, and the air we breathe all influence our chances of developing serious health problems. On the other hand, the same factors can also influence our chances of living a longer, more productive life. Surprisingly, many people still believe that soil and water in a "natural," "pure," or "virgin" state must be "good" and that if human activites have changed or modified them, they have become "contaminated," "polluted," and therefore "bad." This belief is by no means the entire story (5).

Relationships between geology and health are significant research and discussion topics. Although few valid cause-and-effect relationships have been isolated, we are learning more all the time about the subtle ways the geologic environment affects general health. Treatment of the various aspects of medical geology at even the introductory level requires discussion of the natural distributions of elements in the earth's crust and the ways in which natural and artificial processes concentrate or disperse those elements.

Natural Abundance of Elements

There is an inverse relation between atomic number and the abundance of elements in the universe. The lighter elements are encountered more frequently than the heavier ones. In general, this relation holds for both the lithosphere and the biosphere. Table 11.1, the periodic table of the elements, shows the position and atomic number of each element and indicates some of the more abundant elements in the earth's crust and environmentally important trace elements. Table 11.2 shows the most abundant elements in the average composition of the rocks of the continental crust. Note that more than 99 percent of these rocks by weight are concentrated in the first 26 elements of the periodic table. Table 11.3 shows the distribution of the more abundant elements in the average adult human body. Note that more than 99 percent of the body by weight is composed of elements in the first 20 elements of the periodic table.

Living tissue, animal or vegetable, is composed primarily of 11 elements, so-called bulk elements. Five of these are metals: hydrogen, sodium, magnesium, potassium, and calcium. Six are nonmetals: carbon, nitrogen, oxygen, phosphorus, sulfur, and chlorine. For those species that have hemoglobin, iron is added to the list. In addition to the bulk elements, living tissue requires several other elements to function properly. These are trace-element metals (present in minute quantities) that help to regulate the dynamic processes of life. Trace elements that have been studied and shown essential for nutrition include fluorine, chromium, manganese, iron, cobalt, copper, zinc, selenium, molybdenum, and iodine. This list is not complete, and it would not be surprising to learn that many more trace elements are essential or at least active in life processes (6,7). Other elements, such as nickel, arsenic, aluminum, and barium, accumulate as tissues age and are known as "age elements." The physiological consequences of the accumulation of some elements in living tissue are known in some cases but completely unknown or poorly understood in many others (7).

Concentration and Dispersion of Chemical Components

The movement of chemical compounds along various paths through the lithosphere, hydrosphere, atmosphere, and biosphere makes up the geochemical cycle. Natural processes, such as the release of gas by volcanic activity or the weathering of rock and rock debris, release chemical material into the environment. In addition, human use may result in the release of material and substances that lead to pollution or contamination of the environment. In general, but with many exceptions, concentrations of many trace elements tend to increase from rock to soil and water to plants and animals. Figure 11.2 shows some of the paths that trace elements may take to become concentrated in the human body, possibly causing health problems.

Once released by natural or artificial processes, elements and other substances are cycled and recycled by geochemical and rock-form processes. Thus, a certain concentration of a particular trace element in igneous rocks may have quite a different concentration in sedimentary rocks formed from the weathered products of the igneous rock. Whether the concentration has increased or decreased depends on the nature of the geochemical and rock-forming processes. Table 11.4 lists the concentrations of selected elements in igneous and sedimentary rocks. Although this information is not detailed, it is useful because it indicates a change in the relative abundance of elements produced

Key

Atomic Number

Environmentally Important Trace Elements **

* Element Relatively Abundant in Earth's Crust

Element Symbol

Element Name

1																	2
H Hydrogen																	He Helium
3 ** Li Lithium	4 Be Beryllium											5 * B Boron	6 * C Carbon	7 N Nitrogen	8 * O Oxygen	9 ** F Fluorine	10 Ne Neon
11 * Na Sodium	12 * Mg Magnesium											13 * Al Aluminum	14 *,** Si Silicon	15 P Phosphorus	16 S Sulfur	17 Cl Chlorine	18 Ar Argon
19 * K Potassium	20 * Ca Calcium	21 Sc Scandium	22 Ti Titanium	23 ** V Vanadium	24 ** Cr Chromium	25 ** Mn Manganese	26 *,** Fe Iron	27 ** Co Cobalt	28 ** Ni Nickel	29 ** Cu Copper	30 ** Zn Zinc	31 Ga Gallium	32 Ge Germanium	33 As Arsenic	34 ** Se Selenium	35 Br Bromine	36 Kr Krypton
37 Rb Rubidium	38 Sr Strontium	39 Y Yttrium	40 Zr Zirconium	41 Nb Niobium	42 ** Mo Molybdenum	43 Tc Technetium	44 Ru Ruthenium	45 Rh Rhodium	46 Pd Palladium	47 Ag Silver	48 ** Cd Cadmium	49 In Indium	50 ** Sn Tin	51 Sb Antimony	52 Te Tellurium	53 ** I Iodine	54 Xe Xenon
55 Cs Cesium	56 Ba Barium	57 La Lanthanum	72 Hf Hafnium	73 Ta Tantalum	74 W Wolfram	75 Re Rhenium	76 Os Osmium	77 Ir Iridium	78 Pt Platinum	79 Au Gold	80 ** Hg Mercury	81 Tl Thallium	82 *,** Pb Lead	83 Bi Bismuth	84 Po Polonium	85 At Astatine	86 Rn Radon
87 Fr Francium	88 Ra Radium	89 Ac Actinium															

58 Ce Cerium	59 Pr Praseodymium	60 Nd Neodymium	61 Pm Promethium	62 Sm Samarium	63 Eu Europium	64 Gd Gadolinium	65 Tb Terbium	66 Dy Dysprosium	67 Ho Holmium	68 Er Erbium	69 Tm Thulium	70 Yb Ytterbium	71 Lu Lutetium
90 Th Thorium	91 Pa Protactinium	92 U Uranium	93 Np Neptunium	94 Pu Plutonium	95 Am Americium	96 Cm Curium	97 Bk Berkelium	98 Cf Californium	99 Es Einsteinium	100 Fm Fermium	101 Md Mendelevium	102 No Nobelium	103 Lw Lawrencium

TABLE 11.1

Periodic table of the elements with examples of environmentally important trace elements and elements that are relatively abundant in the earth's crust highlighted.

TABLE 11.2
The relative abundance of the most common elements in the rocks of the earth's crust.

Atomic No.		Element	Weight (%)
8	O	Oxygen	46.4
14	Si	Silicon	28.15
13	Al	Aluminum	8.23
26	Fe	Iron	5.63
20	Ca	Calcium	4.15
11	Na	Sodium	2.36
12	Mg	Magnesium	2.33
19	K	Potassium	2.09
		Total	99.34

TABLE 11.3
Distribution of the more abundant elements in the adult human body.

Atomic No.	Element		Weight (%)
8	Oxygen	(O)	65.0
6	Carbon	(C)	18.0
1	Hydrogen	(H)	10.0
7	Nitrogen	(N)	3.0
20	Calcium	(Ca)	1.5
15	Phosphorus	(P)	1.0
16	Sulfur	(S)	0.25
19	Potassium	(K)	0.2
11	Sodium	(Na)	0.15
17	Chlorine	(Cl)	0.15
12	Magnesium	(Mg)	0.05
	Total		99.30

by rock-forming and biological processes, as, for example, the approximately tenfold increase in selenium from the original weathering of igneous rocks to the formation of shale. With the exception of coal and phosphorites (rock that is rich in calcium phosphate), other types of sedimentary rocks do not show the similar increase of selenium. This example and others suggest that geochemical and rock-forming processes such as weathering, leaching, accretion, deposition, and biological activity effectively sort, concentrate, and disperse elements and other substances throughout the environment.

Weathering is the physical and chemical breakdown of rock material and a major process in the formation of soil. Regardless of whether the parent material for soil is bedrock or rock debris transported and deposited by running water, wind, or ice, weathering is a natural process that frees trace elements to be used by the biosphere in life processes.

The artificial counterpart of weathering is pollution or contamination. These processes release trace elements into the environment; for example, lead is released into the environment when lead additives in gasoline are emitted through exhaust systems. Mercury, cadmium, nickel, zinc, and other metals are released into the atmosphere and water through industrial and mining operations.

Because some studies have shown a dramatic increase of lead in the air, water, and soil, and even the Antarctic ice, it is obvious that we need a closer examination of the possible adverse effects of chronic lead exposure. With the exception of human lead poisoning through lead-based paint and plant damage caused by exposure to lead close to highways or by heavy mineral concentrations released in mining areas, however, documented adverse health effects of lead are somewhat speculative. In fact, one author claims there is little difference in the amounts of lead in blood and urine samples from urban and rural residents (8).

FIGURE 11.2
A schematic drawing showing mechanisms by which trace elements may find their way to humans and animals, thus influencing the quality of health or producing disease. (After K. E. Beeson, *Geochemistry and the Environment,* vol. 1, 1974. Reproduced with permission of the National Academy of Sciences.)

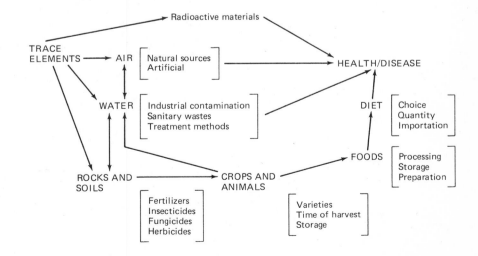

Leaching, accretion, deposition, biologic activity, and other processes may concentrate or disperse elements after they are released by natural and artificial processes. *Leaching* is the natural removal of soluable material (in solution) from the upper to lower soil horizons. Material that leaches out of soil may enter the groundwater system and be dispersed or diluted. If the material is sufficiently abundant or toxic or otherwise harmful, it also may pollute the groundwater. Leaching is most prevalent in warm, humid climates where the soil may be nutrient-poor because the nutrients are removed. Furthermore, trace elements that remain may be concentrated at undesirable levels.

Accumulation refers to processes that cause or increase retention of material in soil. Examples include salts that may accumulate on the surface and the upper zones of soils through evaporation processes, and materials that have been removed by leaching from the A horizon and accumulate in the B horizon. An example of the latter is found in semiarid regions where accumulation of calcium carbonate (caliche) is found in the B horizon of some soils. (See chapter 3 for a discussion of soil horizons.)

Deposition of earth materials has two environmentally important aspects. First, materials such as heavy metals cause biologic disruptions when deposited in streams, lakes, and oceans. Second, problems develop in areas where there is a deficiency of needed trace elements that results because the elements were not originally deposited along with other sediments.

Absorption of mercury to suspended sediments and bottom sediments may lead to high concentrations of mercury in aqueous environments. This absorption may lead to biological disruption because inorganic substances in the water are basic to the life processes of low life-forms in aquatic environments; these organisms assimilate the mercury, which is then passed on at a higher and higher concentration through the food chain.

Deficiencies that result when a certain material is not deposited with sediments moved by water, ice, and wind are less well understood because of the possibility of interactions of other processes such as leaching. Erosion and deposition by wind are particularly susceptible to selective removal. One author attributes the lack of lead, iron, copper, cobalt, and other materials in the sand hills of Nebraska to the fact that these metals occur in grains smaller and heavier than quartz and therefore were not moved with the quartz grains that formed the sand hills (3).

TRACE ELEMENTS AND HEALTH

Every element has a whole spectrum of possible effects on a particular plant or animal. For example, selenium is toxic in seleniferous areas, has no observable effect in most conditions, and is beneficial to animal production (raising cattle, sheep, etc.) in some areas. The apparent contradiction is resolved when we recognize that the first case is one of oversupply of selenium; the second represents a balanced state; and the third case is one of deficiency, which in some cases is rectified by supplementing the animals' food supply with the element (9).

It was recognized many years ago that the effects of a certain trace element on a particular organism depend on the dose or concentration of the element. This *dose dependency* can be represented by a dose-response curve, as shown in Figure 11.3 (9,10). When various concentrations of an element present in a bio-

FIGURE 11.3
Generalized dose-response curve.

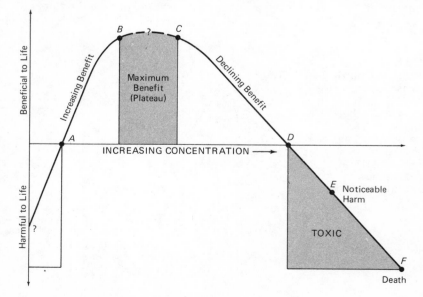

TABLE 11.4

Concentrations of some elements in various natural materials.

Type of Material	Concentrations by Elements (ppm)					Concentrations by Elements (ppm)				
	Cadmium	Chromium	Copper	Fluorine	Iodine	Lead	Lithium	Molybdenum	Selenium	Zinc
Ultramafic igneous[a,b]	0–0.2 0.05(?)	1,000–3,400 1,800	2–100 15	—	0.06–0.3 0.1	— 1	— 0.5	— 0.3	— 0.05	— 40
Basaltic igneous[a,b]	0.006–0.6 0.2	40–600 220	30–160 90	20–1,060 360	— 0.5	2–18 6	3–50 20	0.9–7 1.5	— 0.05	48–240 110
Granitic igneous[a,b]	0.003–0.18 0.15	2–90 20	4–30 15	20–2,700 870	— 0.5	6–30 18	10–120 35	1–6 1.4	— 0.05	5–140 40
Shales and clay[a,b,c]	0–11 1.4	30–590 120	18–120 50	10–7,600 800	212–380 5(?)	16–50 20	4–400 80	— 2.6	— 0.6	18–180 90
Black shales (high C)[d]	0.3–8.4 1.0	26–1,000 100	20–200 70	—	—	7–150 20	—	1–300 10	—	34–1,500 100 (?)
Deep-sea clays[a,b]	0.1–1 0.5	— 90	— 250	— 1,300	11–50 35 (?)	— 80	— 57	— 27	— 0.17	— 165
Limestones[a,b,c,e]	— 0.05	— 10	— 4	0–1,200 220	0.4–29 5	— 9	5–10 7	— 0.4	— 0.08	— 20
Sandstones[a,b,c]	— 0.05	— 35	— 2	10–880 180	— 1.7	1–31 7	7–90 30	— 0.2	— 0.05	2–41 16
Phosphorites[f]	0–170 30	30–3,000 300	10–100 30	24,000–41,500 31,000	—	10–30 10	—	3–300 30	1–100 18	20–300 50
Coals (ash)[a]	— 2	10–1,000 20	2–40 15	40–480 80	1–11 4	2–50 15	2–300 50	0.2–16 5	0.4–3.9 2[g]	7–108 50

Note: The upper figure is the range usually reported; the lower figure, the average.

[a]Turekian and Wedepohl (1961).

[b]Parker (1967).

[c]Becker et al. (1972).

[d]Vine, J. D., and Tourtelot, E. B., Econ. Geol. 65, 223 (1970).

[e]Wedepohl (1970).

[f]Gulbrandsen, R. A., Geochim. Cosmochim. Acta 30, 769 (1966).

[g]U.S. Geological Survey (1972).

Source: M. Fleischer. U.S. Geological Survey. Reproduced with permission of the National Academy of Science.

logical system are plotted against effects on the organism, three things are apparent. First, while relatively large concentrations are toxic, injurious, and even lethal (D–E–F in Figure 11.3), trace concentrations may be beneficial or even necessary for life (A–B). Second, the dose-response curve has two maxima (B–C) forming a plateau of optimal concentration and maximum benefit to life. Third, the threshold concentration, where harmful effects to life begin, is not at the origin (zero concentration), but varies with concentrations less than at point A and greater than at point D in Figure 11.3 (9,10).

Points A, B, C, D, E, and F in Figure 11.3 are significant threshold concentrations. Unfortunately, points E and F are known only for a few substances for a few organisms, including people, and the really important D is all but unknown (10). The width of the maximum-benefit plateau (points B, C) for a particular life form depends on the organism's particular physiological equilibrium (10). In other words, the different phases of activity, whether beneficial, harmful, or lethal, may differ widely both quantitatively and qualitatively for different substances, and, therefore, are completely observable only under special conditions (9).

A complete discussion of the geology and environmental effects of all trace substances could be a textbook in itself. Our objective here is to discuss representative examples to show the possible effects of imbalances of trace elements. For this purpose, we have selected fluorine, iodine, zinc, and selenium. In addition, it is valuable to relate trace-element problems to human use of the land, which we will do with examples of mining activity.

Fluorine

Fluorine is fairly abundant in rocks (Table 11.4) and soils (Table 11.5). Most of the fluorine in soils is derived from the parent rock, but it can also be added by volcanic activity, which deposits fluorine-rich volcanic ash on the land. Industrial activity and application of fertilizers have also, on a limited basis, contributed lo-

cally to an increase in the concentration of fluorine in soils.

Fluorine is an important trace element that forms fluoride compounds such as calcium fluoride which increases the crystallinity of the apatite (calcium phosphate) crystals in teeth. It helps prevent tooth decay by facilitating the growth of larger, more perfect crystals. The same processes occur in bones where fluoride assists in the development of more perfect bone structure that is less likely to fail with old age.

Relations between the concentration of fluoride (a compound of fluorine such as sodium fluoride, NaF) and health indicate a specific dose-response curve, as shown in Figure 11.4. The optimum fluoride concentration (point B) for the reduction of dental caries (DMF index: decayed, missing, and filled) is about 1 part per million (100 parts per million = 0.01 percent) (Figures 11.5, 11.6, and 11.7). Figure 11.8 shows areas in the United States known to have fluoride content in the groundwater greater than 1 part per million and areas with a fluoride content less than 0.4 part per million. Fluoride levels greater than 1.5 parts per million do not significantly decrease the DMF index, but they do increase the occurrence and severity of mottling (discoloration of teeth) (11). In concentration of about 4 to 6 ppm, fluoride may help prevent calcification of the abdominal aorta (Figure 11.9). Finally, fluoride concentrations of 4 to 6 parts per million may also reduce the prevalence of osteoporosis, a disease characterized by reduction in bone mass (Figure 11.10) and collapsed vertebrae (Figure 11.11) (12). The letters N.S. in the figures indicate the differences that are not statistically significant. Based on the study of fluoride in osteoporosis, point C on the dose-response curve for fluoride was located (Figure 11.4). Point E on the curve was located by the fact that fluoride concentrations of 8 to 20 parts per million are associated with excessive bone formation in the periosteum (dense, fibrous, outer layer of bone) and calcification of ligaments that usually do not calcify (7). The position of points A, D, and F are not known precisely, but fluoride in massive doses is the main ingredient of some rodent poisons.

TABLE 11.5
Concentration of fluorine in parts per million in samples of soils and other surficial earth materials from the conterminous United States.

Area	Range	Geometric Mean	Geometric Deviation	Arithmetic Mean[a]
Entire contiguous United States (911)	10–3,680	180	3.57	400
Western United States, west of the 97th medidian (491)	10–1,900	250	2.66	410
Eastern United States, east of 97th meridian (420)	10–3,680	115	4.38	340

[a]Estimated by method of Sichel (1952).

Note: Number of samples is given in parentheses after area.

Source: H. T. Shacklette et al., U.S. Geological Survey Circular 692, 1974.

FIGURE 11.4
Dose-response curve for fluoride.

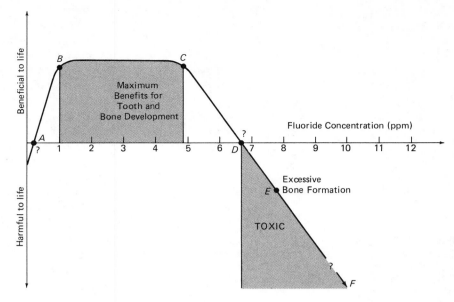

Iodine

Thyroid disease, caused by a deficiency of iodine, is probably the best-known example of the relationship between geology and disease. The thyroid gland, located at the base of the neck, requires iodine for normal function. Lack of iodine causes goiter, a tumorous condition involving enlargement of the thyroid gland

(13). Furthermore, a child born to a mother who suffers iodine deficiency during pregnancy may be a *cretin*, a mentally retarded dwarf. At one time cretinism was fairly common in areas of Mexico and Switzerland that had a high goiter-rate (14). The incidence of goiter is clearly related to deficiencies of iodine, as shown by the relation between iodine-deficient areas and the occurrence of goiter in the United States that appears in

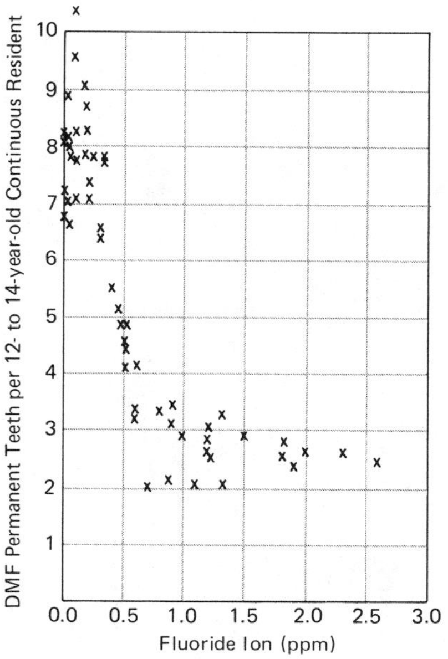

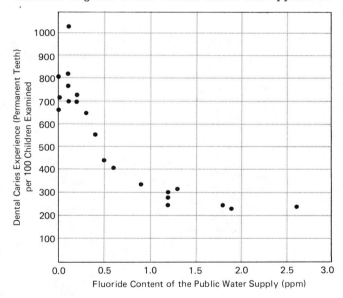

FIGURE 11.5
Relationship between fluoride concentration and prevalence of dental decay. (Reprinted from F. J. Maier, *Water Quality and Treatment,* 3rd ed., by permission of the Association. Copyrighted 1971 by the American Water Works Association, Inc., 6666 West Quincy Avenue, Denver, Colorado 80235. Used with permission of McGraw-Hill Book Company.)

FIGURE 11.6
Relation between number of dental caries (in permanent teeth) observed in 7,257 selected 12- to 14-year-old school children in 21 cities of four states, and the fluoride content of public water supply. (Reprinted from F. J. Maier, *Water Quality and Treatment,* 3rd ed., by permission of the Association. Copyrighted 1971 by the American Water Works Association, Inc., 6666 West Quincy Avenue, Denver, Colorado 80235. Used with permission of McGraw-Hill Book Company.)

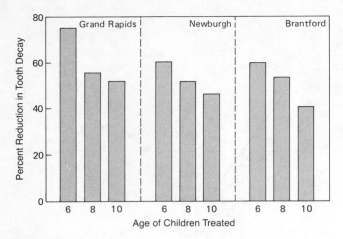

FIGURE 11.7
Reduction in tooth decay at pioneer cities as a result of fluoridation of the water. (Reprinted from F. J. Maier, *Water Quality and Treatment,* 3rd ed., by permission of the Association. Copyrighted 1971 by the American Water Works Association, Inc., 6666 West Quincy Avenue, Denver, Colorado 80235. Used with permission of McGraw-Hill Book Company.)

Figure 11.12. Use of iodized salt is now common in the goiter belt, and it has been found that all adolescent and many adult goiters slowly decrease in size when iodized salt is used. In only four years (1924–28), the use of iodized salt in Michigan reduced the incidence of goiter from 38.6 to 9 percent (13). In spite of this, a recent study indicated that the use of iodized salt is not as prevalent as it was before World War II, and goiter is still endemic to some areas in the United States, especially among the poor (14).

There is considerable speculation over what processes are responsible for concentrating iodine in or removing it from surficial earth materials. The most popular notion is that iodine is released by weathering of rocks. Some of it then enters the rivers and eventually the sea; therefore, the oceans have become great iodine reservoirs, containing possibly about 25 percent of the earth's total iodine (15). Because most iodine compounds are quite soluble, however, it is unlikely that much iodine is residual after long weathering processes. A more likely theory is that iodine is picked up in the ocean in a gaseous state or absorbed onto dust particles, transported by the atmosphere, and deposited by precipitation onto land areas. This idea proposes that iodine has slowly accumulated in the soil over a long period of time. Another possibility is related to the recent (geologically speaking) glaciation that ended just a few thousand years ago in the Great Lakes region. It might be argued that glaciated areas in the United States which roughly coincide with the goiter belt have low iodine concentrations in the soils be-

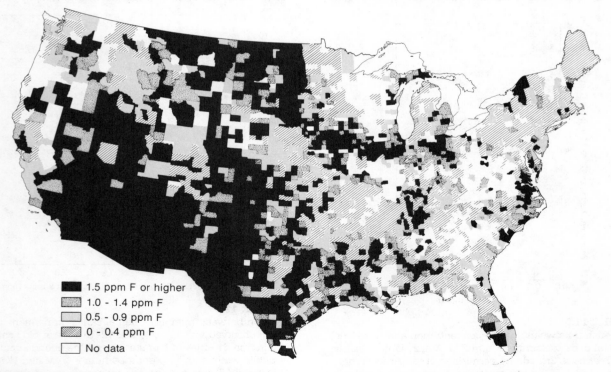

FIGURE 11.8
The fluoride content of groundwater in the coterminous United States. (From M. Fleischer, U.S. Geological Survey, Geological Society of America Special Paper 90, 1967.)

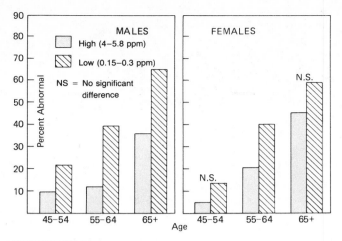

FIGURE 11.9
Percentage of radiologically detectable calcification of abdominal aorta in males and females residing in high- and low-fluoride areas. (Reprinted from the *Journal of the American Medical Association,* October 31, 1966, vol. 198. Copyright 1966, American Medical Association.)

cause glaciation removed and destroyed the older soils that had accumulated iodine for a long period, and the young, post-glaciation soils have not had sufficient time to accumulate large concentrations of iodine from the atmosphere (15). This is unlikely, except from a relative standpoint, considering the solubility of iodine compounds.

Biologic processes are also significant in affecting the amount of iodine available to plants and animals in an area. Plants affect the iodine content in soils by absorption of iodine into their living tissue, and by reten-

tion of iodine in the organic-rich (humus) upper soil horizon. Several analyses have shown that iodine tends to be concentrated in the upper soil horizons. Therefore, biological processes that cycle iodine in soils and plants might be more significant in determining availability of iodine than the amount of iodine in the local bedrock (15).

Zinc

Zinc is a trace element necessary to plants, animals, and people. Although zinc is a heavy metal that in excessive amounts has been associated with disease, it is known primarily from zinc-deficiency studies. The zinc content of surficial materials in the United States is shown in Figure 11.13. Zinc deficiencies are known in 32 states and have resulted in a variety of plant diseases that cause low yields, poor seed development, and even total crop loss (16).

Zinc deficiency in plants is related primarily to three soil conditions: low content of zinc, unavailability of zinc present in the soil, and poor soil management. The amount of zinc is low in areas that are highly leached, as in many coastal areas in the Coastal Plain of North Carolina, Georgia, and Florida (Figure 11.13). Zinc content is also low in areas such as the sand hills of Nebraska (northern Nebraska in Figure 11.13) where wind transport of the sand that formed the hills failed to move the heavy metals along with the lighter quartz grains (16).

Zinc is recognized as essential to all animals and people, especially during early stages of development and growth. Although required concentrations are small, even slight deficiencies can cause loss of fertil-

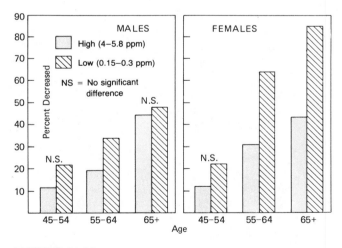

FIGURE 11.10
Percentage of decreased bone density in subjects from a high-fluoride area compared to those from a low-fluoride area (0.15 to 0.3 ppm). (Reprinted from the *Journal of the American Medical Association,* October 31, 1966, vol. 198. Copyright 1963, American Medical Association.)

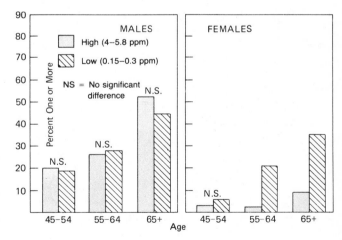

FIGURE 11.11
Percentage of subjects with one or more collapsed vertebrae from high- and low-fluoride areas. (Reprinted from the *Journal of the American Medical Association,* October 31, 1966, vol. 198. Copyright 1966, American Medical Association.)

FIGURE 11.12
Map of the United States showing relationship between goiter
occurrence and iodine deficiencies. (From Armed Forces Institute of Pathology.)

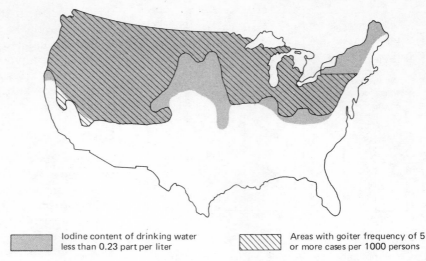

Iodine content of drinking water less than 0.23 part per liter

Areas with goiter frequency of 5 or more cases per 1000 persons

ity, delayed healing, and disorders of bones, joints, and skin. Zinc deficiencies are not unknown in humans and may be associated with some chronic arterial disease, lung cancer, and other chronic diseases (13). A problem is that interrelations between zinc deficiencies in hospital patients and the causes of disease are poorly understood. Do these people have the disease partially because they suffer from a zinc deficiency, or do they have a zinc deficiency because they have the disease? If the former is true, then zinc therapy with its known beneficial effects on tissue repair may be useful in treating some chronic diseases (16).

Adding zinc supplements to the soil may help prevent retardation of plant and animal growth. Care must be taken in adding zinc, however, because if the supplementary zinc is not of high quality, relatively large amounts of cadmium, which is often associated with zinc, may be inadvertently released into the environment. The result might be elimination of the growth problem, but a new hazard from the cadmium. Cadmium has been associated with bone disease, heart disease, and cancer.

Selenium

In concentrated amounts, selenium may be the most toxic element in the environment, and is a good example of the increasing concern over the necessity for controlling health-related trace elements. Selenium is required in the diet of animals at a concentration of 0.04 part per million, beneficial to 0.1 part per million, and toxic above 4 parts per million. Until a few years ago, selenium was of concern to biologists because of its toxicity, but we now know that greater losses have resulted from selenium deficiency than from selenium toxicity (17,18).

The primary source of selenium is volcanic activity. It has been estimated that throughout the history of the earth, volcanoes have released about 0.1 gram of selenium for every square centimeter of the earth's surface (17). Selenium ejected from volcanoes is in a particulate form. Therefore, it is easily removed from volcanic gas by rain and is usually concentrated near the volcano. This explains why the average concentration of selenium in the crust of the earth is about 0.05 parts per million, whereas the soils of Hawaii, which are derived from volcanic material, contain 6 to 15 parts per million (17).

Selenium in soils varies from about 0.1 part per million in deficient areas to as much as 1,200 parts per million in organic-rich soils in toxic areas. It is interesting to note that Hawaii, with a selenium content of 6 to 15 parts per million in the soil, does not produce toxic seleniferous plants, whereas South Dakota and Kansas, with soils containing less than 1 part per million selenium, do produce toxic seleniferous plants. This apparent dichotomy is resolved by examing the availability of selenium. Elemental selenium is relatively stable in soils and not available to plants. Availability of the element is controlled primarily by whether the soil is alkaline or acidic. In acid soils (as in Hawaii), the selenium is in an insoluble form, whereas in alkaline soils, selenium may be oxidized to a form that is extremely soluble in water and thus readily available to plants. This explains how forage crops grown for animals in soil containing soluble selenium are toxic, whereas forage crops grown in soils containing insoluble selenium have selenium deficiencies (17).

Selenium tends to be somewhat concentrated in living organisms. Some plants, called *selenium accumulators,* may contain more than 2,000 parts per million selenium, whereas other plants in the same area may have less than 10 parts per million total selenium (18). Selenium levels in human blood range from 0.1 to 0.34 part per million, which is about 1,000 times that

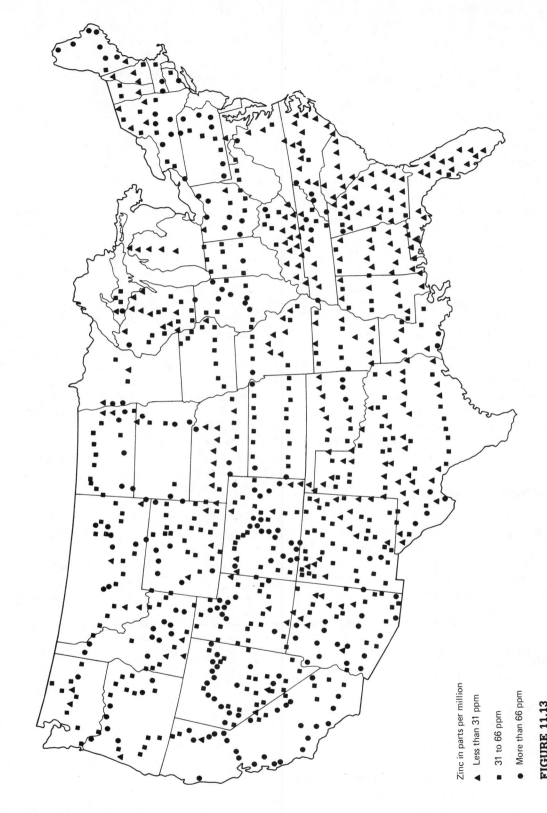

Zinc in parts per million

▲ Less than 31 ppm

■ 31 to 66 ppm

● More than 66 ppm

FIGURE 11.13

Zinc content of surficial earth materials in the coterminous United States. (From H. T. Shacklette, U.S. Geological Survey Professional Paper 574D, 1971.)

found in river water and several thousand times that found in seawater. Marine fish, on the other hand, contain selenium in amounts of about 2 parts per million which is many thousand times that found in seawater (17). Because selenium accumulates in organic material, it may be concentrated in organic sediment and soil. Therefore, fossil fuels such as coal that are developed from organic material also contain selenium. It has been estimated that the annual release of selenium by combustion of coal and oil in the United States is about 4,000 tons (17).

Little is known about selenium deficiency and toxicity in humans. A few cases of selenium poisoning have been reported in people who live on food grown in highly seleniferous soils or who drink seleniferous water in undeveloped countries. In economically developed countries where interregional shipments of food are standard and diets are high in protein, there is probably little problem with selenium (18).

Geographic patterns of selenium distribution are valuable in assessing the possibility that forage and other feed crops contain relatively high or low selenium concentrations. Figure 11.14 and Table 11.6 show the selenium concentrations in surficial materials of the United States. Figure 11.15 shows the general distribution of selenium in plants in the United States. This map can serve as a general guide to locating areas where the addition of selenium to animal feed might be beneficial. In some areas, selenium content cannot be accurately predicted, however, so detailed analysis and mapping remain necessary (18).

Human Use, Trace Substances, and Health

Agricultural, industrial, and mining activities have all been responsible for releasing potentially hazardous and toxic materials into the environment. This risk is part of the price we pay for our life style. For example, unexpected and unanticipated problems have arisen from seemingly beneficial chemicals that control pests and disease. As we become more sophisticated at anticipating and correcting problems, it is expected that environmental disruption from release of trace substances will be reduced.

Geologically significant examples of unexpected problems with trace substances are found in mining processes. Ironically, we spend time, energy, and money to extract resources concentrated by the geological cycle but in doing so, sometimes concentrate and release potentially harmful and toxic trace elements into the environment. Two rather different examples illustrate this situation: in Japan, the occurrence of a serious bone disease related to mining zinc, lead, and cadmium; and in Missouri, a metabolic imbalance of cattle associated with mining clay for the ceramic industry.

A serious chronic disease known as *Itaiitai* in the Zintsu River Basin, Japan, has claimed many lives. This extremely painful disease—the name *Itaiitai* literally means "ouch, ouch"—attacks bones, causing them to become so thin and brittle that they break easily. The disease broke out near the end of World War II when the Japanese industrial complex was damaged and good industrial-waste disposal practices were largely ignored. Mining operations for zinc, lead, and cadmium dumped mining wastes into the rivers. Farmers used the contaminated water downstream for domestic and agricultural purposes. For years, the cause of the disease was unknown, then, in 1960, bones and tissues of the victims of Itaiitai disease were examined and found to contain large concentrations of zinc, lead, and cadmium (19).

Measurement of heavy-metal concentrations in the water, sediment, and plants of the Zintsu River Basin, and subsequent experiments on rats fed diets of heavy metal, established two facts. First, although the water samples generally contained less than 1 part per million cadmium and 50 parts per million zinc, these minerals are selectively concentrated at high rates in

TABLE 11.6
Concentration of selenium in parts per million in samples of soil and other surficial earth materials from the coterminous United States.

Area	Range	Geometric Mean	Geometric Deviation	Arithmetic Mean[a]
Entire contiguous United States (912)	0.1–4.32	0.31	2.42	0.45
Western United States, west of the 97th meridian (492)	0.1–4.32	0.25	2.53	0.38
Eastern United States, east of the 97th meridian (420)	0.1–3.88	0.39	2.17	0.52

[a]Estimated by method of Sichel (1952).

Note: Number of samples is given in parentheses after area.

Source: U.S. Geological Survey Circular 692, 1974.

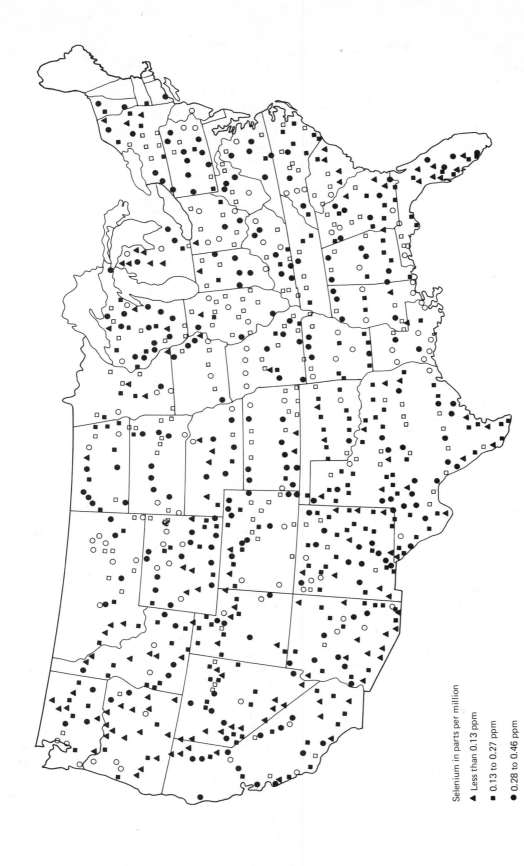

Selenium in parts per million

▲ Less than 0.13 ppm

■ 0.13 to 0.27 ppm

● 0.28 to 0.46 ppm

□ 0.47 to 0.77 ppm

○ More than 0.77 ppm

FIGURE 11.14
Selenium concentrations in surficial earth material of the coterminous United States. (From H. T. Shacklette, et al., U.S. Geological Survey Professional Circular, 1974.)

293

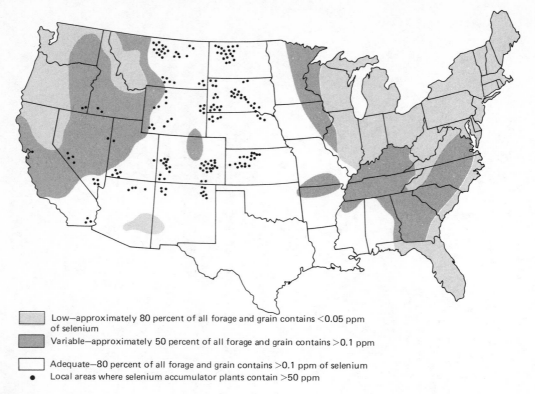

FIGURE 11.15
Map showing the concentration of selenium in plants in the coterminous United States. (From *Micronutrients in Agriculture,* 1972, by permission of the Soil Science Society of America, Inc.)

the sediment and higher yet in plants. One set of data for five samples shows an average of 6 parts per million cadmium in polluted soils. In plant roots, this average increased to 1,250, and in the rice it was 125. Second, rats fed a diet of 100 parts per million cadmium lost about 3 percent of their total bone tissue, and rats fed a diet containing 30 parts per million cadmium, 300 zinc, 150 lead, and 150 copper lost an equivalent of about 33 percent of their total bone tissue (19).

Although measurements of concentration of heavy metals are somewhat variable in the water, soil, and plants of the Zintsu River Basin, the general tendency is clear. Scientists are fairly certain that heavy metals, especially cadmium, in concentrations of a few parts per million in the soil and rice produce Itaiitai disease (20).

Strip mining of clay, coal, and other minerals may bring to the surface materials that contain anomalous concentrations of elements that are hazardous or toxic to an area's plants and animals. An example is a clay pit area in Missouri that was investigated by the U.S. Geological Survey (21). (See Figure 11.16.) Mining clay for use in the ceramic industry has resulted in severe disturbances in the metabolism of beef cattle; these

disturbances have interfered with the cattle's growth, nutrition, and reproduction.

The topography in the western part of the area is characterized by a flat upland surface generally above 244 meters in altitude. The underlying rocks are interbedded clay, shale, sandstone, and coal covered with a variable thickness of windblown silt deposits. The eastern part of the area is generaly below 244 meters in altitude, and the underlying rocks are limestone. Sinkholes in the limestone often contain deposits of fire clay that is mined. Figure 11.17 shows an abandoned clay pit approximately 100 meters wide, about 20 meters deep, and nearly filled with water. The large clay pile consists of clay with numerous fragments of shale, limestone, and pyritic material (iron sulfide). The smaller pit is almost entirely clay. Part of the surface runoff drains into the clay pit, and the remainder drains into the Rocky Branch. Only the cattle located on land downstream with direct access to the Rocky Branch suffered from metabolic disorders.

Although the rock types contain normal concentrations of elements, they are relatively rich in some elements compared to the soils. The occurrence of pyrite in the rocks is a particular problem because when

FIGURE 11.16
Map of a clay pit area in Callaway County, Missouri. The numbers correspond to sample sites. (After Ebens, et al., U.S. Geological Survey Professional Paper 807, 1971.) (700 to 800 ft = 213 to 244 m)

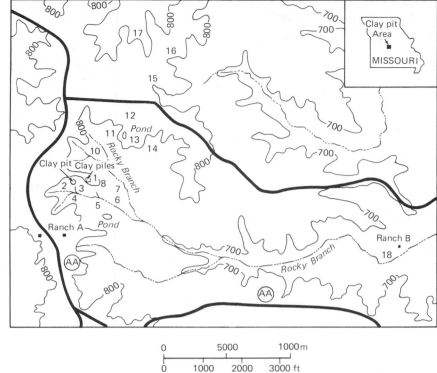

pyrite is exposed to water, it weathers to sulfuric acid, which increases the solubility of other compounds. This is essentially the same problem that is prevalent in coal-mining areas where acidic water from mines contaminates streams. Water that drains from clay piles into the pit and streams is very acidic and contains high concentrations of some trace elements. This material is subsequently deposited in the bed of the stream and on the floodplains. Plants that grow on the clay piles along the floodplains contain some trace elements in concentrations toxic to animals that eat the plants (21).

FIGURE 11.17
Aerial view of a clay pit area, Callaway County, Missouri. Dumptruck at lower right indicates the scale. (From Ebens, et al., U.S. Geological Survey Professional Paper 807, 1973.)

Extensive analysis of more than 30 different elements in the clay, sediment, and plants of the clay pit area revealed that about 20 were present in anomalous concentrations. These fit into one of five general groups:

1 Elements that occur in anomalous amounts in the clay or sediment or both, and are also found in anomalous amounts in many plants; examples include aluminum, copper, molybdenum, nickel, and sodium.
2 Elements that occur in anomalous concentrations in the clay and sediment, but are seldom found in anomalous amounts in plants; examples include barium, beryllium, chromium, and vanadium.
3 Elements that occur in relatively low amounts in the clay and sediment, but are concentrated in some plants; examples include boron, cadmium, calcium, and zinc.
4 Elements that are anomalous in clay or sediment, but are not easily evaluated in plants; examples include carbon, selenium, and silicon.
5 Elements whose concentrations and rates of movement through the local environment are not easily categorized at this time; examples include iron, lead, magnesium, manganese, and strontium (21).

This grouping suggests that some elements may influence metabolic imbalances in cattle because the animals eat plants with a toxic level of a particular element, as in the first and third groups. On the other hand, some elements may influence the metabolism of grazing animals because the animals directly ingest clay and sediment or drink water that contains particularly harmful elements in solution or suspension.

Evaluation of the entire geochemical environment established two items: (1) four elements (beryllium, copper, molybdenum, and nickel) are conspicuously anomalous in the clay, sediment, and plants; and (2) other elements are highly mobile (beryllium, cobalt, copper, and nickel, among others). Of these elements, three (cobalt, copper, and molybdenum) are known to be significant in metabolic processes of animals. In trace amounts they are essential, but in high concentrations they are likely to be toxic (21).

The geologic, hydrologic, and biologic processes responsible for providing, transporting, and concentrating toxic trace elements are indeed complex. Figure 11.18 lists and describes the movement of elements through the geochemical system of the clay pit area. This example is especially valuable because it indicates the interaction of the complex, multidimensional aspects of geochemical systems. Indeed, complexity is much more common than simplicity in the study of anomalous concentrations of trace elements and their effects on the biosphere.

FIGURE 11.18
Movement of elements through the geochemical system of a clay pit area, Callaway County, Missouri. (From Ebens, et al., U.S. Geological Survey Professional Paper 807, 1973.) (700 to 800 ft = 213 to 244 m)

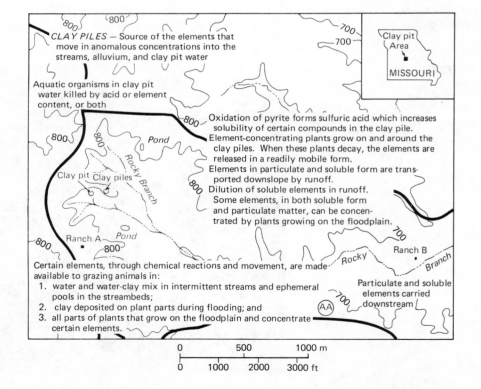

CHRONIC DISEASE AND GEOLOGIC ENVIRONMENT

Health is an organism's state of adjustment to its own internal environment and to its external environment. The relationships between the geologic environment and regional or local variations in chronic disease such as cancer and heart disease in people have been observed for many years. Although evidence suggesting associations between the geochemical environment and chronic disease continues to accumulate, the real significance remains to be discovered. Reasons for the lack of conclusive results are twofold. First, hypotheses for testing relations between the geologic environment and disease have not been specific enough. Basic research and field verification need to be better coordinated. Second, there remain many difficulties in obtaining reliable and comparable data for medical-geologic studies (22).

Although the significance of geologic variations in contributing to disease compared to that of other environmental factors, such as climate, remains an educated guess, the benefit to humankind of learning more about these relationships is obvious. The geographic variations of the incidence of heart disease in the United States may be related to the geologic environment; and it has been estimated that two-thirds of the cancerous tumors in the Western Hemisphere result from environmental causes, while genetic and racial factors are secondary in importance (22).

Heart Disease and the Geochemical Environment

The term *heart disease* here includes coronary heart disease (CHD) and cardiovascular disease (CVD). Variations of heart disease mortality have generally shown interesting relationships with the chemistry of drinking water. For example, studies in Japan, England, Wales, Sweden, and the United States all conclude that communities with relatively soft water have a higher rate of heart disease.

Perhaps the first report of a relationship between water and heart disease came in 1957 from Japan, where the most prevalent cause of death is apoplexy, a sudden loss of body functions caused by the rupture of a blood vessel. The geographic variation of the disease in Japan is related to the ratio of the sulfate to bicarbonate (SO_4 to HCO_3) of river water. Water with low sulfate-to-bicarbonate ratios is relatively hard, whereas high ratios indicate soft (acid) water. Figure 11.19 shows areas in Japan where the readjusted death rates are relatively high, greater than 120 per 100,000 population, compared to areas where the sulfate-to-bicarbonate ratio is relatively high, greater than 0.6 (23).

The abundance of sulfate, especially in the northeastern part of Japan, evidently stems from the sulfur-rich volcanic rock found there. Rivers in Japan that flow through sedimentary rocks are, in contrast, low in sulfate and high in bicarbonate, like most river water in the world. In general, however, river water in Japan is relatively soft, with a hardness less than 40 parts per million, compared to water in the United States where the mean hardness of raw municipal water is 139 parts per million (24).

A general relationship between soft water and high death rates from heart disease is also present in the United States. Figure 11.20 shows the adjusted death rate from heart disease from 1949 to 1951 and the average hardness of the water (25). The figures are consistent with a more recent study which also found a significant negative correlation between hardness of water and death rates from heart disease (24). We em-

FIGURE 11.19
Maps of Japan comparing the ratio of SO_4 to CO_3 in rivers to the death rate of apoplexy in 1950. (Data from T. Kobayashi, 1957.)

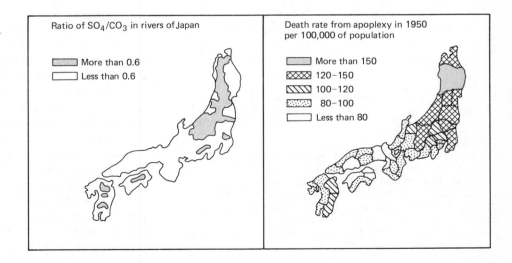

Ratio of SO_4/CO_3 in rivers of Japan

More than 0.6
Less than 0.6

Death rate from apoplexy in 1950 per 100,000 of population

More than 150
120–150
100–120
80–100
Less than 80

phasize, however, that the generally negative correlation between heart disease and hardness of water is not conclusive. A study in Indiana found a small positive correlation, suggesting that many other variables may exert considerable influence on rates of heart disease (26), and a study in Ohio suggests that sulfate (SO_4) and bicarbonate (HCO_3) concentrations possibly influence the incidence of heart disease (27). The Ohio study found a slight positive relation between deaths from heart attack and sulfate concentration in drinking water. Ohio counties with sulfate-rich drinking water derived from coal-bearing rocks in the southeast part of the state tend to have a higher death rate due to heart attack (Figure 11.21). On the other hand, Ohio counties that have drinking water characterized by low-sulfate and high-bicarbonate concentration, derived from

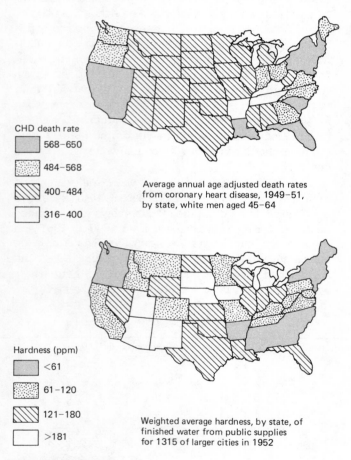

CHD death rate

- 568–650
- 484–568
- 400–484
- 316–400

Average annual age adjusted death rates from coronary heart disease, 1949–51, by state, white men aged 45–64

Hardness (ppm)

- <61
- 61–120
- 121–180
- >181

Weighted average hardness, by state, of finished water from public supplies for 1315 of larger cities in 1952

FIGURE 11.20

Maps of the coterminous United States comparing the average annual adjusted death rate from coronary heart disease (CHD) to the weighted average hardness of the public water supplies. (Reprinted from E. F. Winton and L. J. McCabe, *Journal of American Water Works Association,* Volume 62, by permission of the Association. Copyrighted 1971 by the American Water Works Association, Inc., 6666 West Quincy Avenue, Denver, Colorado 80235.)

young glacial deposits, tend to have a relatively low death rate from heart attack (Figure 11.21) (27).

The meaning of relationships between water chemistry and death rates from heart disease is difficult to surmise, although there are several possibilities. First, the relationships may have nothing to do with heart disease. Second, soft water is acidic and may, through corrosion, attack pipes and release into the water trace elements that cause heart disease. Third, some substances dissolved in hard water or water with low-sulfate and high-bicarbonate concentration may retard heart disease. Or, fourth, some characteristics of soft water or water with high-sulfate and low-bicarbonate concentration may enhance heart disease. Of course, some combination of the second, third, and fourth factors, along with others, is also likely, consistent with our observation that a disease may have several causes. Additional research is needed to prove the benefit of hard water and perhaps treat soft water to reduce heart disease; for now, we can only say that heart disease may indeed be related to the chemistry of drinking water, but the explanation remains obscure.

An interesting study of patterns of heart disease mortality and possible relations to the geologic environment was conducted in Georgia by the U.S. Geological Survey (28,29). The contrast in heart disease mortality among the 159 counties in Georgia is such that the differences between the highest and lowest rates are nearly as great as can be found in any two counties in the United States. Nine counties in northern Georgia with low rates of death caused by heart disease and nine counties in central and south central Georgia with high death rates caused by heart disease were selected for geochemical analysis.

The locations of these counties are shown in Figure 11.22. Table 11.7 lists the death rates from heart disease and other causes in the selected counties. In the nine selected counties with low rates of death from heart disease, the range is from 560 to 682 deaths per 100,000 population for white males 35 to 74 years of age during the period 1950 to 1959. In the nine selected counties with high mortality rates, the range is from 1,151 to 1,446 deaths per 100,000 population for the same age group and time period (28). Mortality rates for the south central part of the state are thus about twice as high as for the northern section.

The nine counties with high heart-disease rates are primarily in southern Georgia on the Atlantic Coastal Plain (Figure 11.22). The landscape characteristically has low relief, sluggish drainage, and swamps. In general, sandy soils overlie Cenozoic marine sedimentary rocks that have undergone intensive weathering. Small-scale agricultural activity is prevalent where there are sufficient relief and drainage.

FIGURE 11.21
Distribution of sulfate-rich and
bicarbonate-rich surface water
by county and occurrence of
heart-attack death by county in
Ohio. (After R. J. Bain, 1979, *Geology* 7: 7–10.)

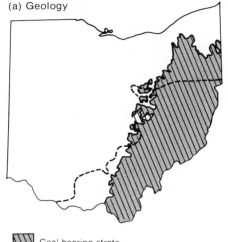

(a) Geology

Coal bearing strata

(b) Water Chemistry (by county)

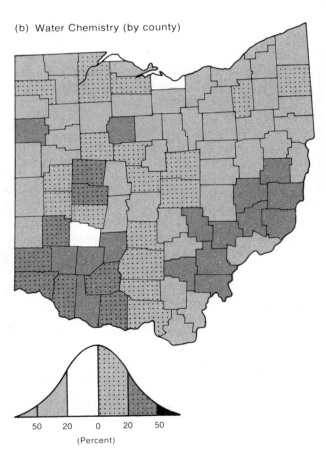

50   20   0   20   50
(Percent)

State average per 100,000 persons is 389.

(c) Heart-attack deaths (by county)

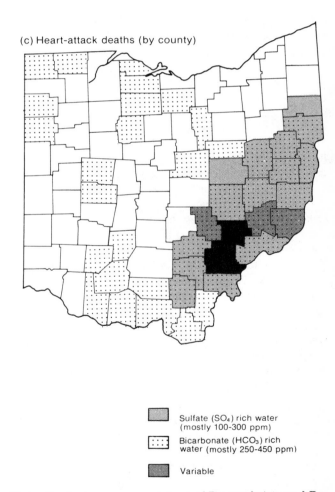

Sulfate (SO₄) rich water
(mostly 100-300 ppm)

Bicarbonate (HCO₃) rich
water (mostly 250-450 ppm)

Variable

The nine counties with low heart-disease rates are located in northern Georgia in the Appalachian Highlands, known as the Piedmont, the Blue Ridge, and the Valley and Ridge (Figure 11.22). The *Piedmont* extends from the Blue Ridge Mountains to the Coastal Plain. Rocks here are a mixture of Precambrian and Paleozoic metamorphic rocks and intrusive igneous rocks of varying ages. Soils are generally mixtures of sand, silt, and clay (loam) with clay subsoils. The *Blue Ridge* topography is mountainous, with narrow valleys and

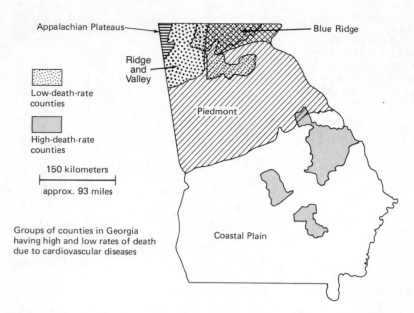

FIGURE 11.22
Map showing physiographic regions and Georgia counties having high and low death rates from cardiovascular disease. (After Shacklette, Sauer, and Miesch, U.S. Geological Survey Professional Paper 574C, 1970.)

Appalachian Plateaus

Blue Ridge

Ridge and Valley

Low-death-rate counties

High-death-rate counties

Piedmont

150 kilometers

approx. 93 miles

Coastal Plain

Groups of counties in Georgia having high and low rates of death due to cardiovascular diseases

turbulent streams. The rocks include metamorphic and sedimentary rocks of Precambrian to early Cambrian age and Paleozoic igneous rocks. Soils are acid with relatively high organic content; high relief is not favorable to agricultural activity. The *Valley and Ridge* is characterized by linear ridges and parallel valleys oriented northeast to southwest; rocks are folded and faulted sedimentary rock of Paleozoic age. Soils vary from rich loam soils in valleys with limestone bedrock to less fertile clay soils on shales and relatively infertile soils derived from sandstone bedrock (28).

A detailed geochemical investigation of the soils and plants of the low-death-rate area in northern Georgia and high-death-rate area in south central Georgia was conducted (28, 29). It was concluded that the geochemical variations in the soil reflected differences in the bedrock. Thirty elements were analyzed, and significant concentrations of aluminum, barium, calcium, chromium, copper, iron, potassium, magnesium, manganese, niobium, phosphorus, titanium, and vanadium were found. Zirconium was the only element that occurred in significantly larger concentrations in soils of the high-death-rate counties (29). The differences probably reflect the fact that soils in the high-death-rate counties of the Coastal Plain derive from unconsolidated sediments (sands and clays) that were previously weathered and leached during deposition.

Several trace elements are known to have beneficial effects on heart disease. Of these, manganese, chromium, vanadium, and copper are more highly concentrated in the low-death-rate areas of Georgia. Therefore, the low death rate may be a result of an abundance of beneficial trace elements in the soils rather than concentrations of harmful elements. In other words, the tendency for higher mortality from

heart disease in Georgia may be a causal relationship facilitated by a deficiency rather than an excess of certain trace elements (29).

Additional information is needed about other elements and their relationship to heart disease. Cadmium, fluorine, and selenium, among other elements, require further study to increase our understanding of how trace elements affect the heart and circulatory system. With respect to cadmium, we know that individuals who die from hypertensive complications generally have greater concentrations of cadmium or higher ratios of cadmium to zinc in their kidneys compared to individuals who die from other diseases. Surprisingly, however, workers who are exposed to cadmium dust and accumulate the element in their lungs do not have abnormal rates of hypertension (30).

Cancer and the Geochemical Environment

Cancer tends to be strongly related to environmental conditions. As with heart disease, however, relationships between geochemical environment and cancer cannot be proven, and undoubtedly the causes of the various types of cancers are complex and involve many variables, some of which may be the presence or absence of certain earth materials that contribute to, or help protect people from, disease.

Relationships between cancer and environment have two aspects: cancer-causing (carcinogenic) substances released into the environment by *human use* of resources; and cancer-causing substances that occur *naturally* in earth materials such as soil and water.

Recent information suggests that the presence of cancer-causing substances in drinking water is ubiquitous. This may or may not be true, but certainly water polluted with industrial waste containing toxic chemicals, some of which are possible carcinogens, is being released into our surficial water supplies. The Mississippi River, particularly, has pollution problems, and, ironically, even present water treatment causes problems. When combined with chlorination in water treatment, some industrial waste turns into cancer-producing material. Other carcinogens escape water treatment because antiquated treatment procedures fail to remove toxic substances.

The occurrence of certain cancers has been related to the natural environment. Examples include iodine deficiency and its relation to breast cancer (14); mineralized drinking water and its relation to stomach cancer (31); organic matter, zinc, and cobalt in soil and their relation to stomach cancer (32,33); and soil salinity, climate, vegetation, and agriculture and their rela-

TABLE 11.7

The death rates from selected causes, white males age 35 to 74, for selected counties of Georgia during the period 1950 to 1959.

| County | Cardiovascular diseases (330–334, 400–468[a]) | | | All Noncardiovascular Causes of Death | All Causes of Death |
	Total	Coronary Heart Disease (420[a])	Other Cardiovascular Diseases		
Low-rate counties					
Cherokee	671.0	344.1	326.9	525.4	1196.4
Fannin	655.2	262.2	393.0	607.9	1263.1
Forsyth	635.7	275.1	360.6	529.6	1165.3
Gilmer	647.5	296.9	350.6	485.4	1132.9
Hall	667.7	328.6	339.1	573.3	1241.0
Murray	680.8	274.8	406.0	593.6	1274.4
Pickens	681.7	332.2	349.5	735.5	1417.2
Towns	587.6	201.2	386.4	480.0	1067.6
Union	560.1	287.1	273.0	477.5	1037.6
High-rate counties that were studied					
Bacon	1302.2	743.1	559.1	789.9	2092.1
Bleckley	1205.5	650.6	554.9	546.0	1751.5
Burke	1167.0	603.7	563.3	834.3	2001.3
Dodge	1206.3	569.4	636.9	736.1	1942.4
Emanuel	1167.8	467.4	700.4	731.5	1899.3
Jeff Davis	1446.4	701.1	745.3	568.6	2015.0
Jefferson	1151.2	679.3	471.9	677.9	1829.1
Jenkins	1330.8	530.1	800.7	802.5	2133.3
Warren	1321.7	915.6	406.1	760.2	2081.9
Other high-rate counties					
Baldwin[b]	1218.8	781.5	437.3	744.7	1963.5
Chatham	1177.2	753.7	423.5	757.2	1934.4
Johnson	1170.3	644.1	526.2	700.1	1870.4
Lee	1170.6	754.5	416.1	927.8	2098.4
Long	1162.3	614.6	547.7	638.1	1800.4
Marion	1155.8	667.2	488.6	686.6	1842.4
Randolph	1252.1	572.2	679.9	618.1	1870.2
Richmond[b]	1203.0	773.5	429.5	771.3	1974.3
Treutlen	1171.7	533.2	638.5	766.2	1937.9
All counties of Georgia	925.2	520.4	404.8	649.0	1274.2

[a]Code number of International Statistical Classification of Diseases, 7th Revision (World Health Organization, 1957).

[b]With adjustments for resident institution populations.

Note: Average annual death rates per 100,000 population, age adjusted by 10-year age groups by the direct method to the entire United States population age 35–74 in 1950. Deaths tabulated by Georgia State Department of Health, by county of usual residence.

Source: Shacklette, Sauer, and Miesch, U.S. Geological Survey Professional Paper 574C, 1970.

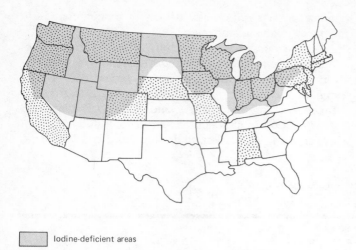

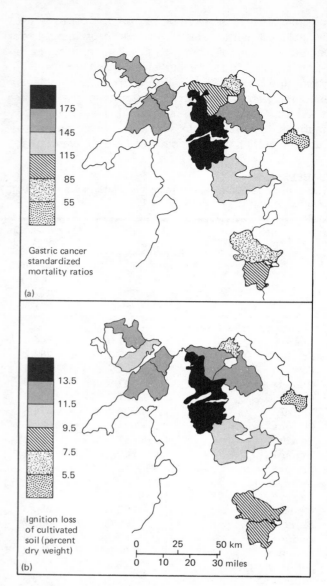

Gastric cancer
standardized
mortality ratios

(a)

Ignition loss
of cultivated
soil (percent
dry weight)

(b)

FIGURE 11.25
Maps of North Wales rural districts comparing cancer mortality ratios and ignition loss of cultivated soils. (After C. D. Legon, *British Medical Journal*, September 27, 1952.)

Iodine-deficient areas

24 states with greatest
incidence of breast cancer

FIGURE 11.23
Map of coterminous United States comparing iodine-deficient areas ("goiter belt") with the 24 states with the greatest incidence of breast cancer. (Data from Bogardus and Finley, 1961, and Spencer, *The Texas Journal of Science*, 1970).

tion to esophageal cancer (34). These studies do not prove a cause-and-effect relationship between certain earth materials and cancer, but they do indicate an area for future research that promises to be significant in solving environmental health problems.

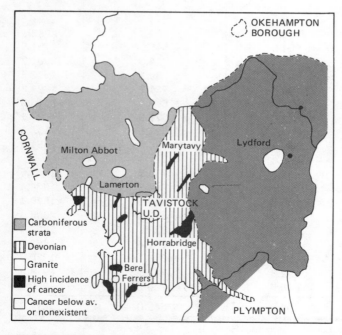

FIGURE 11.24
Map of West Devon, England, showing generalized rock types and areas where the incidence of cancer is high or low. (From Allen-Price, *The Lancet*, vol. 1, 1960.)

The leading cause of death for women between 40 and 44 years of age is breast cancer. It is also the leading cause of death for all types of cancer for all women between the ages 35 and 55 (14). The occurrence of breast cancer in the United States has been related to areas with an iodine deficiency (the goiter belt). Figure 11.23 shows areas with a relatively high incidence of breast cancer superimposed on areas with iodine deficiency. It has been observed that countries with iodine deficiencies and a resultingly high incidence of goiter also have high incidences of breast cancer. The converse also holds; that is, countries with

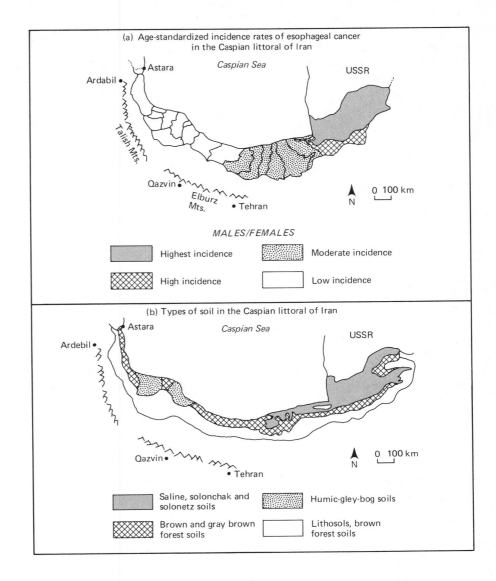

(a) Age-standardized incidence rates of esophageal cancer in the Caspian littoral of Iran

MALES/FEMALES

Highest incidence   Moderate incidence

High incidence   Low incidence

(b) Types of soil in the Caspian littoral of Iran

Saline, solonchak and solonetz soils   Humic-gley-bog soils

Brown and gray brown forest soils   Lithosols, brown forest soils

sufficient iodine have a low incidence of goiter and breast cancer (14). The significance of this observation is not well understood, but is apparently real.

A study into the relationship between the occurrence of cancer, especially stomach cancer, in West Devon, England, established the hypothesis that cancer in that area is connected with water supplies derived from a particular rock type. Figure 11.24 shows the geology of the area and the high incidence of cancer where drinking water is pumped from the rocks of Devonian age (345 to 400 million years old). These sedimentary rocks are highly mineralized compared to the sedimentary rocks of Carboniferous age (280 to 345 million years old) or to the granite that is mineralized but to a lesser extent than the Devonian rocks. Research suggests the cancer occurrence is associated with the mineralization of the Devonian rocks, but no specific cancer-causing substance has been isolated (31).

Two studies in Wales and England have established relationships between stomach cancer and soil characteristics. The earlier study concluded that areas with a high incidence of stomach cancer had soils with a relatively high amount of organic matter (31). The amount of organic material is measured by the percentage of weight loss of a dry soil after burning. Figure 11.25 shows the correlation of areas with high mortality rates of stomach cancer with organic content of cultivated soils. This study was not able to associate disease rates with specific cancer-producing substances in the soils.

A more recent, detailed study of soils in northern Wales and Cheshire, England, combined geochemical analysis of the soil and measurement of organic content (33). It concluded that the abnormal rates of stomach cancer were associated with the amount of organic material in the soil such that abnormally high rates of the disease were related to long residence on soils with an organic content between certain limits. In addition, the study found a positive relationship between the concentration of zinc, cobalt, and chromium and cancer of the stomach. Zinc and cobalt may be especially significant for two reasons. First, zinc is an active part of some essential enzyme systems in the human body and is also active in the processes of gastric digestion. Second, cobalt is known to have carcinogenic properties and is important in plant and animal activity. Although the high rates of cancer in this study are related to excesses of trace elements in the soil, the geographic distribution of such soils is apparently unrelated to that of stomach cancer (33).

Incident rates of esophageal cancer in northern Iran, near the Caspian Sea, offer another opportunity to examine relations between environment and chronic disease. In this area, the rates of esophageal cancer are tremendously variable, and marked differences occur over short distances (34). Research in the area indicates a relationship between cancer and the climate, soils, vegetation, and agricultural practices. There is a great deal of interrelationship among these factors, however, as the climate obviously affects the soils, vegetation, and agricultural practices. The highest correlation between rates of esophageal cancer and environment is for soil types. Figure 11.26 shows relative rates of cancer and general soil types. It is obvious that the highest rates of the disease are associated with the saline (salty) soils in the eastern section. These soils are in an area of little rainfall. Rainfall increases to the west by a factor of 4, and the soils become more and more leached. The lowest incidence of cancer is in the rain belt in the western section (Figure 11.26). Vegetation and agriculture of the east systematically change from grazing on sparse cover of specially adapted plants to lush forests and dry farming of rice, fruit, and tea in the west. Paralleling this change from east to west is the continuously lessening incidence of esophageal cancer (34).

The examples of relations between environment and incidence of cancer further establish the importance of recognizing multiple causes of chronic disease. Regardless of the numerous contradictions and exceptions that can be found, sufficient data are now available to establish beyond a reasonable doubt that hypothesizing a single cause for cancer is invalid. Rather, we must pursue a multifactorial approach, and variations in the environment are certainly among the significant factors.

SUMMARY AND CONCLUSIONS

The death rate in some areas is certainly greater or less than in others, and the cause of a particular disease is not a one-cause, one-effect relationship.

Cultural factors can affect the geographic pattern of a particular disease. Examples include stomach cancer in Japan and lead poisoning in numerous areas.

Temperature, humidity, and precipitation are important factors in the climatic control of disease patterns, of which schistosomiasis and malaria are the best-known examples. In addition, other diseases are associated with a particular climate (for example, Burkett's tumor), but cause-and-effect relations are not clear.

The natural distribution of matter in the universe is such that the lighter elements are more abundant. The first 26 elements of the periodic table are sufficient to account by weight for the vast majority of the lithosphere and biosphere.

The movement of elements along numerous paths through the lithosphere, hydrosphere, biosphere, and atmosphere is known as the *geochemical cycle*. Along these paths, natural processes such as weathering and volcanic activity, combined with human pollution, release all types of substances into the environment. They are then concentrated or dispersed by other natural processes such as leaching, accretion, deposition, and biologic activity.

Every element has an entire spectrum of possible effects on plants and animals. The graph of concentration of a certain element against effects on a particular organism is known as a dose-response curve. For a particular element, trace concentrations may be beneficial or necessary for life. Higher concentrations may be toxic, and still higher concentrations lethal, to a particular plant or animal. Interesting examples of environmentally important elements include iodine, fluorine, zinc, and selenium, among others.

Agricultural, industrial, and mining activities have released hazardous and toxic materials into the environment. These materials have been associated with bone disease in people, metabolic disorders in cattle, and other biological problems.

Relations between chronic disease and geologic environment are complex and difficult to analyze. Furthermore, relationships are not proof of cause and effect. Nevertheless, considerable evidence is being gathered and studied, and preliminary results suggest that the geochemical environment is indeed a significant factor in the incidence of serious health problems such as heart disease and cancer.

Heart disease rates are apparently related to water chemistry such as total hardness and concentration of sulfate or bicarbonate ions. In addition, a study in Georgia suggests that deficiencies of trace elements may also be related to abnormal rates of heart disease.

Studies of cancer rates suggest that abnormal incidence in specific areas may be related to such factors as organic content of the soil and abundance of certain trace elements. In other areas, saline soils may be related to cancer. Although most tumors in the Western Hemisphere are assumed to be caused by environmental factors, evidence of simple cause-and-effect relations is lacking, and multiple causation of disease remains more likely.

REFERENCES

1. SAUER, H. I., and BRAND, F. R. 1971. Geographic patterns in the risk of dying. In *Environmental geochemistry in health,* ed. H. L. Cannon and H. C. Hopps, pp. 131–50. Geological Society of American Memoir 123.

2. HOPPS, H. C. 1971. Geographic pathology and the medical implications of environmental geochemistry. In *Environmental geochemistry in health,* ed. H. L. Cannon and H. C. Hopps, pp. 1–11. Geological Society of America Memoir 123.

3. YOUNG, K. 1975. *Geology: The paradox of earth and man.* Boston: Houghton Mifflin.

4. BYLINSKY, G. 1972. *Metallic menaces.* In *Man, health and environment,* ed. B. Hafen, pp. 174–85. Minneapolis: Burgess Publishing.

5. WARREN, H. V., and DELAVAULT, R. E. 1967. A geologist looks at pollution: Mineral variety. *Western Mines* 40:23–32.

6. FURST, A. 1971. Trace elements related to specific chronic diseases: Cancer. In *Environmental geochemistry in health,* ed. H. L. Cannon and H. C. Hopps, pp. 109–30. Geological Society of America Memoir 123.

7. CARGO, D. N., and MALLORY, B. F. 1974. *Man and his geologic environment.* Reading, Mass.: Addison-Wesley.

8. UNDERWOOD, E. J. 1971. Geographical and geochemical relationships of trace elements to health and disease. In *Trace substances in environmental health—IV,* ed. D. D. Hemphill, pp. 3–11. Columbia, Missouri: University of Missouri Press.

9. MERTZ, W. 1968. Problems in trace element research. In *Trace substances in environmental health—II,* ed. D. D. Hemphill, pp. 163–69. Columbia, Missouri: University of Missouri Press.

10. HORNE, R. A. 1972. Biological effects of chemical agents. *Science* 177: 1152–53.

11. MAIER, F. J. 1971. Fluorides in water. In *Water quality and treatment,* 3d ed., pp. 397–440. New York: McGraw-Hill.

12. BERNSTEIN, D. S., SADOWSKY, N., HEGSTED, D. M., GURI, C. D., and STARE, F. J. 1966. Prevalence of osteoporosis in high- and low-fluoride areas in North Dakota. *The Journal of the American Medical Association* 198: 499–504.

13. GILBERT, F. A. 1947. *Mineral nutrition and the balance of life.* Norman, Oklahoma: University of Oklahoma Press.

14. SPENCER, J. M. 1970. Geologic influence on regional health problems. *The Texas Journal of Science* 21: 459–69.

15. SHACKLETTE, H. T., and CUTHBERT, M. E. 1967. Iodine content of plant groups as influenced by variations in rock and soil type. In *Relations of geology and trace ele-*

ments to nutrition, eds. H. L. Cannon and D. F. Davidson, pp. 31–45. Geological Society of America Special Paper 90.

16. PORIES, W. J., STRAIN, W. H., and ROB, C. G. 1971. Zinc deficiency in delayed health and chronic disease. In *Environmental geochemistry in health and disease,* eds. H. C. Cannon and H. C. Hopps, pp. 73–95. Geological Society of America Memoir 123.

17. LAKIN, H. W. 1973. Selenium in our environment. In *Trace elements in the environment,* ed. E. L. Kothny, pp. 96–111. Advances in Chemistry Series 123, American Chemical Society.

18. ALLAWAY, W. H. 1969. Control of environmental levels of selenium. In *Trace substances in environmental health—II,* ed. D. D. Hemphill, pp. 181–206. Columbia, Missouri: University of Missouri Press.

19. PETTYJOHN, W. A. 1972. No thing is without poison. In *Man and his physical environment,* ed. G. D. McKenzie and R. O. Utgard, pp. 109–10. Minneapolis: Burgess Publishing.

20. TAKAHISA, H. 1971. Discussion in *Environmental geochemistry in health and disease,* ed. H. L. Cannon and H. C. Hopps, pp. 221–22. Geological Society of America Memoir 123.

21. EBENS, R. J., ERDMAN, J. A., FEDER, G. L., CASE, A. A., and SELBY, L. A. 1973. *Geochemical anomalies of a claypit area, Callaway County, Missouri, and related metabolic imbalance in beef cattle.* U.S. Geological Survey Professional Paper 807.

22. ARMSTRONG, R. W. 1971. Medical geography and its geologic substrate. In *Environmental geochemistry in health and disease,* ed. H. L. Cannon and H. C. Hopps, pp. 211–19. Geological Society of America Memoir 123.

23. KOBAYASHI, J. 1957. On geographical relationship between the chemical nature of river water and death-rate of apoplexy. *Berichte des Ohara Institute fur landwirtshaftliche biologie* 11: 12–21.

24. SCHROEDER, H. A. 1966. Municipal drinking water and cardiovascular death-rates. *Journal of the American Medical Association* 195: 125–29.

25. WINTON, E. F., and McCABE, L. J. 1970. Studies relating to water mineralization and health. *Journal of American Water Works Association* 62: 26–30.

26. KLUSMAN, R. W., and SAUER, H. I. 1975. *Some possible relationships of water and soil chemistry to cardiovascular diseases in Indiana.* Geological Society of America Special Paper 155.

27. BAIN, R. J. 1979. Heart disease and geologic setting in Ohio. *Geology* 7: 7–10.

28. SHACKLETTE, H. T., SAUER, H. I., and MIESCH, A. T. 1970. *Geochemical environments and cardiovascular mortality rates in Georgia.* U.S. Geological Survey Professional Paper 574C.

29. SHACKLETTE, H. T., SAUER, H. I., and MIESCH, A. T. 1972. Distribution of trace elements in the occurrence of heart disease in Georgia. *Geological Society of America Bulletin* 83: 1077–82.

30. SCHROEDER, H. A. 1965. Cadmium as a factor in hypertension. *Journal of Chronic Disease* 18: 647–56.

31. ALLEN-PRICE, E. D. 1960. Uneven distribution of cancer in West Devon. *Lancet* 1: 1235–38.

32. LEGON, C. D. 1952. The Aetiological significance of geographical variations in cancer mortality. *British Medical Journal,* 700–702.

33. STOCKS, P., and DAVIES, R I. 1960. Epidemiological evidence from chemical and spectrographic analysis that soil is concerned in the causation of cancer. *British Journal of Cancer* 14: 8–22.

34. KMET, J., and MAHBOUBI, E. 1972. Esophageal cancer in the Caspian littoral of Iran. *Science* 175: 846–53.

FOUR

Minerals, Energy, and Environment

One of the most fundamental concepts of environmental geology is that this earth is our only suitable habitat and its resources are limited. This belief holds even though some resources such as timber, water, air,* and food are renewable, whereas other resources such as oil, gas, and minerals are recycled so slowly in the geologic cycle that they are essentially nonrenewable. It is important to recognize, however, that renewable resources are renewable only as long as environmental conditions remain favorable for their natural or planned reproduction. Careless use of renewable resources, such as air, water, vegetation, and, to a lesser extent, soils, may render these renewable resources less renewable than we would wish.

Because resources are indeed limited, important questions arise. How long will a particular resource last? How much short- or long-term environmental deterioration are we willing to concede to insure that resources are developed in a particular area? How can we make the best use of available resources? These questions have no easy answers. We are now struggling with better ways to estimate the quality and quantity of resources. Unfortunately, it is extremely hard to address the second and third questions without a satisfactory answer to the first question, but determining how long a particular resource will last is further complicated because availability will change as our technological skills and discovery techniques become more sophisticated. Furthermore, the nature and extent of people's and institutions' willingness to conserve resources and

---

*A discussion of the air environment including: air as a renewable resource; the urban air and air pollution; the urban microclimate; and global climate change is presented in Appendix I.

Photo courtesy of Pacific Gas and Electric Company.

protect sensitive environments from mineral exploitation and development depends on factors such as national security, desired standard of living, environmental awareness, and economic incentives, as well as many other factors, all of which are difficult to evaluate and which change with time.

Discussion of mineral resources often centers on elements such as aluminum, copper, iron, lead, and zinc. Many people are surprised to learn that, with the exception of iron, none of these elements is used as much as carbon, sodium, nitrogen oxygen, sulfur, potassium, and calcium. The annual worldwide consumption of carbon is on the order of 10 billion tons, reflecting our dependence on coal, oil, and gas. Sodium and iron are consumed at a rate of about 1 billion tons per year. Iron is the structural metal our civilization most depends on, and sodium is used primarily in chemical industries (chlorine production, glass, soda ash, etc.). Nitrogen, oxygen, sulfur, potassium, and calcium are each consumed at the rate of about 100 million tons per year; of these five elements, four are used primarily as soil conditioners or fertilizers. Elements such as copper, aluminum, and lead have annual worldwide consumption rates of about 10 million tons, and the much discussed metals such as gold and silver have annual consumption rates less than 10,000 tons. Therefore, a significant conclusion concerning the consumption of elements is that the nonmetalics, with the exception of iron, are consumed at much greater rates than elements used for their metallic properties[1].

Basically, our present resource problem (or crisis) is related to a people problem—too many people. As we stated earlier, it is impossible to maintain an ever-growing population on a finite resource base. With this perspective, chapter 12 will explore geologic and environmental aspects of mineral resources, and chapter 13 will discuss energy resources. Although there is considerable overlap between these two subjects, for organizational purposes we will discuss all resources associated with energy production separately. Mineral resources will include a wide variety of earth materials, including metallic and nonmetallic minerals as well as sand, gravel, crushed rock, and ornamental rock (dimension stone) that have commercial value. Some earth materials, such as those used for construction material (sand and gravel, etc.), are a low-value resource and have primarily a "place value." That is, they are extracted economically because they are located near where they are to be used. Long hauls of these low-value, high-bulk materials drastically increase their price and reduce their chances of competing with other, closer supplies. On the other hand, materials such as diamonds, copper, gold, and aluminum are high-value resources. These materials are extracted wherever they are found and transported around the world to numerous markets regardless of distances[2].

[1]SKINNER, B. J. 1969. *Earth resources.* Englewood Cliffs, New Jersey: Prentice-Hall.

[2]FLAWN, P. T. 1970. *Environmental geology.* New York: Harper & Row.

Chapter 12

Mineral Resources and Environment

MINERALS AND POPULATION

Infinite resources, including space, could support an infinite population. Unfortunately, the earth and its resources are finite while the population grows at an ever-faster pace. Although overpopulation has been a problem in some areas for at least several hundred years, it is now apparent that it is a global problem, sometimes called the *population bomb* (Figure 12.1). From 1830 to 1930, the world population doubled from 1 to 2 billion people, an annual growth rate of less than 1 percent. By 1970, it had nearly doubled again, and by the year 2000, it is expected to double once more. These statistics suggest that in the last 40 years the growth rate was a little less than 2 percent per year, and now it is a little greater than 2 percent per year. The problem is that this is *exponential growth,* which is a dynamic process. Consider the simple example of a student who, upon taking a job for one month, requests from the employer a payment of 1 cent for the first day of work, 2 cents for the second day, 4 the third day, and so on. In other words, the payment would double each day. What would be the total? It would take the student eight days to earn a wage of more than $1 per day, and by the eleventh day, earnings would be more than $10 per day. Payment for the six-

teenth day of the month would be more than $300, and on the last day of a 31-day month, the student's earnings for that one day would be more than $10 million!

Consider another hypothetical and completely fictitious example of exponential growth with more environmental significance (1). A farm family owns a pond in which a fast-growing water plant becomes established. As it grows, the plant chokes off all other life in the pond. The family is advised that the plant, which doubles its area each day, will cover the entire pond in about six weeks, or 42 days. As the plant appears to be growing rather slowly the first few days and weeks, the family decides not to worry about it until it covers approximately one-half the pond, at which time they plan to remove it. That time will come on the forty-first day, and they will have only one remaining day to save their pond!

The question often asked is, how safe is our way of life? From a resource standpoint, affluence is indeed in jeopardy (1). Consider that approximately 1000 kilograms of new mineral material (excluding energy resources) are required each year for each person in the United States (Figure 12.2). When resource data are combined with population data, the conclusion is clear—it is impossible in the long run to match exponential population growth with production of useful

FIGURE 12.1
The population bomb. (From U.S.
Department of State.)

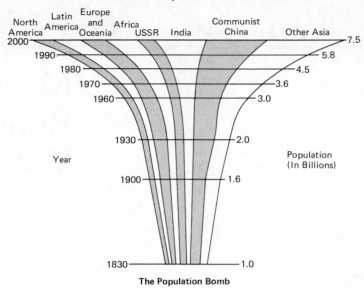

Growth of World Population to the Year 2000

The Population Bomb

materials based on a finite resource base. Considering that many other countries also aspire to affluence, while the world population increases, the availability of necessary resources becomes questionable. World production of commonly used metals such as iron, aluminum, copper, and lead, for example, would have to increase several times to bring the rest of the world up to the standard of the United States, suggesting the possibility of an eventual resource crisis. Historically, however, many expected crises have not developed, and it is hoped that, through conservation, planning, and scientific advancements, there is time to insure that the looming resource crisis will also fail to materialize.

The United States and other affluent societies today are completely dependent on minerals and in fact could not survive without them (2). Table 12.1 lists a

few of the mineral products, including petroleum, in a typical American home. This list shows that minerals are used extensively in almost every activity in our homes, offices, and industries.

A rather simple relationship summarizes the role of minerals in influencing our expected standard of living (3):

$$L \cong \frac{R \times E \times I}{P}$$

where L is the average standard of living as measured by useful consumption of goods and services and is a function of: R, the useful consumption of resources such as metals, nonmetals, soil, water, etc.; E, the useful consumption of energy; I, the useful consumption of all types of ingenuity, including technical, social, economic, and political; and P, the total number of people

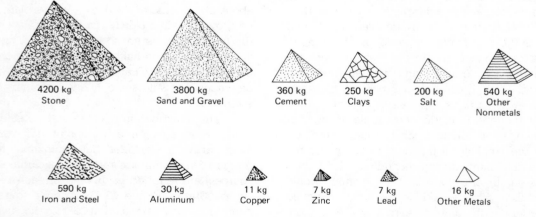

FIGURE 12.2
Amount of new mineral materials required annually by each U.S. citizen. (After U.S. Bureau of Mines, *Mining and Mineral Policy*, 1975.)

TABLE 12.1
A few of the mineral products in a typical American home.

Building materials	Sand, gravel, stone, brick (clay), cement, steel, aluminum, asphalt, glass
Plumbing and wiring materials	Iron and steel, copper, brass, lead, cement, asbestos, glass, tile, plastic
Insulating materials	Rock, wool, fiberglass, gypsum (plaster and wallboard)
Paint and wallpaper	Mineral pigments (such as iron, zinc, and titanium) and fillers (such as talc and asbestos)
Plastic floor tiles, other plastics	Mineral fillers and pigments, petroleum products
Appliances	Iron, copper, and many rare metals
Furniture	Synthetic fibers made from minerals (principally coal and petroleum products); steel springs; wood finished with rotten-stone polish and mineral varnish
Clothing	Natural fibers grown with mineral fertilizers; synthetic fibers made from minerals (principally coal and petroleum products)
Food	Grown with mineral fertilizers; processed and packaged by machines made of metals
Drugs and cosmetics	Mineral chemicals
Other items	Windows, screens, light bulbs, porcelain fixtures, china, utensils, jewelry: all made from mineral products

Source: U.S. Geological Survey Professional Paper 940, 1975.

involved in the system. This relationship clearly shows that any increase in R, E, or I will increase the standard of living, while an increase in P will decrease the standard of living. Of course, an even worse condition arises if R and E decrease while P increases and I cannot increase fast enough to compensate.

A problem facing the United States and many other affluent nations is that domestic supplies of many mineral resources must be supplemented, to a lesser or greater extent, by importing them from other nations. Figure 12.3 and Table 12.2 show that imports supply a significant percentage of the total United States resource demand. That a particular material is imported does not necessarily suggest, however, that it does not exist in quantities that could be mined in the United States; it may suggest instead that for economic, political, or conservation purposes, it is easier and more practical to import certain resources.

RESOURCES AND RESERVES

Mineral resources may be broadly defined as elements, chemical compounds, minerals, or rocks that are concentrated either in the crust of the earth or in the oceans in a form such that a usable commodity can be obtained (4). From a practical viewpoint, this concept of a resource is unsatisfactory because, except in emergencies, a particular commodity will not be extracted unless extraction can be accomplished at a profit. Therefore, a more pragmatic definition of a *resource* is a concentration of a naturally occurring material (solid, liquid, or gas) in or on the crust of the earth in a form such that economical extraction is currently or potentially feasible (2). A *reserve,* on the other hand, is that portion of the total resource that is identified and from which usable materials can be extracted economically and legally *at the time* of the evaluation (2). The distinction between resources and reserves, therefore, is based on current geologic and economic factors (Figure 12.4). Resources include both identified reserves and other materials that are either identified but not currently economically or legally available or are undiscovered (currently hypothetical or speculative) (2).

TABLE 12.2
Ratio of U.S. production to consumption for selected materials in 1978. A low ratio (import dependent) means that domestic production is much less than consumption, whereas a high ratio (export potential) means domestic production exceeds consumption.

Import Dependent		U.S. Production Matches Consumption		Export Potential	
Manganese	0.02	Steel	0.94	Ammonia	1.14
Cobalt	0.03	Aluminum	0.95	Soda Ash	1.23
Bauxite	0.11	Lead	0.96	Phosphates	1.57
Tin	0.20	Copper	1.00	Magnesium	1.63
Nickel	0.25	Iron Ore	1.00	Boron	1.69
Potassium	0.43	Cement	1.03	Molybdenum	2.21
Zinc	0.51	Lime	1.06		

Source: Mining and Mineral Policy, 1979.

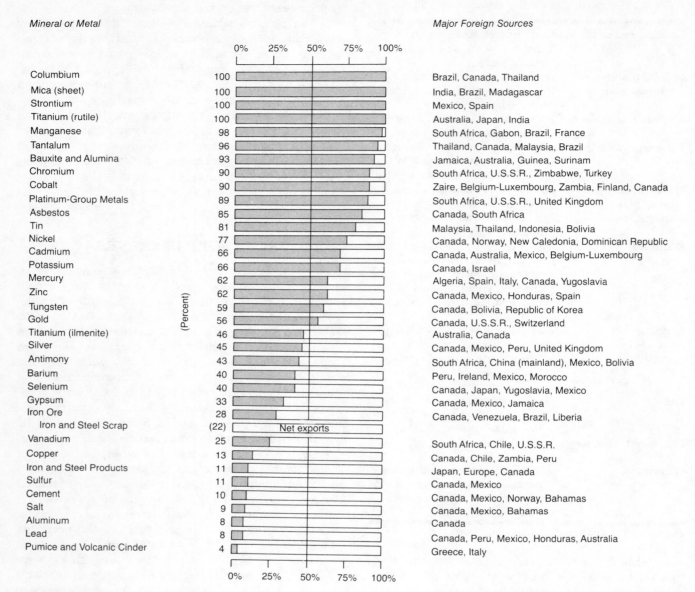

Percentage Imported

Mineral or Metal		Major Foreign Sources
Columbium	100	Brazil, Canada, Thailand
Mica (sheet)	100	India, Brazil, Madagascar
Strontium	100	Mexico, Spain
Titanium (rutile)	100	Australia, Japan, India
Manganese	98	South Africa, Gabon, Brazil, France
Tantalum	96	Thailand, Canada, Malaysia, Brazil
Bauxite and Alumina	93	Jamaica, Australia, Guinea, Surinam
Chromium	90	South Africa, U.S.S.R., Zimbabwe, Turkey
Cobalt	90	Zaire, Belgium-Luxembourg, Zambia, Finland, Canada
Platinum-Group Metals	89	South Africa, U.S.S.R., United Kingdom
Asbestos	85	Canada, South Africa
Tin	81	Malaysia, Thailand, Indonesia, Bolivia
Nickel	77	Canada, Norway, New Caledonia, Dominican Republic
Cadmium	66	Canada, Australia, Mexico, Belgium-Luxembourg
Potassium	66	Canada, Israel
Mercury	62	Algeria, Spain, Italy, Canada, Yugoslavia
Zinc	62	Canada, Mexico, Honduras, Spain
Tungsten	59	Canada, Bolivia, Republic of Korea
Gold	56	Canada, U.S.S.R., Switzerland
Titanium (ilmenite)	46	Australia, Canada
Silver	45	Canada, Mexico, Peru, United Kingdom
Antimony	43	South Africa, China (mainland), Mexico, Bolivia
Barium	40	Peru, Ireland, Mexico, Morocco
Selenium	40	Canada, Japan, Yugoslavia, Mexico
Gypsum	33	Canada, Mexico, Jamaica
Iron Ore	28	Canada, Venezuela, Brazil, Liberia
Iron and Steel Scrap	(22) Net exports	
Vanadium	25	South Africa, Chile, U.S.S.R.
Copper	13	Canada, Chile, Zambia, Peru
Iron and Steel Products	11	Japan, Europe, Canada
Sulfur	11	Canada, Mexico
Cement	10	Canada, Mexico, Norway, Bahamas
Salt	9	Canada, Mexico, Bahamas
Aluminum	8	Canada
Lead	8	Canada, Peru, Mexico, Honduras, Australia
Pumice and Volcanic Cinder	4	Greece, Italy

FIGURE 12.3
Imports supplied a significant percentage of total U.S. demand for minerals and metals in
1979. (From U.S. Bureau of Mines, *Minerals and Materials,* 1980.)

The important point about resources and reserves remains: *resources are not reserves!* An analogy from personal finances might help to clarify this point (2). Financial *reserves* are an individual's liquid assets, such as money in the bank, whereas *resources* include the total income the individual can be expected to earn in his or her lifetime. The distinction between resources and reserves may have serious consequences because, to reiterate, potential resources are not re-

serves. From our financial analogy, resources are "frozen assets," or next year's income, and cannot be used to pay this month's bills (4).

Regardless of the inherent danger, it is important from a long-range planning perspective to compute future resources, which necessitates continual reassessment of all components of total resources by considering new technology, probability of geologic discovery, and shifts in economic and political conditions (2).

This approach to resources and reserves, developed by the U.S. Geological Survey and the U.S. Bureau of Mines, is superior to a simple periodical listing of the total amount of material available or likely to become available. Such a list is misleading when used for planning purposes. A superior method is to list or graph mineral resources in terms of the resource classification (Figure 12.4). United States domestic reserves and resources for selected materials, projected through the year 2000, are shown in Table 12.3. Figure 12.5 is an example of how data from an identified resource can be tabulated; it also shows how identified resources can be further subdivided. "Measured-identified" resources refer to those that are well known and measured and for which the total tonnage or grade is well established. "Indicated-identified" resources are not so well known and measured and therefore cannot be outlined completely by tonnage or grade. "Inferred-identified" resources have quantitative estimates based on broad geologic knowledge of the deposit. The category into which a particular identified resource will fit is obviously a function of available geologic information that involves testing, drilling, and mapping, all of which become more expensive with greater depth. It is therefore not surprising that most of the information for coal is available for identified resources buried less than 300 meters from the surface, which includes 89 percent of the identified resources (Figure 12.5). With increasing depth, less information is available because most detailed mapping is still done from surface exposure (outcrop) of rocks where local relief is seldom more than 100 meters or so. Sampling of subsurface materials is less certain, and indirect geophysical observations are usually not sufficient to make accurate measurements of identified resources that are well below the earth's surface.

GEOLOGY OF MINERAL RESOURCES

The geology of mineral resources is intimately related to the entire geologic cycle, and nearly all aspects and processes of the cycle are involved to a lesser or greater extent in producing local concentrations or occurrences of useful materials. The term *ore* is sometimes used for those useful metallic minerals that can be mined at a profit, and locations where ore is found anomalous concentrations of these minerals. The necessary concentration factor of a particular mineral before the mineral can be classified as an ore varies with technology, economics, and politics, as well as with its natural concentration in the earth's crust. For example, the natural concentration of aluminum in the crust is about 8 percent, and needs to be concentrated only to about 34 percent (concentration factor of 4), whereas the natural concentration of mercury in the crust is only a tiny fraction of 1 percent and must have a concentration factor of about 10,000 to be mined economically. Table 12.4 lists other metallic elements and the current concentration factors necessary before mining becomes profitable. The concentration factors necessary for mining may vary as demand changes.

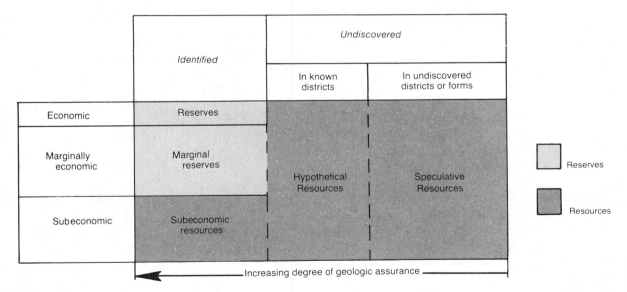

FIGURE 12.4
Classification of mineral resources used by the U.S. Geological Survey and the U.S. Bureau of Mines. (After U.S. Geological Survey Circular 831, 1980.)

TABLE 12.3
A generalized outlook of domestic reserves and resources through the year 2000.

Group 1: *Reserves* in quantities adequate to fulfill projected needs well beyond 25 years.	Coal	Phosphorus
	Construction stone	Silicon
	Sand and gravel	Molybdenum
	Nitrogen	Gypsum
	Chlorine	Bromine
	Hydrogen	Boron
	Titanium (except rutile)	Argon
	Soda	Diatomite
	Calcium	Barite
	Clays	Lightweight aggregates
	Potash	Helium
	Magnesium	Peat
	Oxygen	Rare earths
		Lithium

Group 2: *Identified Subeconomic Resources* in quantities adequate to fulfill projected needs beyond 25 years and in quantities significantly or slightly greater than estimated undiscovered resources.	Aluminum
	Nickel
	Uranium
	Manganese
	Vanadium
	Zircon
	Thorium

Group 3: Estimated *undiscovered resources* (hypothetical and speculative) in quantities adequate to fulfill projected needs beyond 25 years and in quantities significantly greater than identified subeconomic resources. Research efforts for these commodities should concentrate on geologic theory and exploration methods aimed at discovering new resources.	Iron	Platinum
	Copper	Tungsten
	Zinc	Beryllium
	Gold	Cobalt
	Lead	Cadmium
	Sulfur	Bismuth
	Silver	Selenium
	Fluorine	Niobium

Group 4: *Identified subeconomic* and *undiscovered resources* together in quantities probably not adequate to fulfill projected needs beyond the end of the century; research on possible new exploration targets, new types of deposits, and substitutes is necessary to relieve ultimate dependent on imports.	Tin
	Asbestos
	Chromium
	Antimony
	Mercury
	Tantalum

In a broad-brush approach to the geology of mineral resources, there are relations between tectonic plate boundaries and the origin of ore deposits such as iron, gold, copper, and mercury (Figure. 12.6). The basic ideas relate to processes operating at the diverging and converging plate boundaries.

The origin of metallic ore deposits at *divergent* plate boundaries is related to the migration (movement) of ocean water. Basically, cold dense ocean water moves down through numerous fractures in the basaltic rocks at oceanic ridges and is heated by contact with or heat from nearby molten rock (magma). The warm water is lighter and more chemically active and rises up (convects) through the fractured rocks, leaching out metals. The metals are carried in solution and deposited (precipitated) as metallic sulfides (5). Several

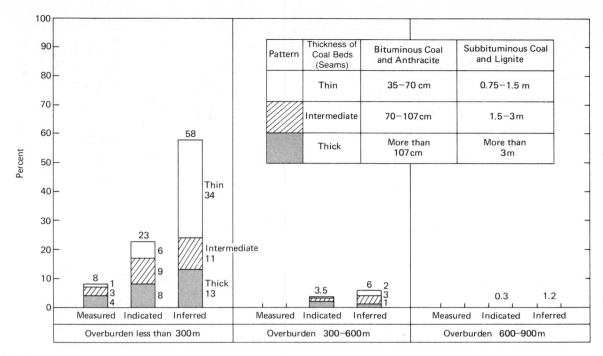

FIGURE 12.5
Major resource categories of identified U.S. coal resources. (After D. A. Brobst and W. P. Pratt, eds., U.S. Geological Survey Professional Paper 820, 1973.)

hot water vents with sulfide deposits have recently been discovered along oceanic ridges, and undoubtedly numerous others will be located.

The origin of metallic ore deposits at *convergent* plate boundaries is hypothesized to be the result of relationships between sea-water-saturated oceanic rocks and partial melting of rocks associated with the descent of oceanic lithosphere at a zone of convergence (subduction zone). The idea is that, as the sea-water-saturated rocks are subjected to elevated temperatures and pressures in the subduction zone, partial melting occurs. The combination of heat, pressure, and the resulting partial melting facilitates the mobilization of metals which are concentrated and ascend as compo-

TABLE 12.4
Approximate concentration factors of selected metals necessary before mining is economically feasible.

Metal	Natural Concentration (Percent)	Percent in Ore	Approximate Concentration Factor
Gold	0.0000004	0.001	2,500
Mercury	0.00001	0.1	10,000
Lead	0.0015	4	2,500
Copper	0.005	0.4 to 0.8	80 to 160
Iron	5	20 to 69	4 to 14
Aluminum	8	35	4

Source: Data from U.S. Geological Survey Professional Paper 820, 1973.

nents of the magma. The metal-rich fluids are eventually released (or escape) from the magma, and the metals are deposited in a host rock (5).

Perhaps the best example of this relation is the global occurrence of known mercury deposits (Figure 12.7). All the belts of productive deposits of mercury are associated with volcanic systems and are located in close proximity to convergent plate boundaries (subduction zones). It has been suggested that the mercury, originally found in oceanic sediments of the crust, is distilled out of the downward-plunging plate and emplaced at a higher level above the subduction zone (2). The significant point, however, is that mercury deposits are intimately associated with volcanic and tectonic processes, so that convergent plate junctions characterized by volcanism and tectonic activities are likely places to find mercury. A similar argument can be made for other ore deposits, but there is danger in oversimplification, since many deposits are not directly associated with plate boundaries (Figure 12.8).

The geology of economic mineral and rock materials is as diversified and complex as the processes responsible for their formation or accumulation in the natural environment. Most deposits, however, can be related to various parts of the rock cycle within the influence of the tectonic, geochemical, and hydrologic cycles. The genesis of mineral resources with commercial value can be subdivided into several categories:

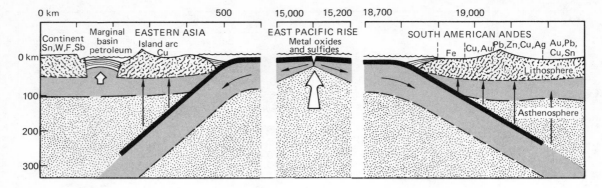

FIGURE 12.6
Idealized diagram showing the relationship between the East Pacific Rise (divergent plate boundary), Pacific margins (convergent plate boundaries), and metallic ore deposits. (After NOAA, *California Geology,* vol. 30, no. 5, 1977.)

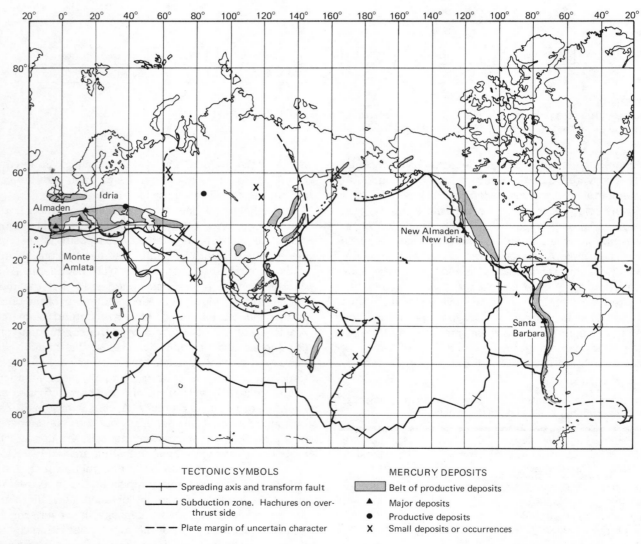

TECTONIC SYMBOLS

─┼─ Spreading axis and transform fault

└┴┴┘ Subduction zone. Hachures on over-
    thrust side

─ ─ ─ Plate margin of uncertain character

MERCURY DEPOSITS

▨ Belt of productive deposits

▲ Major deposits

● Productive deposits

X Small deposits or occurrences

FIGURE 12.7
Relation between mercury deposits and recently active subduction zones. (From D. A. Brobst and W. P. Pratt, eds., U.S. Geological Survey Professional Paper 820, 1973.)

FIGURE 12.8
Occurrence of economic deposits of some selected metals in the United States. (Data from mineral resource maps of the U.S. Geological Survey.)

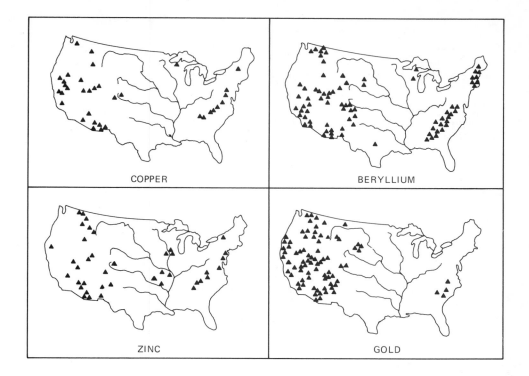

COPPER

BERYLLIUM

ZINC

GOLD

- Igneous processes, including crystal settling and late magmatic and hydrothermal activities
- Metamorphic processes associated with contact or regional metamorphism
- Sedimentary processes, including accumulation in oceanic, lake, stream, wind, and glacial environments
- Weathering processes such as soil formations and in situ (in-place) concentration of insoluble minerals in weathered rock debris

Table 12.5 lists several examples of ore deposits from these categories.

Igneous Processes

Most ore deposits caused by igneous processes are a result of an enrichment process that concentrates an economically desirable ore of metals such as copper, nickel, or gold. In some cases, however, an entire igneous rock mass contains disseminated crystals that can be recovered economically. Perhaps the best-known example is the occurrence of diamond crystals found in a coarse-grained igneous rock called *kimberlite,* which characteristically is found as a pipe-shaped body of rock that decreases in diameter with depth. Almost the entire kimberlite pipe is the ore deposit, and the diamond crystals are disseminated throughout the rock (Figure. 12.9).

Ore deposits can also result from igneous processes that segregate crystals formed earlier from those formed later. For example, as magma cools, heavy minerals that crystallize early may slowly sink or settle toward the lower part of the magma chamber, where they form concentrated layers. Deposits of chromite (ore of chromium) have formed by this process (Figure 12.10).

Late magmatic processes occur after most of the magma has crystallized, and rare and heavy metalliferous materials in water- and gas-rich solutions remain.

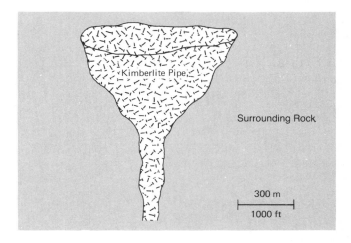

Kimberlite Pipe

Surrounding Rock

300 m

1000 ft

FIGURE 12.9
Idealized diagram showing a typical South African diamond pipe. Diamonds are scattered throughout the cylindrical body of igneous rock known as *kimberlite*. (After Foster, *Physical Geology,* 3d ed. [Columbus, Ohio: Charles E. Merrill, 1979].)

TABLE 12.5
Examples of different types of mineral resources. (Modified from Robert J. Foster, *Physical Geology,* 3d ed. Columbus, Ohio: Charles E. Merrill, 1979.)

Type	Example
Igneous	
Disseminated	Diamonds—South Africa
Crystal settling	Chromite—Stillwater, Montana
Late magmatic	Magnetite—Adirondack Mountains, New York
Pegmatite	Beryl and lithium—Black Hills, South Dakota
Hydrothermal	Copper—Butte, Montana
Metamorphic	
Contact metamorphism	Lead and silver—Leadville, Colorado
Regional metamorphism	Asbestos—Quebec, Canada
Sedimentary	
Evaporite (lake or ocean)	Potassium—Carlsbad, New Mexico
Placer (stream)	Gold—Sierra Nevada foothills, California
Glacial	Sand and gravel—northern Indiana
Deep-ocean	Manganese oxide nodules—central and southern Pacific Ocean
Weathering	
Residual soil	Bauxite—Arkansas
Secondary enrichment	Copper—Utah

This late-stage metallic solution may be squeezed into fractures or settle into interstices (empty space) between earlier-formed crystals. Other late-stage solutions form coarse-grained igneous rock known as *pegmatite,* which is rich in feldspar, mica, and quartz, as well as certain rare minerals (6). Pegmatites have been extensively mined for feldspar, mica, spodumene (lithium mineral), and clay that forms from weathered feldspar.

Hydrothermal (hot-water) mineral deposits are the most common type of ore deposit. They originate from late-stage magmatic processes and give rise to a variety of deposits, including gold, silver, copper, mercury, lead, zinc, and other metals, as well as many nonmetallic minerals. The mineral material is either produced directly from the igneous parent rock, as shown in Figure 12.11, or altered by metamorphic processes as magmatic solutions intrude into the surrounding rock as veins or small dikes. Many hydrothermal deposits cannot be traced to a parent igneous rock mass, and for these cases the origin remains unknown. It is speculated that circulating groundwater, heated and enriched with minerals after contact with deeply buried magma, might be responsible for some of these deposits (6,7).

Essentially, hydrothermal solutions that form ore deposits are mineralizing fluids that migrate through a host rock. Two types of deposits can be recognized: *cavity-filling* and *replacement*. Cavity-filling deposits are formed when hydrothermal solutions migrate along openings in rocks such as fracture systems, pore spaces, or bedding planes, and precipitate ore minerals. Replacement deposits, on the other hand, form as hydrothermal solutions react with the host rock, forming a zone in which ore minerals precipitate from the mineralizing fluids and replace part of the host rock. Although replacement deposits are believed to dominate at higher temperatures and pressures than cavity-filling deposits, actually both may be found in close association as one grades into the other; that is, the filling of an open fracture by precipitation from hydrothermal solutions may occur simultaneously with replacement of the rock lining the fracture (7).

Hydrothermal replacement processes are significant because, excluding some iron and nonmetallic deposits, they have produced some of the world's largest and most important mineral deposits. Some of these

FIGURE 12.10
Idealized diagram showing how chromite layers might form. The chromite crystallizes early, and the heavy crystals sink to the bottom and accumulate in layers. [From Foster, *Physical Geology,* 3d ed. (Columbus, Ohio: Charles E. Merrill, 1979).]

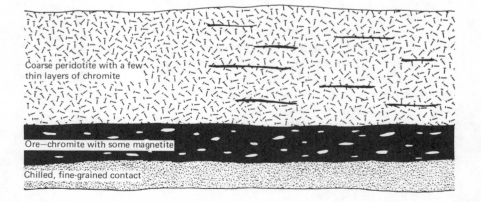

Coarse peridotite with a few thin layers of chromite

Ore—chromite with some magnetite

Chilled, fine-grained contact

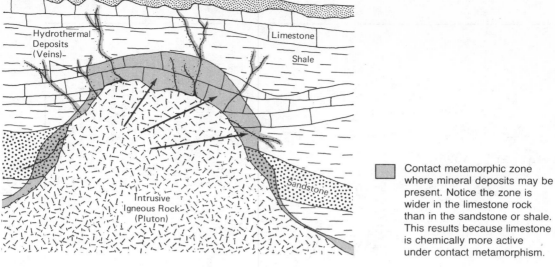

Contact metamorphic zone where mineral deposits may be present. Notice the zone is wider in the limestone rock than in the sandstone or shale. This results because limestone is chemically more active under contact metamorphism.

FIGURE 12.11
Idealized diagram showing how hydrothermal and contact metamorphic ore deposits might form.

deposits result from a massive, nearly complete replacement of host rock with ore minerals that terminate abruptly; others form thin replacement zones along fissures; and still others form disseminated replacement deposits that may involve huge amounts of relatively low-grade ore (7).

The actual sequence of geologic events that lead to the development of a hydrothermal ore deposit is usually complex. Consider, for example, the tremendous, disseminated copper deposits of northern Chile. The actual mineralization is thought to be related to igneous activity, faulting, and folding that occurred 60 to 70 million years ago. The ore deposit is an elongated, tabular mass along a highly sheared (fractured) zone separating two types of granitic rock. The concentration of copper results from a number of factors:

- A source igneous rock supplied the copper
- The fissure zone tapped the copper supply and facilitated movement of the mineralizing fluids
- The host rock was altered and fractured, preparing it for deposition and replacement processes that produced the ore
- The copper was leached and redeposited again by meteoric water, which further concentrated the ore (8)

Metamorphic Processes

Ore deposits are often found along the contact between the igneous rocks and the country rocks they intrude. This area is characterized by *contact metamorphism,* caused by interactions among heat, pressure, and chemically active fluids of the cooling magma in contact with the surrounding (country) rock. The width of the contact metamorphic zone varies with the type of country rock. The zone of limestone is usually thickest because limestone is more reactive; the release of carbon dioxide (CO_2) increases the mobility of reactants. The zone is generally thinnest for shale because the fine-grained texture retards the movement of hot, chemically active solutions; and the zone is intermediate for sandstone, as shown in Figure 12.11. Some of the mineral deposits that form in contact areas originate from the magmatic fluids and some from reactions of these fluids with the country rock.

Metamorphism can also result from regional increase of temperature and pressure associated with deep burial of rocks or tectonic activity. Regardless of its cause, *regional metamorphism* can change the mineralogy and texture of the preexisting rocks, producing ore deposits of asbestos, talc, graphite, and other valuable nonmetallic deposits (7).

Metamorphism has been suggested as a possible origin of some hydrothermal fluids. It is a particularly likely cause in high-temperature, high-pressure zones where fluids might be produced and forced out into the surrounding rocks to form replacement or cavity-filling deposits. For example, the native copper found along the top of ancient basalt flows in the Michigan copper district was apparently produced by metamorphism and alteration of the basalt which released the copper and other materials that produced the deposits (8).

Our discussion of igneous and metamorphic processes has focused primarily on ore deposits. It is also

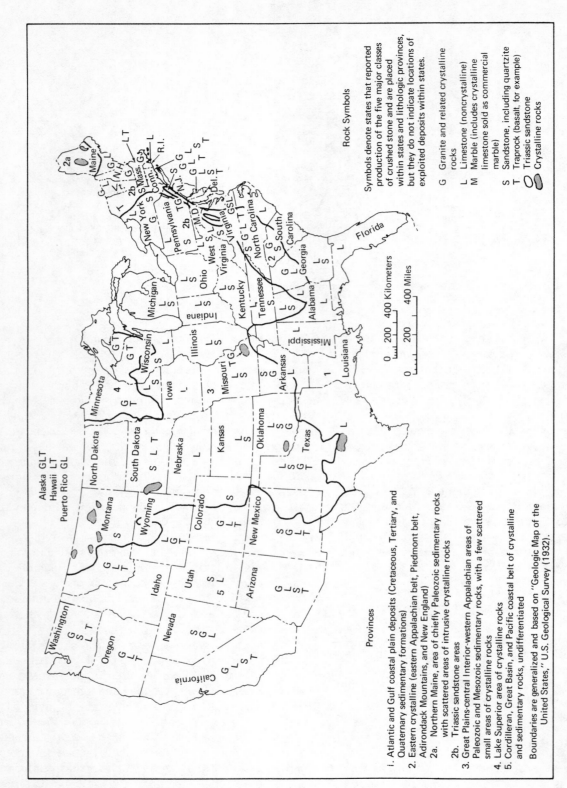

FIGURE 12.12

Map showing provinces of different types of rocks in the United States and related occurrence and production of principal kinds of crushed stone for 1969. (From D. A. Brobst and W. P. Pratt, eds., U.S. Geological Survey Professional Paper 820, 1973.)

Rock Symbols

Symbols denote states that reported production of the five major classes of crushed stone and are placed within states and lithologic provinces, but they do not indicate locations of exploited deposits within states.

G Granite and related crystalline rocks
L Limestone (noncrystalline)
M Marble (includes crystalline limestone sold as commercial marble)
S Sandstone, including quartzite
T Traprock (basalt, for example)
  Triassic sandstone
  Crystalline rocks

Provinces

1. Atlantic and Gulf coastal plain deposits (Cretaceous, Tertiary, and Quaternary sedimentary formations)
2. Eastern crystalline (eastern Appalachian belt, Piedmont belt, Adirondack Mountains, and New England)
  2a. Northern Maine, area of chiefly Paleozoic sedimentary rocks with scattered areas of intrusive crystalline rocks
  2b. Triassic sandstone areas
3. Great Plains-central Interior-western Appalachian areas of Paleozoic and Mesozoic sedimentary rocks, with a few scattered small areas of crystalline rocks
4. Lake Superior area of crystalline rocks
5. Cordilleran, Great Basin, and Pacific coastal belt of crystalline and sedimentary rocks, undifferentiated

Boundaries are generalized and based on "Geologic Map of the United States," U.S. Geological Survey (1932).

322

interesting to learn that igneous and metamorphic processes are responsible for producing a good deal of stone used in the construction industry. Granite, basalt, marble (metamorphosed limestone), slate (metamorphosed shale), and quartzite (metamorphosed sandstone), along with other rocks, are quarried to produce crushed rock and dimension stone in the United States (Figure 12.12 and 12.13). Stone is used in many aspects of construction work; but many people are surprised to learn that, in total value, the stone industry is the largest nonfuel, nonmetallic mineral industry in the United States (9).

Sedimentary Processes

In all aspects of the sedimentary cycle, from transportation to deposition of sediment, sedimentary processes are often significant in concentrating economically valuable earth materials in sufficient amounts that it is advantageous to recover the materials from depositional areas such as floodplains, lakes, or oceans. Throughout all aspects of the transport of sediments, processes of running water and wind help segregate the sediment by size and shape. Thus, for construction purposes, the best sand or sand and gravel deposits are those where water or wind have effectively removed the finer materials and left the larger particles behind. Sand dunes, beach deposits, and deposits in stream channels are good examples.

The sand and gravel industry amounts to more than $1 billion per year, and, by volume mined, is the largest nonfuel mineral industry in the United States. Figure 12.14 shows the production data of sand and gravel compared to those of other construction materials. Currently, most sand and gravel is obtained from river channels and water-worked glacial deposits. The United States now produces more sand and gravel than it needs, but demand is likely to meet supply by the year 2000 (10).

Stream processes transport and sort all types of materials according to size and density. Therefore, if the bedrock in a river basin contains heavy metals such as gold, streams draining the basin may concentrate heavy metals to form "placer deposits" in areas where the turbulence and velocity of the water facilitate deposition. These areas tend to be found in open crevices or fractures at the bottom of pools, on the inside of bends, or on riffles (Figure 12.15).

The settlement of California, Alaska, and other areas in the United States was facilitated by placer mining of gold, known as a "poor man's method" because a miner needed only a shovel, a pan, and a strong back to work the streamside claim. The gold in California attracted miners who acquired the necessary

expertise for locating and developing other resources in the western United States and Alaska.

Placer deposits of gold and diamonds have also been concentrated by coastal processes, primarily wave action; and beach sands and near-shore deposits are mined in Africa and other places.

Rivers and streams that empty into the oceans and lakes carry tremendous quantities of dissolved material derived from the chemical weathering of rocks. From time to time (geologically), shallow marine basins may be isolated by tectonic activity (uplift) which restricts circulation and facilitates evaporation. In other cases, large inland lakes with no outlets essentially dry up. As evaporation progresses, the dissolved materials precipitate out, forming a wide variety of compounds, minerals, and rocks that have important commercial value. Most of these "evaporite deposits" can be grouped into one of three types: *marine evaporites* (solids)—potassium and sodium salt, gypsum, and anhydrite; *nonmarine evaporites* (solids)—sodium and calcium carbonate, sulfate, borate, nitrate, and limited iodine and strontium compounds; and *brines* (liquids derived from wells, thermal springs, inland salt lakes, and seawaters)—bromine, iodine, calcium chloride, and magnesium (10). Heavy metals (such as copper, lead, and zinc associated with brines and sediment in the Red Sea, Salton Sea, and other areas) are important conditional resources that may be exploited in the future. Evaporite materials, are widely used in industry and agriculture (11).

Extensive marine evaporite deposits exist in the United States (Figure 12.16). The major deposits are halite (common salt, NaCl), gypsum ($CaSO_4 2H_2O$), anhydrite ($CaSO_4$), and interbedded limestone ($CaCO_3$). Limestone, gypsum, and anhydrite are present in nearly all marine evaporite basins, and halite and potassium minerals are found in a few (11).

Marine evaporites can form stratified deposits that may extend for hundreds of kilometers with a thickness of several thousand meters. The evaporites represent the product of evaporation of seawater in isolated shallow basins with restricted circulation. Within many evaporite basins, the different deposits are arranged in broad zones that reflect changes in salinity and other factors controlling the precipitation of evaporites; that is, different materials may be precipitated at the same time in different parts of the evaporite basin. Halite, for example, is precipitated in areas where the brine is more saline and gypsum where it is less saline. Economic deposits of potassium evaporite minerals are relatively rare but may form from highly concentrated brines.

Nonmarine evaporite deposits form by evaporation of lakes in a closed basin. Tectonic activity such

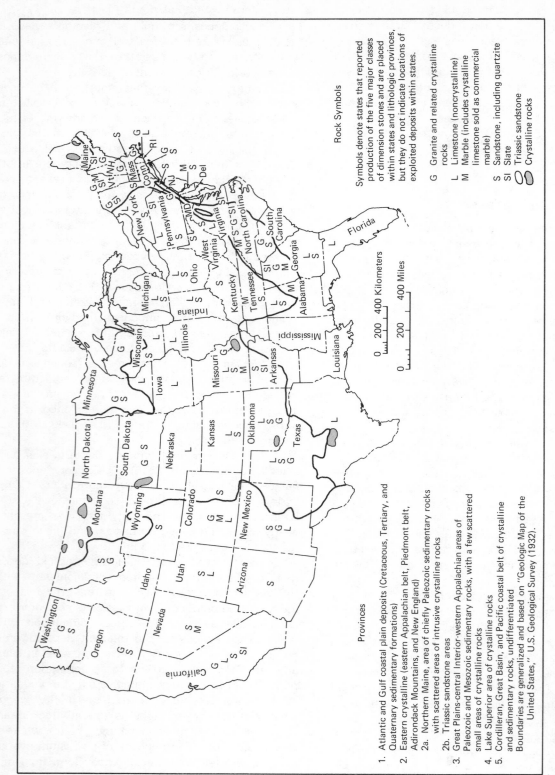

FIGURE 12.13

Provinces of different types of rocks in the United States as related to occurrence and production of principal kinds of dimension stone for 1969. (From D. A. Brobst and W. P. Pratt, eds., U.S. Geological Survey Professional Paper 820, 1973.)

Provinces

1. Atlantic and Gulf coastal plain deposits (Cretaceous, Tertiary, and Quaternary sedimentary formations)
2. Eastern crystalline (eastern Appalachian belt, Piedmont belt, Adirondack Mountains, and New England)
   2a. Northern Maine, area of chiefly Paleozoic sedimentary rocks with scattered areas of intrusive crystalline rocks
   2b. Triassic sandstone areas
3. Great Plains-central Interior-western Appalachian areas of Paleozoic and Mesozoic sedimentary rocks, with a few scattered small areas of crystalline rocks
4. Lake Superior area of crystalline rocks
5. Cordilleran, Great Basin, and Pacific coastal belt of crystalline and sedimentary rocks, undifferentiated

Boundaries are generalized and based on ''Geologic Map of the United States,'' U.S. Geological Survey (1932).

Rock Symbols

Symbols denote states that reported production of the five major classes of dimension stones and are placed within states and lithologic provinces, but they do not indicate locations of exploited deposits within states.

G   Granite and related crystalline rocks
L   Limestone (noncrystalline)
M   Marble (includes crystalline limestone sold as commercial marble)
S   Sandstone, including quartzite
Sl Slate
◯   Triassic sandstone
⬭   Crystalline rocks

324

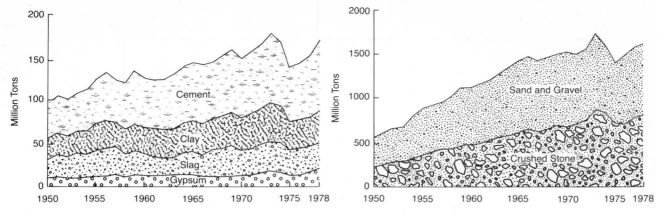

FIGURE 12.14
Supplies of major construction materials in the United States from 1950 to 1978. (From U.S. Bureau of Mines, Mining and Mineral Policy, 1979.)

as faulting can produce an isolated basin with internal drainage and no outlet. However, the tectonic activity must continue to uplift barriers across possible outlets or lower the basin floor faster than sediment can raise it, to maintain a favorable environment for evaporite mineral precipitation. Even under these conditions, economic deposits of evaporites will not form unless sufficient dissolved salts have washed into the basin by

surface runoff from surrounding highlands. Finally, even if all favorable environmental criteria are present, including an isolated basin with sufficient runoff and dissolved salts, valuable nonmarine evaporites such as sodium carbonate or borate will not form unless the geology of the highlands surrounding the basin is also favorable and yields runoff with sufficient quantities of the desired material in solution (11).

Some evaporite beds are compressed by overlying rocks, mobilized, and then pierce or intrude the overlying rocks. Intrusions of salt, called *salt domes,* are quite common in the Gulf Coast of the United States and are also found in northwestern Germany, Iran, and other areas. Salt domes in the Gulf Coast are economically important because

- They are a good source for nearly pure salt
- Some have extensive deposits of elemental sulfur (Figure 12.17) and
- Some have oil reserves on their flanks

They are also environmentally important as possible permanent disposal sites for radioactive waste, although because salt domes tend to be mobile, their suitability as disposal sites for hazardous wastes must be seriously questioned.

Evaporites from brine resources of the United States are substantial (Table 12.6), assuring that no shortage is likely for a considerable period of time. But many evaporites will continue to have a "place value" because transportation of these mineral commodities increases their price, and therefore continued discoveries of high-grade deposits closer to where they will be consumed remains an important goal (11).

Weathering Processes

Weathering processes concentrate some materials to the point that they can be extracted at a profit. For

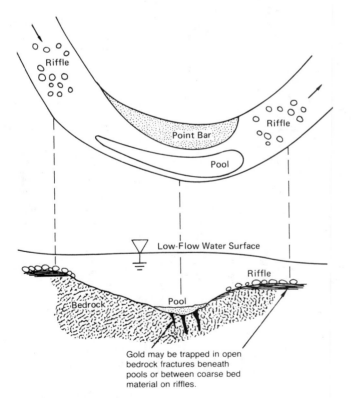

FIGURE 12.15
Idealized diagram of a stream channel and bottom profile showing areas where placer deposits of gold are likely to occur.

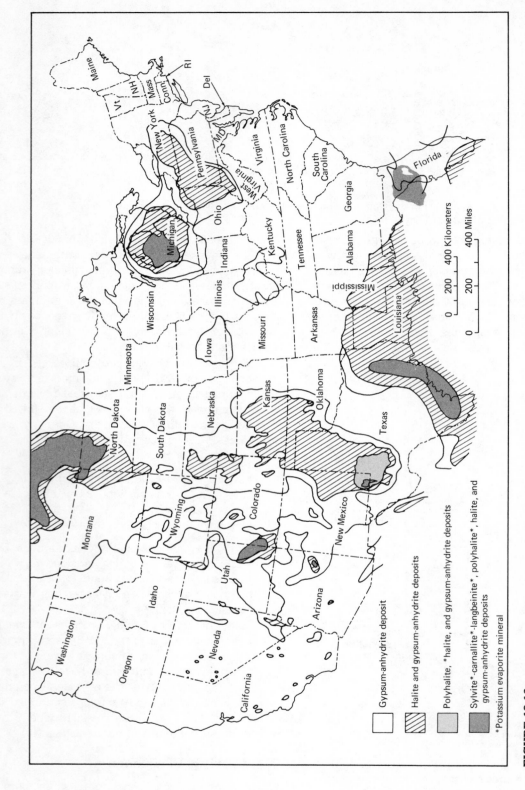

FIGURE 12.16

Marine evaporite deposits of the United States. (After D. A. Brobst and W. P. Pratt, eds., U.S. Geological Survey Professional Paper 820, 1973.)

Legend:

☐ Gypsum-anhydrite deposit

▨ Halite and gypsum-anhydrite deposits

▨ Polyhalite, *halite, and gypsum-anhydrite deposits

▨ Sylvite*-carnallite*-langbeinite*, polyhalite*, halite, and gypsum-anhydrite deposits

*Potassium evaporite mineral

400 Kilometers
400 Miles
200
200
0
0

326

FIGURE 12.17
Idealized diagram showing a cross section through a typical salt dome of the type found in the Gulf Coast of the United States.

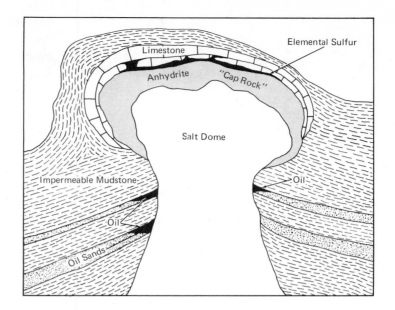

example, intensive weathering of residual soils (laterite) derived from aluminum-rich igneous rocks may concentrate relatively insoluble hydrated oxides of aluminum and iron, while the more soluble elements such as silica, calcium, and sodium are selectively removed by soil and biological processes. If sufficiently concentrated, residual aluminum oxide forms an ore of aluminum known as *bauxite* (Figure 12.18). Important nickel and cobalt deposits are also found in laterite soils developed from ferromagnesium-rich igneous rocks.

TABLE 12.6
Evaporite and brine resources of the United States expressed in years of supply at current rates of domestic consumption.

Commodity	Identified Resources[a] (Reserves[b] and Subeconomic Deposits)	Undiscovered Resources (Hypothetical[c] and Speculative[d] Resources)
Potassium compound	100 years	Virtually inexhaustible
Salt	1,000 + years	Do
Gypsum and anhydrite	500 + years	Do
Sodium carbonate	6,000 years	5,000 years
Sodium sulfate	700 years	2,000 years
Borates	300 years	1,000 years
Nitrates	Unlimited (air)	Unlimited (air)
Strontium	500 years	2,000 years
Bromine	Unlimited (seawater)	Unlimited (seawater)
Iodine	100 years	500 years
Calcium chloride	100 + years	1,000 + years
Magnesium	Unlimited (seawater)	Unlimited (seawater)

[a]Identified resources: Specific, identified mineral deposits that may or may not be evaluated as to extent and grade, and whose contained minerals may or may not be profitably recovered with existing technology and economic conditions.

[b]Reserves: Identified deposits from which minerals can be extracted profitably with existing technology and under present economic conditions.

[c]Hypothetical resources: Undiscovered mineral deposits, whether of recoverable or subeconomic grade, that are geologically predictable as existing in known districts.

[d]Speculative resources: Undiscovered mineral deposits, whether of recoverable or subeconomic grade, that may exist in unknown districts or in unrecognized or unconventional form.

Source: G. I. Smith et al., U.S. Geological Survey Professional Paper 820, 1973.

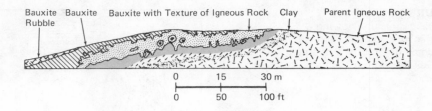

FIGURE 12.18
Cross section of the Pruden bauxite mine, Arkansas. The bauxite was formed by intensive weathering of the aluminum-rich igneous rocks. (After G. Mackenzie, Jr., et al., U.S. Geological Survey Professional Paper 299, 1958.)

Insoluble ore deposits such as native gold are generally residual and, unless removed by erosion, accumulate in weathered rock and soil. Accumulation is favored where the parent rock that contains insoluble ore minerals is relatively soluble, such as limestone (Figure 12.19). Care must be taken in evaluating a residual weathered rock or soil deposit because the near-surface concentration is a much higher grade of ore than that in the parent, unweathered rocks (6).

Weathering is also important in producing secondary-enrichment sulfide-ore deposits from a lower-grade, primary ore. Near the surface, primary ore containing minerals such as iron, copper, and silver sulfides is in contact with slightly acid soil water in an oxygen-rich environment. As the sulfides are oxidized, they dissolve to form solutions that are rich in sulfuric acid and silver and copper sulfate that migrate downward, producing a leached zone devoid of ore minerals (Figure 12.20). Below the leached zone and above the groundwater table, oxidation continues, and sulfate solutions continue their downward migration. Below the water table, if oxygen is no longer available, the solutions are deposited as sulfides, enriching the metal content of the primary ore by as much as ten times. In this way, low-grade primary ore is rendered more valuable, and high-grade primary ore is made even more attractive (7,8).

The presence of a residual iron oxide cap at the surface indicates the possibility of an enriched ore below, but is not always conclusive. Of particular importance to formation of a zone of secondary enrichment is the presence in the primary ore of iron sulfide (for example, pyrite). Without it, secondary enrichment seldom takes place, because iron sulfide in the presence of oxygen and water forms sulfuric acid, which is a necessary solvent. Another factor that favors development of a secondary-enrichment ore deposit is that the primary ore is sufficiently permeable to allow water and solutions to migrate freely downward. Given a primary ore that meets these criteria, the reddish iron oxide cap probably does indicate that secondary enrichment has taken place (7).

Secondary enrichment is important in concentrating dispersed metals, and the economic success of several disseminated copper deposits has resulted from this process. For example, secondary enrichment of a disseminated copper deposit at Miami, Arizona, increased the grade of the ore from less than 1 percent copper in the primary ore to as much as 5 percent in some localized zones of enrichment (8).

Minerals from the Sea

Mineral resources in seawater or on the bottom of the ocean are vast and in some cases, such as magnesium, nearly unlimited. In the United States, magnesium was first extracted from seawater in 1940; by 1972, 80 percent of the domestic production capacity of magnesium was from one company in Texas that used seawater as its raw material source. Companies in Alabama, California, Florida, Mississippi, and New Jersey are also extracting magnesium from seawater.

The deep-ocean floor may be the site of the next big mineral rush. Two types of deposits have been identified: massive sulfide deposits associated with hydrothermal vents, and manganese oxide nodules.

Massive sulfide deposits, containing zinc, copper, iron, and trace amounts of silver, are produced at di-

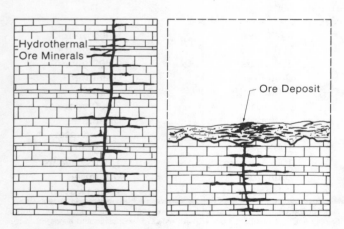

FIGURE 12.19
Idealized diagram showing how an ore deposit of insoluble minerals might form by weathering and formation of a residual soil. As the limestone that contained the deposit weathered, the ore minerals became concentrated in the residual soil. [From Foster, *Physical Geology*, 3d ed. (Columbus, Ohio: Charles E. Merrill, 1979).]

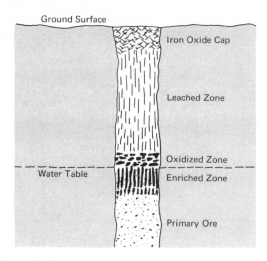

FIGURE 12.20
Idealized diagram showing the typical zones that form during secondary enrichment processes. Sulfide ore minerals in the primary ore vein are oxidized and altered, and then are leached from the oxidized zone and redeposited in the enriched zone. The iron oxide cap is generally a reddish color and may be helpful in locating ore deposits that have been enriched. [After Foster, *Physical Geology,* 3d ed. (Columbus, Ohio: Charles E. Merrill, 1979).]

vergent plate boundaries (oceanic ridges) by the forces of plate tectonics. Pressure created by several thousand meters of water at ridges forces cold seawater deep into numerous rock fractures where it is heated by upwelling magma and emerges at hot springs, as illustrated in Figure 12.21. The circulating water leaches the rocks and removes metals which are deposited when hot mineral-rich water (at temperatures up to 350°C) is ejected into the cold sea. Sulfide minerals precipitate near vents, called "black smokers" because of the color of the ejected mineral-rich water, forming towerlike massive formations rich in metals. The hot vents are of particular biologic significance because they support a unique assemblage of animals, including giant clams, tube worms, and white crabs. Communities of these animals base their existence on sulfide compounds extruded from black smokers through a process called *chemosynthesis* as opposed to photosynthesis, which supports all other known ecosystems on earth.

We do not know the extent of sulfide mineral deposits along oceanic ridges, and although leases to possible deposits are under consideration, it seems unlikely that they will be extracted profitably in the near future. Potential environmental degradation, such as water quality and sediment pollution, will need careful evaluation prior to any proposed mining activity.

Manganese oxide nodules (Figure 12.22) which contain, on a very rough average, dominant man-

ganese (24 percent) and iron (14 percent) with secondary copper (1 percent), nickel (1 percent), and cobalt (0.25 percent) cover vast areas of the deep-ocean floor. Nodules are found in the Atlantic Ocean off Florida, but the richest and most extensive accumulations occur in large areas of the northeastern, central, and southern Pacific where nodules cover 20 to 50 percent of the ocean floor (12).

Manganese oxide nodules are usually discrete, but are welded together locally to form a continuous pavement. Although a few nodules have been found buried in sediment, they are usually surficial deposits on the seabed. The average size of the nodules varies from a few millimeters to several tens of centimeters in diameter. Individual nodules are composed primarily of concentric layers of manganese and iron oxides mixed with a variety of other materials. Deposition is around a nucleus of broken nodules, fragments of volcanic rock, and sometimes fossils. The estimated rate of growth is 1 to 5 millimeters per million years. The nodules are most abundant in those parts of the ocean where sediment accumulation is minimal, generally at depths of 2,500 to 6,000 meters (13).

The origin of the nodules is not well understood, and there are probably several ways they might form. The most probable theory is that they form from material weathered from the continents and transported by rivers to the oceans, where ocean currents carry the material to the deposition site in the deep-ocean basins. The minerals from which the nodules form may also derive from submarine volcanism, or are released during physical and biochemical processes and reactions that occur near the water-sediment interface during and after deposition of the sediments (13).

Expenditures for mining and metallurgical research to recover the nodules have surpassed $190 million, and proposed expenditures through the early 1980s are expected to approach $800 million. At least 20 corporations in several countries are in the race, and metallurgical systems are being examined. Some would produce cobalt, copper, nickel, and manganese, while others would produce combinations of only copper and nickel (14).

Actual mining involves developing a system to lift the nodules off the ocean bottom and up to the mining ship. French and Japanese researchers are examining a system with a continuous-line bucket dredge, a continuous rope with buckets attached at prescribed intervals, strung between two ships. The buckets drag along the bottom as the two ships move, and the nodules are dumped into one of the ships as the rope loop is reeled from one ship to the other. Other methods of recovery that are under study are hydraulic lifting and use of airlift in conjunction with hydraulic dredging

FIGURE 12.21
Idealized diagrams showing (a)
oceanic ridge hydrothermal environment and (b) detail of black
smokers where massive sulfide
deposits form.

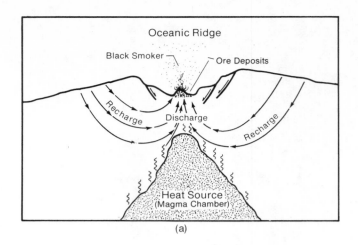

(a)

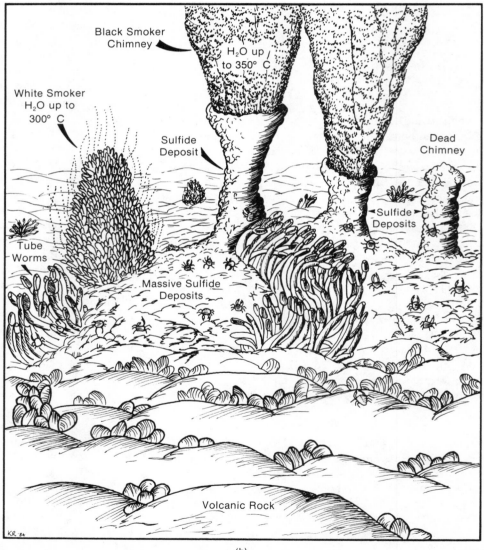

(b)

FIGURE 12.22
Manganese oxide nodules on the floor of the Atlantic Ocean.
(Photo courtesy of R. M. Pratt.)

(14). It is fair to say that the whole manganese oxide nodule industry is in its infancy, and considerable change in methods and technology is likely. Nevertheless, it has been determined from present data that mining of the nodules is technologically feasible and potentially profitable, if prospective interest groups can cooperate and manage the resource and if the environmental impact on the ecology of the ocean bottom is carefully evaluated to minimize seabed degradation.

ENVIRONMENTAL IMPACT OF MINERAL DEVELOPMENT

The impact of mineral exploitation on the environment depends upon such factors as mining procedures, local hydrologic conditions, climate, rock types encountered, size of operation, topography, and many more interrelated factors. The impact varies with the resource's stage of development; for example, the exploration and testing stages have considerably less impact than the mining and processing stages.

Exploration activities for mineral deposits vary from collecting and analyzing remote sensing data gathered from airplanes or satellites to field work such as surface mapping, drilling, and gathering geophysical data. Exploration usually has minimal impact on the environment when care is taken in sensitive areas.

Mining and processing mineral resources, on the other hand, can have a considerable impact on land, water, air, and biologic resources. That impact is part of the price we bear for the benefits of mineral consumption, and it is unrealistic to expect that we can mine our resources without affecting some aspect of the local environment. We must, however, develop those resources with a minimum of adverse impact. Minimizing environmental degradation caused by mining can be very difficult, though, since the demand for minerals continues to grow while the volumes of highly concentrated mineral deposits diminish. Therefore, to provide more and more material, we will need larger and larger operations to mine ever-poorer grades of ore. At the same time, mining ore deposits that are transitional into the surrounding rocks, allowing recovery of ever-lower grades of ore, is not always possible because some ore deposits terminate abruptly along geologic boundaries, as shown in Figure 12.23 (15).

The trend in recent years has been away from subsurface mining and toward large, open-pit (surface) mines, such as the Bingham Canyon copper mine in Utah (Figure 12.24) and Liberty Pit near Ruth, Nevada (Figure 12.25). The Bingham Canyon mine is one of the world's largest artificial excavations, covering nearly 8 square kilometers to a maximum depth of nearly 800 meters.

Surface mines and quarries today cover about 0.25 percent of the total area of the United States. Figure 12.26 shows the percentage of land use for selected activities—notice that nearly half of the land used for mining involves bituminous coal—and Figure 12.27 shows the states that use the greatest amount of land for mining bituminous coal, copper, phosphate rock, and iron ore.

Even though the impact of mining tends to be a local phenomenon, there is growing awareness that numerous local occurrences will eventually constitute a larger problem. The concern is that environmental degradation will tend to go beyond the excavation and surface plant areas. For example, area streams and groundwater may become polluted by particulate or dissolved sediment derived from rapidly eroding, unprotected waste piles exposed to surface erosion and migrating groundwater. In a worst-case situation, the rate of erosion and thus sediment production from waste piles may be up to 1000 times the natural rate. The excess sediment may enter streams and rivers, where it will be deposited, decreasing the capacity of the channel and thus increasing the flood hazard.

Even abandoned mines can cause serious problems; for example, subsurface mining for lead and zinc in the tristate area of Kansas, Missouri, and Oklahoma that began in the late 19th century and ceased in some areas in the 1960s is causing serious water pollution problems in the 1980s. The mines, extending to depths of 100 m below the water table, were kept dry by pumping while the mines were in production. Since the mining stopped, some have flooded and started to overflow into nearby creeks. The water is extremely

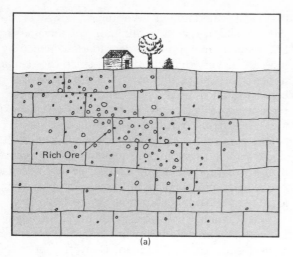 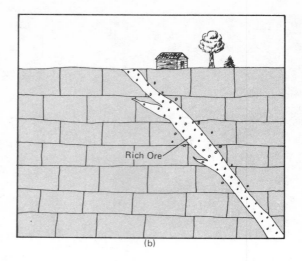

(a)   (b)

FIGURE 12.23
Idealized diagram showing (a) an ore deposit in which the grade of the ore gradually lessens with distance from the greatest concentration, and (b) an ore deposit in which the high-grade ore is concentrated in a definite zone with no outward gradation to lower concentrations of the ore.

acidic because sulfide minerals in the mine react with oxygen and groundwater to form sulfuric acid, a problem known as "acid mine-drainage." The problem was so severe in Oklahoma's Tar Creek area that the Environmental Protection Agency in 1982 designated it as the nation's number one hazardous waste site. Acid mine-drainage problems are particularly prevalent in coal mining areas because of the pyrite (iron sulfide) commonly associated with coal. As we have noted, oxidation of the pyrite in the presence of water produces sulfuric acid, which can pollute streams and other bodies of surface or groundwater.

FIGURE 12.24
The Bingham Canyon copper mine, one of the largest artificial excavations in the world. (Photo courtesy of Kennecott Copper Corporation.)

FIGURE 12.25
The Liberty Pit Copper mine near Ruth, Nevada. (Photo courtesy of Kennecott Copper Corporation.)

Following mining activities, land reclamation is necessary if the mining has had detrimental effects and if the land is to be used for other purposes. Table 12.7 shows, by area of activity, the amount of land the U.S. mining industry used and reclaimed from 1930–1980. It is interesting to note that 69 percent of the total area utilized by the industry was for surface mining, of which only 55 percent has been reclaimed. Figure 12.28 shows the land utilized and reclaimed by stages for the 1930-to-1980 period, and Figure 12.29 illustrates that the percentage of land reclaimed in 1971 (79 percent) and 1980 (54 percent) was significantly greater than the average (47 percent) for the period 1930 to 1980. On the other hand, the apparent reduction in percentage of land reclaimed in 1971 compared to 1980, along with the fact that over 50 percent of the

TABLE 12.7
Land used and reclaimed by the mining industry in the U.S. (1930–1980) by area of activity.

Area of activity[1]	Utilized, km^2	Percent of total land utilized	Reclaimed, km^2	Percent reclaimed
Surface area mined (area of excavation only)...............................	15,920	69%	8,735	55.0%
Area used for disposal of overburden waste from surface mining[2]	3,683	16%	1,910	52.0%
Surface area subsided or disturbed as a result of underground workings[3]......	425	2%	24	5.6%
Surface area used for disposal of underground mine waste....................	770	3%	103	13.4%
Surface area used for disposal of mill or processing waste.....................	2,242	10%	210	9.4%
Total[4]	23,067	100%	10,927	47.4%

[1]Excludes oil and gas operations.
[2]Includes surface coal operations for 1930–71 only.
[3]Includes data for 1930–71 only.
[4]Data may not add to totals shown because of independent rounding.
Source: Johnson, W., and Paone, J., 1982, U.S. Bureau of Mines Information Circular 8862.

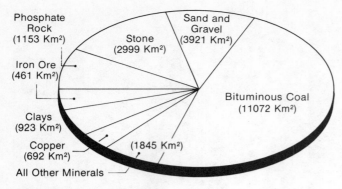

FIGURE 12.26
Amount of land use in the U.S. by selected commodity. Total land utilized is 23067 km². (After W. Johnson and J. Paone, 1982, U.S. Bureau of Mines Information Circular 8862.)

total land utilized from 1930 to 1980 is still not reclaimed, suggests the need for a greater effort to reclaim old mining areas. (We will discuss methods of mine reclamation in chapter 13 when we consider the impact of coal mining.)

Considering the entire resource utilization pattern, it is obvious that the demand for mineral resources will increase. Therefore, the only logical approach to the eventual environmental degradation is to strive to minimize both on-site and off-site problems by controlling sediment, water, and air pollution through good engineering and conservation practices. These practices will raise the cost of mineral commodities, and hence the price of all products produced from these materials, but it will yield other returns of equal or higher value to future generations.

As we mentioned in the previous chapter, a serious problem associated with mineral resource development is the possible release into the environment of trace elements in hazardous concentrations. Trace elements such as cadmium, cobalt, copper, lead, molybdenum, and others, when leached from mining wastes and concentrated in water, soil, or plants, can be toxic or can cause diseases in people and animals who drink the water, eat the plants, or use the soil.

RECYCLING RESOURCES

In the United States alone, more than 180 million metric tons of household and industrial waste are collected each year. Of this, about 27 million metric tons are incinerated, which generates more than 6 million metric tons of residue. The tremendous tonnage of waste that is not incinerated and the residues from burning are usually disposed of at sanitary landfill sites or open dumps. These materials are sometimes referred to as

urban ore because they contain many materials that could be recycled to provide energy or useful products (16,17).

The concept of "real" urban ore was realized recently when it was discovered that ash from the incineration of sewage sludge in Palo Alto, California (much of which comes from metal-using, high technology industries) contains large concentrations of gold (30 parts per million), silver (660 parts per million), copper (8,000 parts per million) and phosphorus (6.6 percent). Each metric ton of the ash contains approximately 1 troy ounce of gold and 20 ounces of silver (worth about $500 in 1984). The gold is concentrated above natural abundance by a factor of 7,500 times, making the "deposit" double the average grade that is mined today. Silver in the ash has a concentration factor of 9,400, similar to rich deposits in Idaho. Copper is concentrated in the ash by a factor of 145 times above natural concentrations and is thus of common ore grade. Commercial phosphorus deposits vary from 2 to 16 percent; the ash with 6.6 percent phosphorus has the potential of a high-value resource. Therefore, as of 1978, ash in the Palo Alto dump is a silver and gold deposit with a 1980 value of about $10 million, and approximately $2 million of gold and silver are being concentrated and delivered each year (18).

The most likely sources of the metals in the Palo Alto sewage are the large electronics and photographic industries in the area. Gold in significant amounts has been found in the sewage of only one other city, and silver is usually present in much smaller concentrations than at Palo Alto. Thus Palo Alto and its unique urban ore offers an unusual opportunity to study and develop methods for recycling valuable materials concentrated in urban waste (18). The city has now contracted for a private company to extract the gold and silver.

The notion of reusing waste materials is not new—metals such as iron, aluminum, copper, zinc, and lead have been recycled for many years (Figure 12.30). Of the millions of automobiles discarded annually, nearly 90 percent are now dismantled by auto wreckers and scrap processors for recyclable metals (17). Recycling metals from discarded automobiles is a sound conservation practice, considering that nearly 90 percent by weight of the average automobile is metal (Figure 12.31).

Recycling appears to be one way to delay or partly alleviate a possible crisis caused by convergence of rapidly rising population and a limited resource base. Recycling the wide variety of materials in urban waste is not an easy task, however, and before it can become a widespread practice, we will need improved technology and more economic incentives. Figure 12.32 shows a possible flowchart with necessary equipment to recycle urban refuse. Experiments conducted by the U.S. Bureau of Mines at a pilot plant using these techniques have been encouraging. Although refinements are necessary, urban refuse has been successfully separated into concentrates of light-gauge iron, massive metals, glass, paper, plastics, organic waste, and other combustible wastes (16).

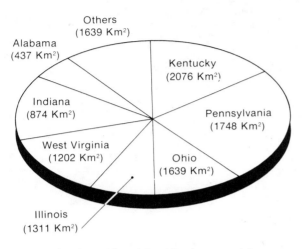

(a) Land used for mining bituminous coal, by state 1930-80. Total land is 10,927 Km².

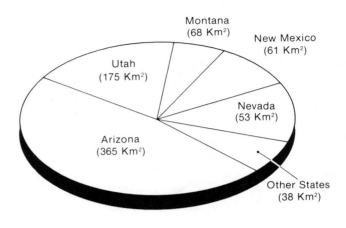

(b) Land used for mining copper, by state, 1930-80. Total land is 761 Km².

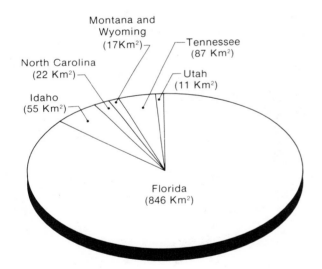

(c) Land used for mining phospate rock, by state, 1930-80. Total land is 1,093 Km².

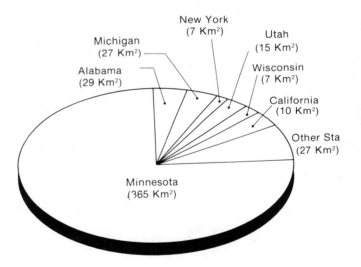

(d) Land used for mining iron ore, by state, 1930-80. Total land is 486 Km².

FIGURE 12.27
Amount of land (by state) mined for selected materials. (Data from W. Johnson and J. Paone, 1982, U.S. Bureau of Mines Information Circular 8862.)

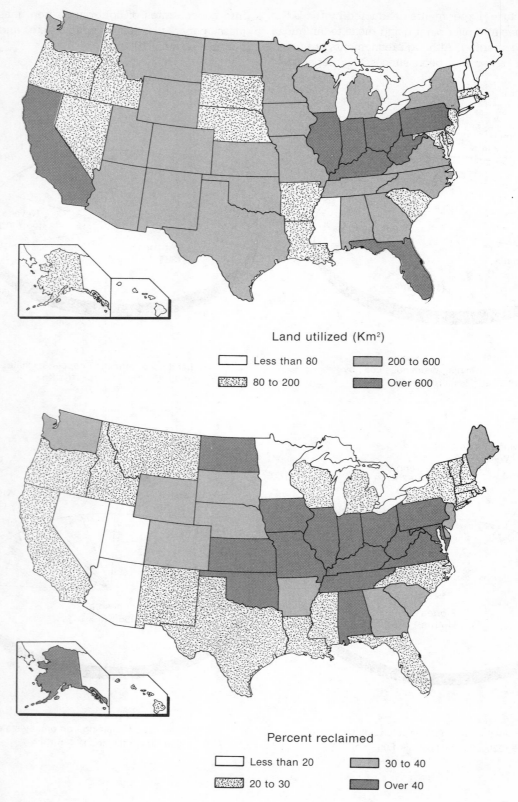

Land utilized (Km²)

☐ Less than 80   ▨ 200 to 600

▨ 80 to 200   ▨ Over 600

Percent reclaimed

☐ Less than 20   ▨ 30 to 40

▨ 20 to 30   ▨ Over 40

FIGURE 12.28
Geographic distribution of land mined and reclaimed from 1930 to 1980. (Data from W. Johnson and J. Paone, 1982, U.S. Bureau of Mines Information Circular 8862.)

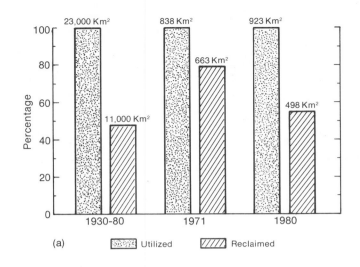

(a)   Utilized   Reclaimed

(b)

FIGURE 12.29
Relationship between land mined and reclaimed in the U.S. from 1930–1980, compared to the same relationship for 1971 and 1980 (a) and photograph of a Missouri strip mine reclaimed for agricultural use (b). (Data from U.S. Bureau of Mines; photograph courtesy of Jerry D. Vineyard and Missouri Geological Survey.)

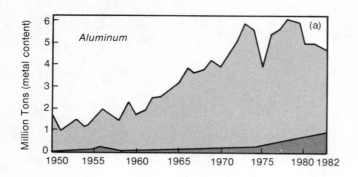

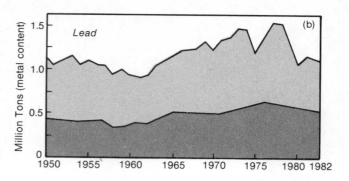

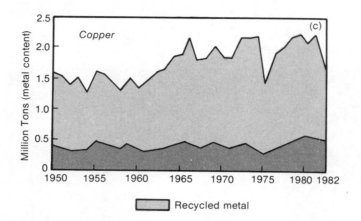

Recycled metal

FIGURE 12.30
United States consumption and recycling of selected metals. (Data from U.S. Bureau of Mines, 1979 and 1983.)

FIGURE 12.31
Content of a typical discarded automobile. (From F. F. Davis, "Urban Ore," *California Geology*, May 1972.)

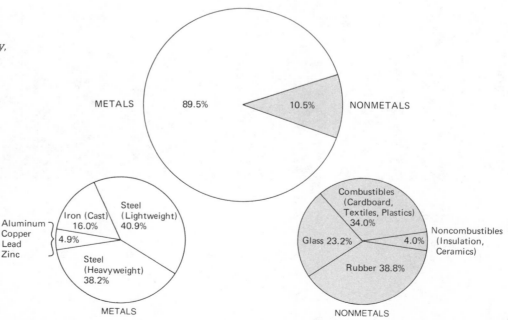

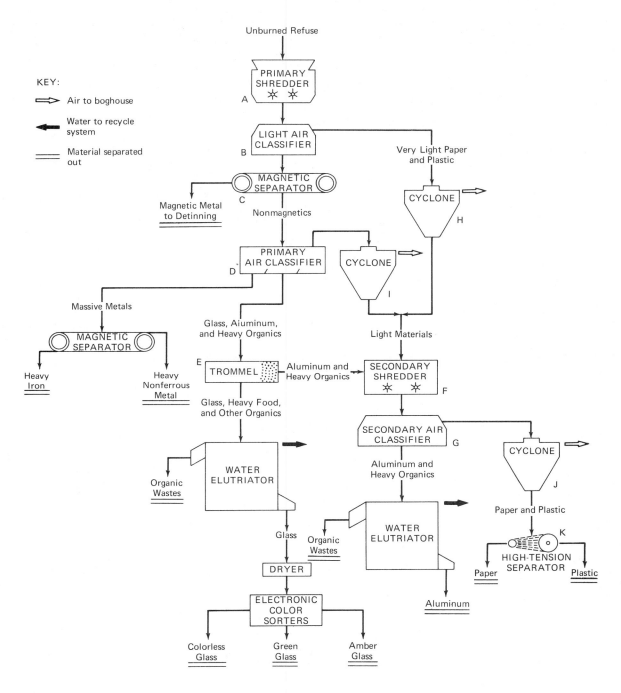

FIGURE 12.32
Flow chart showing how raw refuse may be separated for recycling of resources. (From
P. M. Sullivan, et al., U.S. Bureau of Mines, R.I. 7760, 1973.)

SUMMARY AND CONCLUSIONS

The exponential increase of a population whose needs are filled by a finite resource base may eventually explode as a food, energy, or other resource crisis. Based on present technology, there are simply too many people for worldwide affluence similar to that of the United States and other industrial nations. Therefore, affluence in many developed areas of the world may have to be reduced to a more conservation-oriented, less materialistic scale (but not necessarily a lower quality of life), while underdeveloped nations increase the availability of goods and services to their people.

An important concept in analyzing resources and reserves is that resources are not reserves! They are "birds in the bush" or "fish in the pond," and unless we discover and capture them, we cannot use them to solve present shortages.

The geology of mineral resources is complex and intimately related to various aspects of the geologic cycle. In a broad-brush approach, many, but certainly not all, metallic mineral deposits may correlate with dynamic earth processes that occur at junctions of lithospheric plates. On a more practical scale, mineral resources for construction and industrial uses are concentrated in the geologic environment by: first, igneous or magmatic processes such as crystal settling or hydrothermal activity; second, metamorphic processes such as contact and regional metamorphism; third, sedimentary processes including oceanic and lake processes, running water, wind, and moving ice; and fourth, weathering processes including soil-forming activity and in situ concentrations of insoluble minerals.

Recycling of resources is technically feasible; for conservation purposes, conversion of waste to energy or useful resources should be a high-priority goal of urban areas. Before it becomes practical on a large scale, however, increased economic incentives and better procedures will be necessary.

REFERENCES

1. PARK, C. F., JR. 1968. *Affluence in jeopardy.* San Francisco: Freeman, Cooper & Company.

2. U.S. GEOLOGICAL SURVEY. 1975. *Mineral resource perspectives 1975.* U.S. Geological Survey Professional Paper 940.

3. McKELVEY, V. E. 1973. Mineral resource estimates and public policy. In *United States mineral resources,* D. A. Brobst and W. P. Pratt, eds., pp. 9–19. U.S. Geological Survey Professional Paper 820.

4. BROBST, D. A., PRATT, W. P., and McKELVEY, V. E. 1973. *Summary of United States mineral resources.* U.S. Geological Survey Circular 682.

5. NOAA. 1977. Earth's crustal plate boundaries: Energy and mineral resources. *California Geology* (May 1977): 108–109.

6. FOSTER, R. J. 1973. *General geology.* Columbus, Ohio: Charles E. Merrill.

7. BATEMAN, A. M. 1950. *Economic ore deposits.* 2d ed. New York: John Wiley & Sons.

8. PARK, C. F., JR., and MacDIARMID, R. A. 1970. *Ore deposits.* 2d ed. San Francisco: W. H. Freeman.

9. LAWRENCE, R. A. 1973. Construction stone. In *United States mineral resources,* ed. D. A. Brobst and W. P. Pratt, pp. 157–62. U.S. Geological Survey Professional Paper 820.

10. YEEND, W. 1973. Sand and gravel. In *United States mineral resources,* ed. D. A. Brobst and W. P. Pratt, pp. 561–65. U.S. Geological Survey Professional Paper 820.

11. SMITH, G. I., JONES, C. L., CULBERTSON, W. C., ERICKSON, G. E., and DYNI, J. R. 1973. Evaporites and brines. In *United States mineral resources,* ed. D. A. Brobst and W. P. Pratt, pp. 197–216. U.S. Geological Survey Professional Paper 820.

12. CORNWALL, H. R. 1973. Nickel. In *United States mineral resources,* ed. D. A. Brobst and W. P. Pratt, pp. 437–42. U.S. Geological Survey Professional Paper 820.

13. VAN, N., DORR, J., CRITTENDEN, M. D., and WORL, R. G. 1973. Manganese. In *United States mineral resources*, ed. D. A. Brobst and W. P. Pratt, pp. 385–99. U. S. Geological Survey Professional Paper 820.

14. SECRETARY OF THE INTERIOR. 1975. *Mining and mineral policy, 1975.* Washington, D. C.: U.S. Government Printing Office.

15. FAGAN, J. J. 1974. *The earth environment.* Englewood Cliffs, New Jersey: Prentice-Hall.

16. SULLIVAN, P. M., STANCZYK, M. H., and SPENDBUE, M. J. 1973. *Resource recovery from raw urban refuse.* U.S. Bureau of Mines, R.I. 7760.

17. DAVIS, F. F. 1972. Urban ore. *California Geology* (May 1972): 99–112.

18. GULBRANDSEN, R. A., RAIT, N., KRIER, D. J., BAEDECKER, P. A., and CHILDRESS, A. 1978. *Gold, silver, and other resources in the ash of incinerated sewage sludge at Palo Alto, California—a preliminary report.* U.S. Geological Survey Circular 784.

Chapter 13

Energy and Environment

ENERGY AND PEOPLE

Our discussion of mineral resources established that resources are not infinite, and that it is impossible to support an exponential increase in population on a finite resource base. The same is true for energy derived from mineral resources.

United States citizens have recently encountered the effects of energy shortages for the first time, including increases in the prices of energy and products produced from petroleum. Nevertheless, to many people energy still seems unlimited. The United States, whose total population is only a small percentage of the world's population, continues to consume a disproportionate share of the total electric energy produced in the world. The appetite of the American people for a higher standard of living and its accompanying energy consumption are shown in Figure 13.1.

Nearly 90 percent of the energy consumed in the U.S. today is produced from coal, natural gas, and petroleum (Figure 13.2), with small amounts of hydropower and, more recently, nuclear power. We still have huge reserves of coal, but there are restrictions: major new sources of natural gas and petroleum are becoming scarce; few new large hydropower plants can be expected; and planning and construction of new nu-clear power plants have become uncertain for a variety of reasons. On the brighter side, alternative energy sources such as solar power for homes, farms, and offices are becoming economically feasible and thus more common. It is likely, however, that people in industrialized countries will have to realize that the quality of life is not directly related to ever-expanding energy needs.

Projections of supply and demand for energy are at best difficult, because technical, economic, political, and social assumptions that underlie the projections are constantly changing. Thus recent predictions state that energy consumption in the U.S. in the year 2010 may exceed 100 exa (10^{18}) joules or be as low as 63 exa joules (energy consumption in 1982 was about 72 exa joules); one exa joule is approximately equal to one quad (10^{15} Btu). The higher value assumes no change in energy policies, whereas the lower value assumes aggressive energy conservation policies.

Evaluation of all potential energy sources and conservation practices is necessary to insure that the flow of energy will be sufficient to maintain our industrial society and a quality environment. This difficult task is expected to become even more so, because more innovations in energy production are likely to be forthcoming, and each innovation requires a new eval-

FIGURE 13.1
(a) Relation between energy consumption and standard of living for various countries. (From Pennsylvania Department of Education, *The Environmental Impact of Electrical Power Generation: Nuclear and Fossil,* 1973.)
(b) Oil consumption for free world countries, 1950–1978. (From Department of Energy, Annual Report to Congress, 1979.)

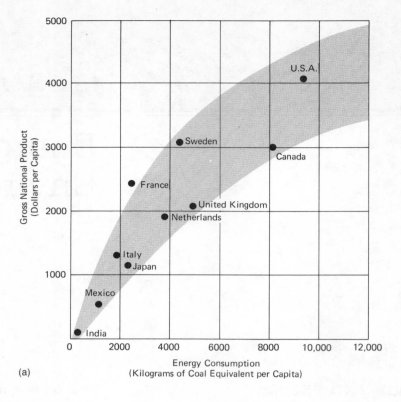

(a)

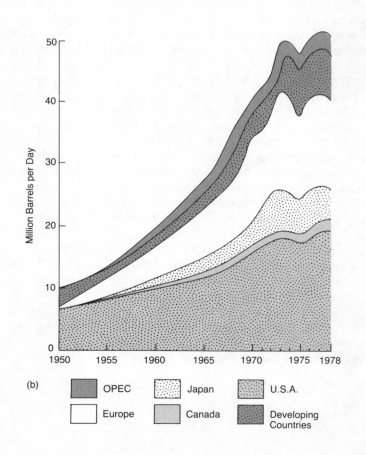

(b)

344

FIGURE 13.2
United States consumption of energy, 1949 to 1982 (From Department of Energy.)

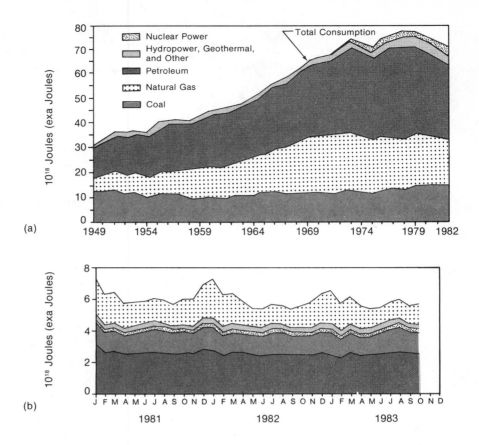

(a)

(b)

uation of the total available data (1). Of particular importance will be energy uses with applications below 100° C, because a large portion of the total energy consumption (for uses below 300° C) in the United States is for space heating and water heating (see Figure 13.3). With these ideas in mind, we will cautiously ex-

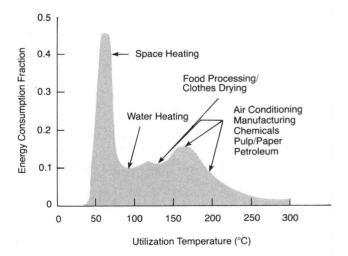

FIGURE 13.3
Spectra of energy use below 300° C in the United States. [From Los Alamos Scientific Laboratory (L.A.S.L. 78–24), 1978.]

plore some selected geologic and environmental aspects of well-known energy resources such as coal and petroleum, nuclear sources, and other possibly important sources including oil shale, tar sands, and geothermal resources. We will also discuss briefly the expected increase in demand for water as a result of energy production and several other alternative energy sources such as hydropower (river and tidal) and solar power.

COAL

Geology of Coal

Like other fossil fuels, coal is made up of organic materials that have escaped oxidation in the carbon cycle. Coal is essentially the altered residue of plants that have flourished in ancient freshwater or brackish-water swamps, typically found in estuaries, coastal lagoons, and low-lying coastal plains or deltas (2).

Processes that form coal (Figure 13.4) involve the development of a swamp rich in plants that are partially decomposed in an oxygen-deficient environment and that accumulate slowly to form a thick layer of peat. These swamps and accumulations of peat may then be inundated by a prolonged slow rise of sea level

(a) *Coal swamp forms.*

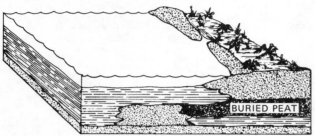

(b) *Rise in sea level buries swamp in sediment.*

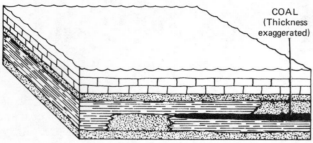

COAL
(Thickness
exaggerated)

(c) *Compression of peat forms coal.*

FIGURE 13.4
Idealized diagram showing the processes by which buried plant debris (peat) is transformed into coal. Considerable lengths of geologic time must elapse before the transformation is complete.

(a relative rise, as the land may be sinking) and covered by sediments such as sand, silt, clay, and carbonate-rich material. As more and more sediment is deposited, water and organic gases (volatiles) are squeezed out, and the percentage of carbon increases in the compressed peat. As this process continues, the peat is eventually transformed to coal. Because there are often several layers of coal in the same area, scientists believe the sea level may have risen and fallen alternately, allowing development and then drowning of coal swamps.

Classification and Distribution of Coal

Coal is commonly classified according to rank and sulfur content. The rank is generally based on the per-

TABLE 13.1
Distribution of United States coal resources according to their rank and sulfur content.

| | Sulfur Content (Percent) | | |
| | Low | Medium | High |
Rank	0–1	1.1–3.0	3 +
Anthracite	97.1	2.9	—
Bituminous coal	29.8	26.8	43.4
Subbituminous coal	99.6	.4	—
Lignite	90.7	9.3	—
All ranks	65.0	15.0	20.0

Source: U.S. Bureau of Mines Circular 8312, 1966.

centage of carbon, which increases from lignite to bituminous to anthracite, as shown in Figure 13.5. This figure also shows that heat content is maximum in bituminous coal, which has relatively few volatiles (oxygen, hydrogen, and nitrogen) and low moisture content. Heat content is minimum in lignite, which has a high moisture content. The distribution of the common coals (bituminous, subbituminous, and lignite) in the contiguous United States and Alaska is shown in Figure 13.6.

The sulfur content of coal may be generally classified as low (zero to 1 percent), medium (1.1 to 3 percent), or high (greater than 3 percent). Most coal in the United States is of the low-sulfur variety (Table 13.1), and by far the greatest part of the low-sulfur coals is a

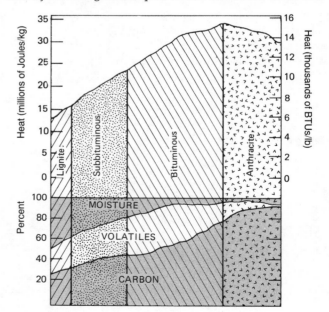

FIGURE 13.5
Generalized classification of different types of coal based upon their relative content (in percentage) of moisture, volatiles, and carbon. The heat values of the different types of coal are also shown. (After D. A. Brobst and W. P. Pratt, eds., U.S. Geological Survey Professional Paper 820, 1973.)

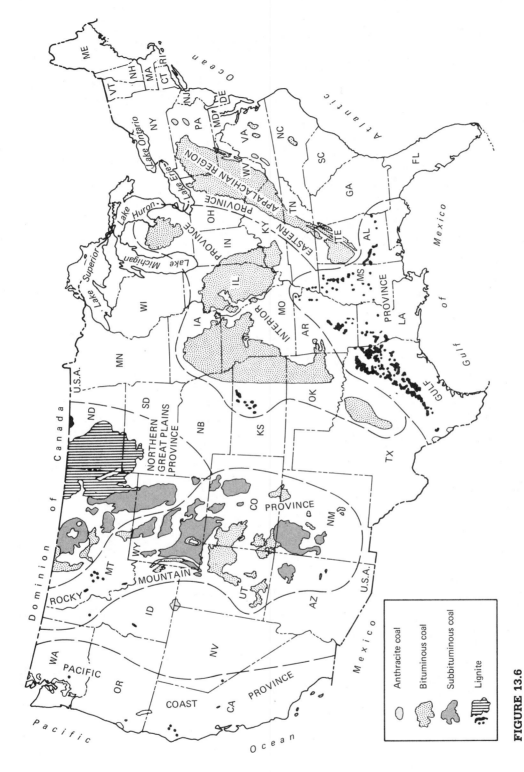

FIGURE 13.6
Coal fields of the United States. (After U.S. Bureau of Mines Information Circular 8531, 1971. Adapted from U.S.G.S. Coal Map of Alaska, 1960.)

347

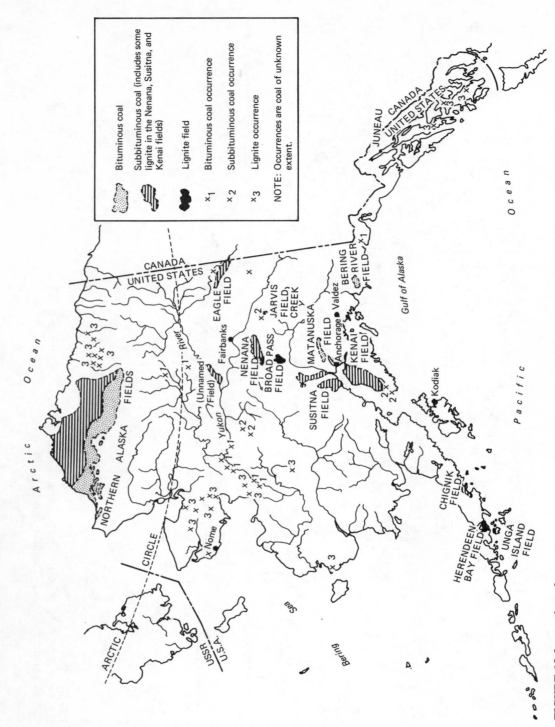

FIGURE 13.6, *continued*

348

relatively low-grade, subbituminous variety found west of the Mississippi River (Figure 13.7). The location of coal reserves has environmental significance because, with all other factors equal, the use of low-sulfur coal as a fuel for power plants causes less air pollution. Therefore, to avoid air pollution, thermal power plants

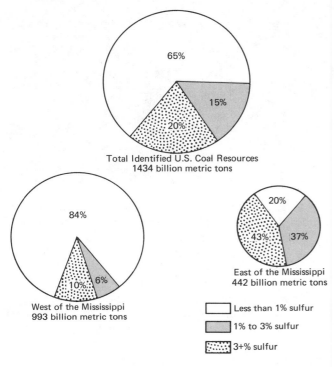

FIGURE 13.7
Geographical distribution of identified U.S. coal resources by sulfur content. (After Federal Energy Administration, *Monthly Energy Review,* July 1975.)

FIGURE 13.8
Coal strip mine, Henry County, Missouri. (Photo by J. D. Vineyard, courtesy of Missouri Geological Survey.)

on the highly populated East Coast will have to continue to treat some of the local coal to lower its sulfur content. This treatment increases the cost, but may be more economical than shipping low-sulfur coal long distances.

Impact of Coal Mining

Much of the coal mining in the United States is still done underground, but strip mining (Figure 13.8), which started in the late nineteenth century, has steadily increased, whereas production from underground mines has stabilized. The trend toward strip mining has developed because the method is in many cases technologically and economically more advantageous than underground mining. The increased demand for coal will lead to more and larger strip mines to extract the estimated 40 billion metric tons of coal reserves that are now accessible to surface mining techniques. In addition, approximately another 90 billion metric tons of coal within 50 meters of the surface is potentially available for stripping if need demands.

The impact of large strip mines varies from region to region depending on topography, climate, and, most importantly, reclamation practices. In humid areas with abundant rainfall, mine drainage of acid water is a serious problem (Figure 13.9). Surface water infiltrates the spoil banks (material left after the coal or other minerals are removed) where it reacts with sulfide minerals such as pyrite (FeS_2) to produce sulfuric acid. The sulfuric acid then runs into and pollutes streams and groundwater resources. Acid water also drains from underground mines and road cuts and areas where coal and pyrite are abundant, but the problem is magnified

FIGURE 13.9
The stream in the foreground is flowing through waste piles of a Missouri coal mine. The water reacts with sulfide minerals and forms sulfuric acid. This is a serious problem in coal mining areas. (Photo by J. D. Vineyard, courtesy of Missouri Geological Survey.)

when large areas of disturbed material remain exposed to surface waters. Acid drainage can be minimized through proper use of water diversion practices that collect surface runoff and groundwater before they enter the mined area and divert them around the potentially polluting materials. This practice reduces erosion, pollution, and water-treatment cost (3).

In arid and semiarid regions, water problems associated with mining are not as pronounced as in wetter regions, but the land may be more sensitive to mining activities such as exploration and road building. In some arid areas, the land is so sensitive that even tire tracks across the land survive for years. Soils are often thin, water is scarce, and reclamation work is difficult.

Common methods of strip mining include *area mining*, which is practiced on relatively flat areas (Figure 13.10), and *contour mining*, which is used in hilly terrain (Figure 13.11). The width of the cut in contour mining depends on the ease of excavation and topography. As the width increases, more and more overburden must be removed, which increases the cost. Topography thus limits the total width of the cut. The objective is to confine the cut to a given elevation (contour) and work around the hill while maintaining that elevation. Both methods of strip mining severely disturb the landscape by removing indigenous vegetation, overburden, and desired minerals. Unless the mined land is reclaimed, unsightly conical piles or ridges of waste remain as a source material to pollute groundwater and surface-water. Area mining is used only on relatively flat ground where the potential velocity and erosion of runoff are low. Therefore, the potential to

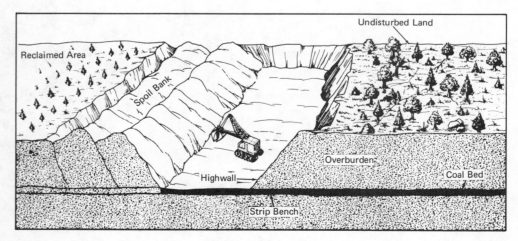

FIGURE 13.10
Idealized diagram showing area strip mining.

FIGURE 13.11
Idealized diagram showing contour strip mining.

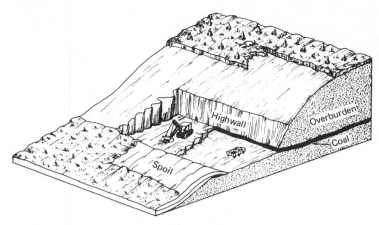

pollute streams by siltation is less than for contour mining. Potential groundwater pollution is greater for area mining, however, because there is more area and more time for precipitation to infiltrate and slowly migrate through the spoil piles (3).

All methods of strip mining have the potential to pollute or destroy scenic, water, biologic, or other land resources, but good reclamation practices can minimize the damage (Figure 13.12). One potentially good method is to segregate the overburden for eventual replacement of the topsoil that is removed. This method has been used rather widely in the coal fields of the eastern United States, and experience has shown that it is a successful way to control water pollution when combined with regrading and revegetation (3).

Low-sulfur coal deposits in the western United States, particularly in Montana, Wyoming, Colorado, Utah, New Mexico, and Arizona, will probably be mined by the area method of strip mining. Reclamation may be difficult even if the overburden is segregated,

because soils are thin, and revegetation is extremely difficult because water to establish new vegetation is scarce. Additional research into methods of reclamation for arid and semiarid regions is essential (3,4).

Underground mining of coal and other resources has caused considerable environmental degradation from mine drainage of acid water and has produced serious hazards such as subsidence over mines or fires in mines (Figure 13.13). In the past, waste material (spoil) from underground mines has been piled on the surface, producing aesthetic degradation and sediment and chemical pollution of water resources resulting from exposure to surface water and groundwater.

Federal guidelines now govern strip mining of coal in the United States. They require, basically, that mined land be restored to support its pre-mining use. Restoration includes disposing of wastes, contouring the land, and replanting vegetation. The hope is that after reclamation, the mined land will appear and function as it did prior to extraction of the coal; however,

FIGURE 13.12
A small open-pit mine before (left) and after (right) reclamation. (Photos by J. D. Vineyard, courtesy of Missouri Geological Survey.)

FIGURE 13.13
Escaping gases from holes drilled
into a burning abandoned coal
mine in western Pennsylvania.
(Photo courtesy of U.S. Bureau of
Mines.)

as previously mentioned, this will be a difficult task, and probably not completely successful. The new regulations also prohibit mining on prime agricultural land and give farmers and ranchers the opportunity to restrict or veto mining on their land, even if they do not own the mineral rights.

Trapper Mine Near Craig, Colorado— A Case History

Trapper Mine on the western slope of the Rocky Mountains in northern Colorado is a good example of a new generation of large coal strip mines. The main operation is designed to minimize environmental degradation during mining and to enable reclaiming the land for dry land farming and grazing of livestock and big game without artificial application of water.

The mine will produce 68 million metric tons of coal over a 35-year period, to be delivered to an 800 megawatt power plant adjacent to the mine. To meet this commitment, approximately 20–24 km^2 of land will have to be strip mined.

Four coal seams, varying from about 1 to 4 m thick, each separated by various depths of overburden, will be mined. Depth of overburden to the coal varies from zero to about 50 meters. The steps in the actual mining are as follows:

1 Vegetation and top soil are removed with dozers and scrapers, and the soil is stockpiled for reuse
2 Overburden along a cut up to 1.6 km long and 53 m wide is removed with a 23-cubic-meter dragline bucket (Figure 13.14)
3 Exposed coal beds are drilled and blasted to fracture the coal, which is removed with a backhoe and loaded into trucks (Figure 13.15)

FIGURE 13.14
Removing overburden at the Trapper Mine, Colorado. The large drag line shown here has a 39 cubic meter bucket.

FIGURE 13.15
Large backhoe at the Trapper Mine, Colorado, removing the coal which is then loaded in large trucks and delivered to a power plant just off the mining site.

4 The cut is filled, top soil is replaced, and the land is either planted in a crop or returned to rangeland (Figure 13.16)

At the Trapper Mine the land is reclaimed without artificially applying water. Precipitation (mostly snow) is about 35 cm per year, which is sufficient to reestablish vegetation provided there is adequate top soil to help hold the soil water. This factor emphasizes that reclamation is site specific; what works at one location or region may not apply to other areas.

Water and air quality are closely monitored at the Trapper Mine. Surface water is diverted around mine pits and groundwater is intercepted while pits are open. Settling basins constructed downslope from pits trap suspended solids before discharging water into local streams. Air quality at the mine is degraded by dust produced from blasting, hauling, and grading of the coal. Dust is minimized by regularly watering or otherwise treating roads and other surfaces that may produce dust.

Reclamation of mined land at the Trapper Mine has been very successful during the first years of operation. Although the environmental protection techniques increase the cost of the coal by as much as 50 percent, the payoff will come in the long-range productivity of the land after termination of mining. Some might argue that the Trapper Mine is unique in that the fortuitous combination of geology, hydrology, and topography allow for successful reclamation; on the other hand, the mine shows that with careful site selection and planning, strip mining is not incompatible with other land uses.

Future Use of Coal

Limited resources of oil and natural gas are greatly increasing the demand for coal, and a significant changeover from burning oil and gas to burning coal in thermoelectric power plants and industrial heat-generating units is forthcoming. About 60 plants have been, or shortly will be, converted at a savings of about 250,000 barrels of oil per day. Total savings on the order of one million barrels per day could be realized if all thermoelectric power plants burn coal. There are sufficient coal reserves to meet the increased demand from

FIGURE 13.16
Reclaimed land at the Trapper Mine, Colorado. The site in the foreground has just had the soil replaced following mining whereas the vegetated sites have been entirely reclaimed.

such conversion and to supply coal for new plants for many hundreds of years (5).

The real crunch on oil and gas is still a few years away, but when it does come, it will put tremendous pressure on the coal industry to open more and larger mines in both the eastern and the western coal beds of the United States. This may have tremendous environmental impacts for several reasons. First, more and more land will be strip mined and will thus require careful restoration. Second, unlike oil and gas, burned coal leaves ash (5 to 20 percent of the original amount of the coal) that must be collected and disposed of. Some ash can be used for landfill or other purposes, but about 85 percent, is presently useless. Third, handling of tremendous quantities of coal through all stages—mining, processing, shipping, combustion, and final disposal of ash—will have potentially adverse environmental effects, such as aesthetic degradation, noise, dust pollution, and, most significant, release of trace elements that are likely to cause serious health problems into the water, soil, and air (5).

The transport of large amounts of coal, or energy derived from coal, from production areas with low energy demand to large population centers is a significant environmental issue. Coal can be converted on site to electricity, synthetic oil, or synthetic gas, all of which are relatively easy to transport, but with few exceptions these alternatives present problems. Transmission of electricity over long distances is expensive, and power plants in semiarid, coal-rich regions may have trouble finding sufficient water for cooling. Conversion of coal to synthetic oil or gas, although possible, is expensive and the technology is primitive. Furthermore, conversion requires a tremendous amount of water and thus, as with the generation of electrical power, will place a significant demand on local water supplies in the coal regions of the western United States (1).

Methods of transporting large volumes of coal for long distances include freight trains and coal slurry pipelines. Trains have the advantage of a relatively low cost for new capital expenses and thus will continue to be used. Coal slurry pipelines, designed to use water to transport pulverized coal, have an economic advantage over trains if (1):

- Transport distance is long and a large volume of coal is shipped
- Inflation rates are high and interest rates low
- Mines are large and customers will purchase large volumes of coal over a long period of time
- Sufficient, low-cost water is available

The economic advantages of the slurry pipeline are thus rather tenuous, especially in the western United States where large volumes of water will be difficult to obtain. For example, a pipeline that transports 30 million tons of coal requires about 20 million cubic meters of water per year, which is enough to meet the water needs for a city of about 85,000 people or to irrigate up to 40 square kilometers of farmland. Despite these problems, some slurry pipelines will probably be constructed. Figure 13.17 shows an idealized description of the slurry pipeline system.

Environmental problems associated with coal, while significant enough to cause concern, are not necessarily insurmountable, and careful planning could minimize them. At any rate, there may be few alternatives in the future to mining tremendous quantities of coal to feed thermoelectric power plants and to providing oil and gas by gasification and liquefaction processes.

OIL AND GAS

Geology of Oil and Gas

Oil and natural gas (methane) are hydrocarbons. Like coal, they are fossil fuels in that they form from organic material that has escaped complete decomposition after burial. Next to water, oil is probably the most abundant fluid in the earth's crust, yet the processes that form it are only partly understood. Most earth scientists accept that oil and gas are derived from organic materials that are buried with marine or lake sediments. Favorable environments where organic debris might escape oxidation include *near-shore areas* characterized by rapid deposition that quickly buries organic material or *deeper-water areas* characterized by a deficiency in oxygen at the bottom which promotes an anaerobic decomposition. Beyond this, the locations where oil and gas form are generally classified as subsiding, depositional basins in which older sediment is continuously buried by younger sediments, thus progressively subjecting the older, more deeply buried material to higher temperatures and pressures (6).

The major source material for oil and gas is a fine-grained, organic-rich sediment that is buried to a depth of at least 500 meters and subjected to increased heat and pressure that physically compress the source rock. The elevated temperature and pressure, along with other processes, start the chemical transformation of organic debris into hydrocarbons (oil and gas). As the pressure increases, the porosity of the source rock is reduced, and the higher temperatures thermally energize the hydrocarbons and induce them to begin an upward migration to a lower-pressure environment with increased porosity. The initial movement of the hydrocarbons upward through the source rock is

PLATE 17
Erosion of soil from urban sites during construction (top) often produces a pollution problem for nearby areas (bottom). Sediment control ordinances in many areas now minimize such problems.

PLATE 18

Off-road vehicle tracks in "devil's basin" near Ventura, California (top) and Pismo Dunes, near Santa Barbara, California (bottom). Use of ORVs is incompatible with most other land uses and is destructive in environmentally sensitive areas.

RARE
PLANTS
LITTLE
COREOPSIS HILL
DUNE 2 - USMS

PLATE 19
Two extremes of channelization. Concrete ditch, Carpinteria Creek, California (top) and channel restoration of Briar Creek near Charlotte, North Carolina (bottom).

PLATE 20
A new type of channelization undergoing experimentation in England, similar to that recommended in Figure 9.43. A small meandering channel is located within the larger flood channel, which lacks vegetation in this photograph taken shortly after construction in 1983. Vegetation will eventually cover the flood channel and provide wildlife habitat. The channel is part of the River Riding near London, England.

Missouri River. This upstream view shows a series of permeable pile dikes used to train the river. (Photo courtesy of the U.S. Army Corps of Engineers, Omaha District)

PLATE 22
Overview of part of the Long Valley Caldera (top). The snow-covered Sierra Nevada forms the western boundary; the area of recent uplift and earthquake activity is in the central portion of the photograph. Close-up of hot spring activity within the Caldera (bottom). Note the plume of steam at the center of the photograph. The construction is for a small geothermal energy plant (7000 kw) scheduled for completion in 1985.

1. Bingham copper mine
2. Great Salt Lake
3. Lake Utah
4. Tailing pond from copper smelters
5. Commercial salt evaporating pond
6. Wasatch front along Wasatch fault that uplifts snow-covered Mts.

0   20 km

N

PLATE 23A

Salt Lake City, Utah. This area contains the Bingham copper mine as well as evaporite resources associated with the Great Salt Lake, which is only a remnant of the once much larger Lake Bonneville. (ERTS photo)

PLATE 23B

Lavendar Pit copper mine, Bisbee, Arizona, (Photo courtesy of Peter L. Kresan)

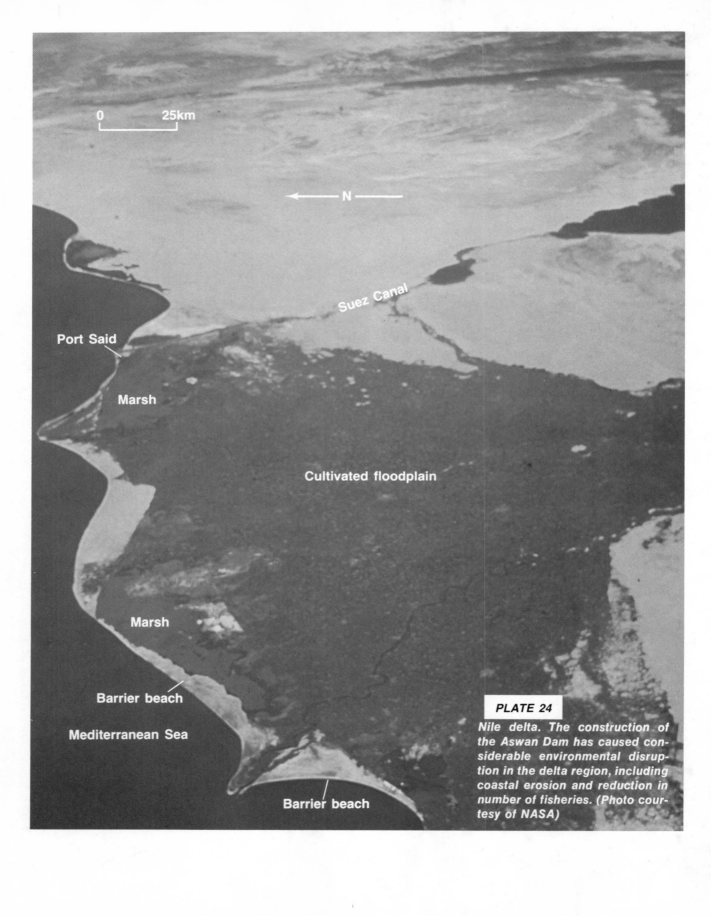

0   25km

N

Suez Canal

Port Said

Marsh

Cultivated floodplain

Marsh

Barrier beach

Mediterranean Sea

Barrier beach

PLATE 24

Nile delta. The construction of the Aswan Dam has caused considerable environmental disruption in the delta region, including coastal erosion and reduction in number of fisheries. (Photo courtesy of NASA)

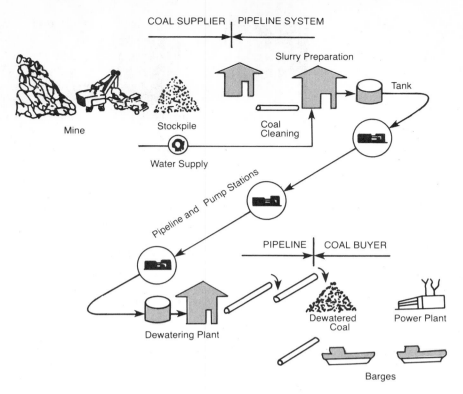

FIGURE 13.17
Idealized diagram showing the coal slurry pipeline system. (From Council on Environmental Quality, 1979.)

termed *primary migration,* which merges into *secondary migration* as the oil and gas move more freely into and through coarse-grained, more permeable rock such as sandstone or fractured limestone. These porous, permeable rocks into which the oil and gas migrate are called *reservoir rocks.* If the path is clear to the surface, the oil and gas will migrate and escape there; this perhaps explains why most of the oil and gas is found in geologically young rocks. That is, hydrocarbons in older rocks have had a longer period of time in which to reach the surface and leak out (6). However, the lower amounts of hydrocarbon in very old rocks might also be explained by tectonic processes that uplift rocks containing oil and gas, exposing them to erosion.

If in their upward migration, oil and gas are interrupted or stopped because they encounter a relatively impervious barrier, then they may accumulate. If the barrier *(cap rock)* is in a favorable geometry (structure) such as a dome or an anticline, the oil and gas will be trapped in their upward movement at the crest of the dome or anticline below the cap rocks (6). Figure 13.18 shows an anticlinal trap and two other possible traps caused by faulting or by an unconformity (buried erosion surface). These are not the only possible types of traps. Any rock that has a relatively high porosity and permeability and that is connected to a source rock containing hydrocarbons may become a reservoir, provided that the upward migration of the oil and gas is impeded by a cap rock oriented such that the hydrocarbons are entrapped at a central high point (6).

Petroleum Production

Production wells in an oil field recover petroleum through primary or enhanced recovery methods. Primary recovery uses natural reservoir pressure to move the oil to the well, but normally, this delivers, by pumping, only up to 25 percent of the total petroleum in the reservoir. To increase the recovery rate to 50 to 60 percent or more, enhanced (secondary or tertiary) recovery methods are necessary. The purpose of the enhancement is to manipulate reservoir pressure and enable the reservoir to better transmit fluids by carefully injecting natural gas, water, steam, and/or chemicals into the reservoir. Enhancement pushes petroleum to wells where it can be lifted to the surface by means of the familiar "horse head" bobbing pumps, submersible pumps, or other lift methods.

Petroleum production always brings to the surface a variable amount of salty water along with oil. After separating the oil and water, the latter must be disposed of, because it is toxic to the surface environment. Disposal can be accomplished by injection as part of enhanced (secondary) recovery, evaporation in lined open pits, or deep-well disposal outside the field.

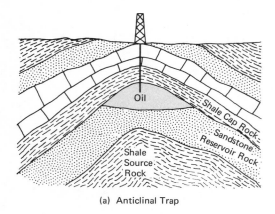

(a) Anticlinal Trap

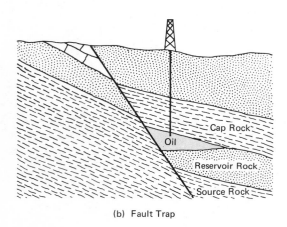

(b) Fault Trap

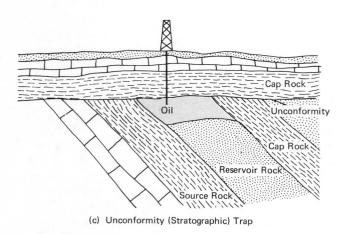

(c) Unconformity (Stratigraphic) Trap

FIGURE 13.18
Idealized diagrams showing several types of oil traps.

Distribution and Amount of Oil and Gas

The distribution of oil and gas in time (geologic) and space is rather complex, but in general, three principles apply. First, commercial oil and gas are produced almost exclusively from sedimentary rocks deposited during the last 500 million years of the earth's history (6). Second, although there are many oil fields in the world, approximately 85 percent of the total production plus reserves occurs in less than 5 percent of the producing fields; 65 percent occurs in about 1 percent of the fields; and, as remarkable as it seems, 15 percent of the world's known oil reserves is in two accumulations in the Middle East (6). Third, the geographic distribution of the world's giant oil and gas fields (Figure 13.19) shows that most are located near tectonic belts (plate junctions) that are known to have been active in the last 60 to 70 million years.

It is difficult to assess the petroleum and gas reserves of the United States, much less those of the entire world. Recent estimates of proven oil and gas reserves of the United States suggest that, at present production rates, the oil and gas will last only a few decades. Using estimates of inferred and undiscovered resources, one can extend these projections; however, even the best data are only estimates subject to large errors. In 1975, for example, the undiscovered recoverable oil resources were estimated at 105 billion barrels, but from a probability standpoint, there is a 95 percent chance that at least 73 billion barrels will be recovered and only a 5 percent chance that as much as 150 billion barrels will be recovered, giving a possible range of 77 billion barrels (7). The significance of these figures and projections is their suggestion that it will be extremely difficult for the United States to become economically self-sufficient in oil and gas (6). The last time the United States was essentially self-sufficient in oil was about 1950, and we have been importing oil since then, despite a peak period of exploratory drilling in 1956 (Figure 13.20). We have also become more successful in drilling, and in 1979, the success ratio increased to a record high of 29 percent, reflecting more careful and better exploration.

What is the future of oil and gas? Unless large, new accumulations of petroleum are discovered in the United States—there is always this possibility, and in fact large fields were recently discovered near Santa Barbara, California—known reserves and projected recoverable resources will last only a few more years before a shortage is evident. When a significant shortage does occur, we will most likely turn to large-scale gasification and liquefaction of our tremendous coal reserves, extract oil and gas from oil shale, perhaps rely more on atomic energy, and certainly rely more on solar

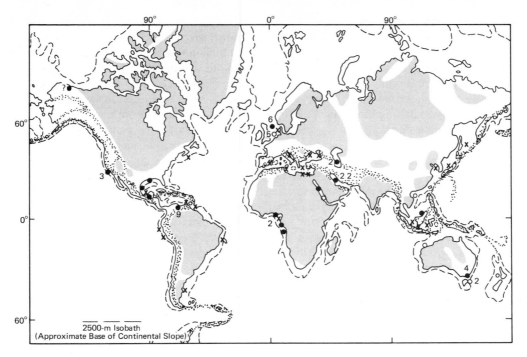

FIGURE 13.19
Giant oil and gas fields of the world relative to generalized tectonic belts. Active tectonic areas are shown by the stippled pattern, and regions not subjected to tectonic activity for the last 500 million years are shaded. Giant oil fields are denoted by the solid circles; giant gas fields by the open circles; and general areas of discovery either of less than giant size or areas under field development are indicated by an X. The numbers indicate how many fields are in a particular location. (After U.S. Geological Survey Circular 694, as adapted from C. L. Drake, *American Association of Petroleum Geologists Bulletin,* vol. 56, no. 2, 1972, with permission.)

FIGURE 13.20
Exploration wells drilled for oil and gas (1949–1979). Notice that since the peak in 1956, the number of wells drilled declined until 1971. From 1971 to 1979, the number of wells drilled increased, and the success ratio also increased to 29 percent in 1979. Also shown is the trend of offshore development to work at increasingly greater water depths. (From U.S. Department of Energy, Annual Report to Congress, 1979.)

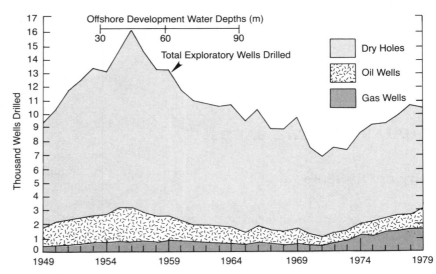

energy. These changes will strongly affect our petroleum-based society, but there appears to be no insurmountable problem if we implement meaningful short- and long-range plans to phase out oil and gas and phase in alternative energy sources. Unfortunately, phasing in alternative energy sources will require many years of research, exploration, and development, so it is crucial that we begin the task now. We are living in an interesting time from an energy standpoint, and it will be exciting to see new innovations, technology, and standards of living. Right now, no one knows the answer, or even all the questions.

Impact of Oil and Gas Exploration and Development

The environmental impact of exploration and development of oil and gas varies from negligible—for remote sensing techniques in exploration—to significant, unavoidable impact for projects such as the Trans-Alaska Pipeline. The impact of exploration for oil and gas can include building roads, exploratory drilling, and building a supply line (for camps, airfields, etc.) to remote areas. These activities, except in sensitive areas such as some semiarid to arid environments and some permafrost areas, generally cause few adverse effects to the landscape and resources compared to development and consumption activities. Development of oil and gas fields involves *drilling* wells on land or beneath the sea; disposing of waste water brought to the surface with the petroleum; *transporting* the oil by tankers, pipelines, or other methods to refineries; and *converting* the crude oil into useful products. All along the way, the possibility for environmental disruption from problems associated with disposal of waste water, accidental oil spills, shipwreck of tankers, air pollution at refineries, and other impacts is well documented. Tragic oil spills have affected the coastlines of Europe and America, spoiling beaches, estuaries, and harbors, killing marine life and birds, and causing economic problems for coastlines that depend on tourist trade. We need additional research and legislation to minimize these occurrences.

The most familiar serious impact associated with oil and gas use is air pollution, which is produced in urban areas when fossil fuels are burned to produce energy for electricity, heat, and automobiles. The adverse effects of smog on vegetation and human health are well documented and need not be restated here.

OIL SHALES AND TAR SANDS

Geology of Oil Shale

Oil shale is a fine-grained sedimentary rock containing organic matter (kerogen). On heating (destructive distillation), oil shale yields significant amounts of hydrocarbons that are otherwise insoluble in ordinary petroleum solvents. (8).

As with other fossil fuels, the origin of oil shale involves the deposition and only partial decomposition of organic debris. Favorable environments for oil shale are lakes, stagnant streams or lagoons in the vicinity of organic-rich swamps, and marine basins (9).

The best-known oil shales in the United States are those in the Green River Formation, which is about 50 million years old and underlies approximately 44,000 square kilometers of Colorado, Utah, and Wyoming (Figure 13.21). The Green River Formation consists of oil shale interbedded with variable amounts of sandstone, siltstone, claystone, and compacted volcanic ash (tuff). Variable amounts of halite (rock salt), trona (hydrous sodium carbonate), nahcolite (sodium bicarbonate), dawsonite (sodium, aluminum, hydroxl, carbonate), and other materials or minerals are also found as nodules, lenses, thin beds, or disseminated crystals (8). Nahcolite is a potentially valuable source of sodium carbonate, and dawsonite is a possible source of aluminum. Recovering these materials with the oil could reduce the cost of mining.

The geology of the Green River Formation and the depositional environment in the large lakes obviously varied significantly as the nature and pattern of deposition varied. The Green River Formation varies in thickness from a few meters to more than 1,000 meters, and some beds are rich in organic materials while others are not. Generally, the thickest accumulation of organic material is in the central parts of the large, shallow lakes where microscopic algae and other microorganisms flourished, died, and were buried with a mixture of sediment and other land-derived organic material. Toward the margin of the basins, less organic material accumulated; the oil shale there is of a lower grade (Figure 13.21) and is interbedded with land-derived, inorganic sediments (sand, silt, etc.) transported into the lake by streams, wind, and other surface processes. Few of these inorganic sediments reached the center of the lake, facilitating the deposition there of rich accumulations of organic material from which the oil shale formed (8).

The large, shallow, ancient lakes of Wyoming, Utah, and Colorado, in which the organic material accumulated, varied from fresh to alkaline water. The lake basins subsided slowly and irregularly, causing the center of deposition of the oil shale to shift slowly during the millions of years the deposition continued. For millions of years after the sediment was buried, little tectonic activity disturbed the sediments. More recently, uplift and local tilting of the rocks have exposed some of the oil shale to erosion (8).

Geology of Tar Sands

Tar sands are rocks that are impregnated with tar oil, asphalt, or other petroleum materials from which recovery of petroleum products by usual methods such as oil wells is not commercially possible. The term *tar sand* is somewhat confusing because it includes several rock

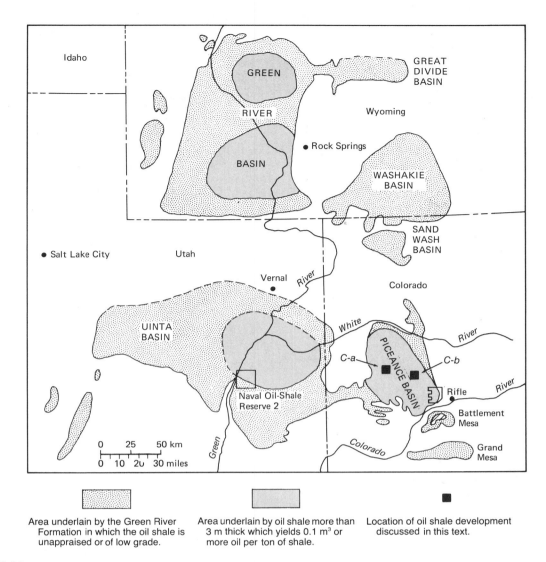

FIGURE 13.21
Distribution of oil shale in the Green River Formation of Colorado, Utah, and Wyoming.
(After D. C. Duncan and V. E. Swanson, U.S. Geological Survey Circular 523, 1965.) (3
meters ≅ 10 feet; 0.1 cubic meter ≅ 25 gallons)

types such as shale and limestone as well as unconsolidated or consolidated sandstone. The one thing all these rocks have in common is that they contain a variety of semiliquid, semisolid, and solid petroleum products, some of which ooze from the rock outcrops and others of which are difficult to remove with boiling water (10).

Oil in tar sands is nearly the same as the heavier oil pumped from wells. The only real difference is that tar-sand oil is much more viscous and therefore more difficult to recover. A possible conclusion concerning the geology of tar sands is that they form essentially the same way that the more fluid oil forms, but much more of the volatiles and accompanying liquids in the reservoir rocks have escaped, leaving the more viscous materials behind.

Distribution of Oil Shale and Tar Sands

Many countries have deposits of oil shale (Figure 13.22), but there is a tremendous range in thickness of the rock, quality of the oil, and areal extent (11).

Total identified shale-oil resources of the world land areas are estimated to contain about 3 trillion barrels of oil, but thorough evaluation of the grade and feasibility of economic recovery with today's technology and economic situation is incomplete. Shale-oil resources in the United States amount to about 2 trillion

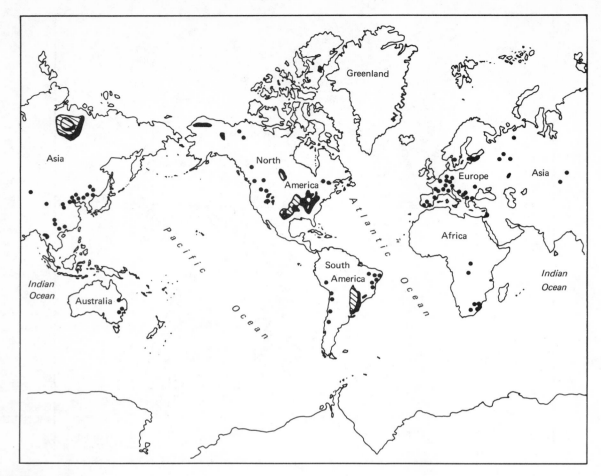

FIGURE 13.22
Oil-shale deposits of the world. Areas denoted by the diagonal lines are those where some major deposits may be extended. (From D. C. Duncan and V. E. Swanson, U.S. Geological Survey Circular 523, 1965.)

barrels of oil, or two-thirds of the total identified in the world; of this, 90 percent, or 1.8 trillion barrels, is located in the Green River Oil Shales (Table 13.2). Even these rich deposits are not equally distributed: the Unita Basin contains approximately two-thirds (1.2 trillion barrels) of the identified oil in the Green River Oil Shales (8), and recoverable resources may be as high as 600 billion barrels (about equal to the known world reserves of crude oil).

Unfortunately, identified resources of oil shale are not all recoverable with today's economics. This situation could change rapidly, however, and probably around 80 billion barrels of oil can now be extracted at a profit (9). We do know of large accumulations of tar sands; for example, the Athabasca Tar Sands of Alberta, Canada, cover an area of approximately 78,000 square kilometers and contain an estimated reserve of 300 billion barrels of oil that might be recovered (12).

TABLE 13.2
Oil Shale Resource in the Green River Formation: Colorado, Utah, and Wyoming (billions of barrels)

Nature of the deposit	Colorado	Utah	Wyoming	Total
At least 100 ft. (30 m) thick with oil yields averaging at least 30 gal/ton	355	50	13	418
At least 15 ft. (5 m) thick with oil yields averaging at least 15 gal/ton, excluding the deposits shown above	840	270	290	1,400
Totals (rounded)	1,200	320	300	1,820

Source: W. C. Culbertson and J. K. Pitman, "Oil Shale," in *United States Mineral Resources*, U.S. Geological Survey Professional Paper 820, 1973.

In addition, smaller tar-sand resources are known in Utah (1.8 billion barrels), California (100 million barrels), Texas, Wyoming, and other areas (10).

Impact of Exploration and Development of Shale Oil and Tar Sands

Recovering petroleum from surface or near-surface oil shale and tar sands involves use of well-established exploration techniques. Numerous deposits are known— what we need are reliable techniques of developing the oil-shale resources that will cause a minimum of environmental disruption. The really significant problem associated with mining oil shale is insufficient water supplies, which we will discuss in a later section on energy and water demand.

The Athabasca Tar Sands in Alberta are now yielding about 50,000 barrels of synthetic crude oil per day from one large open-pit mine operated by the Great Canadian Oil Sands Limited. The average thickness of the mined tar sands is about 42 meters; the amount of overburden removed varies considerably, but is generally less than 50 meters. Approximately 126,000 metric tons of tar sand and 117,000 metric tons of overburden are removed in each day's activity (12).

The methods of surface mining at the Canadian Tar Sands are atypical for four reasons. First, the indigenous vegetation is a water-saturated, organic matte (muskeg swamp) of decayed and decaying vegetation that can be removed easily only in the winter when it is frozen. Second, the muskeg must be drained before excavation, which takes about two years. Third, once the vegetation has been removed, the tar sands are extremely harsh on equipment and difficult to remove; furthermore, the overburden and other unwanted material with the tar and sand occupy considerably more volume than before they were removed, creating a disposal problem and a final land surface that may rise more than 21 meters above the original surface. Fourth, restoration of the land after mining is a serious problem in this fragile environment because it is so difficult to work with (12).

Fortunately, the actual process of recovering oil from tar sands is relatively easy and involves washing the viscous oil out of the sand with hot water. The oil then flows to the surface and is removed (10). Only about 10 percent of the total oil from the tar sands can be economically recovered by open-pit mining, however, so recovery of the oil in place (without removal by surface or subsurface mining) should become a necessary goal for obtaining additional oil (12). It is hoped that in-place recovery techniques will disturb the surface only minimally.

The environmental impact of developing oil-shale resources will vary according to the recovery technique. Presently, surface and subsurface mining as well as in-place (in situ) techniques are being considered.

Surface mining, either open pit or strip mine, is attractive because nearly 90 percent of the oil shale can be recovered, as opposed to less than 60 percent for underground mining. But waste disposal is a major problem with any mining, surface or subsurface, that requires that oil shale be processed at the surface for retorting (crushing and heating raw oil shale to about 540°C to obtain crude shale oil). This disposal problem results because the volume of waste will exceed the original volume of mined shale by 20 to 30 percent. Therefore, the mine from which the shale was removed will not be able to accommodate the waste, which will have to be piled up or otherwise disposed of. The impact of the waste disposal can be considerable: if surface mining produces 100,000 barrels of shale oil per day for 20 years, the operation will produce 570 million cubic meters of waste. If 50 percent of the waste is disposed of on the surface, it could fill an area 8 to 16 kilometers long, 600 meters wide, to a depth of 60 meters. Thus, with large-scale mining, we will have to determine ways to contour and vegetate shale-oil waste to minimize the visual and pollutional impacts (11, 13). Experiments to learn more about how to accomplish this are now going on.

True in situ shale oil development, in which the rock is fractured by explosives then retorted underground, is being tested, but at present has several disadvantages, including these: the technology is not advanced to the point of being ready for application; recovery of shale oil is expected to be low; and there is a serious potential for surface subsidence. On the other hand, true in situ processing has the advantage that mining is not required and spent oil shale doesn't present a waste disposal problem (13).

A process of oil shale recovery that is being seriously tested is known as *modified in situ*, or MIS, in which part of the oil shale (about 20 percent) is mined, and the remainder is highly fractured or rubbled to increase the permeability. A block of rubbled shale, the retort block, is then ignited, and the released oil and gas are recovered through wells (Figure 13.23).

Two tracts known as *C-a* and *C-b* northwest of Rifle, Colorado (Figure 13.21) are presently under development. It is hoped that both tracts will eventually produce about 50,000 barrels of oil per day, but several billion dollars will be invested before any potential return occurs. Tract *C-a* (Figure 13.24) will produce shale oil at a depth of about 130 meters to 300 meters below the surface. A central shaft about 3 meters in diameter has been excavated to a depth of about 300 meters.

Other excavations, lateral and horizontal, will isolate blocks of oil shale to be rubbled and retorted in place by burning. A substantial amount of relatively good quality groundwater has been encountered at Tract *C-a*. The water is pumped out and then injected back into the ground at another location. Poor quality ground-water or groundwater degraded by recovery of the shale oil will be treated on site in lined ponds (Figure 13.25).

Tract *C-b* (Figure 13.21) will also use the modified in situ procedure to recover shale oil. The plan there involves three concrete-lined shafts 5 to 9 m in diame-

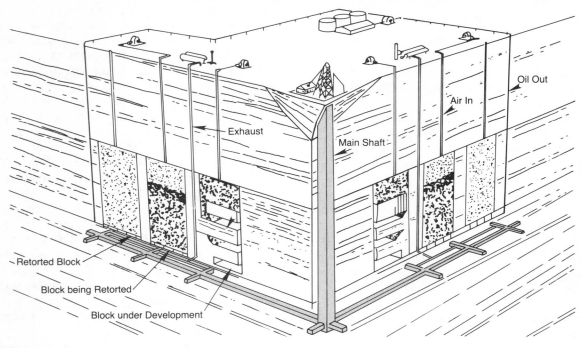

FIGURE 13.23
Idealized diagram showing the modified in situ method of development of oil shale resources. (From U.S. Department of Energy.)

FIGURE 13.24
Oil shale tract *C-a* located northwest of Rifle, Colorado. The mining method is to be modified *in situ,* and it is hoped this location will eventually produce about 50,000 barrels of oil per day.

FIGURE 13.25
Water treatment pond for waste water at oil shale tract *C-a* located northwest of Rifle, Colorado.

ter to a depth of about 600 meters. The operation is zoned to four levels and lateral excavations, or drifts, will eventually extend out about 3 kilometers if development proceeds as planned. Groundwater is also encountered at Tract *C-b*. As of the summer of 1980, the water was pumped out, treated, and stored in ponds, then put into local streams or sprinkled on natural vegetation to observe the effects. By 1984, the future development of Tracts *C-a* and *C-b* had become uncertain as work ceased because of a generally poor economic outlook. Another significant increase in the price of oil, however, would start the project rolling again.

The human environment will change significantly if oil shale is mined in Colorado, Wyoming, or Utah. The mining would necessitate modifications for increased transportation, including roads, pipelines, and airports, as well as augmenting economic activity, construction of various industrial facilities, and rapid urbanization as the population increases—all of which affect the physical environment.

FOSSIL FUEL AND ACID RAIN

Combustion of huge quantities of fossil fuels such as coal and oil is producing one of the most significant environmental problems for the coming decade. In the United States alone, the annual discharge of sulfur and nitrogen oxide into the atmosphere is approximately 50 million metric tons. By a complex set of chemical reactions, these pollutants are often converted into acids

that return to earth with rain or snow. The term for this acid precipitation is "acid rain." There is concern that acid rain will have a severe impact on widespread areas of the United States, including: increased leaching of nutrients and minerals from soils resulting in lower productivity of crops and forests; loss of plant and fish life in lakes located in areas particularly sensitive to acid rain; and accelerated weathering and destruction of stone statues and buildings in urban areas.

All rainfall is slightly acidic (Figure 13.26) because water reacts with atmospheric carbon dioxide to produce a weak carbonic acid. Thus pure rain has a

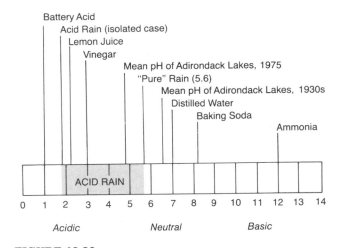

FIGURE 13.26
pH scale. (Modified after U.S. Environmental Protection Agency, 1980.)

pH (numerical value describing the strength of an acid) of about 5.6. Acid rain is defined as precipitation in which the pH is below 5.6. The pH scale is logarithmic, so a pH value of 3 is ten times more acidic than a pH value of 4 and 100 times more acidic than a pH value of 5. A pH value of 1 is extremely acidic, as, for example, battery acid. It is alarming to learn that in Wheeling, West Virginia, rainfall has been measured with a pH value of only 1.5, nearly as acidic as stomach acid and far more acidic than lemon juice or vinegar. More important, perhaps, than isolated cases of very acid rain is the recent growth of the problem. Not too many years ago, acid rain was believed to be primarily a European problem, but in recent years, acid rainfall has spread from a relatively small area in the northeastern United States to nearly all of eastern North America—an ominous trend. Furthermore, the problem is not restricted to the east coast of the United States, since urban centers on the west coast such as Seattle, San Francisco, and Los Angeles are now beginning to record acid rainfall.

The source of acid rain is air pollution, which releases sulfur and nitrogen oxides to the atmosphere as a by-product of burning industrial materials, as well as fossil fuels such as coal, oil, and gas. In 1977 alone, 27.4 million metric tons of sulfur oxide and 23 million metric tons of nitrogen oxide were released into the atmosphere in the United States (Figure 13.27). The oxides of sulfur and nitrogen are primary contributors to the acid rain problem, but other acids such as hydrochloric acid, which can be emitted from coal-fire power plants, also contribute. Sulfur oxide is emitted primarily from stationary sources such as power plants that burn coal or oil. On the other hand, nitrogen oxide is emitted from both stationary and transportation-related sources such as automobiles. In 1977, approximately 56 percent of the nitrogen oxide discharged into the atmosphere

resulted from the combustion of fossil fuels at power plants and other stationary sources, whereas 40 percent was released by transportation-related sources. Because it is expected that there will be a dramatic increase in the burning of fossil fuel (especially coal) over the next several decades, the acid rain problem will undoubtedly also increase rapidly during this period (14).

Geology and Acid Rain

Geology, as well as other factors, including climatic patterns, types of vegetation, and composition of soils all affect the potential impact of acid rain. Figure 13.28 shows areas of the United States that are sensitive to acid rain and is based on an examination of some of these factors. Particularly sensitive areas are those in which the bedrock offers little chance to buffer the acid rain, including terrain dominated by granitic rocks, as well as areas in which the soils have little buffering action. Areas least likely to suffer damage will be those in which the bedrock contains an abundance of limestone or other carbonate material, or in which the soils contain a calcium carbonate-rich horizon. It is expected that soils in sensitive areas will be damaged as nutrients are leached out by the acid. That acid rain can remove material was dramatically demonstrated recently when a block wall, part of a greenhouse, in Lyme, New Hampshire, actually effervesced as acid rain dissolved holes in the blocks.

As soils become depleted of nutrients and other minerals, plant productivity can be adversely affected. In some instances, emission of sulfur oxide and the resulting acid rain actually kills plants over a wide area. For example, acid rain and particulate pollution has occurred over a wide area in the Sudbury region of Ontario, Canada. Large smelting operations there have re-

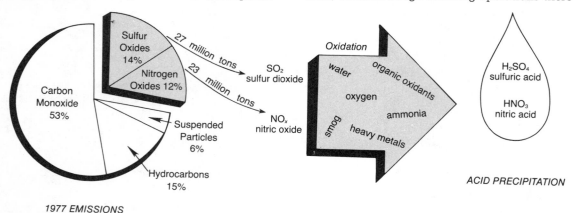

1977 EMISSIONS

FIGURE 13.27

Emission of sulphur and nitrogen oxide and the information of acid rain. (Modified after U.S. Environmental Protection Agency, 1980.)

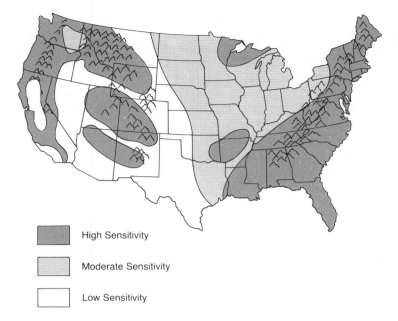

FIGURE 13.28
Areas in the United States sensitive to acid rain. (From U.S. Environmental Protection Agency, 1980.)

High Sensitivity

Moderate Sensitivity

Low Sensitivity

leased several million tons of sulfur oxide into the atmosphere each year by way of smokestacks that are now as high as 365 meters. Released particulates include hundreds of tons of heavy metals, including nickel, copper, iron, and cobalt. As a result of the combination of acid rain and particulate air pollution, the forests that once surrounded the mining town of Sudbury have been devastated over the last 50 years, including an area of approximately 250 square kilometers that is nearly devoid of vegetation. In addition, damage to forests in the region is visible over an area of approximately 3,500 square kilometers. Attempts to minimize the pollution problem close to the smelting operation by increasing the smokestack height have spread the problem even further. In 1977, the pH of rainfall up to 19 kilometers east was less than 4.3 and at a distance of 3 kilometers from the stacks it was often less than 3.0. Secondary effects of the loss of vegetation have been soil erosion and drastic changes in the soil chemistry caused by the influx of the heavy metals (15). The Sudbury situation, while perhaps extreme, may tell us something about the effects we can expect in the future if we are unsuccessful in curbing the acid rain problem.

Lakes in areas that are sensitive to acid rain are also experiencing major problems. As the lake water becomes more and more acidic, plants and fish often suffer until they eventually die, and the lake becomes blue and clear, devoid of life. In the Adirondack Mountains of the northeastern United States, for example, the pH of the rainfall is now just above 4, a forty-fold increase in the last 50 years. Ironically, this area is designated as a wilderness area, yet it clearly suffers from urban and industrial problems that originate many ki-

lometers away. Lakes that will probably continue to have problems in the future are those less able to buffer the effect of acid rain. If this is true, then more eastern lakes as well as lakes in the Rocky Mountains, southern California, and the Pacific Northwest are in for troubled times.

Tragically, classical buildings on the Acropolis in Athens, Greece, have shown considerably more rapid decay in this century than in the past because of the city's air pollution levels that result in acid rainfall. In the United States, cities in the east are more susceptible at this time because emissions of sulfur and nitrogen oxides are more abundant there, but the problem is expected to move westward. Rock types such as marble and limestone are particularly susceptible to damage since they will dissolve in weak acids. Interestingly, the geologic aspect of the problem may also shed light on predictions for future effects of acid rainfall. Since 1875, the Veteran's Administration has provided more than 2.5 million tombstones to various national cemeteries. These tombstones have come from only three rock quarries and are standardized with respect to size and shape. These stones, located in various parts of the country, will thus be valuable in assessing damages caused by acid rainfall since they provide a variety of dates and locations to work from, and may give us valuable data concerning air pollution and the meteorological patterns that facilitate damage from acid rainfall (14).

It is hoped that the acid rainfall problem will be short lived. New standards to control sulfur oxide emissions from future power plants are expected to help alleviate the problem by 1995. Unfortunately, this program does not address the continued emissions from existing power plants during the next 20 years. The technology to clean up coal so that it will burn cleanly is already available, although the cost of removing the sulfur makes the coal more expensive. If nothing is done, however, the consequences of burning sulfur-rich coal during the next 20 years will probably be very costly to us and future generations. The state of New York is currently attempting to save the trout fishing in several Adirondack lakes by treating the lakes every year or so with lime ($CaCO_3$) to reduce the pH of the water. Although commendable, this project is expensive and must eventually fail if emissions of sulfur and nitrogen oxides are not reduced.

NUCLEAR ENERGY: FISSION

The first controlled nuclear fission was demonstrated in 1942 and led the way to the development of the primary uses of uranium in explosives and as a heat source to provide steam for generation of electricity. It now appears that the energy from uranium (one kilogram of uranium oxide produces a heat equivalent of approximately 16 metric tons of coal) will continue as an important source of energy in the United States (16).

Fission is the splitting of uranium (U^{235}) by neutron bombardment (Figure 13.29). The reaction produces three more neutrons released from uranium, fission fragments, and heat. The released neutrons each strike other U^{235} atoms, releasing more neutrons, fission products, and heat. The process continues in a chain reaction—as more and more uranium is split, it releases ever more neutrons.

Three types of uranium occur in nature: U^{238}, which accounts for approximately 99.3 percent of all natural uranium; U^{235}, which makes up about 0.7 percent; and U^{234}, which makes up about 0.005 percent. Uranium-235 is the only naturally occurring fissionable material, and is therefore essential to the production of nuclear energy. However, uranium is processed to increase the amount of U^{235} from 0.7 percent to about 3 percent before it is used in a reactor. The processed fuel is called *enriched uranium*. Uranium-238 is not naturally fissionable, but is "fertile material" because upon bombardment by neutrons, it is converted to plutonium-239, which is fissionable (16).

Most reactors today consume more fissionable material than they produce and are known as *burner reactors*. The reactor itself (Figure 13.30) is part of the nuclear steam supply system which produces the steam to run the turbine generators that produce the electricity (17). The main components of the reactor shown in Figure 13.31 are the core, control rods, coolant, and reactor vessel. Fuel pins consisting of enriched uranium pellets placed into hollow tubes with a diameter less than about 1 cm are packed together (40,000 or more in a reactor) into fuel subassemblies in the core (Figure 13.32). A stable fission chain reaction

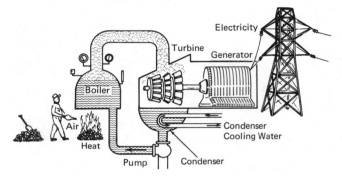

FOSSIL FUEL POWER PLANT

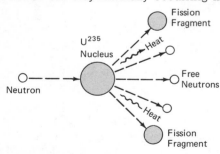

FIGURE 13.29
Idealized diagram showing fission of U^{235}. A neutron strikes the U^{235} nucleus, producing fission fragments and free neutrons and releasing heat. The released neutrons may then each strike another U^{235} atom, releasing more neutrons, fission fragments, and energy. As the process continues, a chain reaction develops.

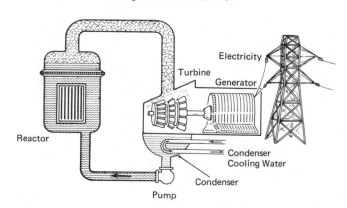

FIGURE 13.30
Idealized diagram comparing a fossil fuel power plant and nuclear power plant. Notice that the nuclear reactor has exactly the same function as the boiler in the fossil fuel power plant. (Reprinted, by permission, from *Nuclear Power and the Environment,* American Nuclear Society, 1973.)

in the core is maintained by controlling the number of neutrons that cause fission as well as the fuel concentration. A minimum fuel concentration is necessary to keep the reactor critical (achieve a self-sustaining chain reaction).

The control rods contain materials that capture neutrons, and can thus be used to regulate the chain

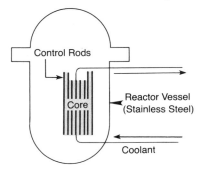

FIGURE 13.31
Idealized drawing showing the main components of a reactor. (From ERDA–76–107, 1976.)

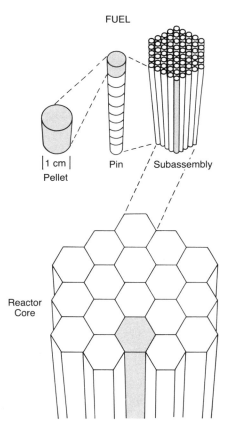

FIGURE 13.32
Fuel pellets of enriched uranium are placed in hollow tubes forming fuel pins that are bundled into fuel subassemblies, which are placed in the reactor as part of the core. (Modified from ERDA–76–107, 1976.)

reaction. If the rods are pulled out, the chain reaction speeds up; if they are inserted into the core, the reaction slows down (18).

The function of the coolant is to remove the heat produced by the fission reactions. When the coolant is water, besides removing heat, the coolant also acts as a moderator, slowing down the neutrons and facilitating efficient fission of uranium-235 (18).

The core of the reactor is contained in a heavy stainless steel reactor vessel. Then, for extra safety and security, the entire reactor is contained in a reinforced concrete building (18).

In addition to the reactor, other parts of the nuclear steam supply system are the *primary coolant loops* and pumps that circulate a coolant (usually water) through the reactor, extracting heat produced by fission, and *heat exchangers* or *steam generators* that use the fission-heated coolant to make steam. Figure 13.33 shows how these combine in a pressurized water reactor (PWR), which is a *light water reactor* (a type of burner reactor) so designated because the coolant is water. Water heated by the reactor core is circulated in the primary coolant loop (a closed system) through a steam generator (heat exchanger) turning the water in the secondary loop to steam which produces electricity (17).

A second type of light water reactor in use is the *boiling water reactor* (BWR), which is a direct cycle system because there is no heat exchanger. The pri-

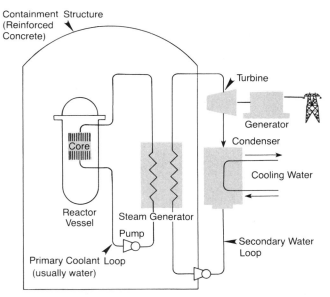

FIGURE 13.33
Pressurized water reactor (PWR) showing the three main parts of the nuclear steam supply system: reactor, primary coolant loop, and steam generator (heat exchanger). (Modified after Energy Research and Development Administration, 1976, ERDA–76–107.)

mary coolant loop goes through the reactor core directly to a turbine, producing electricity (Figure 13.34). Examination of this figure reveals a major disadvantage of the BWR. The steam that turns the turbine comes directly from the reactor core rather than first going through a heat exchanger, as in the PWR system. Therefore, particular care must be taken to keep hazardous radiation from leaking into the turbine system that is outside the main containment structure. The primary coolant (water) will have significantly induced radioactivity, requiring that the turbines be heavily shielded, increasing construction and maintenance costs (17).

To conserve U^{235}, work is progressing on a reactor that actually produces more fissionable material (nuclear fuel) than it uses. These *breeder reactors*, using a fuel core of fissionable material (plutonium-239) surrounded by a blanket of fertile material (U^{238}), will produce or breed additional plutonium-239 from the U^{238}. The transformation from U^{238} to P^{239} occurs in the breeder reactor at the same time that the plutonium nuclei in the core are undergoing fission, providing heat that produces steam to drive turbines and generators that provide electricity. Development of the breeder reactor will greatly extend the limited supply of natural U^{235}. In fact, breeder reactors that use plutonium as a fuel can use uranium-238 tailings from enriched uranium or uranium recovered from spent light water reactor fuel. It has been estimated that the ura-

nium now stockpiled as tailings and spent fuel, if used in a breeder reactor, could supply the total electrical energy demands in the United States for up to 100 years (18).

In 1982, the demand for electric power decreased for the first time in many years. It will increase in the future, however, as will the demand for nuclear power, which produced about 13 percent of our electricity in 1984. By the year 2000, it is expected that heat from nuclear reactors may have the capacity to produce about 20 percent of the electrical power in the United States. About 82 reactors are now in operation (88 percent of which are in the eastern USA), but many more will be needed to realize the projected energy from uranium. Probably because of increased costs, environmental considerations, and other factors, the projected generating capacity is too high and will have to be revised. Thus, the full impact of what began in 1942 is still to be determined.

Geology and Distribution of Uranium

The natural concentration of uranium in the earth's crust is about 2 parts per million. Uranium originates in magma and is concentrated to about 4 parts per million in granitic rock, where it is found in a variety of minerals. Some uranium is also found with late-stage igneous rocks such as pegmatites. To be mined at a profit, uranium with a concentration factor of 400 to 2,500 times the natural concentration must be used (16).

Fortunately, uranium forms a large number of minerals, many of which can be found in high-grade deposits being mined today. Three types of deposits have produced most of the uranium in the last few years: sandstone impregnated with uranium minerals, veins of uranium-bearing materials localized in rock fractures, and placer deposits in river or delta deposits (now coarse-grained sedimentary rock) more than 2.2 billion years old (19).

Uranium in *sandstone* is often found in thin lenses interbedded with mudstone. It is hypothesized that the uranium in these deposits was derived by leaching from volcanic glass associated with the sedimentary rocks or from granitic rocks exposed along the margins of the sedimentary basins. Supposedly, the uranium was transported by groundwater and then precipitated into the pore spaces of the sandstone under reducing (oxygen-deficient) conditions (16).

Most of the uranium mined in the United States has been from sandstone deposits of two types: roll-type and tabular deposits. *Roll-type* uranium deposits

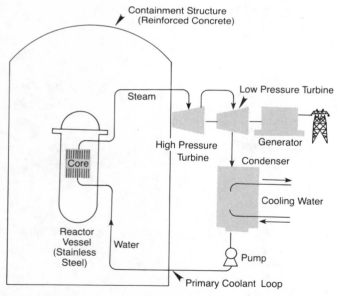

FIGURE 13.34
Boiling water reactor (BWR) showing the two main parts of the nuclear steam supply system: reactor, and primary coolant loop. (Modified after Energy Research and Development Administration, 1976, ERDA–76–107.)

form at the interface between oxidizing and reducing conditions. Characteristically, the ore deposits are elongated bodies scattered like interconnected beads along kilometers of interface between altered (oxidized) sandstone and unaltered sandstone (Figure 13.35). *Tabular* uranium deposits are discrete masses completely enclosed by altered (reduced) sandstone which is itself enveloped in oxidized sandstone, as shown in Figure 13.36 (16,19).

Uranium deposits in *veins* generally fill fissures (rock fractures) in many types of rocks of varying geologic age. The veins usually vary from several centimeters to a few meters wide and several hundred meters long. Veins may coalesce to form vein systems that can extend for several thousand meters (16).

Uranium in ancient river or delta sediment was deposited before the occurrence of abundant free oxygen in the atmosphere. Therefore, it is believed, the uranium mineral was deposited as stream-rounded particles along with gold, pyrite, and other typical placer material (19).

The distribution of uranium deposits mined in the United States is shown in Figure 13.37. Although sandstone ores have been the main source of uranium in this country, vein deposits are a significant source in Australia, Canada, France, and Africa, and ancient placer deposits are being mined in Africa and Canada (19).

The amount of uranium from the known deposits is sufficient to last until the 1980s. Beyond that, additional exploration for deposits and research of techniques to locate new reserves will be required (16). If the breeder reactor program is successful and shown to be environmentally safe, the supply of fissional material will be greatly extended.

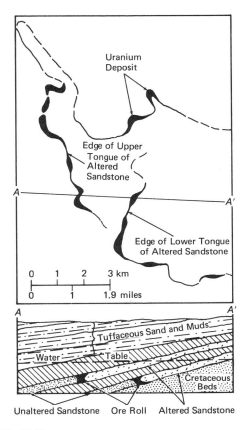

FIGURE 13.35
Geologic map and cross section of typical roll-type uranium deposits. (After U.S. Geological Survey, INF–74–14, 1974.)

FIGURE 13.36
Geologic map and cross section of tabular uranium deposits. (After U.S. Geological Survey, INF–74–14, 1974.)

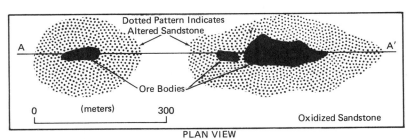

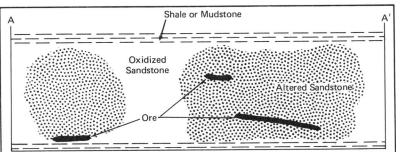

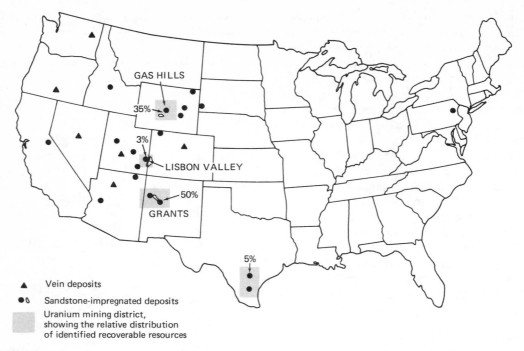

- ▲ Vein deposits
- ●◖ Sandstone-impregnated deposits
- ▨ Uranium mining district, showing the relative distribution of identified recoverable resources

FIGURE 13.37
Distribution of major uranium deposits in the United States. (After A. P. Butler, Jr., U.S. Geological Survey Circular 547, 1967.)

Nuclear Energy (Fission) and the Environment

Nuclear energy and the possibly adverse effects associated with it have been subjects of vigorous debate. The debate is healthy because the number of reactors that produce heat to drive generators is expected to increase in the future, and we should examine the consequences carefully.

A first approach to evaluating the environmental effects of nuclear power might be to compare the effects of all common methods of generating electrical power by power source: coal, oil, gas, and uranium (Table 13.3). All methods affect the environment in various ways during normal operation. Disregarding the magnitude of a particular effect (as, for example, thermal pollution, which is greater for nuclear power because the generator capacity is generally larger and no heat is directly dissipated to the atmosphere), the real difference of nuclear power is radioactivity. The possible radiation hazard from nuclear energy production and waste disposal really concerns people.

Throughout the entire nuclear cycle, from mining and processing uranium to controlled fission, to reprocessing spent nuclear fuel, to final disposal of radioactive waste, various amounts of radiation enter and, to a lesser or greater extent, affect the environment.

The chance of a disastrous nuclear accident is estimated to be very low; nevertheless, the chance of an accident increases with every reactor put into operation. The 1979 accident at the Three Mile Island nuclear power plant near Middletown, Pennsylvania, involved a chain of what were believed to be highly improbable events resulting from both mechanical failure and human error. Although a major disaster was avoided, the incident raised important questions about reactor safety; five other reactors were shut down temporarily as a result of potential design-safety problems, reducing the total amount of nuclear power produced in the United States for the first time since 1960.

There are additional serious hazards associated with transporting and disposing of nuclear material (see Chapter 10) as well as supplying other nations with reactors. Terrorist activity and the possibility of irresponsible actions by governments add a risk that is present in no other form of energy production. Nuclear energy may indeed be the answer to our energy problems, and perhaps someday will provide unlimited cheap energy. But along with nuclear power comes the responsibility of insuring that nuclear power is used for, not against, people, and that future generations will inherit a quality environment free from worry about hazardous nuclear waste.

NUCLEAR ENERGY: FUSION

The long-range energy strategy for the United States is to develop and provide sufficient energy resources to

TABLE 13.3
Comparison of environmental effects of coal, oil, gas, and uranium electrical power generation.

Energy Source	Effects on Land	Effects on Water	Effects on Air	Biological Effects	Supply
Coal	Disturbed land; large amounts of solid waste; leaching of soil and rocks by acid rain	Acid mine drainage; increased water temperature; acid rain	Sulfur oxides; nitrogen oxides; particulates; some radioactive gases	Respiratory problems from air pollutants; drainage from acid rain	Large reserves
Oil	Wastes in the form of brine; pipeline construction; leaching by acid rain	Oil spills; increased water temperature; acid rain	Nitrogen oxides; carbon monoxide; hydrocarbons	Respiratory problems from air pollutants; damage from acid rain	Limited domestic reserves
Gas	Pipeline construction	Increased water temperature	Some oxides of nitrogen	Few known effects	Extremely limited domestic reserves
Uranium	Disposal of radioactive waste	Increased water temperature; some radioactive liquids	Some radioactive gases	None detectable in normal operation	Large reserves if breeders are developed

Source: Pennsylvania Department of Education. *The Environmental Impact of Electrical Power Generation: Nuclear and Fossil,* 1973.

meet energy demand in the twenty-first century and beyond. Energy sources capable of meeting long-term needs are those that are inexhaustible: *solar, breeder reactor,* and *fusion reactor.* Our discussion here will focus on fusion.

In contrast to fission, which involves splitting heavy atoms such as uranium, fusion involves combining light elements such as hydrogen to form a larger element such as helium. As fusion occurs, heat energy is released (Figure 13.38). Similar reactions are the source of energy in our sun and other stars. In a hypothetical fusion reactor, two isotopes of hydrogen (atoms with variable mass resulting from a different number of

neutrons in the nucleus), deuterium (D) and tritium (T), are injected into the reactor chamber where necessary conditions (temperature, time, and density) for fusion are maintained. Products of the D-T fusion include helium, carrying 20 percent of the energy released, and neutrons, carrying 80 percent of the energy released (Figure 13.38) (20).

These are the conditions necessary for fusion: **(1)** extremely high temperature (approximately 100 million degrees Centigrade for D-T fusion); **(2)** sufficiently high density of the fuel elements (at the necessary temperature for fusion nearly all atoms are stripped of their electrons forming a *plasma,* an electrically neutral ma-

FIGURE 13.38
Deuterium-Tritium (D-T) fusion reaction. (Modified from U.S. Dept. of Energy, 1980, DOE/ER–0059.)

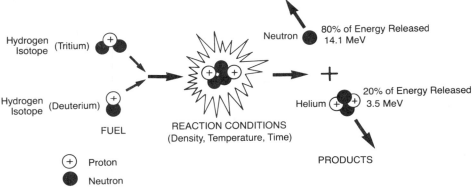

terial consisting of positively charged nuclei, ions, and negatively charged electrons); and **(3)** confinement of the plasma for a sufficient time to insure that the energy released by the fusion reactions exceeds the energy supplied to maintain the plasma (20, 21).

With development of fusion reactor-power plants, such as that shown in Figure 13.39, the potential available energy is nearly inexhaustible. One gram of D-T fuel (from a water and lithium fuel supply) has the energy equivalent of 45 barrels of oil. Deuterium can be extracted economically from ocean water, tritium can be produced in a reaction with lithium in a fusion reactor, and there are no problems obtaining lithium, as it is abundant and can be extracted economically.

Many problems remain to be solved before nuclear fusion is commercially available. The research is still in the first stage, which involves basic physics, testing possible fuels (mostly D-T), and magnetic confinement of plasma. Progress in fusion research has been steady in recent years, and there is optimism that useful power will eventually be produced from controlled fusion.

Energy from fusion, once released, has a variety of applications, including heating and cooling buildings and producing synthetic fuels, but producing electricity is probably the most important. It is expected (but not proven) that fusion power plants will be economically competitive with other sources of electric energy.

From an environmental view, fusion certainly appears attractive: land use and transportation impacts are small compared to fossil fuel or fission energy sources; and compared to fission breeders, fusion produces no fission products, little radioactive waste, and has a much lower risk of potential hazard resulting from an accident (22). On the other hand, fusion power plants will probably use materials that are harmful to humans, such as lithium, which is toxic when inhaled or ingested. Other potential hazards are the strong magnetic fields and microwaves used in confining and heating plasma, and short-lived radiation emitted from the reactor vessel (20).

GEOTHERMAL ENERGY

The useful conversion of natural heat from the earth's interior (*geothermal energy*) to heat buildings and generate electricity is an exciting application of geologic knowledge and engineering technology. The idea of harnessing the earth's internal heat is not new. As early as 1904, geothermal power was developed in Italy using dry steam, and natural internal heat is now used to generate electricity in the USSR, Japan, New Zealand, Iceland, Mexico, and California. The existing geothermal facilities probably use only a small portion of the total energy that might eventually be tapped from the earth's reservoir of internal heat.

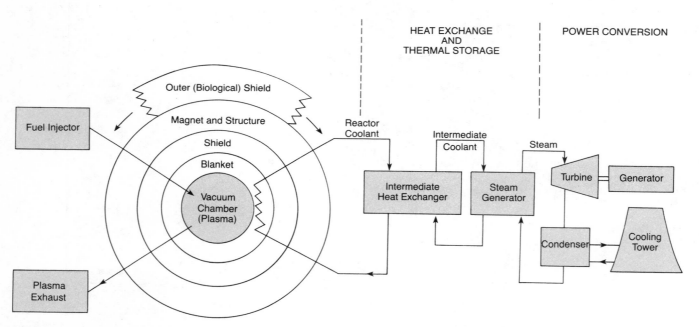

FIGURE 13.39
Schematic diagram of a fusion reactor and power plant. The vacuum chamber is where plasma is confined by strong magnetic fields and fusion reactions take place. (Modified after U.S. Department of Energy, 1979, DOE/EDP-0052.)

Geology of Geothermal Energy

Natural heat production within the earth is only partly understood. We do know that some areas have a higher flow of heat from below than others, and that for the most part these locations are associated with the tectonic cycle. Oceanic ridge systems (divergent plate boundaries) and convergent plate boundaries, where mountains are being uplifted and volcanic island arcs are forming, are areas where this natural heat flow from the earth is anomalously high.

The relative distribution of heat flow from within the earth for the United States is shown in Figure 13.40. The data for this generalized classification of hot, normal, and cold crustal areas are not sufficient to accurately define the limits of the regions (23), so the map has limited value for locating specific sites where geothermal energy resources could be developed.

The region of high heat flow is concentrated in the western United States where tectonic and volcanic activity have been recent. The wide width of the belt is somewhat of an anomaly, and for years geologists have vigorously debated the origin of the rock structure and tectonic activity of the Basin and Range region

(Figure 13.40), which is characterized by tensional stress and volcanic activity. Some geologists believe the region represents a place where the plate on which North America rides (American plate) is overlapping the Pacific plate, including the East Pacific Rise (oceanic ridge). With this interpretation, the Basin and Range is the surficial expression of oceanic ridge activity. Other geologists believe the East Pacific Rise has not been overridden by the American plate but is separated along the San Andreas fault. Regardless of the actual tectonic picture, the entire western region is generally characterized by recent uplift and volcanic activity, and is therefore, with limited exceptions, a good prospect for geothermal energy exploration.

Within the region of relatively high heat flow, commercial development of geothermal power is most likely where a heat source such as convecting magma is relatively near the surface (3 to 10 kilometers) and in thermal contact with circulating groundwater. Thus, the likely sites for exploration are areas with naturally occurring hot springs and geysers that reflect near-surface hot spots; areas of recent volcanic activity, particularly those characterized by high-silica magma, because they are more likely to have stored heat near the

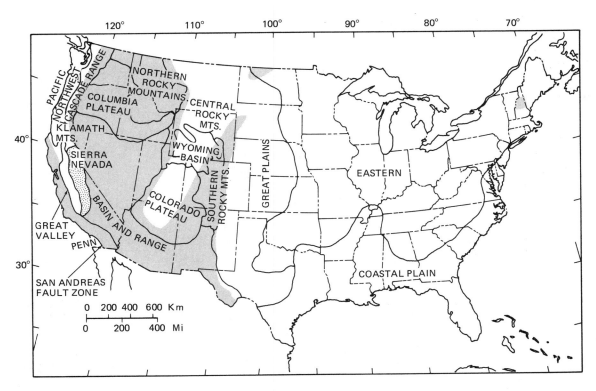

FIGURE 13.40
Map showing the generalized extent of hot (shaded), normal (white), and cold (stippled) crustal regions of the United States. Also shown are the major physiographic provinces such as the Colorado Plateau. (After D. F. White and D. L. Williams, eds. U.S. Geological Survey Circular 726, 1975.)

surface where it might be accessible to drilling (23); and other localized hot spots with little or no near-surface expression that are discovered by direct and indirect (geophysical) subsurface exploration. The best location may not necessarily be that with the most prominent surface expression of geyser and hot-spring activity, however, because geothermal systems with a high rate of upflow, feeding geysers, and hot springs might not be as well insulated as a system with little or no leakage, as shown in Figure 13.41 (24). There is thus likely to be more stored heat at shallow depths in the well-insulated geothermal reservoir.

Based on geologic criteria, several geothermal systems may be defined: hydrothermal convection systems, hot igneous systems, and geopressured systems (25, 26). Each system has a different origin and different potential as an energy source.

Hydrothermal convection systems are characterized by a permeable layer in which a variable amount of hot water circulates. They are of two basic types: *vapor-dominated* systems and *hot-water* systems. Va-

por-dominated hydrothermal convection systems are geothermal reservoirs in which both water and steam are present at depth. Near the surface, where pressure is less, the water flashes to superheated steam, which can be tapped and piped directly into turbines to produce electricity. These systems characteristically have a sufficiently slow recharge of groundwater so the hot rocks can convert the water to steam. That is, the heat supply is great enough to boil off more water than can be replaced by natural recharge (25).

Vapor-dominated systems are not very common. Only three have been identified in the United States: The Geysers, 145 kilometers north of San Francisco, California; Mt. Lassen National Park, California; and Yellowstone National Park, Wyoming. The parks are not available for energy development, but steam from The Geysers in California has been producing electric energy for years (Figure 13.42).

In the United States, hot-water systems are about twenty times more common than vapor-dominated systems (Figure 13.43). These systems, with surface tem-

FIGURE 13.41

Generalized diagrams showing (a) a hot-spring type of geothermal system with a high rate of upflow characterized by vigorous hot-spring or geyser activity at the surface; and (b) an insulated geothermal reservoir with little or no leakage. (After U.S. Geological Survey Publication GPO: 1969 O-339-536.)

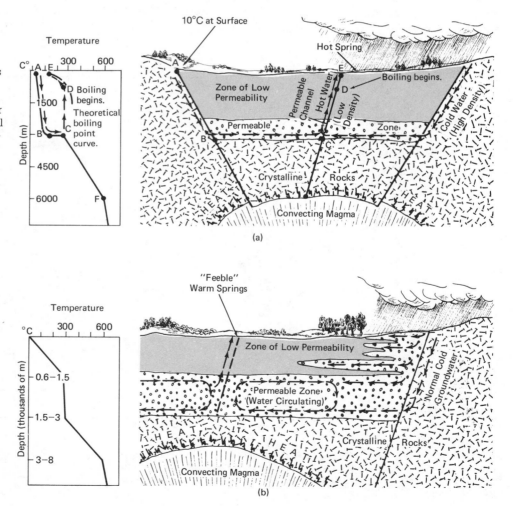

FIGURE 13.42

The Geysers Power Plant north of San Francisco, California, with about 1000 megawatts installed capacity, is the world's largest geothermal electricity development. (Photo courtesy of Pacific Gas and Electric Company.)

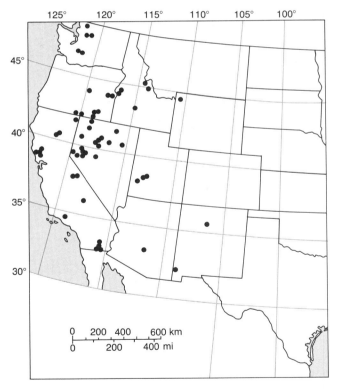

FIGURE 13.43

Distribution of hydrothermal convection systems in the contiguous United States where subsurface temperatures are thought to be in excess of 150 degrees Celsius. (After D. F. White and D. L. Williams, eds., U.S. Geological Survey Circular 726, 1975.)

peratures greater than 150 degrees Celsius, have a zone of circulating hot water (no steam) that when tapped moves up with reduced pressure to yield a mixture of steam and water at the surface. The water must be removed from the steam before the steam can be used to drive the turbine (26). One problem with this system is disposal of the water. However, as ideally shown in Figure 13.44 the water could be injected back into the reservoir to be reheated.

Hot igneous systems may involve the presence of molten magma at temperatures of 650 degrees Celsius to about 1,200 degrees Celsius depending on the type of magma. Even if the igneous mass is not still molten, it may involve a large quantity of hot, dry rocks. These systems contain more stored heat per unit volume than any of the other geothermal systems; however, they lack the circulating hot water of the convection systems, and innovative methods to use the heat will have to be developed (25). Some of these geothermal reservoirs have hot, dry rocks accessible to drilling, in which case the rocks might first be drilled and then fractured with explosives or hydrofracturing techniques. Then cold water might be injected into the rock at one location and pumped out at elevated temperatures at another location to recover the heat. The induced circulation of water would effectively mine the heat of the dry rock system (27). A pilot project by the Los Alamos Scientific Laboratory is currently going on in the Jemez Mountains of New Mexico. Approximately 3 to 6 megawatts (thermal energy) have been demonstrated using

FIGURE 13.44

Idealized diagram showing a hot-water geothermal system. At the power plant, the steam will be separated from the water and used to generate electrical power. The water will be injected back into the geothermal system by a disposal well. (Diagram courtesy of Pacific Gas and Electric Company.)

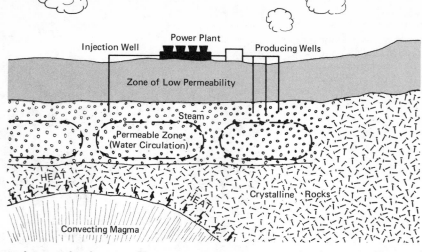

an innovative system that involves a nearly completely closed system of circulating water. Injected water is heated at a depth of about 3,000 meters to about 150°C. Pumped to the surface, the water goes through a heat exchanger, causing rapid expansion of freon which turns a turbine, generating electricity. The Jemez installation eventually may generate 35 megawatts while demonstrating the potential of hot, dry geothermal resources.

Recovery of heat directly from magma accessible to drilling (less than 10 kilometers deep) is an exciting prospect; however, the necessary technology is yet to be developed, and in fact, recovery may not be even remotely possible (28). Nevertheless, the possibility is being studied and evaluated. A generalized diagram of

the concept to tap magma is shown in Figure 13.45. Some of the major problems to be solved include drilling and heat-extraction technology. Some important questions are how heat is transferred in magma, and how the introduction of a heat-exchange device would affect the proportion of crystals, liquid, vapor, and other physical and chemical properties of the magma in the vicinity of the device (29). Extraction of heat directly from magma, even if theoretically possible, is a long way off.

Geopressured systems exist where the normal heat flow from the earth is trapped by impermeable clay layers that act as an effective insulator. The favorable environment is characterized by rapid deposition of sediment and regional subsidence. Deeply buried

FIGURE 13.45

Idealized diagram showing the general concept of a magma tap. (After J. L. Colp, Sandia Laboratories Energy Report SAND–74–0253, 1974.)

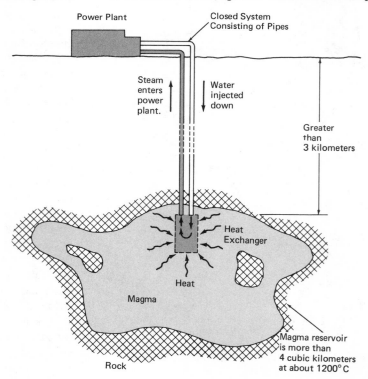

water trapped in the sediments develops considerable fluid pressure and is heated (26).

Perhaps the best-known regions where geopressurized systems develop are in the Gulf Coast of the United States, where temperatures of 150 degrees Celsius to 273 degrees Celsius at depths of 4 to 7 kilometers have been identified. These systems have the potential to produce large quantities of electricity for three reasons:

- They contain hot-water thermal energy that could be extracted
- They contain mechanical energy from the high-pressure water that could be used to turn hydraulic turbines to produce electricity
- As an added asset, the waters contain considerable amounts of dissolved methane gas, up to 1 cubic meter per barrel of water, that could be extracted (26, 27)

Of the three, the energy potential from the heat and methane gas far exceeds that of the high-pressure water (30).

Environmental Impact of Geothermal Energy Development

Although the potentially adverse environmental impact of intensive geothermal energy development is perhaps not as extensive as that of other sources of energy, it is nevertheless considerable. Geothermal energy is developed at a particular site, and environmental problems include on-site noise, gas emissions, and industrial scars. Fortunately, development of geothermal energy does not require the extensive transportation of raw materials or refining that are typical of the fossil fuels. Furthermore, geothermal energy does not produce atmospheric particulate pollutants associated with burning fossil fuels; nor does it produce any radioactive waste. Geothermal development, except for the vapor-dominated systems, does, however, produce considerable thermal pollution from hot waste waters, which can be saline or highly corrosive. The plan is to dispose of these waters by reinjecting them into the geothermal reservoir, but that kind of disposal has problems because injecting fluids may activate fracture systems in the rocks and cause earthquakes. In addition, original withdrawal of fluids may compact the reservoir, causing surface subsidence. It is also feared that subsidence could occur because, as the heat in the system is extracted, the cooling rocks will contract (26).

Future of Geothermal Energy

Geothermal energy appears to have a bright future. Over a thirty-year period, the estimated yield from this vast resource, which is both identified and recoverable at this time (disregarding cost) from hydrothermal convection systems and geopressured systems, is about equivalent to that of 140 modern nuclear power plants. In addition, it is expected that many more systems with presently recoverable energy are yet to be discovered (25). Furthermore, geothermal energy in the form of hot water (not steam) using normal heat flow may be a future source of energy to heat homes and other buildings along the Atlantic Coast, where alternative energy sources are scarce.

At present, geothermal energy supplies only a small fraction of one percent of the electrical energy produced in the United States. With the exception of the unusual vapor-dominated systems such as The Geysers in California, the production of electricity from geothermal reservoirs is still rather expensive, generally more so than from other sources of energy. For this reason commercial development of the energy will not proceed rapidly until the economics are equalized. Even so, the total energy output (electrical) from geothermal sources is not likely to exceed a few percent (10 percent at most) in the near future, even in states like California where it has been produced and where expanding facilities are likely (27).

RENEWABLE ENERGY SOURCES

The energy transition from fossil fuels to renewable energy sources has begun. In the United States today, oil and gas still supply approximately three-fourths of our energy needs, but as oil and gas become scarce and/or more costly, we will slowly change to new energy sources. The three major sources of energy likely to be used are fossil fuels such as coal, oil shale, and tar sands; nuclear energy (fission today and perhaps fusion sometime in the future); and renewable energy, broadly defined to include solar energy, hydropower, biomass, and wind. Geothermal energy might also be added to this list, but in many instances is better thought of as a nonrenewable resource even though in specific situations the natural heat flow may be high enough that it can be replenished relatively quickly. An important point concerning the renewable energy sources is that in many parts of the world they may be the only energy sources indigenous to a given region.

The necessity for seriously considering renewable energy sources is emphasized by examining the environmental impact associated with other possible sources of energy. That is, the fossil fuels such as coal, although abundant, are often damaging to the environment throughout their entire fuel cycle, from mining to processing to eventual consumption. Fossil fuels also carry the threat of global climate modification through increased discharge of carbon dioxide, particulates, and other materials. Nuclear energy, while imposing no threat to climate modification, is associated with serious problems such as waste disposal, accidents, and weapons proliferation. Nuclear energy also releases waste heat into the environment through on-site cooling processes and when the electricity produced at the power plant is transported and used. Finally, we must also recognize that all fossil fuels and nuclear energy from fission are ultimately exhaustible, again emphasizing the need for eventual transition to the renewable energy sources (31).

Broadly defined, solar energy includes a number of different sources, as shown in Figure 13.46. The advantages of these sources are that they are inexhaustible and are generally associated with minimal environmental degradation. With the exception of burning biomass or urban waste, they do not pose the threat of adding carbon dioxide to the atmosphere and thus causing potential climate modification. One major disadvantage is that most forms of solar energy, with the possible exceptions of hydropower, biomass, and ocean thermal conversion, are intermittent and spatially variable. Some of these sources, such as solar cells (pho-

tovoltaics) that produce electricity directly from solar energy, are presently much more expensive than fossil or nuclear energy sources.

Another important aspect of renewable resources is the lead time necessary to implement technology. Table 13.4 summarizes the technology, availability, and lead time necessary for several of the renewable resources. We emphasize that many of these are commercially available, and the construction lead time is often short relative to development of new sources or power plants using fossil or nuclear fuels.

Direct Solar Energy

The direct use of solar energy is not new. Nearly 2,500 years ago Greeks designed homes to capture sunlight, and ancient cliff-dwelling Indians in the southwestern United States used asymmetry of valleys and exposure to sunlight for natural winter heating and summer shading (Figure 13.47). Amory Lovins points out that throughout history there has been a steady evolution in solar architecture and technology, but progress has been periodically interrupted by the influx of apparently plentiful, cheap fuels such as new forests or large deposits of oil, natural gas, coal, and uranium. Earlier en-

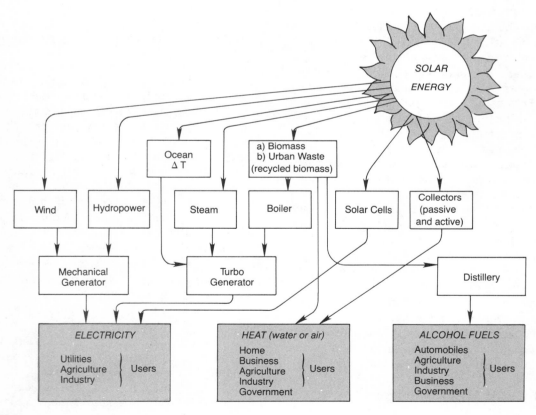

FIGURE 13.46
Idealized diagram showing the routes of the various types of renewable solar energy.

TABLE 13.4
Commercial status and lead times of renewable energy technologies.

Technology	Commercially Available	Regulatory Plus Construction Lead Time (years)
Biomass		
Direct combustion, electric	Yes	2–4.5
Direct combustion, non-electric	Yes	2
Gasification	Early 1980s	2
Anaerobic digestion	Yes	1–2
Municipal solid waste combustion	Mid 1980s	5–8
Wood stoves	Yes	Less than 1
*Wind turbines**		
Large	Mid 1980s	0.5–2
Small	Mid 1980s	0.5
Solar		
Passive	Yes	Less than 1
Water heating	Yes	Less than 1
Active space heating	Yes	Less than 1
Photovoltaics	Dispersed—mid 1980s	0.5–2
	Central station—after 2000	5
Thermal electric	Late 1980s or 1990s	9
Industrial heat	Some applications	2

*Lead times given do not include resource verification, which requires a minimum of one year.

Source: California Energy Commission 1980, Comparative Evaluation of Non-traditional Energy Resources.

(a)

(b)

FIGURE 13.47
Frijole Canyon, Bandelier National Monument, New Mexico. Photograph (a) shows the asymmetric profile of the canyon with the steep south-facing slope and more gentle north-facing slope. Early cliff-dwelling Indians used the south-facing cliff (b) for homesites. This is an example of the use of passive solar energy, as the homes provided partial shade from the summer sun and allowed the lower incident angle winter sun to enter the caves, helping to warm them.

ergy crises, beginning with shortages of wood during the Greek and Roman eras to coal strikes in the United States at the end of the nineteenth century, led to the rediscovery of earlier knowledge concerning practical applications of solar energy. These lessons should not be neglected in the current energy situation, since it appears that, with the possible exception of nuclear fusion, solar energy, broadly defined as shown in Figure 13.46, is the only long-term alternative at this time (32).

The total amount of solar energy that reaches the earth's surface is tremendous. On a global scale, two weeks' worth of solar energy is roughly equivalent to the energy stored in all known reserves of coal, oil, and natural gas on earth. In the United States, on the average, 13 percent of the sun's original energy entering the atmosphere arrives at the ground. The actual amount at a particular site is quite variable, however, depending upon the time of year and the cloud cover. The average value of 13 percent is equivalent to approximately 177 watts per square meter per hour (33).

Solar energy may be used directly through either passive or active solar systems. Passive solar systems often involve architectural design, without implementation of mechanical power, to enhance and take advantage of natural changes in solar energy that occur throughout the year. Many homes and other buildings in the southwestern United States as well as in other parts of the country now use passive solar systems for at least part of their energy needs. Active solar systems require mechanical power, usually pumps and other apparatus to circulate air, water, or other fluids from

solar collectors to a heat sink, where the heat is stored until used. Solar collectors are usually flat panels, a glass cover plate over a black background upon which water is circulated through tubes. Short-wave solar radiation enters the glass and is absorbed by the black background. As longer-wave radiation is emitted from the black material, it cannot escape through the glass, so the water in the circulating tubes is heated, typically to temperatures of 38 to 93°C (100–200°F) (33). In the U.S., the number of solar systems that use collectors like these exceeds 100,000 and is rapidly growing. Figure 13.48 shows the increase in production of solar collectors from 1974 through 1978. Figure 13.49 lists the types of collectors and uses for 1978 and clearly illustrates the dominance of low- and medium-temperature collectors used for heating swimming pools and domestic water heating.

Another potentially important aspect of direct solar energy involves solar cells or photovoltaics that convert sunlight directly into electricity. Although these cells may be feasible today in unique situations, they are expensive, costing about twenty times as much as electricity produced from traditional fossil fuel sources (34). On the other hand, the field of solar technology is changing quickly, and low-cost solar cells may become widely available in the future.

Questions often asked about direct solar energy are: where can it be used and how much energy will it save? These questions are not easy to answer. From an availability standpoint, Figure 13.50 shows in a general way the estimated year-round usability of solar energy

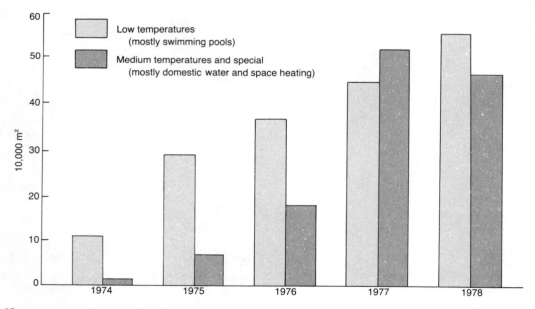

FIGURE 13.48
Production of solar collectors from 1974 through 1978. (From Department of Energy, Annual Report to Congress, 1979.)

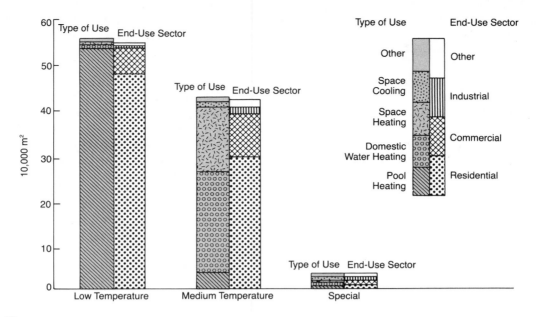

FIGURE 13.49
Types of solar collectors shipped by producers in 1978. Low-temperature collectors are primarily used for swimming pool heating. Medium-temperature and special collectors are used predominantly for domestic water heating and space heating. In 1978 approximately 77 percent of the total shipments went to residential users. (From Department of Energy, Annual Report to Congress, 1979.)

in the United States. But this view of solar energy availability is analogous to trying to paint a picture with a paint roller. Many sites for solar energy are site specific and detailed observation in the field is necessary to evaluate the solar energy potential in a given area. As to how much energy might be saved by converting to direct solar power, this again is quite variable. Optimists hope that by the year 2000 direct solar energy may supply up to 20 percent of the total energy demand. Considering how fast the solar energy industry is growing and the short lead time necessary for conversion to solar energy, these projections may prove accurate. On the other hand, conversions may be much slower if other sources of energy at cheaper prices be-

FIGURE 13.50
Estimated year-round usability of solar energy for the contiguous United States. (Modified after National Wildlife Federation, 1978.)

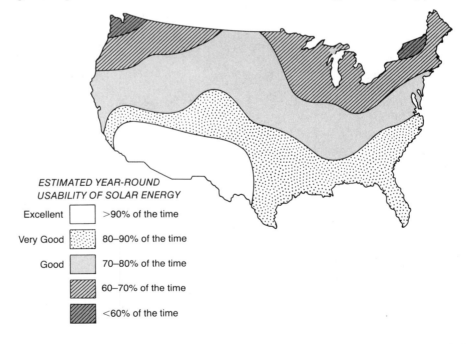

come available in the next few years. Since this possibility seems unlikely, the future of solar energy looks secure.

Two other interesting types of solar energy are the solar power tower and solar ponds. The concept of a power tower is shown in Figure 13.51. The system works by collecting solar energy as heat and delivering this energy in the form of steam to turbines that produce electric power. An experimental 10-megawatt power tower is underway near Barstow, California. The tower is approximately 100 meters high, surrounded by approximately 2,000 mirror modules, each with a reflective area of about 40 square meters. The mirrors adjust continually as the sun moves to reflect as much sunlight into the tower as possible. When excess steam is available it will be stored for extraction during periods when no sun is available. The Japanese are also interested in power towers and currently have a project to produce one megawatt of electricity from a tower approximately 75 meters high. If the Japanese experiment proves successful, they hope to eventually build a 60-megawatt plant. Although there is certainly room for optimism concerning power towers, the electricity generated will be very expensive, at least during the first few years when the technology is under development. It is probably safe to say that power towers may not prove economically viable in the near future; nevertheless, research to develop the technology is worthwhile and should continue.

Use of shallow solar ponds to generate relatively low-temperature water of about 68°C (180°F) is an interesting prospect for sources of commercial, industrial, and agricultural heat. Presently two types of ponds are being developed. One type is relatively deep (3 meters)

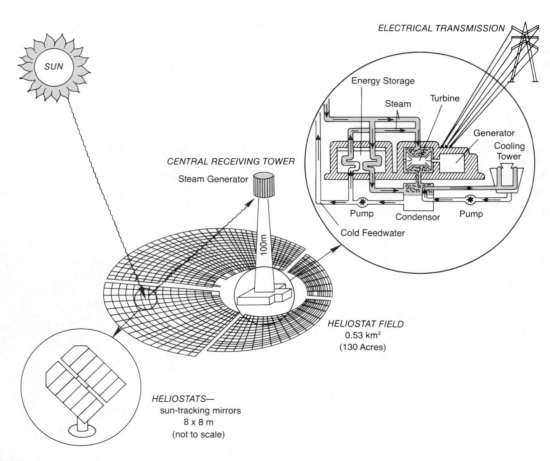

FIGURE 13.51
Idealized diagram showing the solar power tower being tested near Barstow, California in the Mojave Desert. This power plant is expected to produce as much as 10 megawatts of electricity and is a cooperative effort between the U.S. Department of Energy and the Southern California Edison Company. (Modified after a drawing prepared by Southern California Edison Company.)

and approximately 20 meters by 10 meters wide. The pond is designed to collect heat during the summer, building up a bottom water temperature of about 68°C in the fall. During the winter, heat can then be extracted from the bottom. The pond keeps the heated water on the bottom by the addition of salt, which makes the bottom water heavier. Circulation is restricted so the dense bottom water tends to remain on the bottom and not mix with the water above. Reportedly the pond works well even in areas with rather harsh winter. Up to 25 centimeters of ice and snow may accumulate on the surface of the pond and the bottom water will still remain approximately 38°C (100°F) throughout the winter (34).

The second experimental type of solar pond resembles a large water bed approximately 3.5 meters wide by 60 meters long and 5 to 10 centimeters deep. The top of the bed is transparent to solar energy while the bottom is an energy-absorbing black material. The pond is insulated from below to prevent heat loss to soil and rock. Essentially these ponds are just large solar collectors much like the smaller panels used for heating water in homes. So far several ponds have been tested, and the results are promising, since the cost is presently competitive with burning fossil fuel oil for heat. In particular, this type of solar pond seems applicable to small-scale industrial and agricultural use (34).

A last example of direct use of solar energy involves using part of the natural oceanic environment as a gigantic solar collector. The surface temperature of ocean water in the tropics is often about 28°C (82°F). At the bottom of the ocean, at a depth as shallow as about 600 meters, however, the temperature of the water may be 1–3°C (35–38°F). Low-efficiency heat engines can be designed to exploit this temperature differential either by using the seawater directly or by using an appropriate heat exchange system in a closed cycle in which a fluid such as ammonia or propane is vaporized by the warm water. The expanding vapor is then used to propel a turbine and generate electricity. After generation of electricity, the vapor is cooled and condensed by means of the cold water. Whether or not large-scale ocean thermal plants are ever built will depend primarily upon whether locations are discovered close to potential markets and whether the plants are economically feasible (33). Answers to these questions are uncertain so the future of ocean thermal plants remains speculative.

Water Power

Water power is really a form of stored solar energy, and since at least the time of the Roman Empire, the use of water as a source of power has been a successful venture. Waterwheels that harness water power and convert it to mechanical energy were turning in Western Europe in the seventeenth century, and during the eighteenth and nineteenth centuries, large waterwheels provided the energy to power grain mills, sawmills, and other machinery in the United States. Today, hydroelectric power plants provide about 15 percent of the total electricity produced in the United States. Although the total amount of electrical power produced by running water will increase somewhat in the coming years, the percentage may be reduced as other energy sources such as nuclear, solar, and geothermal increase faster.

Water power is clean power. It requires no burning of fuel, does not pollute the atmosphere, produces no radioactive or other waste, and is efficient. There is an environmental price to pay, however. Water falling over high dams may pick up nitrogen gas, which enters the blood of fish, expands, and kills them. Nitrogen has killed many migrating game fish in the Pacific Northwest. Furthermore, dams trap sediment that would otherwise reach the sea and replenish the sand on beaches. In addition, many people do not want to turn all the wild rivers into a series of lakes. Therefore, since many good sites for dams are already used, others are extremely limited, and many rivers will be saved by legislation. Thus, the growth of hydroelectric power from rivers is likely to be curtailed somewhat in the future.

Another form of water power might be derived from ocean tides in a few places where there is favorable topography, such as the Bay of Fundy region of the northeastern United States and Canada. The tides in the Bay of Fundy have a maximum rise of about 15 meters. A minimum rise of about 8 meters is necessary to even consider developing tidal power.

The principle of tidal power is to build dams across the entrance to bays, creating a basin on the landward side so as to create a difference in water level between the ocean and the basin. Then, as the water in the basin fills or empties, it can be used to turn hydraulic turbines that will produce electricity (35).

Wind Power

Wind power, like solar, has evolved over a long period of time, from early Chinese and Persian civilizations to the present. Wind has propelled ships and driven windmills for grinding grain or pumping water. More recently, wind has been used to generate electricity. The potential energy that might eventually be derived from the wind is tremendous; yet there will be problems because winds tend to be highly variable in time, place, and intensity (33).

Wind prospecting is becoming an important endeavor. On a national scale, regions with the greatest potential for development of wind energy are the Pacific Northwest coastal area, the coastal region of the northeastern United States, and a belt extending from northern Texas northward through the Rocky Mountain states and the Dakotas. There are also many other good sites, such as in mountain areas in North Carolina and the northern Coachella Valley in southern California.

At a particular site, the direction, velocity and duration of the wind may be variable, depending on local topography and regional to local magnitude of temperature differences in the atmosphere (36). For example, wind velocity often increases over hilltops or mountains, or may be funneled through a broad mountain pass (Figure 13.52). The increase in wind velocity over a mountain is caused by a vertical convergence of the wind, whereas through a pass, it is caused partly by horizontal convergence as well. Because the shape of a mountain or pass is often related to the local or regional geology, prospecting for wind energy is a geologic as well as geographic and meterologic problem.

Improvements in the size of windmills and the amount of power they produce occurred from the late 1800s through approximately 1950, when many countries in Europe and the United States became interested in larger-scale generators driven by the wind. In the United States, thousands of small wind-driven generators have been used on farms. Most of the small windmills generated about one kilowatt of power, which is much too small to be considered for central power generation needs. Interest in wind power declined during the several decades prior to the 1970s because of the abundance of cheap fossil fuels, but interest has revived in building larger windmills that might each supply a few megawatts of electrical power. Pilot projects in the Blue Ridge Mountains of North Carolina and deserts of southern California are ongoing. Figure 13.53 shows a large windmill near Palm Springs, California. The three-bladed rotor of the windmill is approximately 55 meters in diameter and is designed to generate as much as three megawatts of electricity, or enough to supply nearly 1,000 homes. The windmill is mounted on a tower approximately 33 meters high, in an area northwest of Palm Springs, California, where the winds average 24–27 kilometers per hour. It is expected that the windmill will operate about two-thirds of the time, but will not run at full power for

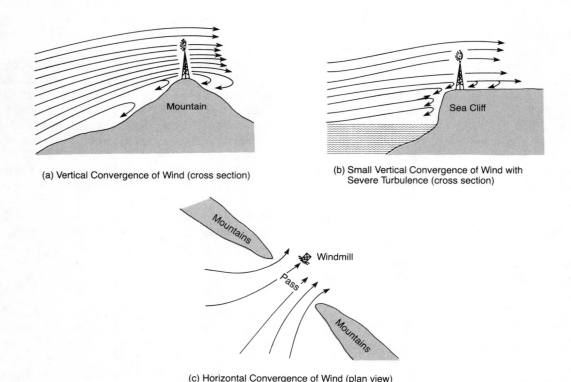

(a) Vertical Convergence of Wind (cross section)

(b) Small Vertical Convergence of Wind with Severe Turbulence (cross section)

(c) Horizontal Convergence of Wind (plan view)

FIGURE 13.52
Idealized diagrams showing how wind may be converged (and velocity increased) vertically (a & b) or horizontally (c). Tall windmills are necessary on hilltops or on the top of a sea cliff to avoid rear surface turbulence.

FIGURE 13.53
Giant windmill near Palm Springs, California. This windmill is designed to generate as much as 3 megawatts of electricity, sufficient to supply nearly 1,000 homes. The tower is approximately 100 feet or 33 meters high. More common are "wind farms" consisting of groups of smaller windmills. (Photograph courtesy of Mel Manalis.)

more than about 6 percent of the time. "Wind farms" consisting of rows of small windmills can also be used to generate power, and such farms are becoming more and more common in southern California.

Wind power will not solve all our energy problems, but it is yet another alternative energy source that is renewable and can be used in particular sites to reduce our dependency on foreign oil. Wind energy has a few detrimental qualities: first, demonstration projects suggest that vibrations from windmills may be a source of problems; second, windmills may cause interference with radio and television broadcast; and finally, windmills, particularly if there are many of them, may degrade scenic resources of an area. Still, everything considered, wind energy has a very low environmental impact and its continued use should be carefully researched and evaluated. A number of large demonstration units are now being tested in Ohio, New Mexico, Hawaii, North Carolina, Washington, and Rhode Island, as well as California.

Energy from Biomass

As illustrated in Figure 13.46, energy from biomass can take several routes: direct burning of biomass to produce either electricity or heat, water, or air, and distillation of biomass to produce alcohol for a fuel. Biological materials are certainly one of the oldest sources of energy. Today in the United States various biomass sources supply nearly 2 percent of the entire energy consumption (approximately 1.5 exa joules per year), primarily from the use of wood in the forest products

industry and for home heating (37). More than 1 billion people in the world today still use wood as their primary source of energy for heat and cooking.

It has been estimated that by the year 2000, biological processes could supply up to 15–20 percent of the current energy consumption in the United States. This projection depends upon a number of factors, however, including availability of cropland, increased yields, good resource management, and, most important, development of efficient conversion processes. Most of the increase is expected to result from burning a greater amount of wood. Figure 13.54a summarizes U.S. energy use in 1979 and Figure 13.54b shows two possible scenarios, one high and one low, for potential bioenergy supplies.

Biomass can be recycled, and today there are 20 facilities in the United States that process urban waste for use in generating electricity or producing fuel. Presently, only about 1 percent of the nation's municipal solid wastes are being recovered for energy; however, if all plants operated at full capacity and if the additional 20 or so plants under construction were completed and operating, about 10 percent of the country's waste, or 18 million tons per year, could be used to extract energy. At processing plants such as those in Baltimore County, Maryland; Chicago, Illinois; Milwaukee, Wisconsin; Tacoma, Washington; and Akron, Ohio, municipal waste is burned and the heat energy is used to make steam for a variety of purposes ranging from space heating to industrial generation of electricity. The United States has been slower to use urban waste as an energy source than other countries. In western Europe a number of countries now use between one-third to one-half of their municipal waste for energy production. With the end of cheap available fossil fuels, more and more energy recovery systems that use urban waste will be forthcoming (31).

Biomass in the form of organic waste that has been buried in landfills is also a potential source of energy. As the waste decomposes, it produces methane gas that can be recovered through shallow wells. Gas is now obtained from several landfills in California, Illinois, and Minnesota, and landfills that were once considered a nuisance are now viewed as a potentially valuable resource. As discussed in chapter 10, however, methane gas produced from landfills can also become a hazard if not carefully managed.

Biomass in its various forms appears to have a bright future as an energy source. The only question is how much can be produced and how quickly. As the price of fossil fuels, particularly oil and gas, continues to rise, these renewable energy sources certainly will become more attractive. The transition from fossil fuels to renewable energy sources is just now underway;

FIGURE 13.54
Energy use in the United States for 1979 (a), and two scenarios of potential bioenergy supplies, excluding speculative sources or municipal waste (b). (From: Office of Technology Assessment, 1980, *Energy from Biological Processes*, OTA-E-124.)

Total = 78.7 exa Joules/year

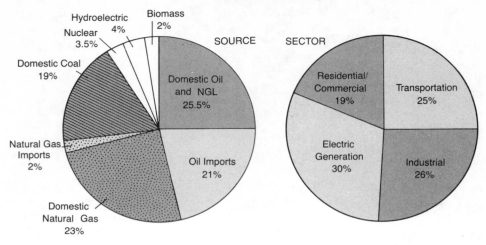

(a) *U.S. ENERGY USE IN 1979*

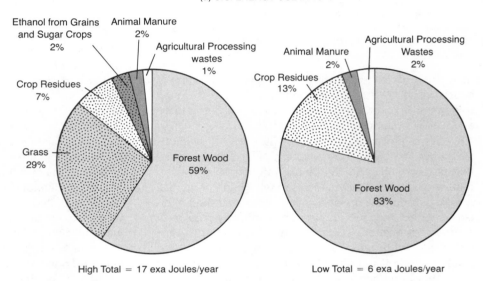

High Total = 17 exa Joules/year   Low Total = 6 exa Joules/year

(b) *POTENTIAL BIOENERGY SUPPLIES*

only the future will tell us how complete the transition will be.

ENERGY AND WATER DEMAND

The recent push to accelerate energy development rapidly has raised an important question: Is there enough water to support the projected development? This complex question has recently been addressed by a U.S. Geological Survey evaluation, and our discussion here summarizes some of the conclusions from that report (38).

The withdrawal of water for major uses is shown in Figure 13.55. Note that in 1955, the largest withdrawal was for irrigation. But by 1970, water used for thermoelectric power generation (fossil fuel, nuclear, or geothermal heat sources) had increased by almost 150 percent and exceeded irrigation demands for water. Most of the water for thermoelectric power generation was withdrawn to cool and condense the steam on the outlet side of the turbine. Most of that water is withdrawn from a river or lake, circulated once through the condenser, then carries the waste heat to a point of discharge back in the river or lake. Where thermal pollution is unacceptable or there is no large lake or river from which to obtain abundant cooling water, cooling towers, evaporation ponds, and water-sprayed evaporation ponds can be used for the cooling process (38).

We can thus see that a great deal of water is used in the thermoelectric generating industry. However, water is needed in all stages of energy development, from mining and land reclamation, to on-site processing, refining, and converting fuels to other energy forms. Figure 13.56 compares water consumption in refining and conversion processes for present commercial energy sources and those likely to be used in the future.

The amount of water required for mining and reclamation varies for different fuels and different climatic conditions. The water demand for mining coal or uranium is generally low, but in arid regions, obtaining the water necessary to establish vegetation as part of the reclamation program may be a serious problem. On the other hand, reclaiming a surface mine in a humid climate would probably not have a problem of lack of water. Another example can be drawn from the oil industry. Little water is used in drilling for oil, but during the recovery phase, a tremendous amount of water may be pumped down into the well to float the oil as part of a secondary recovery technique (38).

Mining oil shale, and particularly disposal of spent shale and surface reclamation in Colorado, Utah, and Wyoming, are also associated with high water demand. In fact, water requirements will probably limit production of shale oil to about one million barrels per day. A much larger production would require purchasing and transferring water that is now used for agricultural purposes (38).

The conclusion concerning the demand for water and production of energy in the United States is that in the entire nation, there is enough water to support accelerated energy development. In arid regions, how-

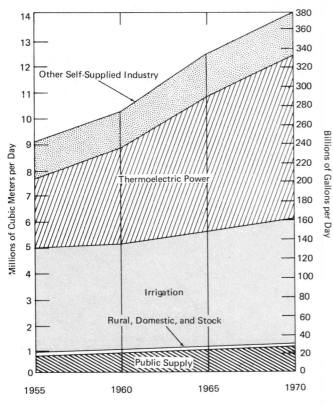

FIGURE 13.55
Withdrawal of water for major uses from 1955 to 1970 in the United States. (After G. H. Davis and L.A. Wood, U.S. Geological Survey Circular 703, 1974.)

FIGURE 13.56
Consumption of water for various energy-related refining and conversion processes. (After G. H. Davis and L. A. Wood, U.S. Geological Survey Circular 703, 1974.)

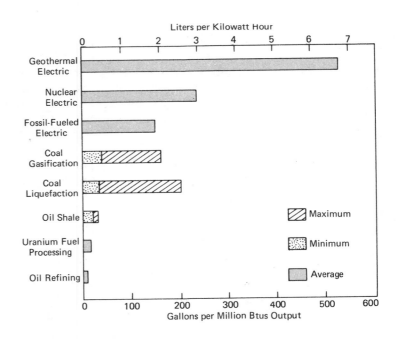

ever, limited water affects site location and design of energy recovery and conversion facilities (38).

ENERGY FOR TOMORROW

Although accurate predictions of future energy supplies are extremely difficult, it is interesting to speculate on the availability of energy in the United States to the year 2020. Even more interesting is the idea of a low-energy scenario based on assumptions of reduced supplies of oil, gas, and nuclear energy and increasing dependence on coal, solar, wind, and geothermal power.

A recently published low-energy scenario for the United States calls for a 40-percent reduction in total energy use by the year 2020. The authors of this scenario (Figure 13.57) emphasize that the reduction need not be associated with a lower quality of life, but rather with increased conservation of energy and a more energy-efficient distribution of urban populations, agriculture, and industry. They offer these alternatives: new settlement patterns that encourage "urban villages" with populations of about 100,000; constrained and minimized private transportation, and maximum accessibility of services, agricultural practices that emphasize locally grown and consumed food consisting largely of grains, beans, potatoes, and vegetables; and new industrial guidelines designed to conserve energy and minimize production and consumer waste to help reduce the total energy demand (39). To an extent, these alternatives are already taking form; that is, in some instances, highway construction has a lower priority than development of mass transit systems, agricultural lands near urban centers are preserved, and industry is more receptive to recycling and decreasing production of consumer wastes (such as unnecessary packaging).

The low-energy picture of the future also assumes the eventual demise of nuclear power by the year 2015 and a peak in coal production in the year 2010; both assumptions are highly speculative. It may be comforting to learn that this scenario is an estimate of the lowest energy use deemed plausible and was developed by eliminating possible future energy supplies that would require unlikely technology, drastic changes in human behavior, or a new political system (39).

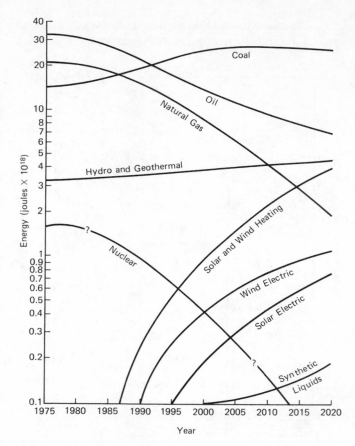

FIGURE 13.57
Energy supply for a low-energy scenario of the United States. (Data from J. S. Steinhart et al., "A low energy scenario for the United States: 1975–2050." In *Perspectives on Energy*, 2nd ed., ed. L. C. Ruedisili and M. W. Firebaugh [New York: Oxford University Press, 1978] p. 573.)

From an energy planning viewpoint, the next thirty years will be crucial to the United States and the rest of the industrialized world. The energy decisions we make in the near future will greatly affect both our standard of living and quality of life. Optimistically, we have the necessary information and technology to insure a bright, warm, lighted, and moving future—but time is short and we need action now! Essentially, we can either continue to take things as they come and live in the year 2000 with the results of our present inaction, or we can build for the future an energy picture of which we can be proud (39).

SUMMARY AND CONCLUSIONS

The ever-increasing population and appetite of the American people and the world for energy is staggering. It is time to seriously question the need and desirability of an ever-increasing demand for electrical and other sources of energy in industrialized societies. Quality of life is not necessarily directly related to greater consumption of energy.

The origin of fossil fuels (coal, oil, and gas) is intimately related to the geologic cycle. These fuels are, essentially, stored solar energy in the form of organic material that has escaped total destruction by oxidation. The environmental disruption associated with exploration and development of these resources must be weighed against the benefits gained from the energy; but this development is not an either-or proposition, and good conservation practices combined with pollution control and reclamation can help minimize the environmental disruption associated with fossil fuels.

Nuclear fission will remain an important source of energy. The growth of fission reactors in the United States will not be as rapid as projected, however, because of concern for environmental hazards and increasing costs to produce large nuclear power plants. To avert an eventual nuclear fuel shortage, work should continue on the breeder reactor.

Nuclear fusion is a potential energy source for the future, and may eventually supply a tremendous amount of energy from a readily available and nearly inexhaustible fuel supply.

Use of geothermal energy will become much more widespread in the western United States where natural heat flow from the earth is relatively high. Although the electric energy produced from the internal heat of the earth will probably never exceed 10 percent of the total electric power generated, it nevertheless will be significant. Geothermal energy also has an environmental price, however, and the possibility of causing surface subsidence through withdrawal of fluids and heat, as well as possibly causing earthquakes from injection of hot waste water back into the ground, must be considered.

Renewable sources of energy depend upon solar energy and can take a variety of forms, including passive and active solar, wind and hydropower, and energy from biological processes (including recycled biomass from urban waste). These energy sources have varying attributes, but generally cause little environmental disruption and are indigenous to the land. These energy sources will not be depleted, so are dependable in the long term. There is an energy transition going on in the world today, and sooner or later we must make a complete transition from oil to other sources. Certainly we can mine more coal, but this will be expensive and could have unacceptable environmental consequences. A mixture of energy options, including coal and nuclear energy, along with renewable energy resources, seems a probable alternative. Of particular importance in the next few years will be the continued growth in direct solar energy as well as continued growth in recycling of urban waste to produce energy.

Hydropower will undoubtedly continue to be an important source of electricity in the future, but it is not expected to grow much because of lack of potential sites and environmental considerations. Wind power and use of solar cells along with various biomass alternatives are more speculative, so it is difficult to predict their growth. We are in an interesting time from an energy standpoint, and many changes and innovations will occur.

REFERENCES

1 U.S. COUNCIL ON ENVIRONMENTAL QUALITY. 1978. *Environmental quality.* Ninth Annual Report of the Council on Environmental Quality.

2 AVERITT, P. 1973. Coal. In *United States mineral resources,* ed. D. A. Brobst and W. P. Pratt, pp. 133–42. U.S. Geological Survey Professional Paper 820.

3 U.S. ENVIRONMENTAL PROTECTION AGENCY. 1973. *Processes, procedures and methods to control pollution from mining activities.* EPA-430/9-73-001.

4 U.S. BUREAU OF MINES. 1971. *Strippable reserves of bituminous coal and lignite in the United States.* U.S. Bureau of Mines Information Circular 8531.

5 COMMITTEE ON ENVIRONMENT AND PUBLIC PLANNING. 1974. Environmental impact of conversion from gas or oil to coal for fuel. *The Geologist,* Newsletter of the Geological Society of America. Supplement to vol. 9, no. 4.

6 McCULLOH, T. H. 1973. Oil and gas. In *United States mineral resources,* ed. D. A. Brobst and W. P. Pratt, pp. 477-96. U.S. Geological Survey Professional Paper 820.

7 FEDERAL ENERGY ADMINISTRATION. 1975. FEA reports on proven reserves. *Geotimes,* August 1975, pp. 22–23.

8 CULBERTSON, W. C., and PITMAN, J. K. 1973. Oil shale. In *United States mineral resources,* ed. D. A. Brobst and W. P. Pratt, pp. 497–503. U.S. Geological Survey Professional Paper 820.

9 DUNCAN, D. C., and SWANSON, V. E. 1965. *Organic-rich shale of the United States and world land areas.* U.S. Geological Survey Circular 523.

10 OFFICE OF OIL AND GAS, U.S. DEPARTMENT OF THE INTERIOR. 1968. *United States petroleum through 1980.* Washington, D.C.: U.S. Government Printing Office.

11 COMMITTEE ON ENVIRONMENTAL AND PUBLIC PLANNING. 1974. Development of oil shale in the Green River Formation. *The Geologist,* Newsletter of The Geological Society of America. Supplement to vol. 9, no. 4.

12 ALLEN, A. R. 1975. Coping with oil sands. In *Perspectives on energy,* ed. L. C. Ruedisili and M. W. Firebaugh. New York: Oxford University Press, pp. 386–96.

13 OFFICE OF TECHNOLOGY ASSESSMENT. 1980. *An assessment of oil shale technologies.* OTA-M-118. 517P.

14 U.S. Environmental Protection Agency. 1979. *Acid rain.* EPA-600/8-79-028.

15 HUTCHINSON, T. C., and WHITBY, L. M. 1977. The effects of acid rainfall and heavy metal particulates on a boreal forest ecosystem near the Sudbury smelting region of Canada. *Water, Air and Soil Pollution* 7: 421–438.

16 FINCH, W. I. et al. 1973. Nuclear fuels. In *United States mineral resources,* ed. D. A. Brobst and W. P. Pratt, pp. 455–76. U.S. Geological Survey Professional Paper 820.

17 DUDERSTADT, J. J. 1977. Nuclear power generation. In *Perspectives on energy,* ed. L. C. Ruedisili and M. W. Firebaugh. New York: Oxford Univ. Press, pp. 249–273.

18 ENERGY RESEARCH AND DEVELOPMENT ADMINISTRATION. 1978. *Advanced nuclear reactors: An introduction.* ERDA-76-107. 24P.

19 U.S. GEOLOGICAL SURVEY. 1973. *Nuclear energy resources: A geologic perspective.* U.S.G.S. INF-73-14.

20 U.S. DEPARTMENT OF ENERGY. 1980. *Magnetic fusion energy.* DOE/ER-0059.

21 U.S. DEPARTMENT OF ENERGY. 1979. *Environmental development plan for magnetic fusion.* DOE/EDP-0052.

22 U.S. DEPARTMENT OF ENERGY. 1978. *The United States magnetic fusion energy program.* DOE/ET-0072.

23 SMITH, R. L., and SHAW, H. R. 1975. Igneous-related geothermal systems. In *Assessment of geothermal resources of the United States—1975,* ed. D. F. White and D. L. Williams, pp. 58–83. U. S. Geological Survey Circular 726.

24 WHITE, D. F. 1969. *Natural steam for power.* U.S. Geological Survey Publication, GPO-19690-339-S36.

25 WHITE, D. F., and WILLIAMS, D. L. 1975. In *Assessment of geothermal resources of the United States—1975,* ed. D. F. White and D. L. Williams, pp. 1–4. U.S. Geological Survey Circular 726.

26 MUFFLER, L. J. P. 1973. Geothermal resources. In *United States mineral resources,* ed. D. A. Brobst and W. P. Pratt, pp. 251–61. U.S. Geological Survey Professional Paper 820.

27 WORTHINGTON, J. D. 1975. *Geothermal development.* Status Report—Energy Resources and Technology, A Report of The Ad Hoc Committee on Energy Resources and Technology, Atomic Industrial Form, Incorporated.

28 COLP, J. L. 1974. *Magma tap—the ultimate geothermal energy program.* Sandia Laboratories Energy Report, SAND74-0253.

29 COLP, J. L. et al. 1975. *Magma workshop: assessment and recommendations.* Sandia Laboratories Energy Report, SAND75-0306.

30 PAPADOPULUS, S.S. et al. 1975. Assessment of onshore geopressured-geothermal resources in the northern Gulf of Mexico basin. In *Assessment of geothermal resources of the United States,* ed. D. F. White and D. L. Williams, pp. 125–46. U.S. Geological Survey Circular 726.

31 COUNCIL ON ENVIRONMENTAL QUALITY. 1979. *Environmental quality,* Tenth Annual Report of the Council on Environmental Quality.

32 LOVINS, A. B. 1979. Foreword in K. Butti and J. Perlin, *A golden thread.* Palo Alto, Calif.: Cheshire Books.

33 EATON, W. W. 1976. Solar energy. In *Perspectives on energy,* 2d ed., ed. L. C. Ruedisili and M. W. Firebaugh. New York: Oxford University Press, pp. 418–50.

34 RALOFF, J. 1978. Catch the sun. *Science News* 113, no. 16.

35 COMMITTEE ON RESOURCES AND MAN, NATIONAL ACADEMY OF SCIENCE. 1969. *Resources and man.* San Francisco: W. H. Freeman.

36 NOVA SCOTIA DEPARTMENT OF MINES AND ENERGY. 1981. *Wind power.*

37 OFFICE OF TECHNOLOGY ASSESSMENT. 1980. *Energy from biological processes,* OTA-E-124.

38 DAVIS, G. H., and WOOD, L. A. 1974. *Water demands for expanding energy development.* U.S. Geological Survey Circular 703.

39 STEINHART, J. S., HANSON, M. E., GATES, R. W., DEWINKEL, C. C., BRIODY, K., THORNSJO, M., and KABALA, S. 1978. A low energy scenario for the United States: 1975–2050. In *Perspectives on energy,* ed. L. C. Ruedisili and M. W. Firebaugh, pp. 553–88. New York: Oxford University Press.

FIVE

Land Use and Decision Making

P erhaps the most controversial environmental issue today is determining the most appropriate uses of our land and resources. The controversy is a human issue that involves responsibility to future generations as well as compensation for landowners today.

Before we decide how to plan the use of the land and its resources, we must develop reliable methods to evaluate the land and develop a legal framework within which we can make sound environmental decisions. Chapter 14 discusses several aspects of landscape evaluation such as land-use planning, site selection, landscape aesthetics, and environmental impact. Chapter 15 introduces the relatively new and expanding field of environmental law. Although environmental law is obviously not geological, there is a real need for applied earth scientists to know something about our legal system and the laws and legal theory that affect the natural environment.

Photograph courtesy of U.S. Army Corps of Engineers.

Chapter 14

Landscape Evaluation

Evaluating the landscape for purposes such as land-use planning, site selection, construction, and environmental impact is a common practice. Although they are in various stages of quantification and testing for validity, the methodology and techniques to accomplish evaluation are a significant part of applied environmental work.

The role of the geologist in landscape evaluation is to provide geologic information and analysis before planning, design, and construction. Table 14.1 shows some of the possible interrelationships among geology and other professional disciplines during the planning phase of landscape evaluation. This table clearly demonstrates that the geologist is an important member of the planning team (1).

The specific information geologists apply as part of the landscape evaluation process varies from area to area and project to project, but generally includes these factors:

- Physical and chemical properties of earth materials, including seismic stability, potential for building materials, and acceptability for waste disposal
- Slope stability
- Presence of active or possibly active faults (at the centers of historical earthquakes) and fracture systems

- Depth to the water table and groundwater flow characteristics
- Depth to bedrock
- Extent of any floodplains present

Table 14.2 lists the importance of some of the geologic conditions for typical land uses (1).

LAND-USE PLANNING

Figure 14.1 shows traditional land use in the contiguous United States from 1900 to 1974. Although the increase in conversion of rural land to nonagricultural uses appears to be very slow, it currently amounts to about 8,100 square kilometers per year (Figure 14.2). The intensive conversion of rural land to urban development, transportation networks and facilities, and reservoirs is nearly matched by the extensive conversion of rural land to wildlife refuges, parks, and wilderness and recreation areas.

The primary contribution of earth scientists to land-use planning is in emphasizing that all land is not the same and that particular physical and chemical characteristics of the land may be more important to society than geographic location. Earth scientists recognize that there is a limit to our supply of land, and that we must therefore strive to plan so that suitable

TABLE 14.1
Professional interrelationships during planning.

Major Study Groups	Typical Study Topics	Geologists	Geographers	Civil Engineers	Sanitary Engineers	Architects	Landscape Architects	Planners	Recreation Planners	Conservationists	Lawyers	Public Administrators	Sociologists
Regional economy	Economic base		■					●	■		■	●	
	resource potential	●	■	■		■	■	■	■	■	■	■	■
Regional population	Population studies,				■			●	■				■
	socio-economic studies							■	●		■	■	●
Transportation	Transport facilities,	●	■	●		■	■	●		●		■	
	public transport,	■		●		■		●			■	■	
	parking facilities	■		●		■	■	●		■		■	
Natural environment and public utilities	Natural resources,	●	●	●	■		■	●	■	●			
	hazards protection,	●		●	■			■		■	■		
	land reclamation,	●	■	●	●			■		●	■	■	
	public utilities	●		●	●			●			■	●	
Community facilities	Schools, libraries,	■				●	●	●				●	●
	police and fire,		■					●		■	●	●	■
	parks, recreation	■	■	■		■	●	●	●	●		●	■
Land use	General plans,	■	■	■	■		●	●	■	●			■
	neighborhood plans,	■	■	●	■	■	●	●	■	■			●
	commercial developments,	■	■	●		●	●	●			■	■	
	industrial developments	■	■	●		●	●	●			■	■	
Housing and public buildings	Private dwellings,	■				■	●	■				■	■
	public buildings	■		■		●	●	●	■			●	
Aesthetics	History, cultural values,	■	■			●	●	●	●	●			●
	community improvement,	■		●		■	●	●	●	■		●	■
	legislative controls					■	■	■	■			●	●
Administration and legislation	Legislation		■					■			■	●	●
	administration							●	■		●	●	■
Finance	Capital improvements,					■		●	■		●	●	
	federal-aid programs		●	■	■			●		■	●	●	
Other planning studies	Defining goals,	■	●			●	●	●	●	●	■	■	●
	urban renewal programs,	■	●	■	●	●	●	●	●	●		●	●
	waste disposal,	●		●	●			■	■	●		■	
	public health,	■		■	●			■	■			●	■
	civil defense	■		■				■				●	

- ● Primary interrelationship
- ■ Secondary interrelationship
- ☐ No interrelationship

Source: Turner and Coffman, "Geology for Planning: A Review of Environmental Geology," *Quarterly of the Colorado School of Mines,* vol. 68, no. 3, 1973.

FIGURE 14.1
Land use from 1900–1974 for the contiguous United States. The category "specialized land" includes land used for urban purposes, which is about 3 percent in the United States. (After Council on Environmental Quality, 1979.)

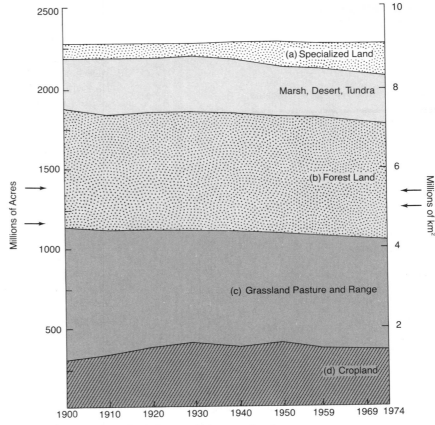

ᵃUrban and other built areas, highways, railroads, airports, parks, and other land.
ᵇExcludes forested areas reserved for parks and other special uses.
ᶜIncludes grassland pasture and range, private and public.
ᵈCropland planted, cropland in summer fallow, soil-improvement crops, and land being prepared for crops and idle.

FIGURE 14.2
Approximate annual conversion of rural land in the United States to nonagricultural uses. (After Council on Environmental Quality.)

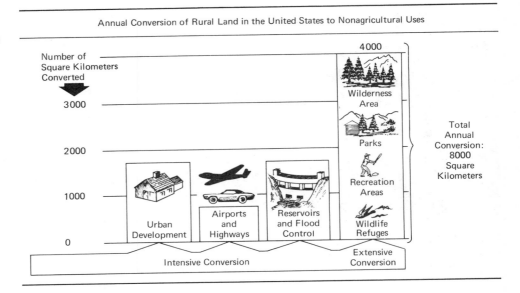

land is available for specific uses for this generation and those that follow (2).

TABLE 14.2
Geologic conditions that affect typical land uses.

Geologic Conditions	Typical Land Uses						
	Light structure construction	Heavy structure construction	Waste disposal	Building material resources	Ease of excavation	Road performance	Agriculture
Physical properties of soils and rocks	•	•	•	•	•	•	•
Slope stability	•	•				•	
Thickness of surficial materials	•	•		•			
Depth to groundwater	•	•	•	•			•
Supply of Surface water				•			•
Danger of flooding	•	•	•	•	•	•	•

▪   Primary importance

☐   Secondary importance

Source: Turner and Coffman, "Geology for Planning: A Review of Environmental Geology," *Quarterly of the Colorado School of Mines*, vol. 68, no. 3, 1973.

The need for land near urban areas in recent years has led to the concepts of *multiple* and *sequential* land use rather than permanent, exclusive use. Multiple land use, as, for example, active subsurface mining below urban land, is probably less common than sequential land use, or changing use with time. The concept of sequential use of the land is consistent with the fundamental principle we discussed earlier—the effects of land use are cumulative, and, therefore, we have a responsibility to future generations. The basic idea is that after a particular activity (perhaps mining or a sanitary landfill operation) has been completed, the land is reclaimed for another purpose.

There are several examples of sequential land use. Sanitary landfill sites in California, North Carolina, and other locations are being planned so that when the site is completed, the land will be used for a golf course. The city of Denver used abandoned sand and gravel pits for sanitary landfill sites that today are the sites of a parking lot and the Denver Coliseum (Figure 14.3) (3). Enormous underground limestone mines in Kansas City, Springfield, and Neosho, Missouri, have been profitably converted to warehousing and cold-storage sites, offices, and manufacturing plants (3). Other possibilities also exist, such as the use of abandoned surface mines for parking below shopping centers, chemical storage, and petroleum storage (Figure 14.4) (3).

The land-use planning process shown in Figure 14.5 includes five specific steps (4):

FIGURE 14.3
Examples of sequential land use in Denver, Colorado. Gravel pits were used as a sanitary landfill site as long ago as 1948. Today the Denver Coliseum and parking lots cover the landfill site. (Photos courtesy of Missouri Geological Survey.)

(a)

(b)

(c)

FIGURE 14.4
Examples of sequential land use of underground mines in Missouri, for (a) freezer storage and (b and c) warehousing. (Photos by J. D. Vineyard, courtesy of Missouri Geological Survey.)

1 Identify and define problems, goals, and objectives
2 Collect and interpret data
3 Formulate land-use plans
4 Review and adopt plans
5 Implement plans

The role of the earth scientist in the land-use planning process is most significant in the data collection and analysis stage. Depending on the specific task or plan, the earth scientist will use available earth science information, collect necessary new data, pre- pare pertinent technical information such as interpretive maps and texts, and assist in the preparation of land-capability maps that, ideally, should match the natural capability of a land unit to specific potential uses (4).

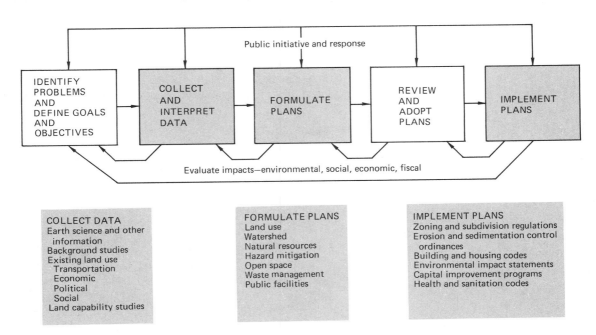

FIGURE 14.5
Diagram showing the land-use planning process. (After U.S. Geological Survey Circular 721, 1976.)

As an example, consider the role of the earth scientist in planning for reduction of natural hazards. Many natural hazards are also natural processes, and in a hazards reduction program the first task is to identify problem areas. Table 14.3 summarizes many of the natural processes that continue to be potentially hazardous to human use of the land. For a particular site or area, individual potentially hazardous processes must be evaluated carefully in light of intended land use. Siting of critical facilities, such as power plants, dams, or highway overpasses, should obviously be prohibited in high-risk areas, as should hospitals, schools, high-rise buildings, and disaster relief facilities.

The applied earth scientist is also qualified to advise about developing hazard-mitigation methods. There are several ways that hazardous risks from natural processes can be reduced (4). First, land-use can be regulated to restrict occupancy of potentially hazardous areas, such as floodplains and unstable slopes. Second, attempts can be made to partially control the potentially hazardous processes; for example, by constructing retaining walls to stabilize slopes or on-site storm water holding ponds in urban areas to minimize flooding. Third, rather than trying to control the process directly, efforts can be initiated to reduce its impact; for example, special construction methods for foundations and buildings may reduce the expansive soils or earthquake hazard. Fourth, hazardous processes can be monitored and studied in detail for help in developing early warning systems to decide when a hazardous area should be evacuated. It is particularly important to develop procedures for releasing predictions of hazardous events such as earthquakes. Table 14.4 lists terms that have been suggested for earthquake prediction, ranging from long-term, low-level forecast and watch to short-term, high-level alert. Finally, in helping to select a particular set of procedures or plans to minimize natural hazards, the earth scientist is an active member of a responsible decision-making team that must consider the physical, environ-

TABLE 14.3
Hazardous natural processes significant in land-use planning.

Process	Description of Hazard
Flooding	Overtopping of river and stream banks by water produced by sudden cloudbursts, prolonged rains, tropical storms, or seasonal thaws; breakage or overtopping of dams; ponding or backing up of water because of inadequate drainage
Erosion and sedimentation	Removal of soil and rock materials by surface water and depositing of these materials on floodplains and deltas
Landsliding	Perceptible downslope movement of earth masses
Faulting	Relative displacement of adjacent rock masses along a major fracture in the earth's crust
Ground motion	Shaking of the ground caused by an earthquake
Subsidence	Sinking of the ground surface caused by compression or collapse of earth materials; common in areas with poorly compacted, organic, or collapsible soils and commonly caused by withdrawal of groundwater, oil, or gas; or collapse over underground openings, such as mine workings or natural caverns
Expansive soils	Soils that swell when they absorb water and shrink when they dry out
High water tabel	Upper level of underground water close to ground surface causing submergence of underground structures, such as septic tank systems, foundations, utility lines, and storage tanks.
Seacliff retreat	Recession of seacliffs by erosion and landsliding
Beach destruction	Loss of beaches owing to erosion and/or loss of sand supply
Migration of sand dunes	Wind-induced inland movement of sand accelerated by disturbance of vegetative cover
Salt-water intrusion	Subsurface migration of seawater inland into areas from which fresh water has been withdrawn, contaminating fresh water supplies
Liquefaction	Temporary change of certain soils to a fluid state, commonly from earthquake-induced ground motion causing the ground to flow or lose its strength

Source: U.S. Geological Survey Circular 721, 1976.

TABLE 14.4
Suggested terms for earthquake prediction with specific time window shorter than a few decades

Long-term Prediction	Few years to a few decades
1. Forecast	
2. Watch	
Intermediate-term Prediction	Few weeks to few years
(no subdivisions)	
Short-term Prediction	Up to a few days
1. Alert	Three days to a few weeks
2. Imminent Alert	Up to three days

Source: R. E. Wallace, J.F. Davis, and K. C. McNally, "Terms for expressing earthquake potential, prediction and probability" (manuscript in preparation).

mental, social, political, and economic aspects of hazard mitigation.

Land-Use Plans

Land-use plans contain four basic elements:

- A statement of land-use issues, goals, and objectives
- A summary of data collection and analysis
- A land classification map
- A report that describes and indicates appropriate development of areas of special environmental concern (5)

The *statement* of major land-use issues to help plan future development over at least a ten-year period should be prepared in cooperation with citizens and public agencies. The statement should include information on issues such as the impact of economic and population trends, housing and services, conservation of natural resources, protection of important natural environments, and protection of cultural, historical, and scenic resources (5).

Data collection and *analysis* include analyses of present population, economy, and land use, and current plans, policies, and regulations; constraints, such as land potential, capacity of community facilities (water and sewer service, schools, etc.); and estimated demand from changes in population and economy, future land needs, and demand for facilities. Table 14.5 lists the required and optional data-analysis items for land-use plans submitted under the North Carolina Coastal Management Act of 1974.

The *land classification map* is the heart of a land-use plan. It serves as a statement of land-use policy and has five aims:

- To achieve and encourage coordination and consistency between local and state land-use policies
- To provide a guide for public investment in land
- To provide a useful framework for budgeting and planning construction of facilities, such as schools, roads, and sewer and water systems
- To coordinate regulatory policies and decisions

- To help provide guidelines for development of equitable land tax

An example of a current land classification system is shown in Table 14.6

The *report* accompanying the land classification map gives special attention to areas of environmental concern—those areas in which uncontrolled or incompatible development might produce irreparable damage. Examples in coastal areas include coastal marshland, estuaries, renewable resources (watersheds or aquifers), and fragile, historic, or natural resource areas (Outer Banks, Barrier Islands, etc.). The report should be precise regarding the permissible land uses of these areas (5).

We can see that the preparation of a land-use plan is complex and requires a team approach. The input of the applied earth scientist is especially valuable in evaluating natural resources and suitability of land for a particular purpose and in defining areas of environmental concern. To this end, it is important to make maximum use of soils information and to develop environmental geology maps to aid planners.

Soil Surveys and Land-Use Planning

When properly used the information from detailed soil maps can be extremely helpful in land-use planning. Soils can be rated according to their limitations for land uses, such as housing, light industry, septic-tank systems, roads, recreation, agriculture, and forestry. Soil characteristics that help determine possible limitations for particular land use include slope, water content, permeability, depth to rock, susceptibility to erosion, shrink-swell potential, bearing strength (ability to support a load, such as a building), and corrosion potential.

The limitations can be mapped from detailed soil maps and the accompanying description of the individual soil types. As an example, a soil evaluation to determine limitations for buildings in recreation areas will emphasize the value of a soil survey (6). Figure 14.6 is a detailed soil map and brief description of soils for 4.1 square kilometers in Aroostock County, Maine. Table 14.7 lists soil limitations for buildings in recreational

TABLE 14.5
Required and optional data analysis items for a land-use plan.

Element	Required	Optional
1. Present conditions		
a. Population and economy	Brief analysis, utilizing existing information	More detailed analyses relating to human resources (population composition, migration rates, educational attainment, etc.) and economic development factors (labor force characteristics, market structure, employment mix, etc.)
b. Existing land use	Mapped at generalized categories	Mapped with more detailed categories, including more detailed analyses, building inventory, etc.
c. Current plans, policies, and regulations		
1) Plans and policies	1) List and summary	1) Detailed impact analysis of plans and policies upon land-development patterns
2) Local regulations	2) List and description of enforcement mechanisms	2) Detailed assessment of adequacy and degree of enforcement
3) Federal and stage regulations	3) List and summary (to be provided by N. C. Dept. of Natural and Economic Resources)	
2. Constraints		
a. Land potential		
1) Physical limitations	1) Analysis of following factors (maps if information available): Hazard areas; Areas with soil limitations; Sources of water supply; Steep slopes	1) Detailed analysis and mapping of required items. Analysis and mapping of additional factors: Water-quality limited areas; Air-quality limited areas; Others as appropriate
2) Fragile areas	2) Analysis of follwing factors (maps if information available): Wetlands; Frontal dunes; Beaches; Prime wildlife habitats; Scenic and prominent high points; Unique natural areas; Other surface waters; Fragile areas	2) Detailed analysis and mapping of required items; analysis and mapping of additional factors
3) Areas with resource potential	3) Analysis of following factors (maps if information available): Areas well suited for woodland management; Productive and unique agricultural lands; Mineral sites; Publicly owned forests, parks, fish and game lands, and other outdoor recreational lands; Privately owned wildlife sanctuaries	3) Detailed analysis and mapping of required items. Analysis and mapping of additional factors: Areas with potential for commercial wildlife management; Outdoor recreation sites; Scenic and tourist resources
b. Capacity of community facilities	Identification of existing water and sewer service areas. Design capacity of watertreatment plant, sewagetreatment plant, schools, and primary roads. Percent utilization of the above	Detailed community facilites studies or plans (housing, transportation, recreation, water and sewer, police, fire, etc.)
3. Estimated demand		
a. Population and economy		
1) Population	1) 10-yr estimates based upon Dept. of Administration figures as appropriate	1) Detailed estimate and analysis, adapted to local conditions using Department of Administration model
2) Economy	2) Identification of major trends and factors in the economy	2) Detailed economic studies
b. Future land needs	Gross 10-yr. estimate allocated to appropriate land classes	Detailed estimates by specific land-use category (commercial, residential, industrial, etc.)
c. Community facilites demand	Consideration of basic facilities needed to service estimated growth	Estimates of demands and costs for some or all community facilities and services

Source: North Carolina Coastal Resources Commission, 1975.

402

TABLE 14.6
Example of a land classification system.

Classification	Characteristics
Developed	Lands where existing population density is moderate to high and where there are a variety of land uses that have the necessary public services.
Transition	Lands where local government plans to accommodate moderate- to high-density development and to provide basic public services that will accommodate that growth during the following ten-year period.
Community	Lands where low-density development is grouped in existing settlements or will occur in such settlements during the following ten-year period, and will not require extensive public services now or in the future.
Rural	Lands whose highest use is for agriculture, forestry, mining, water supply, etc., based on their natural resource potential. Also includes lands for future needs not currently recognized.
Conservation	Fragile, hazard, and other lands necessary to maintain healthy natural environment and necessary to provide for public health, safety, or welfare.

Source: North Carolina Coastal Resources Commission, 1975.

areas. Information from Figure 14.6 and Table 14.7 is combined to produce the map in Figure 14.7 showing where buildings could be located. The standard limitation classes are defined as follows: *none to slight*— soils are relatively free of limitations that affect the planned use, or have limitations that can readily be overcome; *moderate*—soils with moderate limitations resulting from effects of soil properties (these limitations can normally be overcome with correct planning, careful design, and good construction); and *severe*— soils with severe limitations that make the proposed use doubtful. Careful planning and more extensive design and construction measures are required, possibly including major soil reclamation work (5).

Environmental Geology Mapping

The primary goal of environmental geology mapping is to help the planner (1), so geologic information and engineering properties of earth materials must be presented in a way that is easy to interpret. That is, environmental geology maps must contain important information necessary for planning and eliminate extraneous data (1).

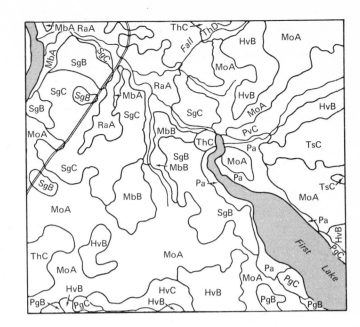

HvB	Howland very stony loam, 0 to 8 percent slopes
HvC	Howland very stony loam, 8 to 15 percent slopes
MbA	Madawaska fine sandy loam, 0 to 2 percent slopes
MbB	Madawaska fine sandy loam, 2 to 8 percent slopes
MoA	Monarda and Burnham silt loams, 0 to 2 percent slopes
Pa	Peat and Muck
PgB	Plaisted gravelly loam, 2 to 8 percent slopes
PgC	Plaisted gravelly loam, 8 to 15 percent slopes
PvC	Plaisted and Howland very stony loam, 8 to 15 percent slopes
RaA	Red Hook and Atherton silt loams
SgB	Stetson gravelly loam, 2 to 8 percent slopes
SgC	Stetson gravelly loam, 8 to 15 percent slopes
ThC	Thorndike shaly silt loam, 8 to 15 percent slopes
ThD	Thorndike shaly silt loam, 15 to 25 percent slopes
TsC	Thorndike and Howland soils, 8 to 15 percent slopes

FIGURE 14.6
Detailed soil map of a 4-square-kilometer tract of land, Aroostock County, Maine. (Reproduced from *Soil Surveys and Land Use Planning*, 1966, by permission of the Soil Science Society of America.)

From a pragmatic view, the best environmental geology maps are those that combine geologic and hydrologic information in terms of engineering properties. The landscape may then be mapped in terms of a specific land use, as is currently done for soils. The idea is to produce a series of maps, one for each possible land use. Thus, an *environmental geology map* is essentially a combination of geologic and hydrologic data expressed in less technical terms to facilitate general understanding by a large audience.

The Illinois State Geological Survey has been a leader in developing methods for mapping for environmental geology. The mapping in DeKalb County, Illinois, illustrates the procedure. The county was glaciated during the Pleistocene Ice Age, and surficial deposits include water-transported sands and gravels

TABLE 14.7
Soil limitations for buildings in recreational areas.

Soil Items Affecting Use	Degree of Soil Limitation[a]		
	None to Slight	Moderate	Severe[b]
Wetness	Well to moderately well drained soils not subject to ponding or seepage. Over 1.2 meters to seasonal water table.	Well and moderately well drained soils subject to occasional ponding or seepage. Somewhat poorly drained, not subject to ponding. Seasonal water table 0.6–1.2 meters[c]	Somewhat poorly drained soils subject to ponding. Poorly and very poorly drained soils.
Flooding	Not subject to flooding	Not subject to flooding	Subject to flooding
Slope	0%–8%	8%–15%	15% +
Rockiness[d]	None	Few	Moderate to many
Stoniness[d]	None to few	Moderate	Moderate to many
Depth to hard bedrock	1.5 meters[b]	0.9–1.5 meters[c]	Less than 1 meter

[a]Soil limitations for septic-tank filter fields; hillside slippage, frost heave, piping, loose sand, and low bearing capacity when wet are not included in this rating, but must be considered. Soil ratings for these items have been developed.

[b]Soils rated as having severe soil limitations for individual cottage sites may be best from an aesthetic or use standpoint, but they do require more preparation or maintenance for such use.

[c]These items are limitations only where basements and underground utilities are planned.

[d]*Rockiness* refers to the abundance of stones or rock outcrops greater than 10 inches in diameter. *Stoniness* refers to the abundance of stones 3 to 10 inches in diameter.

Source: Reproduced from *Soil Surveys and Land Use Planning* (1966) by permission of the Soil Science Society of America.

as well as glacial till (mixture of clay and larger particles).

Interpretive maps showing suitability for a particular land use generally include a color code: green for "go" (favorable conditions); yellow for "caution"; and red for "stop" (unfavorable conditions or problem area). Three shades for each color represent different kinds of limitations. Figure 14.8 is a suitability analysis for solid-waste disposal sites (sanitary landfills) in DeKalb County. In this case, the letters *G, Y,* and *R* are used in place of colors, and the numerals 1, 2, and 3 are used in place of the shades of each color. In theory, G-1 (Green-1) indicates areas with the least limitation, and R-3 indicates problem areas with the most unfavorable conditions (7).

Evaluation of a site for solid-waste disposal requires information on topography, type and thickness of unconsolidated material, present and potential sources of surface water and groundwater, and type of bedrock. The best sites have at least 9 meters of im-

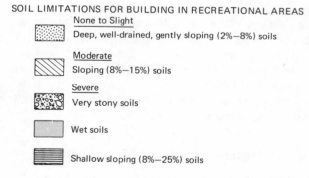

SOIL LIMITATIONS FOR BUILDING IN RECREATIONAL AREAS

None to Slight
Deep, well-drained, gently sloping (2%–8%) soils

Moderate
Sloping (8%–15%) soils

Severe
Very stony soils

Wet soils

Shallow sloping (8%–25%) soils

FIGURE 14.7
Soil limitations for buildings in recreational areas for the 4-square-kilometer tract of land shown in Figure 14.6. (Reproduced from *Soil Surveys and Land Use Planning*, 1966, by permission of the Soil Science Society of America.)

FIGURE 14.8

Suitability of DeKalb County, Illinois, area for solid-waste disposal. (After Gross, Illinois State Geological Survey, Environmental Geology Notes No. 33, 1970)

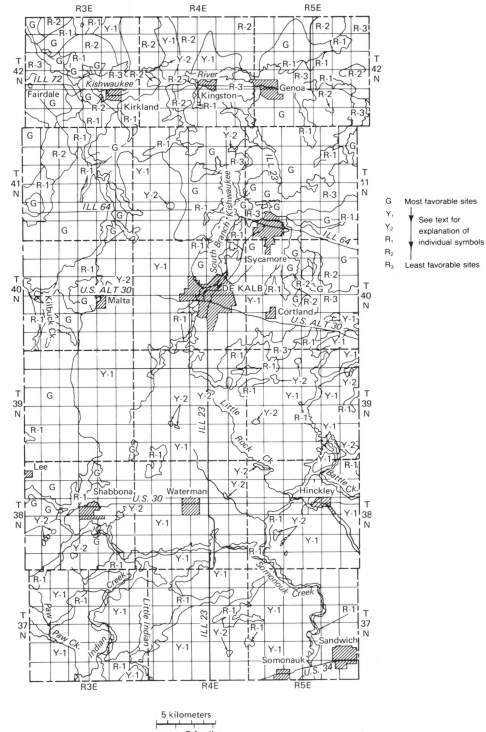

G   Most favorable sites

Y₁

Y₂   See text for explanation of individual symbols

R₁

R₂

R₃   Least favorable sites

5 kilometers

approx. 3.1 miles

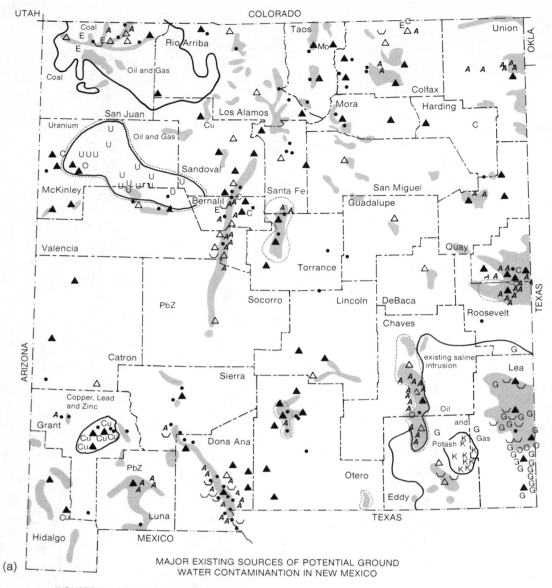

(a)

MAJOR EXISTING SOURCES OF POTENTIAL GROUND
WATER CONTAMINANTION IN NEW MEXICO

INDUSTRIAL/COMMERCIAL OPERATIONS

Cu-Copper
C-Misc. commercial
G-Natural gas refining
O-Oil refining
U-Uranium
I-Misc. industrial

Mo-Molybdenum
PbZ-Lead and Zinc
E-Electricity generation
K-Potash
⌣-Unspecified

A ANIMAL CONTAINMENT SYSTEMS

AGRICULTURE-irrigation operations involving increased
salinity, nitrogen-containing fertilizers, pesticides and
herbicides

SALINE INTRUSION-movement of naturally mineralized
water into fresh-water aquifer

● ON-SITE SYSTEM-discharge from septic tank,
cesspool, and other liquid waste system

MINING -primary resource labelled

MUNICIPAL WASTE WATER

▲ Potential groundwater discharge
△ Discharge into perennial stream

Type and location of major existing sources of potential ground-water contamination in New Mexico.
Note that some industrial and commercial operations are unidentified on this map.

FIGURE 14.9

Major sources of potential groundwater contamination (a) and relative vulnerability of aquifer contamination (b) in New Mexico. (After L. Wilson, 1981, New Mexico Geological Society Spec. Pub. No. 10, pp. 47–54.)

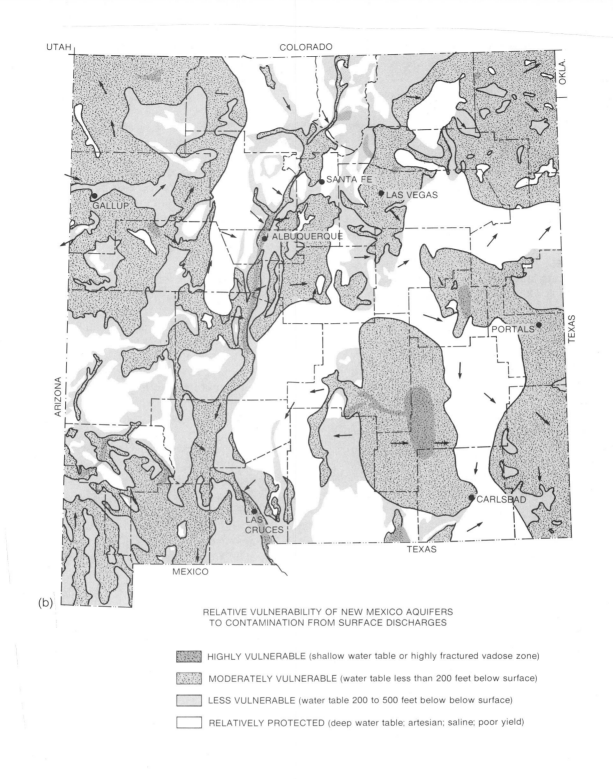

(b)

RELATIVE VULNERABILITY OF NEW MEXICO AQUIFERS
TO CONTAMINATION FROM SURFACE DISCHARGES

▨ HIGHLY VULNERABLE (shallow water table or highly fractured vadose zone)

▨ MODERATELY VULNERABLE (water table less than 200 feet below surface)

▨ LESS VULNERABLE (water table 200 to 500 feet below below surface)

☐ RELATIVELY PROTECTED (deep water table; artesian; saline; poor yield)

Relative vulnerability of New Mexico aquifers to contaminationfrom various surface discharges.
See Figure 1 for potential contaminant sources and names of counties.
Arrows on map indicate gerneral ground-water flow direction.

FIGURE 14.9, *continued*

permeable material between the base of the fill material and the top of any existing or potential groundwater aquifer. The areas labeled *G* in Figure 14.8 are those where it is believed that at least 15 meters of relatively impermeable till overlies any aquifer. Several test borings at a particular proposed site would be necessary to verify this. Less favorable areas, *Y-1* and *Y-2*, primarily in the southern part of the map, are characterized by layers of sand and gravel interbedded with glacial till. There are acceptable sites within these areas, but they require careful preliminary investigation. Areas designated *R-1*, *R-2*, or *R-3* are the least favorable sites for solid-waste disposal. *R-1* areas have surficial deposits of silt, sand, or sand and gravel underlain by till less than 6 meters thick. These areas lie along the streams and rivers, and surface-water pollution from landfill leachate would be a problem. *R-2* areas are those in which the thickness of the glacial deposits over bedrock is thought to be less than 15 meters. The bedrock supplies groundwater to residences, and, therefore, at least 9 meters of relatively impermeable material above the rock is necessary. Test borings will be required before siting of a landfill in these areas. *R-3* areas have severe limitations in that the upper 6 meters of surficial material is silt, sand, or sand and gravel which are likely to cause a pollution problem if landfill leachate is present. Even within the *R-3* areas, however, there are probably areas with thick deposits of silt that would make a satisfactory landfill site (7).

Regional maps that delineate an area's relative vulnerability to environmental disruption are also valuable in planning future development and land use. Figure 14.9, for example, shows major existing sources of potential groundwater contamination and the relative vulnerability of aquifers to pollution from surface dis-

charge in New Mexico. Relative vulnerability to groundwater pollution was predicted by considering aquifer type, depth to the water table, permeability, and presence or absence of natural barriers to migration of pollutants (8). The maps thus present geologic and hydrologic data as well as sources of potential groundwater pollution in a format useful to regional planners.

Environmental geology maps such as those in Illinois and New Mexico are valuable to the planner, but a larger-scale, more detailed map that shows accurate geologic boundaries is preferable, although not always practical—map scale is based on the accuracy and amount of available geologic data, and natural geologic boundaries are often gradual rather than sharp (7).

Another interesting approach to environmental mapping that is likely to become more significant in the future is the concept of an *environmental resource unit* (ERU). This method uses a multidisciplinary team approach to define and analyze the total natural environment—geologic, hydrologic, and biologic. A portion of the environment with a similar set of physical and biological characteristics is called an environmental resource unit. Ideally, every ERU is a natural division characterized by specific patterns or assemblages of structural components (rocks, soils, vegetation, etc.) and natural processes (erosion, runoff, soil genesis, productivity of wildlife, etc.) (1).

A study site of 10.4 square kilometers near Morrison, Colorado, illustrates the concept of defining and working with environmental resource units. Figure 14.10 shows the topography and the area, and Figure 14.11 shows the four ERUs. The first is the *mountain forest* ERU—ridge and ravine topography supporting a pine and Douglas fir forest. Running-water and mass-

FIGURE 14.10
Physiographic diagram of the Morrison Test Site. (From Turner and Coffman, ''Geology for Planning: A Review of Environmental Geology,'' *Quarterly of the Colorado School of Mines*, vol. 68, no. 3, 1973.)

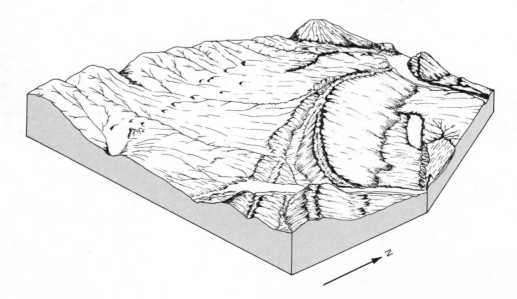

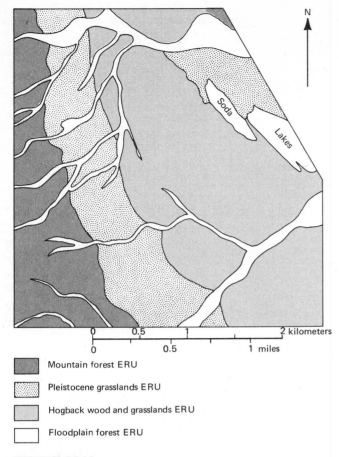

Mountain forest ERU

Pleistocene grasslands ERU

Hogback wood and grasslands ERU

Floodplain forest ERU

FIGURE 14.11
Environmental resource units for the Morrison Test Site. (After Turner and Coffman, "Geology for Planning: A Review of Environmental Geology," *Quarterly of the Colorado School of Mines*, vol. 68, no. 3, 1973.)

wasting processes are important in modifying the land, and the dominant ecological control is the moisture regime that varies with elevation and exposure. The second is the *floodplain forest* ERU—recent alluvial (stream-deposited) material deposited by channel and overbank stream flow. This ERU supports a cottonwood-willow forest that needs abundant water. The third is the *hogback wood* and *grassland* ERU—a series of tilted sedimentary rock units with steep scarp slopes, gentle dip slopes (Figure 14.10), and gentle debris slopes or level bottoms. Mass-wasting processes are controlled by slope and exposure. The scarp slopes and bottomlands support grass, and the dip slopes support junipers. The fourth is the *Pleistocene grasslands* ERU—a complex of older alluvial and mass-wasting material deposited by processes that are no longer operating. Natural vegetation is grass, but in the study area, much of the ERU is used for agricultural purposes (1).

Environmental resource units are used to establish patterns of structure and process and are valuable in establishing land capabilities and suitabilities. The ERU concept is most valuable in defining areas in a systems approach that involves the entire environment. These units may then be analyzed separately or collectively for a specific land use.

The ultimate value of ERUs will be realized when multidisciplinary teams, including geologists, soil scientists, biologists, landscape architects, engineers, and planners, use remote sensing along with field survey techniques to identify different units and then analyze the data for planning purposes by means of computer-drawn maps for rapid data analysis and display.

The Franconia area in Fairfax County, Virginia, less than 15 kilometers southwest of the District of Columbia, offers a good example of the integration of environmental resource units with earth science maps to plan land use that is sensitive to the natural setting (Figure 14.12). The synthesis resulting in the final land-use capability map is an example of computer composite mapping using input of topographic, geologic, and hydrologic data. This map has had a significant impact on decisions to rezone existing land or allow new development in the Franconia area.

Planning Following Emergencies

In recent years, two types of planning have emerged: projects in which engineering design and environmental impact is an integral part; and emergency projects following catastrophic events such as hurricanes or volcanic eruptions that cause widespread damage. Emergency planning is always in response to pressing needs and an influx of emergency money. Too often, the authorized work is either overzealous and beyond what is necessary, or is not carefully thought out. As a result, emergency projects can cause further environmental disruptions instead of help. Acknowledgement of these problems does not imply that all emergency work is poor or unnecessary; nevertheless, when millions of dollars of emergency money arrives, care must be taken to insure that it is used for the best possible purposes. We will discuss two examples: Hurricane Camille in 1969 and the eruption of Mt. St. Helens in 1980.

Following severe storms and floods that struck Virginia in the aftermath of Hurricane Camille, emergency federal aid was used to channelize streams in hopes of alleviating future floods. Although much necessary work was performed, in some instances local bulldozer operators with little or no instruction or knowledge of streams were contracted to clear and straighten stream channels. Results of this unplanned and unsupervised channel work have been disastrous to many kilometers of streams. Since catastrophic storms can seriously damage roads, farms, and homes,

FIGURE 14.12
Flow chart showing the use of
earth science information and
land use planning for the Franconia area, Virginia. (Modified
from A. J. Frolich, and A. D.
Garnaas, U. S. Geological Survey
Professional Paper 950, 1978.)

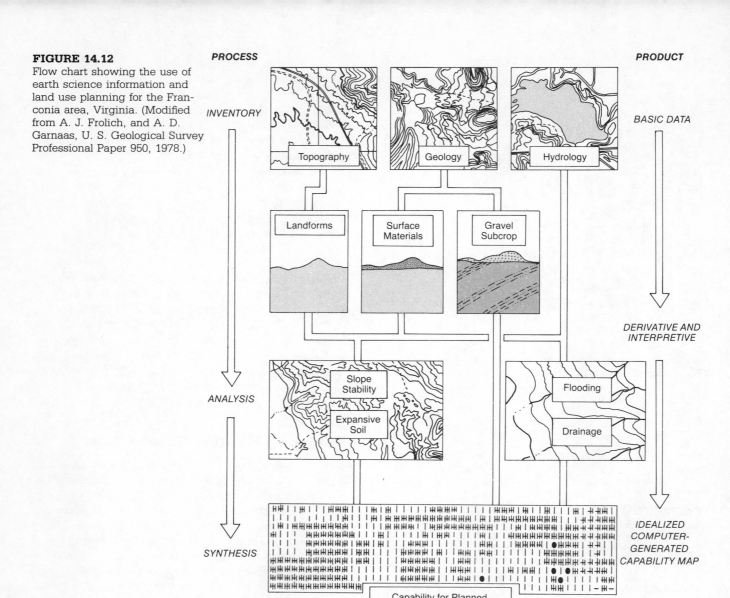

PROCESS

INVENTORY

ANALYSIS

SYNTHESIS

Topography

Geology

Hydrology

Landforms

Surface Materials

Gravel Subcrop

Slope Stability

Expansive Soil

Flooding

Drainage

PRODUCT

BASIC DATA

DERIVATIVE AND INTERPRETIVE

IDEALIZED COMPUTER-GENERATED CAPABILITY MAP

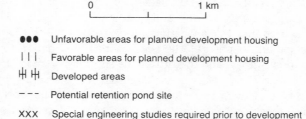

Capability for Planned Development Housing

0   1 km

●●●   Unfavorable areas for planned development housing

| | | Favorable areas for planned development housing

╫╫ ╫   Developed areas

– – –   Potential retention pond site

XXX   Special engineering studies required prior to development emergency channel work is often needed; however, such emergency work should be confined to stream channels that require immediate attention. Emergency funds should not be considered a license for wholesale modification of any stream in the damaged area at the request of property owners or others.

The eruption and catastrophic landslide-debris avalanche of Mt. St. Helens delivered 2.5 cubic kilometers of debris into the Toutle River. Fearing continued downstream sediment pollution, two emergency catch dams were constructed to filter the water through rockfill and gravel barriers, thus maintaining the flow of water in the river while trapping the sediment in the catch dams. Unfortunately, the dams were constructed before reliable estimates could be made of the volume of sediment likely to be delivered to the dams. On the North Fork of the Toutle River, the catch dam had the capacity to hold 0.0065 cubic kilometers of debris, and the Corps of Engineers estimated that the sediment load would be 0.01 to 0.02 cubic kilometers per year, thus requiring 2 to 3 dredgings per year for the dam to remain in service. The U.S. Geological Survey, on the other hand, estimated, after a small August flood, that the sediment yield would be closer to 0.3 to 0.38 cubic kilometers per year; therefore, the catch dams were about 100 times too small to survive expected winter storms and floods. This estimate proved true during a storm in late October when the dams failed. Thus, while the idea of catch dams was worthwhile, they were constructed too quickly and were too small.

Planning following emergencies is important, and even if there is not enough time to prepare an environmental impact statement, all projects should be considered carefully to determine if they are really necessary and will not cause future problems.

SITE SELECTION AND EVALUATION

Site selection is the process of evaluating a physical environment that will support human activities. It is a task shared by professionals in engineering, landscape architecture, planning, earth science, social science, and economics, and thus involves a multidimensional approach to landscape evaluation.

In recent years, a philosophy of site evaluation known as *physiographic determinism* has emerged (9). The thrust of this philosophy is "design with nature." That is, rather than laying down an arbitrary design or plan for an area, it may be more advantageous to find a plan that nature has already provided (8). With this method, planners take advantage of a site's physical conditions in such a way as to maximize the amenities of the landscape while minimizing social and economic expenditures whenever possible. For example, if the soils of an area vary considerably in their ability to corrode, then the best corridor for an underground pipeline might not be the shortest route. Rather, the route might be designed so that the greatest possible length lies in the soils with the least corrosive potential. Although such a route is slightly longer and requires a higher initial cost, it may, by taking advantage of nature, produce tremendous savings in maintenance and replacement costs (10).

Although the philosophy of working in harmony with nature is obviously advantageous, it is often overlooked in the siting process. People still purchase land for various activities without considering whether the land use they have in mind is compatible with the site they have chosen. There are well-known examples of poor siting that have resulted in increased expense, limited production, or even abandonment of partially completed construction. Construction of a West Coast nuclear power facility was terminated when fractures in the rock (active faults) and possible serious foundation problems arose (Figure 14.13). The productivity of a large chicken farm in the southeast was greatly curtailed because the property was purchased before it was determined whether there was sufficient groundwater to meet the projected needs. A housing developer in northern Indiana purchased land and built country homes in one of the few isolated areas where bedrock (shale) is at the surface. Thus, septic-tank systems had to be abandoned and a surface sewage treatment facility had to be built. The rock also made it much more expensive to excavate for basements and foundations.

The goal of site selection for a particular land use is to insure that site development is compatible with both the possibilities and limitations of the natural environment. The earth scientist's role may be significant in providing crucial geologic information to help accomplish this goal. The type of information geologists provided includes: soil and rock types, rock structure (especially fractures), drainage characteristics, groundwater characteristics, landform information, and estimates of possible hazardous earth events and processes, such as floods, landslides, earthquakes, and volcanic activity. The engineering geologist also takes samples, makes tests, and predicts the engineering properties of the earth materials.

Process of Site Selection

The process of site selection begins with a thorough understanding of what the general purposes and specific requirements of the designed activity are and for whom the site is being selected. The next step is an

FIGURE 14.13
Nuclear power development was abandoned due to an earthquake hazard at this site at Bodega Bay, California, after expenditure of millions of dollars and years of time in site preparation. This waste of time and money could have been avoided had data on the fault been available earlier. (From U.S. Geological Survey Circular 701, 1974.)

analysis of possible alternative sites in terms of physical limitations and cost. The final step is to recommend which site or sites are most advantageous for the desired land use (11).

The geologic aspect requires these steps:

- Collection of existing data, such as geologic reports, topographic maps, aerial photographs, soil maps, water surveys, engineering reports, climate records, etc.
- Reconnaissance of likely sites to note potential problems and possibilities
- Collection of necessary new data, including soil, rock, and water analyses
- Determination of the magnitude and importance of geologic limitations and possibilities of the site or sites

The *magnitude* of geologic limitations and possibilities in site selection and evaluation refers to the degree, extensiveness, or scale, while the *importance* is a measure of the weight a particular limitation or possibility should be assigned. For example, if a site evaluation for the foundation of a large dam or nuclear facility reveals that the bedrock has several small fractures, the *magnitude* of the limitation is small. However, the *importance* may be great if the faults are active and less if they prove to be inactive. The presence of an active fault is obviously significant because it could cause the site to be abandoned, whereas inactive fractures can be treated by specific engineering procedures.

METHODS OF SITE SELECTION AND EVALUATION

Although specific methods of determining a site and its suitability obviously vary with the intended purpose for the particular land use, a few methods are common for evaluating the desirability of a particular site for relatively large-scale land uses, such as construction of reservoirs, highways, and canals, and development of parks and other recreational areas. Two common methods are *cost-benefits analysis* and *physiographic determinism*, determining the best site by maximizing physical and social benefits while minimizing social costs. The first approach is strictly economical, while the second balances economics with less tangible aspects, such as aesthetics.

Cost-Benefits Analysis

Cost-benefits analysis is a way to assess the long-range desirability of the particular project, such as construction of a highway or development of a recreational site. The idea is to develop the benefits-to-cost ratio; that is, the total benefits in dollars over a period of time are compared to the total cost of the project, and the most desirable projects are those for which the benefits-to-cost ratio is greater than one (12).

Cost-benefits analysis assumes that all relevant costs and benefits can be determined. There are three difficult aspects: which costs and which benefits are to be evaluated; how they are to be evaluated; and how the intangible costs and benefits, such as aesthetic degradation or improvement, are to be evaluated (12).

Cost-benefits analysis is common for three general types of projects: *water resource* projects, such as irrigation systems, flood control, and hydroelectric power systems, as well as multipurpose schemes, such as reservoirs used for flood control, power generation, and recreation; *transportation* projects, such as highways, roads, railways, and river and canal work to facilitate navigation; and *land-use* projects, such as urban renewal and recreation (12).

Examining some of the costs and benefits for a hypothetical flood-control project will further illustrate the method. The initial tangible costs of a flood-control project are relatively straightforward, but evaluation of repair and maintenance costs and intangible costs over the life of the project is much more difficult. The main benefit from flood control is averting loss of property. (*Loss* here refers primarily to different types of assets, such as property, furnishings, and crops.) The general principle is to evaluate the annual flood loss and damage based on the most probable flood levels for several recurrence intervals, say, the 10-year and 25-year floods. This estimate of annual loss and damage is then taken as the maximum yearly amount that people would be willing to pay for flood control. Other benefits to consider include reducing the number of deaths by drowning and eliminating recurring costs of evacuating flood victims, emergency sandbag work, and incidence of sanitary failure (12).

Physiographic Determinism

An interesting graphic method of site selection based on physiographic determinism, or designing with nature, is structured such that the appropriate site literally selects itself (13). The method uses physical, social, and aesthetic data, and the objective is to maximize social benefit while minimizing social cost.

The actual methodology in selecting a site or a route for a highway, regional development plan, or other land use involves selecting a group of factors that reflect physical and social values. For each factor, three or more grades of the value are mapped on transparencies in tones from clear to dark grey. The darker tones indicate areas of conflict, where physical conditions are not appropriate or social values conflict with the desired land use. When the transparencies (one for each factor) are superimposed on one another, the darkest areas represent those places with the greatest physiographic obstructions and social costs, and the lighter areas represent places where the physical limitations (construction costs and social costs) are least. By this method, the best sites appear as the lightest areas on the set of superimposed transparencies. No matter how dark the entire area may look, there will always be relatively lighter areas (13).

The method of determining lowest social cost in site selection is innovative in addressing social values as well as traditional physical-economic values, but it is not above criticism. It is extremely subjective, in that the person who is evaluating chooses the factors and thus essentially builds the formula used in the evaluation. For example, in evaluating the siting of a highway route, the evaluator may decide that property values are important and that the darker tone should be assigned to high property values. This decision essentially prescribes that the highway should be routed away from people, and most of all from wealthy people, a point some planners would vigorously argue against (9). Another criticism is that all factors, whether physical or social values, are weighted the same; that is, a geologic factor such as suitability of soil for foundation material or susceptibility to erosion is weighted equally with social values, such as scenic value or recreational property value. Regardless of these shortcomings, the method of physiographic determinism in evaluating both physical-economic factors and social values is a step in the right direction (9).

SITE EVALUATION FOR ENGINEERING PURPOSES

The geologist has a definite role in the planning stages of site development for engineering projects, such as construction of dams, highways, airports, tunnels, and large buildings.

Dams

Construction of either masonry (concrete) dams or earth dams (Figure 14.14) requires careful and complete geologic mapping and testing early in the planning process. Testing should include evaluation of the valley and the dam site to determine the present and expected stability of the slopes; identification of possible problems, such as fracture zones in rocks or adverse soil conditions where the dam structure would be in contact with rock or soil (foundation); prediction of the rate of sediment accumulation; and assessment of the availability of building material for the construction of the dam. Continuous geologic mapping during construction may be important because the geologic history of a river valley often involves several periods of alternating erosion and scour that produce irregular or unexpected deposits that can cause problems in construction, operation, and maintenance of the dam and reservoir. We emphasize, however, that major irregularities or problems that might affect the dam or reservoir should be recognized by direct and indirect and indirect geologic investigation before the construction phase of the project.

Foundations for dams vary according to what rock types are encountered. Generally, igneous rocks, such as granite, and some pyroclastic rocks are satisfactory foundations for a dam site. Leakage, if any, will occur along fractures, and fractures can be grouted, that is, filled with a mixture of cement, sediment, and water. Metamorphic rocks may also be good foundation

FIGURE 14.14

Garrison Dam and Lake Saka-kawea on the Missouri River in North Dakota. The earth fill dam is 61 meters high and 3,390 meters long. The width at the base is 1,020 meters. The construction required 50.54 million cubic meters of earth fill and 1.14 million cubic meters of concrete for the spillway, power plant, etc. The power plant generates 400,000 kilowatts. (Photo courtesy of U.S. Army Corps of Engineers, Omaha District.)

material, and the best sites are those in which the foliation of the rocks is parallel to the axis of the dam. Sedimentary rocks such as limestone and particularly compaction shale can be troublesome. Limestone may have large underground cavities, and compaction shales are noted for problems stemming from their tendency to deform and settle when loaded (14,15). Particularly important is the stability of sedimentary rocks following wetting and drying or freezing and thawing. Failure of the St. Francis Dam in California was caused in part by the deterioration upon wetting of the sedimentary rocks in the foundation.

Highways

Site evaluation for highways requires considerable geologic input, including mapping the geology along the proposed route and isolating problems of slope stability, topography, flooding, or weak earth materials that would cause foundation problems. It is also helpful to locate and estimate the amount of possible construction materials in close proximity to the proposed route (15).

Airports

Important geologic aspects of site selection for airports are threefold:

- Topography as it relates to necessary grading, drainage, and surfacing
- Soils—the best are coarse-grained with high bearing capacities and good drainage
- Availability of construction materials

Good surface drainage is important, so areas subject to flooding should be avoided (14).

Tunnels

Construction of tunnels requires detailed knowledge of geology, and evaluation of geologic information is basic to selection of the route and determination of proper construction methods. Geology is a major factor that determines the economic feasibility of the project; therefore, a geologic profile along the center line of the proposed tunnel route is essential before tunneling. This profile is developed from mapping the surface geology and using subsurface data whenever possible (14,15). As with dam investigations, it is important to continue the geologic mapping as construction proceeds to insure that unexpected hazards do not produce problems.

Two distinct types of tunnel conditions are defined by whether the tunnel is in solid rock or in cohesionless, or plastic, earth (soft ground) that may tend to flow and fill the tunnel (15). Both conditions may be present in one tunnel.

Potential problems in rock tunnels are determined primarily by the nature and abundance of fracturing or other partings in the rock. Fracturing greatly affects how much overbreak (excess rock that falls into the tunnel) is present. In general, the closer the fractures and other partings, the greater the overbreak (14). Tunnels in granite or horizontally layered rocks generally develop a symmetrical, naturally arch-shaped overbreak, whereas inclined rocks produce an irregular (asymmetric) roof (Figure 14.15). When possible, it is advisable to select tunnel routes through rock that has a minimum of fractures.

Rock structure, particularly folds and faults, is also significant in locating and orienting tunnels. When

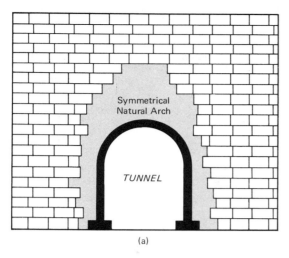 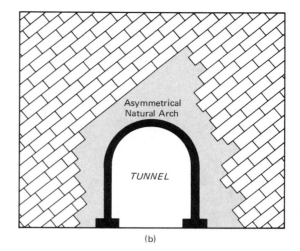

FIGURE 14.15
Idealized diagram showing the overbreak that develops in (a) horizontally layered rocks and (b) inclined rocks.

tunnels are built through synclines, water problems are likely, because the natural inclination of the rocks facilitates drainage into the tunnel (Figure 14.16). In addition, rock fractures in synclines tend to form inverted keystones, and, thus, the roof of the tunnel may require substantial support. On the other hand, tunnels through anticlines are less likely to have water problems, and the fractures tend to form normal keystones that are more nearly self-supporting (14).

Faults in tunnels usually cause water problems, because water often migrates along open fractures associated with fault systems. Therefore, whenever possible, tunnels should be driven at right angles to faults to minimize the length of tunnel that will be in contact with the fault (14). Tunneling in soft ground is generally

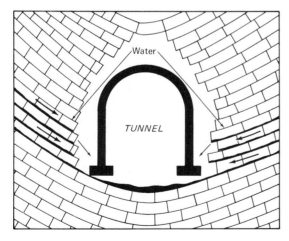

FIGURE 14.16
Idealized diagram showing possible groundwater hazard associated with tunnels driven through synclines.

done at shallower depths, thus facilitating the collection of more detailed subsurface information before construction than is possible in deep rock tunnels (14,15).

Most problems in earth tunneling are compounded when groundwater saturates the material through which the tunnel is being driven. When saturated, sandy material may slowly flow and fill a tunnel or suddenly develop a hazardous condition known as a "blow or boil," in which the material quickly fills the tunnel. Thorough evaluation of soil and hydrogeology is necessary when tunneling in soft ground (14,15).

Large Buildings

The geologist's role in evaluating a site for a large building is primarily to provide geologic information that can be used in designing the building's foundation. The site investigation includes drilling for soil or rock samples (Figure 14.17) to be evaluated for engineering properties to help engineers recognize unanticipated geologic problems that might arise during construction of the foundation. The fact that field investigation is in most cases very simple does not mean that detailed investigation can be routinely shortened (14,15).

Geologic investigation of foundation sites for nuclear facilities may involve detailed geologic mapping of rock structure. Even small fracture systems must be carefully evaluated to insure that there are no active or potentially active faults. In addition, the physical characteristics of old fractures, including their length, width, and degree of rock alteration (weathering) to clay and other material along the fractured planes, must be measured and evaluated for possible engineering treatment before construction. Figure 14.18 shows

FIGURE 14.17
Drilling for soil samples as part of site investigation. (Photo courtesy of Williams and Works.)

the McGuire Nuclear Power Plant during construction; the square boards bolted to the rock, along with dewatering points, stabilize the slope during excavation.

Hazardous Waste Disposal

Determining sites where hazardous waste may be safely disposed of requires careful geologic and hydrologic consideration. Because absolute containment is unlikely, a pragmatic objective is to design, construct, and regulate landfills so as to minimize potential problems associated with release of toxic materials into the environment. Siting hazardous wast disposal facilities

to meet the objective of minimal environmental disruption begins with developing appropriate geologic and hydrologic criteria. Specific criteria vary regionally, according to environmental conditions such as climate and geology, and locally, according to variation in soils, hydrology, geology, and geochemistry. General site criteria for New Mexico, for example, include these:

- Major aquifers must not be present adjacent to or below the site, and the water table must be at least 50 m below the site
- Rocks below the site should be clay-rich, with low permeability and high chemical adsorption properties
- Landscape surrounding the site should be stable, with little evidence of wind or water erosion, and at least 10,000 years old
- Climate at the site should be such that evaporation greatly exceeds precipitation
- Site should be in an area of low seismicity and low potential for landsliding, flooding, or volcanic activity
- Future land or other resource development should not be affected by disposal activities (16)

LANDSCAPE AESTHETICS: SCENIC RESOURCES

Scenery in the United States has been recognized as a natural resource since 1864, when the first state park, Yosemite Valley in California, was established. The early recognition and subsequent action primarily related scenic resources to outdoor recreation, focusing

(a)

(b)

FIGURE 14.18
Construction of the McGuire Nuclear Power Plant near Charlotte, North Carolina (a). Stability of the slope was maintained during excavation by use of rock bolts and dewatering points, shown by the arrows in photograph (b).

on preservation and management of discrete parcels of unique scenic landscapes. Public awareness of and concern for scenic values over a broader range of resource issues is relatively recent. Society now fully recognizes scenery as a valuable resource; recognition that the "everyday" nonurban landscape also has scenic value that enhances a region's resource base is also emerging, based on the assumption that there are varying scenic values just as there are varying economic values relating to other, more tangible resources (17).

Quantitative evaluation of tangible natural resources, such as water, forests, or minerals, is standard procedure before economic development or management of a particular land area. Water resources for power or other uses can be evaluated by flow in the rivers and storage in lakes; forest resources can be evaluated by the number, type, and size of the trees, and their subsequent yield in lumber; and mineral resources can be evaluated by estimating the number of tons of economically valuable mineral material (ore) at a particular location. We can state the quality and quantity of each tangible resource in comparison to some known low quality or quantity. Ideally, we would like to make similar statements about the more intangible resources, such as scenery; that is, we would like to compare scenery to specific standards (18). Unfortunately, this is a difficult task for which few standards are available.

A region's or area's *scenic resources* can be defined as the visual portion of an aesthetic experience; that is, the scenic base consists of those aspects of the landscape that produce visual amenities. Because areas with recognized visual amenities are generally considered unique, they are of particular public concern and more likely to be protected.

Landscape evaluation of scenic resources as part of land-use planning or assessing environmental impact generally rests on a rather subjective methodology. Figure 14.19 shows how a general quantitative evaluation might proceed from a recognition of scenic resources to relative or absolute indices to define their scenic values.

The general philosophical framework for evaluating scenic resources includes certain concepts. First, scenic resources are visual amenities that can be evaluated in terms of an aesthetic judgment. Second, scenic resources, like soil and other resources, vary in quality from place to place. Third, topographic relief, presence of water, and diversity of form and color are three of the significant positive characteristics of scenic quality, whereas artificial change is the most negative characteristic. Fourth, unique landscape is more significant than common landscape. Fifth, although de-

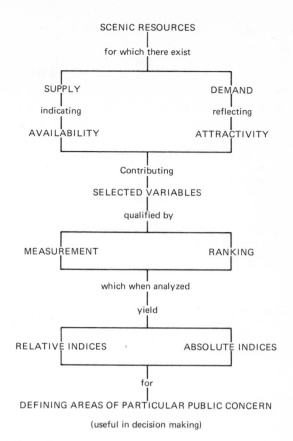

FIGURE 14.19
Idealized diagram showing how a quantitative evaluation of scenic resources might proceed from recognition of scenic resources to development of indices to rank various landscapes hierarchically. (After Coomber and Biswas, 1972, Ecological Systems Branch, Canada.)

termination of what is aesthetically pleasing varies from person to person, the judgment of what is really displeasing is more universal.

Virtually all schemes to evaluate scenic resources involve measuring or observing specifically selected factors. Observation is thus admittedly subjective, but not necessarily a bad practice, as long as relevant factors based on sound judgment are chosen. A completely objective method of analyzing scenic resources, based on our present knowledge of how people perceive the landscape, is probably not possible and certainly not practical. Examining current methods and procedures for evaluating scenic resources suggests that three general categories of factors or variables are most appropriate. The first is *landform characterization*, which assumes that, as the terrain becomes steeper and has greater altitudinal differences (relief) and diversity, the scenic quality increases. The second is *natural characterization*, which assumes that scenic quality increases as both the percentage of an area in natural surfaces (water, forest, marshland, or swamp-land) and the diversity of natural surface types increase. The third is *artificial characterization*—as the amount of area covered by artificial features increases, scenic quality decreases. Agricultural land and nonforested rural land can be considered neutral. With this very general model, a scenic quality index can be developed,

$$SQI = LC + NC - AC$$

where *SQI* is the scenic quality index, *LC* is the landform characterization, *NC* is the natural characterization, and *AC* the artificial characterization. Determining the numbers and possible weights for each factor is a subjective decision that must be clearly stated before evaluation.

Realizing the importance of quantitative and at least procedural-objective methodology for evaluating scenic resources, we will discuss two rather different approaches. The first is a regional study that attempted to evaluate the scenic resources of Scotland (18). Because of the scale, a broad-brush approach was necessary. The method included mapping and assigning numerical values to landforms such that high-relief forms were rated higher than low-relief forms, as shown in Figure 14.20. Land use was also assigned numerical values such that wild or wilderness areas were rated high and urban/industrial areas low (see Figure 14.21). The two maps were then combined to give a final rating of scenic resources, shown in Figure 14.22. Although it uses a general approach, the map clearly shows rather abrupt changes in scenic quality. This is

FIGURE 14.20
Subjective rating of landforms in Scotland. (After Linton, *Scottish Geographical Magazine*, vol. 84, 1968.)

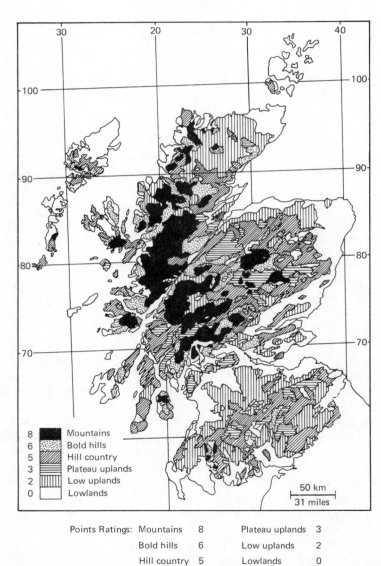

8	▉	Mountains
6	⣿	Bold hills
5	▨	Hill country
3	▤	Plateau uplands
2	▥	Low uplands
0	☐	Lowlands

50 km
31 miles

Points Ratings:	Mountains	8	Plateau uplands	3
	Bold hills	6	Low uplands	2
	Hill country	5	Lowlands	0

FIGURE 14.21
Point rating for land use in Scotland. (After Linton, *Scottish Geographical Magazine*, vol. 84, 1968.)

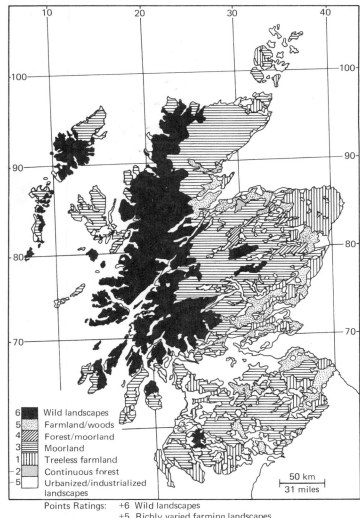

6	Wild landscapes
5	Farmland/woods
4	Forest/moorland
3	Moorland
1	Treeless farmland
-2	Continuous forest
-5	Urbanized/industrialized landscapes

50 km
31 miles

Points Ratings:
+6 Wild landscapes
+5 Richly varied farming landscapes
+4 Varied forest and moorland with some farms
+3 Moorland
+1 Treeless farmland
−2 Continuous forests
−5 Urbanized and industrialized landscapes

not unusual, as changes in relief in landforms tend to be rather abrupt in Scotland, emerging, for example, from forests to open moorland (hills, mires, and peat bogs).

The second approach to evaluating scenic resources is based on the premise that unique landscapes are more significant to society than common landscapes (19,20). Explanation of how a unique landscape is mathematically determined is beyond the scope of this discussion; nevertheless, we can explain the major aspects. The method is most appropriate for corridors such as river valleys, coasts, and highway routes, and involves measuring or observing variables that describe the physical, biological, and human use

of the landscape under evaluation. The data are then analyzed to determine uniqueness indices showing how different, in a relative sense, one landscape is from others (20). Up to this point, the only subjective part of the method is selecting the variables to be used. Figure 14.23 shows uniqueness indices for five stream valleys in Indiana. This type of graph is valuable because it also shows what part of the total uniqueness results from the physical, biological, or human-use variables.

If only "relative uniqueness" were determined, the method would have little application. For example, the most unique landscape might be the only polluted and altered river valley, or it might be the most scenic. The second step of the evaluation therefore, requires

FIGURE 14.22
Rating of scenic resources for
Scotland, developed from Figures
14.20 and 14.21. (After Linton,
Scottish Geographical Magazine,
vol. 84, 1968.)

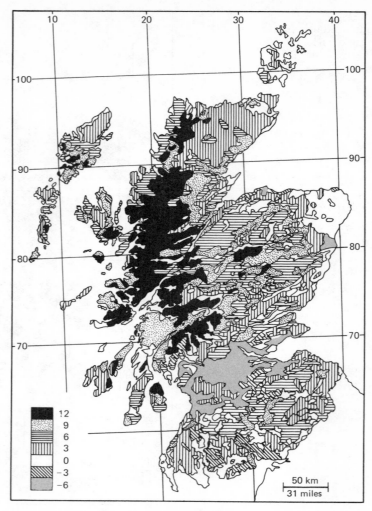

Resource Ratings: More than 12 Points (Black)

+9 to +12 (Cross-hatched)   0 to +3   (Blank)

+6 to +9   (Solid rule)   0 to −3   (Light stipple)

+3 to +6   (Broken rule)   −3 to −6 (Heavy stipple)

determining why a particular landscape is unique (20). With scenic resources, we are most interested in determining whether the most unique landscapes are also the most scenic. To do so, we use our premise that the judgment of what is pleasing may vary, but what is really ugly to one person is usually ugly to many people. In the case of streams, the Wild and Scenic River Act, Public Law 90-542 of 1968, defines scenic rivers as "those rivers or sections of rivers that are free from impoundments, with shorelines or watersheds still largely primitive and shorelines largely undeveloped, but accessible in places by roads." This definition implies that the stream is also clear-running, unpolluted, and unlittered. Using this information, we can compare the data for determining uniqueness with the idealized scenic river and lower a river valley's original unique-

ness in proportion to how much of the original uniqueness is antithetical to our "ideal" scenic river (20). By this philosophy, even a "very unique" stream, if it is heavily developed or polluted, will rank low in terms of a scenic river index. Figure 14.24 shows the scenic river indices for the same five streams in Indiana; the upper line for each stream valley is the uniqueness index, and the lower is the scenic river index. This graph succinctly summarizes the uniqueness and scenic river indices.

Quantifying scenic resources is valuable because the results can be visually displayed and different landscapes hierarchically ranked. Quantification is basically another tool to help make decisions about land-use planning or analyze environmental impact. It also tends to separate facts from emotion while providing a way

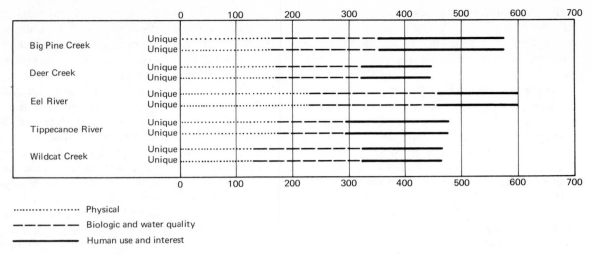

Physical
Biologic and water quality
Human use and interest

FIGURE 14.23
Bar graph of uniqueness indices for five Indiana streams. (From Melhorn, Keller, and Mc-Bane, Purdue University Water Resources Research Center, Technical Report No. 37, 1975.)

to balance intangibles with the more readily envaluated tangible aspects of landscape evaluation.

ENVIRONMENTAL IMPACT

Philosophy and Methodology

The probable effects of human use of the land are generally referred to as *environmental impact*. This term became popular in 1969 when the National Environmental Policy Act (NEPA) required that all major federal actions that could possibly affect the quality of the human environment be preceded by an evaluation of the project and its impact on the environment. To carry out both the letter and the spirit of NEPA, the Council on Environmental Quality prepared guidelines to help in preparing environmental impact statements. The major components of the statement, revised in 1979 (21), include these:

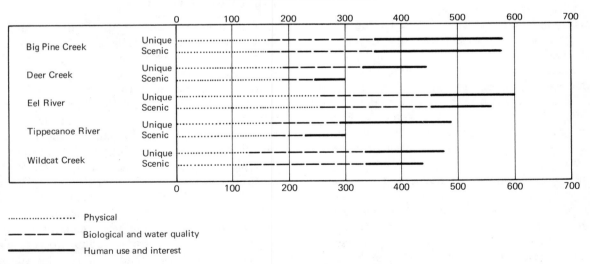

Physical
Biological and water quality
Human use and interest

FIGURE 14.24
Bar graph of scenic river indices for the same five Indiana streams shown in Figure 14.22. (From Melhorn, Keller, and McBane, Purdue University Water Resources Research Center, Technical Report No. 37, 1975.)

1 A statement concerning the purpose and need for the project.

2 A rigorous comparison of the reasonable alternatives.

3 A succinct description of the environment of the area to be affected by the proposed project.

4 A discussion of the environmental consequences of the proposed project and of the alternatives. This must include direct and indirect effects, energy requirements and conversation potential, possible depletion of resources, impact on urban quality and cultural and/or historical resources, possible conflicts with state or local land use plans, policies, and controls, and mitigation measures.

5 A list of the names and qualifications of the persons primarily responsible for preparing the environmental impact statement and a list of agencies to which the statement has been sent.

6 An index.

The environmental impact statement process was criticized during the first ten years under NEPA because it initiated a tremendous volume of paperwork by requiring detailed reports that tended to obscure important issues. In response, the revised regulations (1979) introduced two other important changes: scoping and record of decision. *Scoping* is the process of identifying important environmental issues that require detailed evaluation early in the planning of a proposed project. The *record of decision* is a concise statement by the agency that is planning the proposed project as to the alternatives considered and, specifically, which alternatives are environmentally preferable. The agency then has the responsibility to monitor the project to see that its decision is carried out. This is a significant point; for example, if, in an environmental impact statement for a proposal to construct an interstate highway, the agency commits itself to specific design and locations for the right-of-way to minimize environmental degradation, then the contracts authorizing the work must be conditional on incorporation of those designs and locations.

There is no accepted methodology for assessing the environmental impact of a particular project or action. Furthermore, because of the wide variety of topics, such as hunting of migratory birds or construction of a large reservoir, no one method of impact assessment is appropriate for all situations. What is important is that those responsible for preparing the statement strive to minimize personal bias and maximize objectivity. The analysis must be scientifically, technically, and legally defensible and prepared according to highly objective scientific inquiry standards (22).

Environmental-impact analysis for major projects or actions generally benefits from the combined effort of a task force or a team of investigators, each of whom is assigned specific resource topics or disciplines. It is important to remember that the specific function of the task force is to *evaluate*, not to decide the issue. The work of a task force is to provide information to enable those with authority to make just decisions. At this time in the development of our environmental awareness, knowledge of the nature and extent of information required to make decisions is still in the formative stage. As more and more work is done and more decisions are made, however, our concept of the critical elements of environmental information will be better understood, and eventually some will be acknowledged as requirements in the same way as cost and profit data are now accepted practice in economic analysis (22).

The spirit of NEPA arose in response to the recognized need to consider the environmental consequences of an action before its implementation. The objective is to recognize potential conflicts and problem areas as early as possible so as to minimize regrettable and expensive environmental deterioration. Two examples, the closing of a soda-ash operation in Saltville, Virginia, and the Aswan Dam in Egypt, emphasize the extremes in scale of environmental degradation that it is hoped can be avoided by careful environmental impact analysis before initiation of projects.

Saltville, Virginia

Saltville is a small, historic Appalachian mountain town on the north fork of the Holston River. Before 1972, the community was strictly a company town—the Olin Chemical Works of the Olin Corporation built and maintained almost everything in the town. The plant, which produced soda ash (sodium carbonate), prospered for about 75 years before it finally closed down because it could not economically meet new water-quality standards of the Virginia Water Control Board. The plant disposed of the chloride wastes into the north fork of the Holston River (Figure 14.25), and reportedly up to 5,000 parts per million of chlorine was measured in the river water. The salts from which the sodium was derived came from solution mining, a procedure in which subsurface salts are dissolved and pumped up from deep solution wells. The wells in Saltville were up to 900 meters deep, and the combination of available sodium and carbonate rock facilitated production of the sodium carbonate. The plant started pro-

duction when the dilution and dispersion philosophy of waste disposal was accepted, and, tragically for the town, ended production when this philosophy was no longer acceptable. Seventy-five years ago, people seldom worried about long-range environmental effects. Perhaps requirements of environmental-impact analysis will help alleviate future problems that might not be considered problems today.

High Dam at Aswan

The situation with the High Dam at Aswan and Lake Nasser in Egypt is a dramatic example of what can happen when environmental impact is ignored or inadequately evaluated. The High Dam project is causing many serious problems, including health problems that could one day be catastrophic, as the following case study, reported by Claire Sterling, demonstrates (23).

The High Dam at Aswan, completed in 1964, is one of the biggest and most expensive dams in the world. Its functions are to store a sufficient amount of the Nile's yearly floodwaters to irrigate existing land, reclaim additional land from the desert, produce electricity, and protect against drought and famine.

Unexpected water loss in the reservoir of the Aswan Dam, which can present problems in dry years, results from two factors: first, a 50-percent error in computing the evaporation loss, and second, tremendous water losses to underground water systems. The error in computation of evaporation loss, which takes an unanticipated 5 billion cubic meters of water per year, resulted from overlooking the evaporation loss induced by high-velocity winds traveling over the tremendous expanse of water in this very hot, dry region

of the earth. A chain of smaller and less expensive dams would have prevented this waste of water resources; however, even this loss is small compared to the water lost in the geologic environment (23).

It has long been known that, for hundreds of kilometers upriver from Aswan, the Nile cut across an immense sandstone aquifer that fed the Nile an incalculable amount of water. When the first and much smaller Aswan Dam was built in 1902, the flow of groundwater was reversed; the water pressure caused by the reservoir forced the water to move elsewhere through numerous fissures and fractures in the sandstone. From 1902 to 1964, when the High Dam was completed, the Aswan Reservoir stored about 5 billion cubic meters of water per year, but caused the loss of 12 billion cubic meters of water per year through reversed groundwater flow. The amount of water escaping from Lake Nasser is unknown, but because the new resevoir is designed to store 30 times more water than the old Aswan Reservoir, and because seepage tends to vary directly with lake depth, the amount of water that is being lost must be tremendous (23).

It might be argued that in time the clay settling out from the lake will plug fractures in the rock and the lake will fill, but if the fractures are very large and numerous, the water could essentially escape forever, and Egypt could end up with less water than it had before the dam was constructed (23).

Unfortunately, the direct water problems with the High Dam and Lake Nasser are only the tip of the iceberg when the total impact of the project is analyzed. There are five factors to the impact. First, the High Dam lacks sluices to transport sediment through the reservoir, and therefore the reservoir traps the Nile's sediment, 134 million tons per year, which historically has produced and replenished the fertile soils along the banks of the Nile; no practical, artificial substitute is available to counteract this. Second, marine life in the eastern Mediterranean is deprived of the nutrients in the Nile's sediment, resulting in a one-third reduction of the plankton, the food base for sardine, mackerel, lobster, and shrimp; as a result, the fishing industry in the region of the Nile delta has suffered a tremendous setback. Supposedly, an emerging fisheries industry in Lake Nasser was to compensate for the setback, but this wish has not yet developed. Third, the lake and associated canals are becoming infested with snails that carry the dreaded disease schistosomiasis (snail fever). This disease has always been a problem in Egypt, but the swift currents of the Nile floodwaters flushed out the snails each year. The tremendous expanse of waters in Lake Nasser and the irrigation canals now provide a home for these snails. Fourth, there is an increased threat of a killer malaria carried by a particular mosquito often found only 80 kilometers from the southern shores of Lake Nasser. Malaria has historically migrated down the Nile to Egypt. The last epidemic in 1942 killed 100,000 Egyptians. Authorities fear that the larger reservoir and the irrigation canal system might be invaded on a more permanent basis by the disease-carrying insects. Fifth, salinity of soils is increasing at rather alarming rates in middle and upper Egypt. Soil salinity has been a long-standing problem on the Nile delta but was alleviated upstream by the natural flushing of the salt by the floodwaters of the Nile River. Millions of dollars will have to be spent to counteract the rising soil salinity that threatens the productivity of the land.

On the positive side, the High Dam and lake have converted 2,8000 square kilometers from natural floodwater irrigation to canal irrigation, and allowed double cropping—but at tremendous cost. In the future, tremendous amounts of money will have to be spent on fertilizers to replenish soil nutrients to counteract the rise in soil salinity and to control water- and insect-borne disease.*

The Trans-Alaska Pipeline—An Example of Environmental-Impact Analysis

The controversy surrounding the Trans-Alaska Pipeline, which went into the construction phase in 1975

---

*Claire Sterling, "The Aswan Disaster." Condensed and reprinted by permission from *National Parks and Conservation Magazine* 45, August 1971, pp. 10–13. Copyright © 1971 by National Parks and Conservation Association.

and was completed in 1977, provides a good example of environmental-impact analysis. Experience gained from this project has helped set precedents and establish procedures for subsequent impact analyses.

In 1968, vast subsurface reservoirs of oil and gas were discovered near Prudhoe Bay in Alaska. Since the Arctic Ocean is frozen much of the year, it is impractical to ship the oil by tankers, and thus a 1,270-kilometer pipeline from Prudhoe Bay to Port Valdez, where tankers can dock and load oil the entire year, was suggested (22, 24).

The general route of the Trans-Alaska Pipeline is shown in Figure 14.26. More than 80 percent of the total length is across federal lands. Because of the sensitive nature of the Arctic environment and the certainty of an unknown amount of irreversible environmental degradation, a comprehensive impact analysis was required to evaluate both the negative and positive aspects of the project. A summary of the evaluation published by the U.S. Geological Survey emphasizes the critical aspects of the natural and social-economic impacts of the project (22).

The corridor for the Trans-Alaska Pipeline (Figure 14.26) traverses rough topography, large rivers, areas with extensive permafrost, and areas with a high earthquake hazard.

The pipeline crosses three *major mountain ranges:* the Brooks Range (which is the Alaskan continuation of the Rocky Mountains), the Alaska Range, and the very active (tectonically Chugach Mountains. The pipeline also crosses the Denali fault on the north side of the Alaska Range, an active fault zone that has experienced recent displacement of the ground.

The corridor also traverses a number of *large rivers,* and it is feared that scour or lateral erosion at meander bends might damage the pipeline. A study was conducted to evaluate the river crossing (25), to predict areas that might experience excessive bank erosion or scour during the expected 30-year life of the pipeline. Figure 14.27 shows the Gulkana River crossing in the Copper River Basin between the Alaska Range and the Chugach Mountains. The map shows that the lateral bank erosion of the meander bend upstream from the pipeline crossing has been rapid (over 30 meters) in the last 20 years. This meander bend is expected to be cut off during the life of the pipeline. The cutoff would shorten the course of the Copper River by about 800 meters, increasing the channel slope and causing a meter or so of scour at the pipeline crossing (25). Even smaller rivers (such as Hess Creek, approximately 104 kilometers northwest of Fairbanks) that are actively eroding laterally at a rapid rate will need bank protection to protect the pipeline crossing (Figure 14.28).

FIGURE 14.26
Approximate corridor for the Trans-Alaska Pipeline. (After U.S. Geological Survey Circular 695, 1974.)

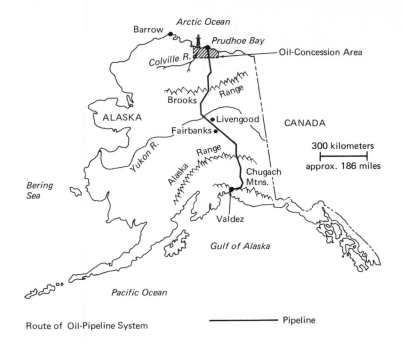

Route of Oil-Pipeline System

——————— Pipeline

The *permafrost* is also a potential hazard, because once frozen ground melts, it is extremely unstable. This aspect of the project has been thoroughly engineered, and it is hoped that problems with permafrost will be minimized. Nevertheless, the many miles of a large hot-oil pipeline crossing hundreds of kilometers of permafrost is cause for concern.

Analysis of possible physical, biological, and social-economic impacts associated with the pipeline establishes three areas of concern (22): the construction, operation, and maintenance of the pipeline system, including access roads and highways; development of the oil fields; and operation of the marine tanker system at Port Valdez. Furthermore, some of the effects

FIGURE 14.27
Gulkana River pipeline crossing near Sourdough, Alaska. Recent lateral erosion on the bends of the river suggest that a meander cutoff may occur in the next 20 years just upstream from the pipeline crossing. (After Brice, U.S. Geological Survey Publication, Alaska District, 1971.)

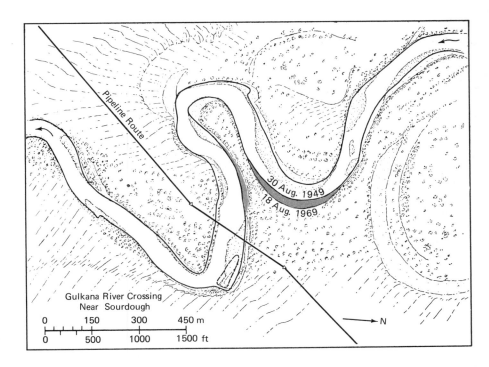

FIGURE 14.28
Tortuous meander bends of Hess
Creek, Alaska, downstream from
the pipeline crossing. (Photo
courtesy of James Brice.)

would result in predictable, unavoidable disruptions, while other effects are more speculative and present a threat, such as the impact of an oil spill caused by a break in the pipeline. These threatened effects are not easy to evaluate and predict (22).

A list of primary and secondary effects of the trans-Alaska hot-oil pipeline, gas pipeline, and tanker transport system that the task force considered to evaluate environmental impact is shown in Table 14.8. Besides these effects, it was necessary to evaluate possible links between the effects that might have an impact on the environment. For example, there are obvious possible links between oil spills or unavoidable release of oil in the tanker operation into the marine environment and the effect on marine resources. This is difficult to evaluate, however, because of the variable or unknown aspects of the hazard, the dynamic hydrologic environment, and seasonal and other changes in fish and marine resources. In general, the task force concluded that the impact linkage data were hard to obtain and evaluate and so were considered the largest problem area in quantitatively determining possible impacts (22).

Based on the environmental analysis, both unavoidable and threatened impacts of the Trans-Alaska Pipeline were determined (22). The main unavoidable effects were threefold:

1   Disturbances of terrain, fish and wildlife habitats, and human use of the land during construction, operation, and maintenance of the entire project, including the pipeline itself and access roads, highways, and other support facilities;

2   Discharge of effluents and oil into Port Valdez from the tanker transport system; and

3   Increased human pressures on services, utilities, and many other areas, including cultural changes of the native population.

Major threatening effects are accidental loss of oil from the oil field, pipeline, or tanker system. Accidental loss of oil from the pipeline could be caused by slope failure (landslides), differential settlement in permafrost areas, streambed or bank scour at river crossings, ground rupture during earthquakes, or destructive sea waves damaging the pipeline, causing a leak or rupture. Oil loss from tankers may be caused by shipwreck or accidental loss during transfer operations at Port Valdez. The potential loss of oil from the pipeline and tanker systems involves many variables, making predictions of oil loss difficult—but some loss is inevitable. Estimates place maximum oil loss at 1.6 to 6 barrels per day at Valdez for tanker operations, and 384 barrels per day from tanker accidents. The latter, of course, could be either a series of small spills or several large spills at unknown times, locations, and intervals (22).

Before the final impact statement was completed early in 1972, until construction began in 1975, the

TABLE 14.8
Primary and secondary impacts associated with the Trans-Alaska Pipeline, arctic gas pipelines, and proposed tanker system.

A. Primary effects associated with arctic pipelines:
1. Disturbance of ground
2. Disturbance of water (including treated effluent discharge into water)
3. Disturbance of air (including waste discharged to air and noise)
4. Disturbance of vegetation
5. Solid waste accumulation
6. Commitment of physical space to pipeline system and construction activities
7. Increased employment
8. Increased utilization of invested capital
9. Disturbance of fish and wildlife
10. Barrier effects on fish and wildlife
11. Scenery modification (including erosional effects)
12. Wilderness intrusion
13. Heat transmitted to or from the ground
14. Heat transmitted to or from water
15. Heat transmitted to or from air
16. Heat to or from vegetation
17. Moisture to air
18. Moisture to vegetation
19. Extraction of oil and gas
20. Bypassed sewage to water
21. Man-caused fires
22. Accidents that would amplify unavoidable impact effects
23. Small oil losses to the ground, water, and vegetation
24. Oil spills affecting marine waters
25. Oil spills affecting freshwater lakes and drainages

26. Oil spills affecting ground and vegetation
27. Oil spills affecting any combination of the foregoing

B. Secondary effects associated with arctic pipelines:
1. Thermokarst development
2. Physical habitat loss for wildlife
3. Restriction of wildlife movements
4. Effects on sports, subsistence, and commercial fisheries
5. Effects on recreational resources
6. Changes in population, economy and demands on public services in various communities, including native communities, and in native populations and economies
7. Development of ice fog and its effect on transportation
8. Effects on mineral resource exploration

C. Primary effects associated with tanker system:
1. Treated ballast water into Port Valdez
2. Vessel frequency in Port Valdez, Prince William Sound, open ocean, Puget Sound, San Francisco Bay, Southern California waters, and other ports
3. Oil spills in any of those places

D. Secondary effects associated with tanker system:
1. Effects on sports and commerical fisheries
2. Effects on recreational resources
3. Effects on population in Valdez and other communities

Source: U.S. Geological Survey Circular 695, 1974.

pipeline generated, and continues to generate, controversy. This controversy results because of the conflicts in balancing the need for resource development and the known or predicted environmental degradation. Although alternative routes and transport systems, including trans-Canada routes and railroads, other Alaska routes, and marine routes, were extensively evaluated (Figure 14.29), the earlier proposed route to Port Valdez was ultimately approved, perhaps because this route led to the most rapid resource development while maintaining national security. We emphasize, however, that no one route is superior in all respects to the others (22). Comparison of the alternative routes suggests these conclusions: first, all the trans-Alaska routes have less unavoidable adverse impact on the abiotic (nonliving) systems than the trans-Alaska-Canada routes. Second, the trans-Alaska route to the Bering Sea would probably have the least unavoidable impact on

terrestrial-biologic and social-economic systems. The trans-Alaska-Canada coastal route would be next in minimizing unavoidable impact on these systems. Third, the trans-Alaska-Canada routes would have the least unavoidable impact on marine environments because no direct marine transport of oil is involved (22).

Based on this information and many other evaluations and analyses (22), it was concluded that, to avoid the marine environment and earthquake zones, and to place both an oil and a gas pipeline in one corridor, the trans–Alaska-Canada route to Edmonton, Canada, would cause the least environmental impact. This is past history, since the route is to Port Valdez, but it shows the alternatives that were considered and the obligation of scientists to state their opinion based on sound scientific information, even though that opinion may be either unpopular or likely to be overridden in the final balancing of alternatives; and it illustrates

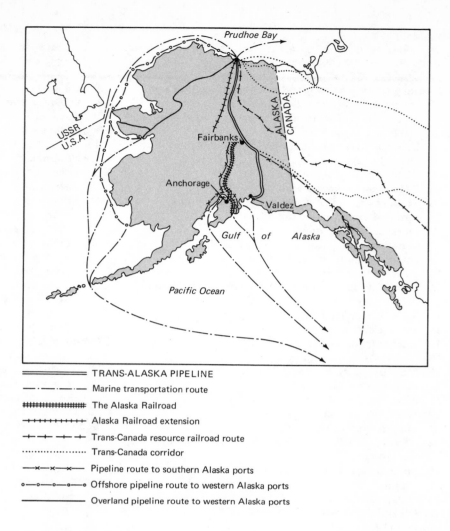

FIGURE 14.29
Alternative routes for transporting oil from the North Slope of Alaska. (From Brew, U.S. Geological Survey Circular 695, 1974.)

=========== TRANS-ALASKA PIPELINE

—·—·—·— Marine transportation route

╫╫╫╫╫╫╫╫╫╫ The Alaska Railroad

+++++++++ Alaska Railroad extension

—+——+——+—— Trans-Canada resource railroad route

··················· Trans-Canada corridor

—×—×—×— Pipeline route to southern Alaska ports

○—○—○—○—● Offshore pipeline route to western Alaska ports

——————— Overland pipeline route to western Alaska ports

the significant power of diverse political maneuvering at all levels in deciding among alternatives.

Cape Hatteras National Seashore—An Example of Environmental-Impact Analysis

The Outer Banks of North Carolina has for generations been inhabited by people living and working in a marine-dominated environment. Until recently, the way of life had depended on raising livestock, fishing, hunting, boat building, and other marine pursuits (26).

The landscape of the Outer Banks characteristically can change in a very short time in response to major storms such as hurricanes and northeasters that periodically strike the islands. Of the two types of storms, the more frequent northeasters probably cause the most erosion. On the other hand, infrequent hurricanes can cause major changes, including extensive overwash and formation of new inlets.

Historically, the people of the Outer Banks have philosophically lived with and adjusted to a changing landscape. In recent times, however, this philosophy has changed because of economic pressure to develop coastal property. A new philosophy of coastal protection has arisen in an attempt to stabilize the coastal environment. Stabilization or a constant-appearing landscape is a prerequisite to commercial development.

Congress approved in 1937 and amended in 1940 an act establishing the Cape Hatteras National Seashore as the first national seashore. The park consists of 115 square kilometers along a 120-kilometer portion of the Outer Banks (Figure 14.30) and includes portions of three islands of the more than 240 kilometers of the Barrier Island system. The Barrier Islands essentially bound and protect the largely undeveloped coastal plains lowland of North Carolina (26).

Eight unincorporated villages are bounded by the Cape Hatteras National Seashore and are spaced along nearly the entire length of the seashore. Legislation au-

FIGURE 14.30
Map showing the Outer Banks of North Carolina and the Cape Hatteras National Seashore. (After Godfrey and Godfrey, in D. R. Coates, ed., *Publications in Geomorphology* [New York: State University of New York, 1971].)

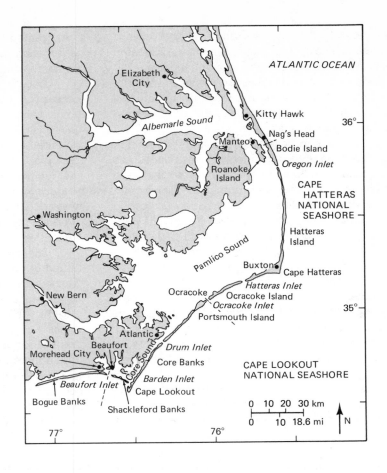

thorizing the park provided for the continued existence of these villages, including the beach in front of each of them facing the Atlantic Ocean. This legislation has been interpreted by many as an obligation of the federal government to maintain and stabilize beach access and frontage of these villages. Unfortunately, stabilization has been difficult, and in some villages, less than 60 meters of beach remains because of recent coastal erosion. This philosophy of protection is contrary to that during the early development of the islands, which was mainly on the sound (inland) side of the islands. With construction of the first dune systems in the 1930s and opening of the road link in the 1950s, however, the basic configuration of the villages began to change as communities began to spread toward the ocean, away from the more protected sites along the sound. This migration has persisted and has grown at an ever-increasing pace (26). For example, in 1953, there were only 75 oceanside subdivision lots, compared to more than 4,000 today.

At one time it was assumed that much of the Outer Banks was heavily forested and that logging and overgrazing had destroyed these forests. Evidence for this is controversial, and in fact extensive forests may

have been developed on only a few islands with an orientation more protected from storms. The wooded areas were located in an area protected by a natural dune line, so the logical conclusion was to construct an artificial barrier dune system to help the area return to a natural state by inhibiting beach erosion. Of course, the dune line would also protect the roads and communities. As a result, there is now an artificial dune system along the entire Cape Hatteras National seashore at an average dune height of 5 meters and an average distance of 100 meters from the ocean (24). This artificial erosion-control measure, along with programs to artificially keep sand on the beach, could cost about $1 million per year if continued. This expensive alternative, along with an essentially geological argument as to whether artificial dune lines will jeopardize the future of the Barrier Islands, has led to a controversy on how to best manage the park.

The geologic problem and controversy of how dynamic earth processes create and maintain barrier islands are at the philosophical heart of the U.S. Park Service's dilemma in developing a long-range program to maintain the natural environment for the use and enjoyment of present and future generations. Although there are several ideas and known ways that coastal processes can develop barrier islands, there is heated debate on the processes needed to maintain them.

It is believed that the present Barrier Islands developed about 5,000 years ago in response to complex wind and water processes associated with a slowly rising sea level. One belief is that, since many of the capes (protruding barrier islands) apparently coincide with mouths of rivers, development of these capes has been facilitated by sediment deposited as deltas during times of relatively low sea levels during the Pleistocene era (the last several million years). As the Pleistocene sea levels retreated because water was held in the continental glaciers, deltas developed seaward of the present coastline. Then, when the ice melted a few thousand years ago and sea levels began to rise, these deltas became the nucleus of the Barrier Islands and capes. Submergence was accompanied by coastal erosion until it produced the present island system, which is still eroding at a rate of 1.2 to 1.8 meters per year in response to continuing rise in sea level. Figure 14.31 shows this general model for development of barrier islands and capes (26, 27).

Although there is debate about the true role of a rising sea level in forming the barrier islands, most investigators do agree that the rise is closely associated with tremendous changes in the coastline and that this is a real and immediate hazard to future coastal development, which demands a static shoreline.

Debate on the nature and extent of geologic processes that maintain the "natural changing" Barrier Islands centers on the importance of whether periodic overwash of the frontal dune system by storm waves is essential to maintaining the islands. If the overwash is essential, then building a dune line is contrary to natural processes, and the final result over a period of years may be deterioration of the islands as a natural system. This would be contrary to the Park Service's stated objectives to preserve the islands in a natural state.

The argument that all barrier islands are maintained in natural landward migration by periodic overwash of frontal dunes is unconvincing, because most of the energy expended by storm waves is concentrated in the surf zone rather than splashing over low dunes. On the other hand, overwash of dunes on some barrier islands is a real process that does tend to facilitate a more "natural appearing" coastline. However, it has been argued that frequent overwash is atypical and occurs only on islands that are already in an unnatural state from overgrazing and subsequent lowering of dunes resulting from loss of protective vegetation. Re-

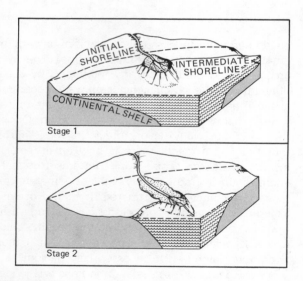
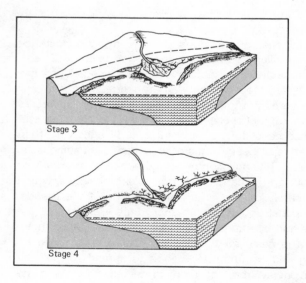

FIGURE 14.31
Idealized diagram illustrating how some capes and barrier islands may form in four stages. Stage 1 is characterized by deposition of sediment at a river's mouth due to a lowering of sea level. Stage 2 involves continued lowering of sea level and subsequent deposition of sediment. Stage 3 is characterized by a rise in sea level, continued submergence of deltaic ridges, and formation of initial sand barriers by wave action and longshore and onshore transport of sand. Stage 4 shows the present cape and barrier island morphology which has resulted from slow submergence and landward migration of earlier formed barrier island systems. (From Hoyt and Henry, *Geological Society of America Bulletin*, vol. 82, no. 1.)

gardless of the cultural influence and historic role of overwash, it is apparent that overwash is a natural process and as such is probably significant in the geologic history of barrier-island migration. This does not mean, however, that it is necessarily always bad to attempt to selectively control rapid coastal erosion by maintaining a protective frontal dune system. Proper placement of dunes and subsequent sound conservation practice in maintaining them remains a feasible alternative to short-range selected coastal erosion problems, particularly near settlements and critical communication lines.

Faced with the problem of selecting a management policy for the Cape Hatteras National Seashore, the Park Service, in 1974, presented five possible alternatives (Table 14.9) ranging from essentially no control of natural coastline processes (Alternative 1) to attempting complete protection (Alternative 5). Each alternative was analyzed to determine the entire spectrum of possible impacts, and although the final decision has not been resolved, few people want complete protection or complete control of the seashore (26). The position of the Park Service is to attempt to strike a compromise in which the Barrier Islands for the most part may be preserved in a natural state, while the need to maintain a transportation link with the mainland is recognized. Thus, the communities will have to live with and adjust to the dynamic high-risk environment they choose to live in, much as the people of the Outer Banks historically have. This is contrary to prevailing trends in coastal development that assume a more stable environment and acknowledge that natural processes play a significant role in preserving the natural environment (26).

A general management plan for the Cape Hatteras National Seashore was recently completed after several years of careful environmental impact work and public review. The plan proposes several actions:

- Natural seashore processes and dynamics will be allowed to occur except in instances when life, health, or significant cultural resources on the major transportation link are jeopardized
- Controlling use of offroad vehicles
- Expanded and easier access to recreation sites on the beach and sound
- Controlling the spread of exotic vegetation species
- Insuring that significant natural and cultural resources are preserved and maintained
- Cooperating with state and local governments in mutually beneficial planning endeavors (28)

For planning purposes, the national seashore was divided into four environmental resource units (ERUs): ocean/beach, vegetated sand flats, interior dunes/maritime forest, and marsh/sound. Table 14.10 summarizes planning objectives for each ERU, and Figure 14.32 illustrates the principle of management zoning for Hatteras Island.

The major impact of the proposed management program will be threefold. First, the Cape Hatteras National Seashore will be preserved in a natural state. Second, residents of the villages will have to live with and adjust to the effects of natural events to a greater event. Third, economic development of the villages will change, and even though the road will be maintained, it will be subject to more periodic damage.

TABLE 14.9
Possible alternative management plans for the Cape Hatteras National Seashore.

	Alternative 1
Management objectives:	To manage Cape Hatteras National Seashore as a primitive wilderness so not to impede natural processes.
Proposed action:	Let nature take its course, and oppose any attempts to control natural forces or stabilize the shoreline.
Legislative changes:	Classify Cape Hatteras as a recreation area, and officially designate the Seashore a wilderness area by administrative decree.
Major impacts:	A natural environment would be restored; the existing highway could not be maintained; property within the villages would be lost; historic structures would be destroyed unless relocated; and the economic base and economic development of the village enclaves would decline.
Costs of alternative:	No direct federal costs except possible relocation of Cape Hatteras Lighthouse and other historic structures. State costs of maintaining the road would be excessive.

TABLE 14.9 *(Continued)*

Alternative 2

Management objectives:	To manage Cape Hatteras National Seashore in accordance with the policies established for natural areas of the National Park Service, and to preserve the area in a natural state.
Proposed action:	Develop an ongoing resource management program, revegetate major overwash areas following destructive storms, and investigate alternative transportation modes.
Legislative changes:	Classify Cape Hatteras as a recreation area, and officially designate the Seashore a natural area by administrative decree.
Major impacts:	The Seashore would be preserved in a natural state, and residents of the villages would have to live with and adjust to natural events. Some property would be lost, and economic development of the villages would be retarded. Historic structures would be relocated or abandoned. The road would be damaged or washed out in places but could be maintained over the next several years.
Costs of alternative:	Estimated at $125,000 annually, or $3,125,000 over a 25-year period.

Alternative 3

Management objectives:	Continue present shore erosion-control practices.
Proposed action:	Management practices of the past—dune building, dune repair, and beach nourishment would continue.
Legislative changes:	No changes would be required.
Major impacts:	The Seashore environment would become increasingly artificial as efforts were concentrated on protecting private development. The village enclaves would continue to grow and the highway would have to be widened to serve these areas.
Costs of alternative:	Estimated at $1 million annually, or $25 million over a 25-year period.

Alternative 4

Management objectives:	To acquire threatened private property within the village enclaves to preserve a public beach fronting the villages and avoid having to protect such property.
Proposed action:	Identify threatened private property, determine fair market value, and purchase the property.
Legislative changes:	Authorize legislation for Cape Hatteras to allow for the acquisition of additional land.
Major impacts:	The economic base of the village enclaves would be reduced, and, as the shoreline continues to recede, future acquisition would be required at much greater cost. This alternative would establish the precedent of compensating unwise development.
Costs of alternative:	Very high—over $25 million for Kinnakeet Township alone (1973 prices). The land involved is the costliest on the Seashore. Loss of tax revenue and revenue derived from visitor expenditures is not included.

TABLE 14.9 *(Continued)*

	Alternative 5
Management objectives:	To protect the development taking place within the village enclaves and protect the highway running through the Seashore.
Proposed action:	Stabilize the shoreline fronting threatened private property and protect exposed sections of highway by relocating or elevating the road or by stabilizing shoreline structures or building dunes.
Legislative changes:	Amend the authorizing legislation for Cape Hatteras, deleting the section pertaining to managing the area as a primitive wilderness.
Major impacts:	The natural setting of the Seashore would be destroyed and the resource base would be seriously degraded. The village enclaves would develop into urban complexes and dominate the landscape. This alternative would acknowledge a Federal responsibility to subsidize unwise economic development.
Costs of alternative:	Overall initial cost of construction would be $40 to $56 million with annual maintenance of $3.2 to $6.4 million.

Source: National Park Service.

TABLE 14.10
Planning objectives for Cape Hatteras National Seashore Environmental Resource Units (ERU)

ERU	Characteristics	Planning Objectives
Ocean/Beach	Shifting sands, frequent overwash, limited vegetation on dunes	Allow natural processes to continue unhampered; allow for wide range of unstructured recreational activities by visitors; no construction allowed
Vegetated sand flats	Located between dune line and edge of salt-water marsh	Continue use as transportation corridor; allow development necessary to support visitor activities and resource protection
Interior dunes/ Maritime forest	Found in relatively few locations; variable topography, remoteness, dense vegetation	Maintain in natural state; allow passive recreation; design any construction to minimize impact on natural processes and systems
Marsh/Sound	Includes the sound, sound shore, and associated marshes	Maintain in natural state; provide limited access to the sound; allow limited development to support passive recreational activities

Source: National Park Service, 1984.

FIGURE 14.32
Proposed management subzones for part of the Cape Hatteras National Seashore (after National Park Service, 1984).

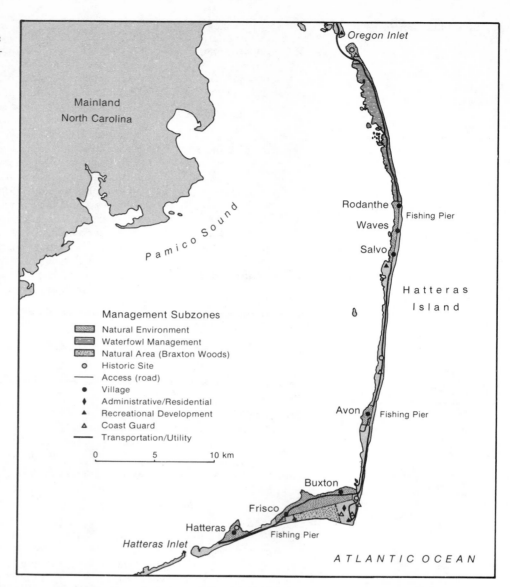

SUMMARY AND CONCLUSIONS

Landscape evaluation, including land-use planning, site selection, evaluation of landscape intangibles, and environmental-impact analysis, is one of the most controversial environmental issues of our times. Before these issues can be resolved satisfactorily, it is necessary to develop a sound methodology to insure that land and water resources are evaluated, used, and preserved in a way that is consistent with the emerging land ethic.

The geologist's role in landscape evaluation is to provide geologic information and analysis before the planning, design, or construction of projects such as reservoirs, large buildings, housing developments, tunnels, pipelines, and parks. In this respect, it is the obligation of the earth scientist to emphasize that all land is not the same and that particular physical and chemical characteristics of the landscape may be more significant to society than geographic location.

Land-use planning in the future will increasingly have to include concepts of sequential or multiple land-use rather than exclusive use. There is a definite limit to the supply of land for specific purposes, so we must strive to plan so that suitable land is available for future generations. The basic elements of land-use planning are development of a statement that includes objectives, issues, and goals; analysis and summary of pertinent data; a land classification map; and a report discussing appropriate developments.

The use of soils and engineering geology is significant in determining possible limitations to land development. This information is best presented in a series of maps that summarize geologic hazards and soil and engineering limitations for specific land uses.

Site selection and evaluation is the process of evaluating the physical environment to determine its capability of supporting human activity and, conversely, the possible effects of human activity on the environment. A philosophy of site evaluation based on "physiographic determinism" or "design with nature" has emerged as a worthy philosophical framework to partially balance the traditional economic aspects of site evaluation issues. From a geologic view, this philosophy essentially requires determining the magnitude and importance of the geologic limitations of a particular site for a particular use.

Site evaluations for engineering purposes, such as construction of dams, highways, airports, tunnels, and large buildings, require careful geologic evaluation before planning and designing the project. The role of a geologist is to work with engineers and indicate possible adverse or advantageous geologic conditions that might affect the project.

Evaluation of scenic resources and other environmental intangibles is becoming more important in landscape evaluation. The significance is in balancing intangibles with the more obvious economic aspects of environmental evaluation. The objective is to be able to quantify and rank alternatives hierarchically, as is done in the evaluation of the more tangible elements of the landscape. All methods that attempt to evaluate environmental intangibles, such as scenery, depend on analysis of landscape factors or variables, such as topographic relief, presence of water, degree of naturalness, and diversity. Thus, all the methods are somewhat subjective, although in this case subjectivity is not necessarily a bad attribute. In fact, a completely objective technique is probably impossible, given our present knowledge of landscape perception by individuals. What is important is that the evaluation be based on sound judgment.

Evaluation of environmental impact is now required by law for all federal actions that could potentially affect the quality of the human environment. The result of the evaluation is an environmental-impact statement that describes the purpose and need of the project; discusses reasonable alternatives, the environment to be affected, and environmental consequences; and considers direct and indirect effects, energy requirements and conservation potential, depletion of resources, impact on urban social systems, and possible conflicts with state or local land-use plans.

There is no one method for determining environmental impact for the wide spectrum of possible actions and projects that might affect the environment, and one method is probably not possible. The objective of the analytic process before design and construction is to minimize the possibility of causing extensive environmental degradation. In the past, rather serious problems of pollution, loss of resources, or creation of a hazard have accompanied certain projects, leading to unfortunate closings of industries and has forced people to adjust to possible hazards and economic loss.

Examples of environmental-impact analysis for the Trans-Alaska Pipeline and the management of the Cape Hatteras National Seashore are significant examples of the way possible impacts and alternatives are evaluated. Both examples stress the importance of considering the geologic aspects of the projects.

REFERENCES

1 TURNER, A. K., and COFFMAN, D. M. 1973. Geology for planning: a review of environmental geology. *Quarterly of the Colorado School of Mines* 68.

2 BARTELLI, L. J., KLINGEBIEL, A. A., BAIRD, J. V., and HEDDLESON, M. R., eds. 1966. *Soil surveys and land use planning*. Soil Science Society of America and American Society of Agronomy.

3 HAYES, W. C., and VINEYARD, J. D. 1969. *Environmental geology in town and country*. Missouri Geological Survey and Water Resources, Educational Series No. 2.

4 WILLIAM SPANGLE AND ASSOCIATES; F. BEACH LEIGHTON AND ASSOCIATES; and BAXTER, McDONALD AND COMPANY. 1976. *Earth-science information in land-use planning—guidelines for earth scientists and planners*. U.S. Geological Survey Circular 721.

5 NORTH CAROLINA COASTAL RESOURCES COMMISSION. 1975. *State guidelines for local planning in the coastal area under the Coastal Area Management Act of 1974.* Raleigh, North Carolina.

6 MONTGOMERY, P. H., and EDMINSTER, F. C. 1966. Use of soil surveys in planning for recreation. In *Soil Surveys and Land Use Planning,* ed. L. J. Bartelli et al., pp. 104–12. Soil Science Society of America and American Society of Agronomy.

7 GROSS, D. L. 1970. *Geology for planning in DeKalb County, Illinois.* Environmental Geology Notes No. 33, Illinois State Geological Survey.

8 WILSON, L. 1981. Potential for ground-water pollution in New Mexico. In *Environmental geology and hydrology in New Mexico,* S. G. Wells and W. Lambert. New Mexico Geological Society Special Publication No. 10, p. 47–54.

9 WHYTE, W. H. 1968. *The last landscape.* Garden City, New York: Doubleday.

10 FLAWN, P. T. 1970. *Environmental geology.* New York: Harper & Row.

11 LYNCH, K. 1962. *Site planning.* Cambridge, Massachusetts: M.I.T. Press.

12 PREST, A. R., and TURVEY, R. 1965. Cost-benefit analysis: a survey. *The Economic Journal* 75: 683–735.

13 McHARG, I. L. 1971. *Design with nature.* Garden City, New York: Doubleday.

14 SCHULTZ, J. R., and CLEAVES, A. B. 1955. *Geology in engineering.* New York: John Wiley & Sons.

15 KRYNINE, D. P., and JUDD, W. R. 1957. *Principles of engineering geology and geotechniques.* New York: McGraw-Hill.

16 LONGMIRE, P. A., GALLAHER, B. M., and HAWLEY, J. W. 1981. Geological, geochemical, and hydrological criteria for disposal of hazardous wastes in New Mexico. In *Environmental geology and hydrology in New Mexico,* ed. S. G. Wells and W. Lambert. New Mexico Geological Society Special Publication No. 10, p. 93–102.

17 ZUBE, E. H. 1973. Scenery as a natural resource. *Landscape Architecture* 63: 126–32.

18 LINTON, D. L. 1968. The assessment of scenery as a natural resource. *Scottish Geographical Magazine* 84: 219–38.

19 LEOPOLD, L. B. 1969. *Quantitative comparison of some aesthetic factors among rivers.* U.S. Geological Survey Circular 620.

20 MELHORN, W. N., KELLER, E. A., and McBANE, R. A. 1975. *Landscape aesthetics numerically defined (land system): Application to fluvial environments.* Purdue University Water Resources Research Center Technical Report No. 37.

21 COUNCIL ON ENVIRONMENTAL QUALITY. 1979. *Environmental quality,* Annual Report.

22 BREW, D. A. 1974. *Environmental impact analysis: The example of the proposed Trans-Alaska Pipeline.* U.S. Geological Survey Circular 695.

23 STERLING, C. 1971. The Aswan disaster. *National Parks and Conservation Magazine* 45: 10–13.

24 NATIONAL RESEARCH COUNCIL. 1972. *The earth and human affairs.* San Francisco: Canfield Press.

25 BRICE, J. C. 1971. *Measurements of lateral erosion at proposed river crossing sites of the Alaskan Pipeline.* U.S. Geological Survey, Water Resources Division, Alaska District.

26 NATIONAL PARK SERVICE. 1974. Cape Hatteras Shoreline Erosion Policy Statement, Department of Interior.

27 HOYT, J. H., and HENRY, V. J., JR. 1971. Origin of capes and shoals along the southeastern coast of the United States. *Geological Society of America Bulletin* 82: 59–66.

28 NATIONAL PARK SERVICE. 1984. *General management plan, development concept plan, and amended environmental assessment, Cape Hatteras National Seashore.*

Chapter 15

Environmental Law

HISTORY AND DEVELOPMENT OF ENVIRONMENTAL LAW

A land, or environmental, ethic is becoming an important part of our government institutions and is being incorporated into our legal and economic principles. No longer is the best use of land that which returns the greatest profit. The new land ethic establishes that the human race is part of a land community that includes trees, rocks, animals, and scenery, and we are morally bound to assure the community's continued existence. Thus, this ethic affirms our belief that this earth is our only suitable habitat (Figure 15.1) and recognizes the rights of people to breathe clean air, drink unspoiled water, and generally exist in a quality environment.

According to the National Environmental Policy Act of 1969, the intention of national environmental policy is to encourage productive and enjoyable harmony between people and their environment, to promote efforts to eliminate or at least minimize degradation of the environment, and to continue investigating relations between ecological systems and important natural resources. This policy suggests that each generation has a moral responsibility to provide the next generation with healthful, productive, and aesthetically pleasing surroundings. It implies a doctrine of trust, as-

serting that all public land and, to a lesser extent, some private land, is essentially held in trust for the general public.

The Trust Doctrine as a theory relates closely to constitutional issues involving people's right to a quality environment. Although nowhere in the Constitution does it state this right, some students of law believe that such a right can be inferred from the Ninth Amendment to the Constitution, which recognizes that the list of specific rights in the Bill of Rights does not deny the existence of other unlisted rights—that is, there are many rights not specifically listed but always held by the people (1).

The fundamental rights entitled to protection under the Ninth Amendment are considered so basic and important to our society that it is inconceivable that they are not protected from unwarranted interference (2). It can therefore be argued that this amendment encompasses the right of people to have clean air, clean water, and other resources necessary to insure a quality environment.

It is important to recognize a certain amount of friction between the Trust Doctrine, which establishes that particularly significant land resources are held in public trust, and the Fifth Amendment, which establishes that land may not be taken from an individual

without due process of law and just compensation. Problems result in interpreting law and determining how to measure just compensation. The real issue is defining at what point the private use of land infringes on the rights of other people and future generations.

Although the term *environmental law* has only recently gained common usage and there are few new laws in this field, it is fast becoming an important part of our jurisprudence (1). The many recent works devoted to environmental law attest to the fact that the subject is becoming increasingly popular among lawyers and environmental law societies, and environmental courses have recently been established at leading law schools (3).

We often take for granted the air, water, and other resources necessary for our survival. In America, we still tend to suffer from the myth of superabundance and to think of resources as inexhaustible. In large cities with accompanying large populations of people, cars, and industries, however, resources are deteriorating, and as a result, regulations and laws are necessary concerning pollution of air and water. This is not new; London's air by 1306 contained large amounts of smoke from burning coal that evidently polluted the air to the extent that a royal proclamation was issued to curtail use of coal. Violations of the law were punishable by death (3).

Air pollution was proclaimed a public nuisance as early as 1611 when an English court ruled, in essence, that property owners have the right to have the air they breathe and smell free of pollutants. The case involved a plaintiff who asked for injunction and damages because his neighbor, the defendant, raised hogs, whose odor the plaintiff considered a nuisance. The defendant was found to be creating a nuisance even though he pleaded that raising hogs was necessary for his subsistence and that neighbors should not have such delicate noses that they could not bear the smell of hogs (4). The English Nuisance Law of 1536 involved a type of common law still used in the United States, the main principle of which is that if other people suffer equally from a particular pollution, an individual cannot bring suit against the polluter. In the case of the hog farmer, it is obvious that the effect of the nuisance varied with proximity to the hog pens (3).

It is beyond the scope of our discussion to consider in detail the processes of law and how they relate to the environment. Suffice it to say that law is a technique for the ordered accomplishment of economic, social, and political purposes, and the most desirable legal technique generally is one that most quickly allows ends to be reached. Law serves primarily the major interest that dominates the culture, however, and in our sophisticated culture, the major concerns are wealth and power. These concerns stress society's ability to use the resource base to produce goods and services, and the legal system provides the vehicle to insure that productivity (5).

The general legal trend has been for the courts to render judgments based upon the greatest good to the greatest number and to use what is commonly referred to as the *Balancing Doctrine*, which asserts that the public benefit from and the importance of a particular action should be balanced against potential injury to certain individuals. The method of balancing is changing, however, and courts are now considering possible long-range injury caused by certain activities to large numbers of citizens other than the immediate complainants. This approach is more likely to insure a true balancing of the equities (5).

CASE HISTORIES

A discussion of particular case histories is useful for understanding some legal aspects of environmental problems. Our selection of cases does not imply a judgment of any particular activity, but rather indicates the considerable variability and possibilities in environmental law. The cases we will discuss include areas of air pollution, aesthetic pollution, and land use.

Two cases, *Madison* v. *Ducktown Sulfur, Copper, and Iron Company,* and *Hulbert* v. *California, Portland Cement Company,* demonstrate different approaches. The equities were balanced in one court; in another, the balancing concept was rejected (4). The *Madison* case involved artificial fumes discharged from smelting and refining companies. The effect of the pollutants in the vicinity of the plants was to kill crops and timber, thereby accelerating soil erosion (Figure 15.2). As a result, people living in the area were prevented from using their farms and homes as they did before construction and operation of the plants. The court chose to balance the equities in favor of the mining industry, declining to grant injunctive relief. The court reasoned that since it was not possible for the industry to reduce the ores in a different way or move to a more remote location, forcing them to stop air pollution would compel the company to stop operating their plants. This would cause 10,000 people to lose their jobs, destroy the tax base of the county, and make the plant properties practically worthless (4). Therefore, although damages were awarded, the plant was allowed to continue operation.

In the *Hulbert* case, the plaintiffs argued for injunction requiring the cement company to stop discharging the cement dust that was falling on their properties. The dust contaminated the plaintiffs' homes and also formed an encrustation on the upper sides of flowers and other plants, especially the plaintiffs' citrus fruit trees. The dust was durable and not removed by natural processes such as strong winds or rain. As a result, the value of the citrus fruit declined, and it was more difficult and expensive to cultivate and harvest the crop. The company argued that they were doing everything possible to stop the discharge of dust and that damages were sufficient to compensate the plaintiffs for their injuries. They also argued that the large company payroll and other benefits to the community must be balanced. Nevertheless, the court rejected the Balancing Doctrine and granted the injunction (4).

When we compare the *Ducktown* and *California* cases, it is important to recognize three facts. First, compared to the cement rock (limestone), the citrus fruit has a relatively high unit value and can be grown only in certain favorable areas. Second, the cement rock has a lower unit value than the Ducktown ores and is found in many more locations. Third, reclamation of an area mined for cement rock is possible and would retain land for other productive uses, an alternative that is neither likely nor very feasible for Ducktown.

Water, land, and air pollution are relatively easy to measure and evaluate compared to the intangible aesthetic pollution of visual, audio, and other senses that affect one's well-being. We do recognize some of

FIGURE 15.2
Fumes from the smelter at Ducktown, Tennessee, killed nearby vegetation and initiated a rapid erosion cycle. (Photo by A. Keith, courtesy of U.S. Geological Survey.)

the possible harmful effects of aesthetic pollution, however, and the courts are forming attitudes on these matters. The precedent that aesthetics is a valid subject of legislative concern was set in New York courts in cases that involved aesthetic degradation, particularly loss of value of property when the view of a lake or mountain was spoiled. It was recognized that reduction in property value from loss of a view that otherwise increased the value of the property should not be borne by the owner whose land was taken for public purposes without permission (3).

Two examples from New York concerning road construction emphasize that the law of aesthetics is becoming more important and that intangible factors that were formerly considered outside the public interest are now as important as other more easily measured factors. In *Clarance* v. *State of New York,* the court awarded the plaintiff $10,000 because a road degraded his view of Seneca Lake and denied him easy access to the lake (3). In a similar case, *Dennison* v. *State of New York,* a property owner was awarded damages because a new highway caused loss of privacy and seclusion, loss of view of forests and mountains, traffic noise, lights, and odors. The court concluded that all these factors tended to cause damage to the plaintiff's property (3).

The Storm King Mountain dispute is a classic example of possible conflict between a utility company and a conservationist. In 1962, the Consolidated Edison Company of New York announced plans for a hydroelectric project approximately 64 kilometers north of New York City in the Hudson River Highlands, an area considered by many to have unique aesthetic value because nowhere else in the eastern United States is there a major river eroded through the Appalachian Mountains at sea level, giving the effect of a fjord (6). Early plans for the project called for a powerhouse to be constructed aboveground, requiring a deep cut into

Storm King Mountain. The project was redesigned to site the powerhouse entirely underground, eliminating the cut on the mountain (Figure 15.3). Regardless, conservationists continued to oppose the project, and the issues have now broadened to include possible damage to fishery industries. The argument is that the high rate of water intake from the river, 31,200 cubic meters per minute, would draw many fish larva into the plant where they would be destroyed by turbulence and abrasion. The most valuable sport fish is the striped bass, and one study showed that 25 to 75 percent of the annual bass hatch might be destroyed if the plant were operating. The fish return from the ocean to tidal water to spawn, and since the Hudson River is the only estuary north of Chesapeake Bay where the striped bass spawn, concern for the safety of the fisheries is real. The problem is even more severe because the proposed plant is located near the lower end of the 13-kilometer reach in which the fish spawn. The project is still under study, and it is the responsibility of the utility company to show that the project will not do unacceptable damage to the fishery resource (6).

After a decade of legal activity, the Storm King Mountain controversy is still unresolved. The case is interesting because it emphasizes the difficulty in making decisions about multidimensional issues. On one hand, a utility company is trying to survive in New York City where there are unbelievably high peak-power demands accompanied by high labor and maintenance costs. On the other hand, conservationists are fighting to preserve our beautiful landscape and fishery resources. Both have legitimate arguments, but in light of their special interests, it is difficult to resolve the conflicts. Present law and procedures are sufficient to resolve the issues, but trade-offs will be necessary. Ultimately, an economic and environmental price must be paid for any decision, and this reflects our desired life-style and standard of living.

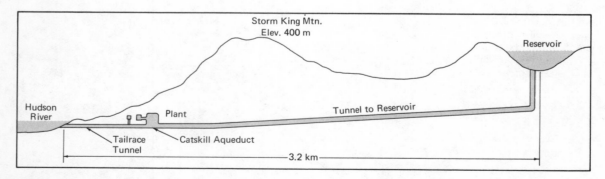

FIGURE 15.3
Idealized diagram showing how the entire Storm King Mountain hydroelectric project might be placed underground. (After L. J. Carter, *Science,* vol. 184, June 1974, pp. 1353–58. Copyright 1974, American Association for the Advancement of Science.)

A case reported by Yannacone, Cohen, and Davison in Colorado emphasizes the signficance of the Trust Doctrine and the Ninth Amendment as they pertain to land management (2). The conflict surrounded the use of 7.3 square kilometers of land near Colorado Springs. The land is part of the Florissant Fossil Beds where insect bodies, seeds, leaves, and plants were deposited in an ancient lake bed about 30 million years ago. Today, they are remarkably preserved in thin layers of volcanic shale. Unfortunately, the fossils are delicate, and unless protected, tend to disintegrate when exposed. Many people consider the fossils unique and irreplaceable. At the time of the controversy, a bill had been introduced into Congress to establish a Florissant Beds National Monument. The bill had passed the Senate, but the House of Representatives had not yet acted on it.

While the House of Representatives was deliberating the bill, a land development company that had contracted to purchase and develop recreational homesites on 7.3 square kilometers of the ancient lake bed announced that it was going to bulldoze a road through a portion of the proposed national monument site to gain access to the property it wished to develop. A citizens' group formed to fight the development until the House acted on the bill. The group tried to obtain a temporary restraining order, which was first denied because there was no law preventing the owner of the property from using that land in any way he wished provided that existing laws were upheld. The conservationists then went before an appeals court and argued that even though there was no law protecting the fossils, they were subject to protection under the Trust Doctrine and the Ninth Amendment. The argument was that protection of an irreplaceable, unique fossil resource was an unwritten right retained by the people under the Ninth Amendment, and that furthermore, since the property had tremendous public interest, it was also protected by the Trust Doctrine. An analogy used by the plaintiffs was that if a property owner were to find the Constitution of the United States buried on the land and wanted to use it to mop the floor, certainly that person would be restrained. After several more hearings on the case, the court issued a restraining order to halt development; shortly thereafter, the bill to establish a national monument was passed by Congress and signed by the president (2).

The court order prohibiting destruction of the fossil beds may have deprived a landowner of making the most profitable use of the property, but it does not prohibit all uses consistent with protecting the fossils. For instance, the property owners are free to develop the land for tourism or scientific research. While this might not result in the largest possible return on the property owner's investment, it probably would return a reasonable profit.*

ENVIRONMENTAL LEGISLATION

The ultimate goal of those who are concerned with how we treat and use our natural environment and resources is to insure that ecologically sound, responsible, socially acceptable, economically possible, and politically feasible legislation is passed (2). To attain this, we need professional people to assist legislators at all levels of government in drafting the needed laws and regulations.

Environmental legislation has already had a tremendous impact on the industrial community. New standards in regulations limiting the discharge of possible pollutants into the environment have placed restrictions on industrial activity. Furthermore, any new activity that directly or indirectly involves the federal government must be preceded by an evaluation of the environmental impact of the proposed activity. Beyond this, federal legislation has set an example that many states are following in passing environmental protection legislation.

Examining all federal legislation that has environmental implications is beyond the scope of our discussion; however, discussing some of the major acts is valuable in understanding the basics of such legislation. For this purpose, we will discuss the Refuse Act of 1899, the Fish and Wildlife Coordination Act of 1958, the National Environment Policy Act of 1969, the Water Quality Improvement Act of 1970, the Resource Conservation and Recovery Act of 1976, the Surface Mining Control and Reclamation Act of 1977, and the Clean Water Act of 1977.

The Refuse Act of 1899

This act states that it is unlawful to throw, discharge, or deposit any type of refuse from any source except that flowing from streets and sewers into any navigable water. Furthermore, the act implies that it is unlawful to discharge refuse into tributaries of navigable water. For all practical purposes, this means that it is against the law to pollute any stream in the United States. However, the Secretary of the Army can allow the discharge of refuse into a stream if a permit is first applied for.

---

*Yannacone et al., *Environmental Rights and Remedies* (San Francisco: Bancroft-Whitney, 1972), pp. 39–46.

The Fish and Wildlife Coordination Act of 1958

This act establishes a national policy that recognizes the important contribution of wildlife resources and specifically provides that conservation of wildlife shall be balanced with other factors in water resources development planning (2).

Wildlife resources in the act are broadly defined to include birds, fish, mammals, and all other types of animals, as well as aquatic and land vegetation upon which the wildlife depends. The act also provides for coordination of wildlife aspects of resource development and requires that projects to develop power, control flooding, or facilitate navigation must first consult with the U.S. Fish and Wildlife Service, as well as with the head of the state agency exercising administrative control of wildlife resources in the particular state where the project is planned. This applies, with the exception of impoundments of less than 4 hectares, to the waters of any stream or other body of water to be impounded, ditched, diverted, or otherwise modified for any purpose by any department or agency of the United States (2).

The object of the Fish and Wildlife Coordination Act is to prevent damage or loss of wildlife resources from water resources development and, at the same time, provide for development and improvement of wildlife and necessary habitat. Reports by the Fish and Wildlife Service, along with those of state agencies, provide details of expectable damage to wildlife resources from a particular project, and include recommendations as to how to reduce the projected damages and develop and improve the fish and wildlife resources. According to the act, agencies that receive the reports must consider these recommendations fully and include in the project plans ways and means to achieve wildlife conservation (2).

The National Environment Policy Act of 1969

The philosophical purposes of the Environment Policy Act are: to declare a national policy that will encourage harmony with our physical environment; to promote efforts that prevent or eliminate environmental degradation, thereby stimulating human health and welfare; and to improve our understanding of relations between ecological systems and important natural resources. To promote interest, research, and authority to achieve these purposes, the act establishes the Council on Environmental Quality. The Council is in the Executive Office of the President and is responsible for preparation of a yearly Environmental Quality Report to the

Nation. It also provides advice and assistance to the president on environmental policies.

The most significant aspect of the act is that it requires an environmental-impact statement before major federal actions are taken that could significantly affect the quality of the human environment. This requirement extends to activities such as construction of nuclear facilities, airports, federally assisted highways, electric power plants, and bridges; release of pollutants into navigable waters and their tributaries; and resource development on federal lands, including mining leases, drilling permits, and other uses. Since enactment of the National Environment Policy, many thousands of environmental impact statements have been prepared.

The Water Quality Improvement Act of 1970

This act, a comprehensive water-pollution control law, essentially gives more power to the Federal Water and Pollution Control Act of 1956. The purpose of the 1956 Act was to enhance the quality and value of our water resources and to establish a national policy to prevent, control, and abate pollution of the country's water resources. The 1970 Act provides for control of oil pollution by vessels and offshore and onshore oil wells, control of hazardous pollutants other than oil, control of sewage from vessels, research and development methods to control and eliminate pollution of the Great Lakes, research grants to universities and scholarships to students to train people in water quality control, and projects to demonstrate methods for eliminating and controlling mine drainage of acid water from both active and abandoned mines (2).

The main regulatory function of the Water Quality Improvement Act of 1970 requires that all facilities or activities that involve discharge into navigable water and that require a federal license must obtain a certificate of reasonable assurance that the proposed activity will not violate the state's water quality standards as approved by the Environmental Protection Agency.

Resource Conservation and Recovery Act of 1976

The purpose of this act is to control hazardous wastes and protect human health. When the act is fully implemented, it will provide for "cradle to grave" control of hazardous wastes. At the heart of the act is the identification of hazardous wastes and their life cycles. Regulations will then require stringent record keeping and reporting to insure that wastes do not present a public nuisance or health problem.

Surface Mining Control and Reclamation Act of 1977

The purpose of the act is to control the environmental effects of strip mining. The act prohibits mining practices that have in the past led to environmental degradation. Reclamation of the land after mining is required. Although the act is making great improvements in the way strip mining is conducted, it has been criticized because it does not sufficiently allow for site-specific conditions in the regulations that control mining and reclamation.

Clean Water Act of 1977

The purpose of this act is to clean up the nation's water. In particular, most municipal sewage treatment plants are to achieve secondary treatment or use the best practicable waste treatment technology. Billions of dollars in federal grants have been awarded to meet these goals. The act clearly encourages innovative and alternative techniques in water treatment and waste disposal, such as land application of sewage sludge, aquifer recharge of treated waste water, and energy recovery. Results of water treatment nationwide have been encouraging. For example, in the 1950s and 1960s, the Detroit River was considered dead. As a result of water treatment, fishermen now catch walleye pike, muskellunge, smallmouth bass, salmon, and trout. Although the Detroit River is still not a really clean river, improvements continue as the discharge of pollutants is reduced.

State and Local Environmental Legislation

The far-reaching federal legislation program is a good example for state and local governments to follow. In fact, a number of states have already enacted legislation analogous to the federal laws, and this trend is expected to continue. Areas of particular public concern at the state and local levels are environmental impact, floodway regulation, sediment control, land use, and water and air quality. The most effective programs and laws, excepting federal activity, will probably be at the local level, because acceptance and enforcement, when combined with education and communication, are probably most effective where environmental degradation is experienced firsthand.

An example of important state and county legislation comes from North Carolina. State legislators recognized that sedimentation of streams, lakes, and ponds is a major pollution problem and that most sediment pollution in urbanizing areas derives from erosion and deposition of sediment associated with con-

struction sites and road maintenance. Thus, they passed the North Carolina Sediment Pollution Control Act of 1973. The act establishes that it is vital to the public interest and necessary for public health and welfare that control of erosion and sedimentation be an important goal. Furthermore, the act provides for creation, administration, and enforcement of a sediment-control program complete with minimal mandatory standards that allow future development with minimal detrimental effects from sediment pollution.

The Sediment Pollution Control Act of 1973 in North Carolina recognizes the need for local participation and encourages local government to draft ordinances consistent with the act. Mecklenburg County, part of the fourth-largest urbanizing area in the southeastern United States, has drafted a sediment-control ordinance that calls for submitting erosion- and sediment-control plans before land-disturbing activities are begun, if more than one contiguous acre is to be uncovered. These plans must be approved by the County Engineer and reviewed by the Soils and Water Conservation District Office for comments and recommendations. An example of such a plan is shown in Figure 15.4.

Communication and education are the keys to success with the sediment-control ordinance. Therefore, a significant number of activities, including urban field trips and workshops on sediment control, are part of an ongoing program. The idea is not to punish developers into sediment control, but to indicate the benefits of sound conservation practices.

WATER LAW

Water is so necessary to all aspects of human use and interest that water resources may be the most legislated and discussed commodity in the arena of environmental law. Struggles for sufficient water by populous areas such as New York City and southern California are examples. Intrastate and legislative contracts were necessary to obtain water for New York City from the Delaware River waters in the Catskill Mountains. In California, there has been a 50-year fight between California and Arizona over Colorado River water (3).

There will always be problems in allocating water resources, and the problem is most severe in areas such as the southwestern United States where water is scarce (Figures 9.6 and 9.7). California is a good example of the type of conflicts that arise when a large population with accompanying industrial and agricultural activity is concentrated in an area with a natural deficiency of water. Approximately two-thirds of the

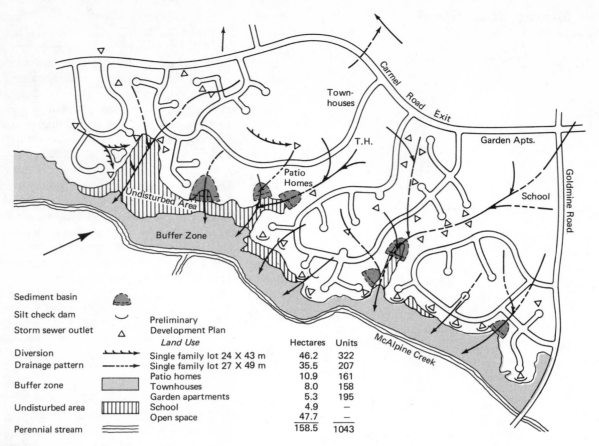

Land Use	Hectares	Units
Single family lot 24 X 43 m	46.2	322
Single family lot 27 X 49 m	35.5	207
Patio homes	10.9	161
Townhouses	8.0	158
Garden apartments	5.3	195
School	4.9	—
Open space	47.7	—
	158.5	1043

FIGURE 15.4
Erosion- and sediment-control plan for a housing development including single-family
homes, patio homes, townhouses, garden apartments, a school, and open space. (Courtesy
of Braxton Williams, Soil Conservation Service.)

state's water supply comes from the northern third of
California, but the greatest need for water is in the
southern two-thirds of the state where the vast majority of people live and where most of the industry and
agricultural activity takes place (7).

The deficiency of water in southern California,
combined with an almost insatiable demand for water,
resulted in construction of the California aqueduct to
move water from the northern and southeastern parts
of the state to the Los Angeles area (Figure 15.5). The
legal grounds that allowed California citizens to vote for
and pass a state bond of nearly $2 billion for the Cali-
fornia aqueduct that now transports water from the
northern part of the state to the southern (Figure 15.6)
was derived from the Constitution of the State of California: "The general welfare requires that water resources of the state be put to beneficial use to the full-
est extent of which they are capable . . . that the
conservation of such water is to be exercised with a
view to the reasonable and beneficial use thereof in the
interest of the people and for the public welfare" (3).

From a philosophical viewpoint, we might raise the
question, "Is continued development in southern California warranted?" Perhaps people should be located
where the water is, rather than moving huge quantities
of water hundreds of kilometers over rough terrain and
active faults, at tremendous cost to an area already over-
crowded and suffering from pollution and other environmental problems. Value judgments in these matters
have little validity, and the fact that the aqueduct was
constructed indicates the price people are willing to
pay to support their standard and style of living. We
should point out, however, that many of the people in
the Feather River and Owens River areas from where
the water transported to Los Angeles is diverted are
not so pleased as those who receive the water.

There is a rather elaborate framework of law surrounding the use of surface water, two major aspects
of which are the Riparian Doctrine and the Appropriation Doctrine.

Riparian rights to water are restricted to owners
of the land adjoining a stream of standing water. The

FIGURE 15.5
Map of California showing the major aqueducts supplying water to Southern California. (From State of California, Department of Water Resources.)

word *riparian* comes from the Latin word meaning *bank,* and traditional Riparian Doctrine is a common law concept. It holds, essentially, that each landowner has the right to make reasonable use of water on his or her land, provided that the water is returned to its natural stream channel before it leaves the property. The property owner also has the right to receive the full flow of the stream undiminished in quantity and quality, but is not entitled to make withdrawals of water that infringe upon the rights of other Riparian owners (8).

The Riparian Doctrine was the prevailing water law in most states before 1850, and is still used in all the states east of the Mississippi and in the first tier of states immediately west of the river (Figure 15.7). The right to use water is considered real property, but the water itself does not belong to the property owner. Riparian water rights are considered natural rights and property that enter into the value of land. They may be transferred, sold, or granted to other people (9).

The Appropriation Doctrine in water law holds that prior usage is a significant factor. That is, the first

to use the water for beneficial purpose is prior in right. This right is perfected by use and is lost if beneficial use ceases (9).

Appropriation water law is common in the western part of the United States, and, generally, states with the poorest water supply manage their water most closely. Arizona is a good example: with an average precipitation of less than 38 centimeters per year, Arizona must, of necessity, manage its water very closely. The state constitution says that riparian water rights are not authorized, and the state's comprehensive water code declares that all water is subject to appropriation. Preferred uses are domestic, municipal, and irrigation. Colorado also has a limited water supply and has declared that all streams are considered public property subject to appropriation (9).

Comparison of the two doctrines suggests that management of water resources is considerably more effective when the principles of appropriation are ap-

FIGURE 15.6
California aqueduct in the San Joaquin Valley, California. (Photo courtesy of State of California, Department of Water Resources.)

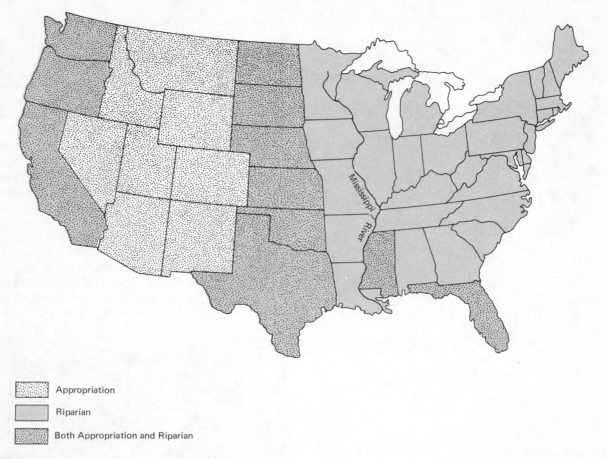

Appropriation

Riparian

Both Appropriation and Riparian

FIGURE 15.7
Surface water laws for the contiguous United States. (Data from New Mexico Bureau of
Mines, Circular 95, 1968.)

plied, because the riparian law requires a judicial decision and is therefore subject to possible variations
and interpretations in different courts. As a result,
property owners are never sure of their position. The
riparian system also tends to encourage nonuse of water and is thus counterproductive in times of shortage.
On the other hand, states with an appropriation system
have the power to make and enforce regulations based
on sound hydrologic principles, which is more likely to
lead to effective management of water resources (9).

LAW OF THE SEABED

For twenty years, there has been a controversy surrounding mining of the seabed. Industrial nations with
international companies interested in mining the
seabed are at odds with developing countries who argue that mineral wealth on the bottom of the sea is a
common heritage to all people on earth, consistent
with a United Nations resolution on the issue of seabed

resources. Five international consortia of companies
have spent hundreds of millions of dollars on research
and exploration in preparation for mining nodules containing manganese, cobalt, nickel and copper (see
chapter 12), and naturally want to be sure that their
investment is returned. Nevertheless, developing nations, wanting a relatively large share of the mineral
wealth, have generally taken a slow-growth-develop-
ment position at Law of the Sea Conferences, a United
Nations' sponsored group of 150 nations.

Lack of action in ratifying a treaty (international
law) concerning mining of the seabed has led to threats
by industrial nations to begin mining without an agreement. Some developing nations have responded with
threats to cut off oil, metals, and other resources to industrial nations that mine the seabed without approval
of the nations in the United Nations group. In spite of
this, the United States Congress has passed legislation
authorizing American companies to mine the seabed.
Although the controversy will go on, it is expected that
plans to mine will continue and developing countries

will eventually modify their demands in light of the tremendous amount of money industrial nations are spending on mining technology and exploration.

LAND-USE PLANNING AND LAW

Few environmental topics are as controversial as land-use planning legislation. The controversy results from several factors. First, unlike air and water pollution that can be measured, evaluated, and possibly corrected, it is very difficult to determine the ''highest and best use'' of the natural environment as opposed to the ''most profitable use.'' Second, landowners fear that land-use planning would take away their right to decide what to do with their property; that is, the idea of individual ownership of property could be converted to a social property ownership in which the individual would have a caretaker role. Third, there is considerable concern over the idea of the federal or state governments' initiating local land-use planning.

People in the United States greatly value private ownership of land, so a law that requires rural land-owners to use their lands in very restricted ways is not likely to be accepted. The topic of private ownership also includes a philosophical question that bothers some people: will enforced land-use planning bring us closer to socialism? If it does, then perhaps it might encourage a basic change in the philosophy of government. Those who state this argument point out that private ownership of property is a sacred right, and only in those countries where private ownership is permitted are there also other personal rights. Everything considered, it is extremely unlikely that private ownership in the United States will be abolished. On the other hand, it is increasingly apparent that private ownership of land does not mean the owner has the right to deliberately or inadvertently degrade the environment. To be effective, land-use planning must leave the property owner certain alternatives of land management from which to choose freely.

It seems inevitable that a federal land-use planning act will eventually become law, with the purpose of assuring that all the land in the nation will be used in such a way as to facilitate harmonious existence with our natural environment, such that environmental, economic, social, and other requirements of present or future generations are not degraded. Legislation will probably require individual states to establish land-use planning agencies to initiate planning or to approve local government planning.

The argument in favor of a federal land-use planning act stresses that such planning is urgently needed to minimize degradation of land resources of statewide or national importance. Those in favor of such an act point out that land use is the most important aspect of environmental quality control that remains to be stated as national policy (10).

Alternative federal roles to a rigidly controlled land-use planning act might include these possibilities: increased federal money to sponsor continued research and dissemination of information programs in the area of land-use planning; federal funds to underwrite selected experiments in guiding growth and land use; and finally, if Congress should determine that it is federal responsibility to recommend land-use planning at the state and local levels then perhaps a special revenue-sharing measure might be appropriate (11).

A number of states are now considering, or have passed, land-use planning legislation. Table 15.1 summarizes land-management programs in the United States. For example, in 1970, the Maine legislature enacted a site law, the objective of which is to maintain the highest and best use of the natural environment. The law was drafted to regulate large developments in such a way that the burden of proof was shifted to the developer; that is, the developer must show that the project will not degrade the land, water, or air. A recent test of the law in court upheld the legislature's policy to regulate land use. This decision was significant because it strongly reaffirmed the right of states to limit property use so as to preserve the environment (12).

In 1970, Vermont also passed a state land-use program that required developers, before they could receive a development permit, to prove to a district environment commission that their project: **(1)** would not cause undue air or water pollution; **(2)** could obtain sufficient water; **(3)** would not result in excessive soil erosion, highway congestion, or municipal service and educational burden; and **(4)** would not have undue detrimental effects on the natural scenery, historical sites, or rare and irreplaceable natural areas. In addition, the development would have to conform to existing local or regional land-use plans as well as any subsequently adopted plan (11). The law also required establishment of an environmental board to hear appeals in the permit procedure and to prepare a state land-use plan consisting of a map showing broad categories of the proper use of all the land in the state. In discussing the Vermont law, one author says the environmental problems that led to the law (that is, shoddy construction, defective sewage systems, poorly designed roads, etc.) could have been solved at the local level, and that the land-use law is unnecessary and in fact produces many more problems (11). This may be correct, for as the author points out, the Vermont legislature in 1974 stripped the environmental board of responsibility to develop a state land-use plan and created a land-use study commission to pursue the matter. The law might have worked better had it involved more local planning at the county and city levels; that is, local planning tends to be more detailed than a broad-brush state plan that may be difficult to interpret.

California citizens concerned about coastal areas (Figure 15.8) voted in 1972 to require a permit for any development taking place in a zone extending from approximately 900 meters inland from the sea's high tide to about 5 kilometers seaward. The law also calls for a master plan for coastal development to be prepared and adopted as state policy.

North Carolina, recognizing the importance of planning for a fragile environment likely to be heavily

TABLE 15.1
State land use programs as of 1976

| State | Type of State Program | | | Coastal Zone Management[4] | Wetlands Management[5] | Power Plant Siting[6] | Surface Mining[7] | Designation of Critical Areas[8] | Differential Assessment Laws[9] | Floodplain Management[10] | Statewide Shorelands Act[11] |
	Comprehensive Permit System[1]	Coordinated Incremental[2]	Mandatory Local Planning[3]								
Alabama				X		X	A			X	
Alaska		X		X		X			B		
Arizona		X				X			A	X	
Arkansas						X	A, B		A	X	
California		X		X		X	X		C	X	
Colorado						X	X	X	A	X	
Connecticut		X		X	X	X			B	X	
Delaware		X		X	X				A		X
Florida	X	X	X	X	X	X	A	X	A, C		
Georgia		X		X	X		A, B				
Hawaii	X	X		X		X	X	X	B	X	
Idaho			X				X		A		
Illinois				X		X	A, B		B	X	
Indiana		X		X			A, B		A	X	
Iowa							A, B		A	X	
Kansas							A, B				
Kentucky						X	A, B		B		
Louisiana				X	X						
Maine	X	X	X (LTD)	X	X	X	A	X	B	X	
Maryland		X		X	X	X	A, B	X	B	X	
Massachusetts				X	X	X			B		
Michigan				X			X		C	X	X
Minnesota		X		X	X	X	X	X	B	X	X
Mississippi				X	X					X	
Missouri						X	X	X	A	X	

[1]State has authority to require permits for certain types of development

[2]State-established mechanism to coordinate state land use-related problems

[3]State requires local governments to establish a mechanism for land-use planning (e.g., zoning comprehensive plan, planning commission)

[4]State is participating in the federally funded coastal zone management program authorized by Coastal Zone Management Act of 1972

[5]State has authority to plan or review local plans or ability to control land use in the wetlands

[6]State has authority to determine siting of power plants and related facilities

[7]State has statutory authority to regulate surface mines—(A) State has adopted rules and regulations; (B) State has issued technical guidelines

developed, passed the Coastal Area Management Act of 1974. This act recognizes the need to preserve for the people an opportunity to enjoy the aesthetic, cultural, and recreational quality of the natural coastlines. The first two objectives of the act are to provide a management system that allows preservation of the estuaries, barrier islands, sand dunes, and beaches such that their natural productivity and biological, economic, and aesthetic value are safeguarded, and to insure that development in the coastal area does not exceed the capabilities of the land and water resources.

The North Carolina act establishes a cooperative program of coastal management between local and state governments such that local government has the

TABLE 15.1,
continued

State	Comprehensive Permit System[1]	Coordinated Incremental[2]	Mandatory Local Planning[3]	Coastal Zone Management[4]	Wetlands Management[5]	Power Plant Siting[6]	Surface Mining[7]	Designation of Critical Areas[8]	Differential Assessment Laws[9]	Floodplain Management[10]	Statewide Shorelands Act[11]
Montana		X	X			X	A, B	X	B	X	X
Nebraska		X				X			B	X	
Nevada		X	X			X		X	B		
New Hampshire				X	X	X			B, C		
New Jersey				X	X	X			B	X	
New Mexico		X				X	A		A		
New York	X	X		X	X	X	X	X	B	X	
North Carolina		X		X	X	X			B	X	
North Dakota						X	A		A		
Ohio				X		X	A		B		
Oklahoma							X		A	X	
Oregon		X	X	X		X	A	X	B		
Pennsylvania				X	X	X	A	X	B		
Rhode Island		X		X	X	X			B		
South Carolina				X		X	A		B		
South Dakota							A	X	A		
Tennessee						X	A, B				
Texas				X	X	X			B		
Utah		X					A		B		
Vermont	X	X			X	X	X		C	X	
Virginia			X	X	X		A, B		B		
Washington		X		X	X	X	A		B	X	X
West Virginia							A, B			X	
Wisconsin		X		X	X	X	X	X		X	
Wyoming		X	X			X	A		A		

[8]State has established rules, or is in the process of establishing rules, regulations, and guidelines for identification and designation of areas of critical state concern (e.g., environmentally fragile areas, areas of historical significance)

[9]State has adopted tax measure designed to give property tax relief to owners of agricultural or open space lands; (A) Preferential Assessment Program—Assessment of eligible land is based on a selected formula, usually use-value; (B) Deferred Taxation—Assessments of eligible land is based on a selected formula, usually use-value, and provides for a sanction, usually payment of back taxes, if the land is converted to a noneligible use; (C) Restrictive Agreements—Eligible land is assessed at its use-value, a requirement that the owner sign a contract, and a sanction, usually payment of back taxes if the owner violates the terms of the agreement

[10]State has legislation authorizing regulation of floodplains

[11]State has legislation authorizing regulation of shorelands of significant bodies of water

Source: Council of State Governors.

FIGURE 15.8
Citizens' concern for coastal areas prompted legislation calling for planning in the coastal environment of California. This oblique aerial photograph is of Morro Bay, California. (Photo courtesy of California Division of Beaches and Parks.)

initiative in the planning. The role of state government is to set standards and to review and support local government in its planning program. The planning has three main steps: development of state planning guidelines for coastal areas; development and adoption of a land-use plan for each county in the coastal area; and use of plans as criteria for issuing or denying permits to develop land or water resources in the coastal area.

SUMMARY AND CONCLUSIONS

A land ethic is emerging in our national environmental policy. Productive and enjoyable harmony with our physical environment is now encouraged, and legal theories based upon the Trust Doctrine or the Ninth Amendment are being used to argue in court for the right of this generation and future generations to an unspoiled land.

The term *environmental law* has only recently gained common usage, but the subject is becoming increasingly popular among lawyers and promises to be more significant in the future. Topics of particular concern to individuals and society are degradation of air, water, scenery, and other natural resources.

Environmental legislation is having a tremendous impact on the industrial community. In addition to regulations controlling emissions of possible pollutants, perhaps the most significant and far-reaching piece of environmental legislation is the National Environmental Policy Act of 1969. The act requires that before any environment-affecting activity that is directly or indirectly involved with federal government can begin, a statement evaluating the environmental impact must be completed. Other important legislation includes the Refuse Act of 1899, the Fish and Wildlife Coordination Act of 1958, the Water Quality Improvement Act of 1970, the Resource Conservation and Recovery Act of 1976, and the Clean Water Act of 1977. Many states are following the federal government's leadership and are developing environmental and land-management legislation.

Areas of particular concern at the state level are environmental impact, floodway regulation, sediment control, land use, and air and water quality.

Water law remains a significant issue, particularly in regions with deficiencies of water. Eastern states generally have what are known as *riparian rights,* whereby owners of land that adjoins water have the right to reasonable use. In the western states, water law is generally governed by the *Appropriation Doctrine,* which holds that prior beneficial usage is the key to water rights. In some states, such as Arizona and Colorado, where water is especially valuable, all water is appropriated on the basis of prior and preferred uses. In comparing the two systems, one can conclude that water appropriation is superior because it leads to better management of water resources.

Perhaps the most controversial and potentially significant of all environmental legislation is land-use planning. Conflicts arise because it is often difficult to determine the highest and best use of land. Also, landowners fear that land-use planning will cause property owners to lose the right to control their property, and there is considerable concern over the idea that federal and state governments should initiate local planning.

The argument in favor of land-use planning legislation stresses that planning is urgently needed to control degradation of the land, and that land use is the remaining important aspect of environmental quality control that has not yet been stated as national policy.

The significant effect of land-management legislation is that it shifts to the developer the burden of proof that an activity will not degrade the land. We will have to live with the concept that the most profitable use of land is not always necessarily the best use. This concept stems directly from the Trust Doctrine and an emerging land ethic.

REFERENCES

1 LANDAN, N. J., and RHEINGOLD, P. D. 1971. *The environmental law handbook.* New York: Ballantine Books.

2 YANNACONE, V. J., JR., COHEN, B. S., and DAVISON, S. G. 1972. *Environmental rights and remedies* 1. San Francisco: Bancroft-Whitney.

3 COATES, D. R. 1971. Legal and environmental case studies in applied geomorphology. In *Environmental Geomorphology,* ed. D. R. Coates, pp. 223–42. Binghamton, New York: State University of New York.

4 JUERGENSMEYER, J. C. 1970. Control of air pollution through the assertion of private rights. *Environmental Law:* 17–46. Greenvale, New York: Research and Documentation Corporation.

5 MURPHY, E. F. 1971. *Man and his environment: Law.* New York: Harper & Row.

6 CARTER, L. J. 1974. Con Edison: Endless Storm King dispute adds to its troubles. *Science* 184:1353–58.

7 CARGO, D. N., and MALLORY, B. F. 1974. *Man and his geologic environment.* Menlo Park, California: Addison-Wesley.

8 Private Remedies for Water Pollution. 1970. *Environmental Law:* 47–69. Greenvale, New York: Research and Documentation Corporation.

9 Legal Approach to Water Rights. 1972. In *Water quality in a stressed environment,* ed. W. A. Pettyjohn, pp. 255–76. Minneapolis: Burgess.

10 HEALY, M. R. 1974. National land use proposal: land use legislation of landmark environmental significance. *Environmental Affairs* 3:355–95.

11 McCLAUGHRY, J. 1974. The land use planning act—An idea we can live without. *Environmental Affairs* 3:595–626.

12 WAINRIGHT, J. K., JR. 1974. Spring Valley: Public purpose and land use regulation in a "taking" context. *Environmental Affairs* 3:327–54.

Renewable Resources: Air

*R*enewable resources are those resources that are recycled or made available by natural or human-induced processes over a period of time that enables human consumption or use. The most important renewable resources are air, water, agricultural products, fish and wildlife, forests and other natural vegetation, and various types of energy derived from interactions between solar radiation and other renewable resources.

An important principle to keep in mind concerning renewable resources is that they are only renewable as long as the environmental factors necessary for reproduction or recycling remain favorable. For example, environmental degradation, whether natural (such as sediment pollution from volcanic ash) or human-induced (as from construction during urbanization), may locally spoil surface water as a renewable resource. Similarly, air pollution, loss of wildlife habitat, or deforestation either by human activity or natural processes may locally render a renewable resource unrenewable. For example, the coastal redwood forest of northern California once extended from south of San Francisco to just north of the Oregon border along the western coast of the United States. Today, only small remnants of a once extensive forest of old growth redwood remain. Following timber harvesting, the redwoods regenerate quite well in some locations, as for

example near Santa Cruz, California. In other areas, the redwoods do not regenerate very well and may be replaced by other species. One reason regeneration may not occur is that soil erosion and microclimatic effects brought on by timber harvesting have removed some of the ingredients necessary for the redwood forest to flourish. In some areas the forest is a relic feature in that it developed during times in the past when the climate was wetter; after timber harvesting, it is not able to regenerate.

On a planetary or global scale, continued assault on renewable resources through environmental degradation is a serious problem: tropical rain forests are being rapidly depleted (15 percent of the world's mature woodlands were lost in one recent decade); the ocean each year continues to receive all sorts of pollutants; and vast areas, as much as 30 percent of the earth's land surfaces, on several continents occupied by one-sixth of the world's population, are being threatened by desertification, a process that over the long term may reduce productivity of the land by the spread of desertlike conditions (Figure A.1) (1).

The lesson to be learned is that renewable resources cannot be casually treated as a commons where all people are allowed equal use or misuse. Rather, renewable resources must be managed care-

453

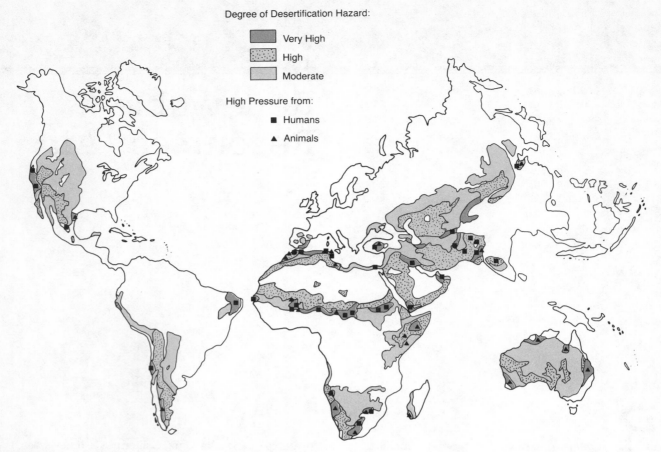

Degree of Desertification Hazard:

⬛ Very High
▨ High
▤ Moderate

High Pressure from:

■ Humans
▲ Animals

FIGURE A.1
Degree of desertification hazards on a global scale. (Council on Environmental Quality, 1978, Annual Report.)

fully to insure that they remain renewable. If this is not the case, then there is little chance to avert a global "tragedy of the commons" and loss of some renewable resources. This is already happening to some extent through the human-induced extinction of species and loss of the coastal redwoods in northern California. What might be the effect of a serious global assault on food or water? Fortunately there is still time to plan and use resources wisely, but the warning time may be growing less each year.

A complete discussion of the renewable biological resources such as fish and wildlife, timber, and agricultural products is beyond the scope of environmental geology. Nevertheless, management of these resources involves an intimate knowledge of the physical environment that in part supports biological systems. Thus, the subjects of earth materials (chapter 3), sediment control (chapter 9), waste disposal (chapter 10) and natural hazards (chapters 4–8) are all important in evaluating and managing renewable biological resources. The remainder of our discussion of renewable

resources will concern air, since we discussed water and energy in chapters 9 and 13.

THE URBAN AIR ENVIRONMENT

The air we breathe is a mixture of nitrogen (78%), oxygen (21%), argon (0.9%), carbon dioxide (0.03%) and minor or trace amounts of numerous elements and compounds, including, among others, methane, ozone, hydrogen sulfide, carbon monoxide, oxides of nitrogen and sulphur, hydrocarbons, and various particulates.

Any substance in the air that occurs in sufficient quantity to adversely affect people, animals, plants, or other earth materials can be considered a pollutant. The primary pollutants that account for more than 90 percent of air pollution problems in the United States are hydrocarbons, particulates, carbon monoxide, nitrogen oxides, and sulphur oxides. Each year, approximately 250 to 300 million metric tons of these materials from human-related processes enter the atmosphere above the United States. About half of this is carbon monoxide; the other four pollutants each account for between 8 to 15 percent. At first glance, several hundred million tons of pollutants appear to be a very large amount; however, if it were uniformly distributed in the atmosphere, it would only amount to about 3 parts per million by weight. Unfortunately, pollutants are not uniformly distributed, because there is little movement of surface air at altitudes greater than about 4,000 meters above the ground. Furthermore, temperature inversions in combination with geologic barriers such as mountains may limit the vertical circulation of air to such an extent that pollutants cannot rise above 600 meters or so (Figure A.2). Therefore, pollutants, particularly in urban areas, may build up beyond trace level concentrations, producing detrimental air quality (2).

In areas such as the Los Angeles basin, nitrogen oxides and hydrocarbons are particularly troublesome because light energy from the sun reacts with them to produce photochemical smog (including oxidants) that irritates eyes, mucous membranes, and lungs; damages certain plants, and retards visibility. Most of the nitrogen oxides and hydrocarbons in Los Angeles are emitted from automobiles, so automobile exhaust is the main producer of smog (3). In other urban areas, such as in Ohio and the Great Lakes region in general, air quality problems result from emissions of sulphur dioxides and particulates into the air from industry and coal-burning power plants that are stationary (point-source), rather than from a dispersed source (or area source) such as automobiles. This is not to say that automobiles are not a problem in areas outside Los Angeles, but the emphasis is on contrasting conditions (see Figure A.3).

The Pollutant Standard Index

A daily air quality report, in terms of whether the air quality is good, moderate, unhealthy, very unhealthy, or hazardous, is becoming a standard item in many urban news broadcasts (Table A.1). These levels or

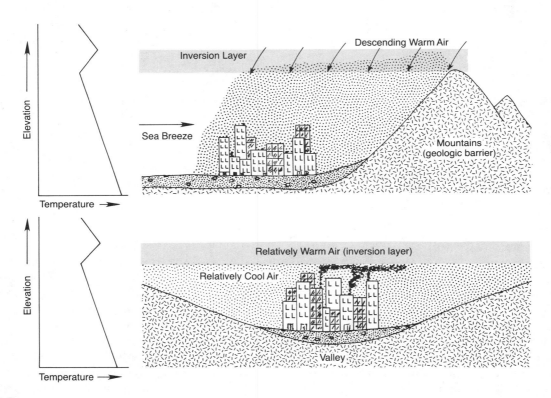

FIGURE A.2
Idealized diagrams showing two ways that temperature inversions can result. The upper diagram shows an inversion associated with a geologic barrier; the lower diagram shows an inversion that has developed over a valley. The graph of temperature vs. elevation shows the inversion. From the ground up, the temperature decreases with elevation to the bottom of the inversion layer, where the slope of the line reverses and temperatures increase through the inversion layer. At elevations higher than the inversion layer, temperatures again decrease with elevation.

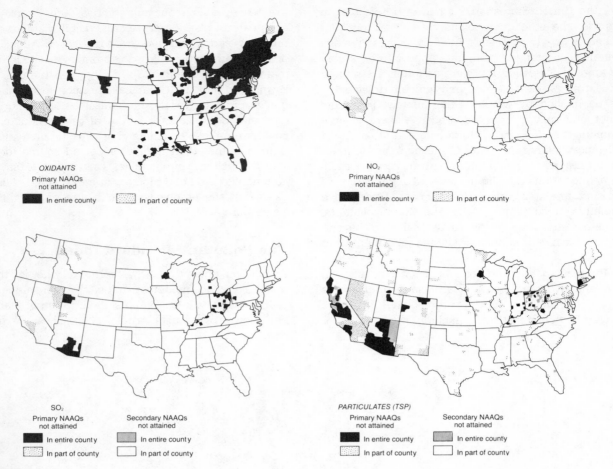

FIGURE A.3
Nonattainment of selected air pollutants for August 1977. NAAQS is National Ambient Air Quality Standard. Notice that nitrogen dioxide (NO$_2$) is primarily a southern California problem whereas sulphur dioxide (SO$_2$) is primarily associated with some parts of the southwestern United States as well as the Great Lakes region. (From Council on Environmental Quality, 1978, Annual Report.)

stages are derived from monitoring the concentration of five major pollutants: total suspended particulates, sulphur dioxide, carbon monoxide, ozone, and nitrogen dioxide. The ozone level is particularly important with reference to photochemical smog since it is formed as oxides of nitrogen combine with hydrocarbons in the presence of sunlight to produce the complicated organic particles known as smog. The ozone level is thus a good indicator of the amount of photochemical smog in urban air. Table A.2 lists national ambient air quality standards as of 1979, which will probably be revised as we learn more about air pollution.

The five major indicators of air quality all suggest that the nation's air quality is improving. Data from 25 major metropolitan areas between 1974 and 1977 indicate that the total number of unhealthful and very un-

healthful days declined (Figure A.4). We must not interpret this to mean, however, that air pollution has been eliminated. On the contrary, in 1977, both the New York and Los Angeles urban areas had unhealthful air about two-thirds of the time. Only about 40 percent of the urban areas in the United States for which good data were available in 1977 reported unhealthful readings for fewer than 10 percent of the days, and in fact, three cities reported an increase in air pollution between 1974 and 1977. The pollutants responsible for sending the pollution index into the unhealthful range in most urban areas were carbon monoxide, hydrocarbons, and their combination to form ozone and photochemical smog. Thus it appears that one way to further improve urban air quality is to obtain a reduction in automobile exhaust pollution.

TABLE A.1

Definition of Pollutant Standard Index (PSI) values

PSI Index Value	Air Quality Level	Pollutant Level					Health Effect	General Health Effects	Cautionary Statements
		TSP (24-hour), μg/m³	SO₂ (24-hour), μg/m³	CO (8-hour), mg/m³	O₃ (1-hour), μg/m³	NO₂ (1-hour), μg/m³			
500	Significant Harm	1,000	2,620	57.5	1,200	3,750			
400	Emergency	875	2,100	46.0	1,000	3,000		Premature death of ill and elderly. Healthy people will experience adverse symptoms that affect their normal activity.	All persons should remain indoors, keeping windows and doors closed. All persons should minimize physical exertion and avoid traffic.
300	Warning	625	1,600	34.0	800	2,260	Hazardous	Premature onset of certain diseases in addition to significant aggravation of symptoms and decreased exercise tolerance in healthy persons.	Elderly and persons with existing diseases should stay indoors and avoid physical exertion. General population should avoid outdoor activity.
200	Alert	375	800	17.0	400[c]	1,130	Very unhealthful	Significant aggravation of symptoms and decreased exercise tolerance in persons with heart or lung disease, with widespread symptoms in the healthy population.	Elderly and persons with existing heart or lung disease should stay indoors and reduce physical activity.
100	NAAQS	260	365	10.0	240		Unhealthful[a]	Mild aggravation of symptoms in susceptible persons, with irritation symptoms in the healthy population.	Persons with existing heart or respiratory ailments should reduce physical exertion and outdoor activity.
50	50% of NAAQS	75[b]	80[b]	5.0	120		Moderate[a]		
0	0	0	0	0	0		Good[a]		

[a]No index values reported at concentration levels below those specified by "Alert Level" criteria.

[b]Annual primary NAAQS.

[c]400 μg/m³ was used instead of the O₃ Alert Level of 200 μg/m³.

Source: U.S. Environmental Protection Agency, "Guidelines for Public Reporting of Daily Air Quality—Pollutant Standard Index."

TABLE A.2
National ambient air quality standards

Pollutant	Averaging Time	Primary Standard Levels	Secondary Standard Levels
Particulate matter	Annual (geometric mean)	75 μg/m³	60 μg/m³
	24 hours[b]	260 μg/m³	150 μg/m³
Sulfur oxides	Annual (arithmetic mean)	80 μg/m³ (0.03 ppm)	—
	24 hour[b]	365 μg/m³ (0.14 ppm)	—
	3 hour[b]	—	1,300 μg/m³ (0.5 ppm)
Carbon monoxide	8 hour[b]	10 mg/m³ (9 ppm)	10 mg/m³ (9 ppm)
	1 hour[b]	40 mg/m³ (35 ppm)	40 mg/m³ (35 ppm)
Nitrogen dioxide	Annual (arithmetic mean)	100 μg/m³ (0.05 ppm)	100 μg/m³ (0.05 ppm)
Ozone	1 hour[b]	235 μg/m³ (0.12 ppm)	235 μg/m³ (0.12 ppm)
Hydrocarbons (nonmethane)[a]	3 hour (6 to 9 a.m.)	160 μg/m³ (0.24 ppm)	160 μg/m³ (0.24 ppm)

[a]A nonhealth related standard used as a guide for ozone control.
[b]Not to be exceeded more than once per year.
Source: U.S. Environmental Protection Agency.

The Urban Microclimate

Air quality in urban areas is primarily a function of the amount of pollutants present or produced and of the ability of the city or urban area to ventilate and thus flush out pollutants. Unfortunately, the air over cities tends to move more slowly than in surrounding areas because buildings and other structures obstruct its flow. In urban areas, it is thus not uncommon that wind velocities are reduced by 20 to 30 percent, and the number of calm days may be 20 percent more abundant relative to nearby rural areas (3).

The combination of lingering air and an abundance of particulates and other pollutants produces the well-known urban dust dome and heat island effects (Figure A.5) (3). Also shown in Figure A.5 is the general circulation pattern of air moving from rural or suburban areas toward the inner city, where it flows up and then disperses laterally near the top of the dust dome. This circulation of air often occurs when a strong heat island develops over the city. Thus, when a dust dome and heat island have accumulated, there is an upward flow of air over the heavily developed Manhattan Island, accompanied by a downward flow over the nearby Hudson and East Rivers which, as green belts, are characterized by cooler air temperatures (4). Figure A.5 also shows the heat profile over the city that delineates the heat island.

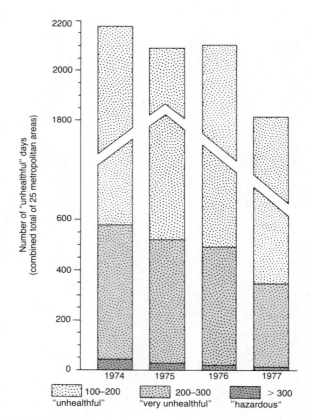

FIGURE A.4
National trend in urban air pollution levels, 1974–1977. (From Council on Environmental Quality, 1979, Annual Report.)

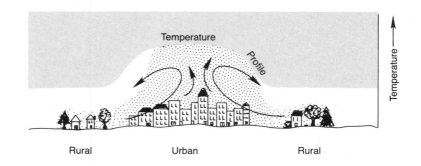

FIGURE A.5
Idealized diagram showing the urban dust dome and heat island effects.

The observable increase in sensible heat (measurable with a thermometer) in urban areas during a calm period is approximately 1–2°C in the winter and 0.5–1.0°C in the summer for midlatitude areas. The increase in temperature results from the dust dome's trapping long wave (infrared) radiation emitted by the earth and artificial heat emitted from the burning of fossil fuels and other sources. In the winter, artificial heat from the city is a prime source for heating the air environment. For example, in Manhattan, the input of artificial heat has been measured in the winter at about 2.5 times the solar energy that reaches the surface of the city; annual average, however, is closer to 33 percent of the solar input. In large urban areas characterized by warmer winters, the heat input from artificial sources is much less (3).

Particulates in the atmosphere over cities are often ten times or more greater than for surrounding areas, and tend to reduce incoming solar radiation by up to 30 percent. The effect is to cool the city; but the cooling effect is small relative to the processes that produce heat in the city (3).

Urban areas are also characterized by large impervious surfaces and lack of surface water. There is thus less exchange (evaporation) of water from the surface to the atmosphere, explaining why midlatitude cities generally record lower relative humidity (2 percent in the winter to 8 percent in the summer) than the surrounding countryside. Concrete, asphalt, and roofs also tend to act as solar collectors and emit heat quickly, helping to increase the sensible heat in cities (4).

Particulates in the dust dome provide condensation nuclei, so urban areas experience 5 to 10 percent more precipitation and considerably more cloud cover and fog than surrounding areas. The formation of fog is particularly troublesome in the winter and can impede air traffic into and out of airports. If the pollution dome moves downwind, then increased precipitation may be reported outside the urban area. For example, in the mid-1960s, effluent from the south Chicago–northern Indiana industrial complex apparently caused a 30 percent increase in precipitation at La Porte, Indiana, 48

kilometers downwind to the south. The effluent from the industries consisted of large amounts of small particulates, heat, and moist air—all of which contributed to the increased precipitation (5). The La Porte case is extreme, but not unique, and there is little doubt that particulate matter in the atmosphere from urban sources has altered local weather at numerous locations.

The urban dust dome and heat island cause problems for many urban areas. Long hot spells with poor air quality can lead to social unrest, characterized by higher crime rates and riots. Although the air quality problems of older cities will be more difficult to minimize, there is some indication that emission of pollutants is beginning to decline. In new urban areas, we can plan more carefully for ways to minimize both the dust dome and heat island effects.

Planning can help reduce air pollution in urban areas. For example, natural ventilation routes in cities, such as valleys and other green belts, should not be blocked by development. Zoning to take advantage of solar energy is also important. Buildings can be constructed in tiers for active and passive solar application, reducing the sensible heat input from burning fuels. Interspersed small parks provide natural filters (from the vegetation) for air pollution and help circulate cooler air toward buildings where heat tends to build up. Finally, stronger controls for automobile and industrial emissions as set forth in the Clear Air Act amendments of 1977 may result in further improvement of urban air quality.

AIR POLLUTION AND GLOBAL CLIMATIC CHANGE

Human use and activity inadvertently changes local, regional, or even the global climate in a number of ways. Table A.3 provides a framework for recognizing the types of changes and probable effects that people can induce by changing the composition of the atmosphere or modifying the surface of the earth. Three items in Table A.3 that have received considerable attention are the increased amount of carbon dioxide

(CO_2) introduced into the atmosphere by burning fossil fuels; the effects of particles in the atmosphere on global temperature; and the effects of ozone destruction in the stratosphere caused by emission of fluorocarbons or nitrogen oxides.

Carbon Dioxide in the Atmosphere

A major controversy now going on in the earth sciences is the debate over the possible effects of increasing carbon dioxide in the earth's atmosphere. It is generally accepted that around the turn of the century, the concentration of carbon dioxide in the atmosphere was about 290 parts per million (0.029 percent); since that time, there has been a steady increase to the present level of approximately 330 parts per million (0.033 percent). By the year 2000 the projected value is 380 parts per million or 0.038 percent, almost 20 percent higher than the present concentration. The increase in carbon dioxide in the atmosphere is generally attributed to the burning of fossil fuels. The actual increase is only half

that expected if all the carbon dioxide remained in the atmosphere, however. It is hypothesized that the ocean acts as a buffer and absorbs about half the carbon dioxide emitted into the atmosphere from fossil fuels. A small amount may also be incorporated into vegetation. The extrapolation of the increase in carbon dioxide until the year 2000 is based on several assumptions that may not be entirely correct. First, it is assumed that the ocean will continue to absorb about 50 percent of the carbon dioxide emitted; and second, it is assumed that large scale and significant atmospheric circulation changes will not take place. Given these assumptions, the temperature increase from atmospheric carbon dioxide should be approximately 0.5°C by the year 2000. This is unlikely to create major problems, but if the CO_2 content continues to build for several hundred more years, a temperature rise of 2–3°C would be possible, and this would cause major problems. A change of a few degrees Centigrade in the mean annual temperature could be sufficient to melt the polar ice caps, raising sea levels around the world and flooding major

TABLE A.3
Examples of possible human-induced climatic changes.

Cause	Probable-Potential Climatic Effects
Changes in atmospheric composition	
Increase in CO_2 (carbon dioxide) by burning fossil fuels	Increase in average temperature 0.2 to 0.3°C for a 10 percent increase in carbon dioxide
Increase number of particulates	Decrease in average surface temperature, increase in precipitation, more urban dust domes, more fog in urban areas
Increase in sulfur	More acid precipitation
Increase in hydrocarbons	More photochemical smog and ozone in urban areas, more acid precipitation
Increase in fluorocarbons	Decrease in ozone in stratosphere; potential tropospheric warming due to both greenhouse effect of fluorocarbons in atmosphere and increase in ultraviolet radiation reaching the surface of the earth
Increase in nitrous oxide (N_2O) from industrial fertilizers	Decrease in ozone in upper atmosphere (20 km); warmer earth surface
Modification of surface of the earth	
More large reservoirs, canals	Increased evaporation, more water in atmosphere, more precipitation in some areas, larger greenhouse effect (due to water in atmosphere)
Urbanization	Dust dome, heat island, local reduction of wind, increased precipitation, lower relative humidity
Deforestation	Increased albedo (reflectivity), decreased evapo-transpiration, increased wind
Overgrazing marginal semiarid lands	Desertification, less evapo-transpiration, increased wind

Source: Modified after R. A. Anthes et al., *The Atmosphere*, 2d ed. Columbus, Ohio. Charles E. Merrill, 1978.

cities. For this reason and others, the carbon dioxide in the atmosphere is being very carefully monitored. Figure A.6 shows the climate of the earth in terms of mean annual temperature for the last 150,000 years. What happens in the next few thousand years may be significantly affected by human-induced changes in the composition of the atmosphere.

The actual mechanism by which carbon dioxide can cause heating of the atmosphere is related to the "greenhouse effect." When incoming short-wave length radiation from the sun enters a greenhouse, it is readmitted as long-wave radiation that is trapped by the glass. Although the main reason a greenhouse heats up relative to outside temperatures is that the air is retarded from circulating with outside air, the effect of trapping long-wave radiation is known as the "greenhouse effect."* In the atmosphere, carbon dioxide effectively traps or absorbs long-wave radiation emitted from the earth, and this trapped radiation heats up the atmosphere. Thus, as the amount of CO_2 in the atmosphere increases, the temperature of the atmosphere has a corresponding increase in temperature. The final statement concerning atmospheric carbon dioxide and its increase because of the burning fossil fuels is that the problem is unresolved. There is no general agreement as to what the effects will be, primarily because we are not certain of all the storage compartments available in the carbon cycle. That is, the ocean may be able to account for a lesser or greater amount of carbon dioxide in the future, and there may be other sinks for carbon dioxide that we have not considered (6,7).

---

*Fluorocarbons and water vapor in the atmosphere may also cause a greenhouse effect.

Particulates in the Atmosphere

The possible effects on global changes of particulate material in the atmosphere in mean annual atmospheric temperature is uncertain. What is more certain is the fact that more particulates are being added to the atmosphere through human activity. Today only about 20 percent of all the particles are from human sources, but this could increase to about 50 percent by the year 2000. The sources of the particulates may be primary, as from burning coal and other fossil fuels, or secondary, as from reactions involving photochemical smog. Regardless of how they are produced, the physical effects of particulates are at least twofold. First, particulates act as condensation nuclei and therefore can cause an increase in precipitation or fog; and second, particulates affect the amount of sunlight that reaches the earth. As the total amount of particulates in the atmosphere increases, a larger percentage of incoming solar radiation is reflected away from the earth. The effect is to cause a decrease in the earth's mean annual temperature. On the other hand, if particles filter out of the atmosphere and are deposited on snow, then a greater portion of the solar radiation will be absorbed, leading to a greater amount of radiation available to heat the atmosphere. There seems little doubt that particulates in the atmosphere can interfere with incoming solar radiation. Catastrophic volcanic eruptions have caused global cooling for several years. This cooling effect was particularly apparent when Krakatoa erupted in 1883, blasting dust into the stratosphere that circled the earth for several years and caused a slight cooling (6). The eruption of Mt. St. Helens in 1980 undoubtedly affected the stratosphere, but will probably not change the global surface weather. As

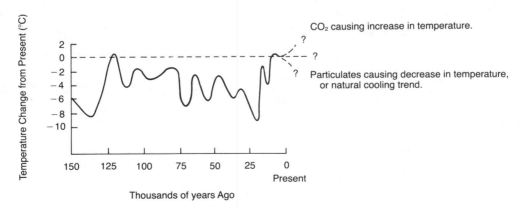

FIGURE A.6
Change in global mean annual temperature for the last 150,000 years with several possibilities for the future. (Modified after National Academy of Science, 1975.)

with carbon dioxide, there is a considerable amount of controversy surrounding the effects that particulates will have on future climates.

Threats to Stratospheric Ozone

Ozone (O_3) is produced in the stratosphere at altitudes of 16–60 kilometers above the earth when two reactions take place. The first reaction is the splitting of an oxygen molecule into atomic oxygen; the second reaction involves a union of a molecule of oxygen (O_2) with an atom of oxygen (O) to make ozone (O_3). The second reaction occurs only when a third molecule (a catalyst) is present. Ozone is destroyed naturally by ultraviolet radiation. Thus, as the ozone is destroyed, it provides a service function to organisms at the surface of the earth by greatly reducing the amount of ultraviolet radiation that reaches the earth. In the stratosphere, ozone is constantly being formed and destroyed and is therefore maintained in a rough equilibrium. Any reduction in ozone could be potentially dangerous because more ultraviolet light would reach the earth and possibly cause an increase in skin cancer. The relationship between ultraviolet radiation and skin cancer is estimated to be such that a 1 percent depletion in ozone would result in a 2 percent increase in skin cancers. Another aspect of the ozone depletion problem stems from the fact that if there were less ozone, more radiation would reach the earth, which might tend to heat up the lower atmosphere.

Controversy arises over the reactions and events that might lead to human-induced changes in the amount of ozone in the stratosphere. It has been suggested that aerosols, particularly fluorocarbons, will destroy some ozone. Others suggest that industrial fertilizers, through interaction with the biosphere, emit nitrous oxide that may rise and eventually destroy stratospheric ozone, producing ozone depletion. As was the case in our discussion of carbon dioxide and particulates, the final potential effects of ozone depletion and the processes involved are not completely understood. Ten years ago it was believed that the proposed American supersonic transport (SST) would release materials into the stratosphere that would deplete ozone, but that theory is now being overturned.

Human activity changes climate, because emissions of carbon dioxide, particulates, and other materials affect the atmosphere. We need more careful monitoring and modeling of various reactions that take place in the atmosphere so that we can make better predictions. This will be an exciting area for earth scientists who are interested in environmental matters and atmospheric physics and chemistry.

REFERENCES

1 COUNCIL ON ENVIRONMENTAL QUALITY. 1978. *Environmental Quality*, Ninth Annual Report of the Council on Environmental Quality.

2 STOKER, H. S., and SEAGER, S. L. 1967. *Environmental Chemistry: Air and Water Pollution*. Glenview, Ill.: Scott, Foresman and Co.

3 DETWYLER, T. R., MARCUS, M. G. et al. 1972. *Urbanization and Environment*. Belmont, Calif.: Duxbury Press.

4 MARSH, W. M., and DOZIER, J. 1981. *Landscape*. Reading, Mass.: Addison-Wesley.

5 LYNN, D. A. 1976. *Air Pollution—Threat and Response*. Reading, Mass.: Addison-Wesley.

6 ANTHES, R. A., PANOFSKY, H. A., CAHIR, J. J., and RANGO, A. 1978. *The Atmosphere*, 2d ed. Columbus, Ohio: Charles E. Merrill.

Common Conversion Factors

LENGTH
1 yard = 3 ft, 1 fathom = 6 ft

	in	ft	mi	cm	m	km
1 inch (in) =	1	0.083	1.58×10^{-5}	2.54	0.0254	2.54×10^{-5}
1 foot (ft) =	12	1	1.89×10^{-4}	30.48	0.3048	
1 mile (mi) =	63,360	5,280	1	160,934	1,609	1.609
1 centimeter (cm) =	0.394	0.0328	6.2×10^{-6}	1	0.01	1.0×10^{-5}
1 meter (m) =	39.37	3.281	6.2×10^{-4}	100	1	0.001
1 kilometer (km) =	39,370	3,281	0.6214	100,000	1,000	1

AREA
1 square mi = 640 acres, 1 acre = 43,560 ft^2 = 4046.86 m^2 = 0.4047 ha
1 ha = 10,000 m^2 = 2.471 acres

	in^2	ft^2	mi^2	cm^2	m^2	km^2
1 in^2 =	1		—	6.4516	—	—
1 ft^2 =	144	1	—	929	0.0929	—
1 mi^2 =	—	27,878,400	1	—	—	2.590
1 cm^2 =	0.155	—	—	1	—	—
1 m^2 =	1,550	10.764	—	10,000	1	—
1 km^2 =	—	—	0.3861	—	1,000,000	1

VOLUME

	in³	ft³	yd³	m³	qt	liter	barrel	gal. (U.S)
1 in³ =	1	—	—	—	—	0.02	—	—
1 ft³ =	1,728	1	—	.0283	—	28.3	—	7.480
1 yd³ =	—	27	1	0.76	—	—	—	—
1 m³ =	61,020	35.315	1.307	1	—	1,000	—	—
1 quart (qt) =	—	—	—	—	1	0.95	—	0.25
1 liter (*l*) =	61.02	—	—	—	1.06	1	—	0.2642
1 barrel (oil) =	—	—	—	—	168	159.6	1	42
1 gallon (U.S.) =	231	0.13	—	—	4	3.785	0.02	1

MASS AND WEIGHT

1 pound = 453.6 grams = 0.4536 kilogram = 16 ounces
1 gram = 0.0353 ounce = 0.0022 pound
1 short ton = 2,000 pounds = 907.2 kilograms
1 long ton = 2,240 pounds = 1,008 kilograms
1 metric ton = 2,205 pounds = 1,000 kilograms
1 kilogram = 2.205 pounds

ENERGY AND POWER

1 kilowatt-hour = 3,413 Btus = 860,421 calories
1 Btu = 0.000293 kilowatt-hour = 252 calories = 1,055 joule
1 watt = 3.413 Btu/hr = 14.34 calories/min
1 calorie = the amount of heat necessary to raise the temperature of 1 gram (1 cm³) of water 1 degree Celsius
1 quadrillion Btu = (approximately) 1 exa joule

TEMPERATURE

F is degrees Fahrenheit
C is degrees Celsius (centigrade)

$$F = \frac{9}{5} C + 32$$

Fahrenheit		Celsius
32	Freezing of H₂O (Atmospheric Pressure)	0
50		10
68		20
86		30
104		40
122		50
140		60
158		70
176		80
194		90
212	Boiling of H₂O (Atmospheric Pressure)	100

OTHER CONVERSION FACTORS

$$1 \text{ ft}^3/\text{sec} = .0283 \text{ m}^3/\text{sec} = 7.48 \text{ gal/sec} = 28.32 \text{ liters/sec}$$
$$1 \text{ acre-foot} = 43,560 \text{ ft}^3 = 1,233 \text{ m}^3 = 325,829 \text{ gal}$$
$$1 \text{ m}^3/\text{sec} = 35.32 \text{ ft}^3/\text{sec}$$
$$1 \text{ ft}^3/\text{sec for one day} = 1.98 \text{ acre-feet}$$
$$1 \text{ m/sec} = 3.6 \text{ km/hr} = 2.24 \text{ mi/hr}$$
$$1 \text{ ft/sec} = 0.682 \text{ mi/hr} = 1.097 \text{ km/hr}$$
$$1 \text{ billion gallons per day (bgd)} = 3.785 \text{ million m}^3 \text{ per day}$$

Strength of common rock types

	Rock Type	Range of Compressive Strength (Thousands of psi)	Comments
Igneous	Granite	23 to 42.6	Finer-grained granites with few fractures are the strongest. Granite is generally suitable for most engineering purposes.
Igneous	Basalt	11.8 to 52.0	Brecciated zones, open tubes, or fractures reduce the strength.
Metamorphic	Marble	6.7 to 34.5	Solutional openings and fractures weaken the rock.
Metamorphic	Gneiss	22.2 to 36.4	Generally suitable for most engineering purposes.
Metamorphic	Quartzite	21.1 to 91.2	Very strong rock.
Sedimentary	Shale	Less than 1 to 33.5	May be a very weak rock for engineering purposes; careful evaluation is necessary.
Sedimentary	Limestone	5.3 to 37.6	May have clay partings, solution openings, or fractures that weaken the rock.
Sedimentary	Sandstone	4.8 to 34.1	Strength varies with degree and type of cementing material, mineralogy, and nature and extent of fractures.

Source: Data primarily from *Handbook of Tables for Applied Engineering Science*, ed. R. E. Bolz and G. L. Tuve (Cleveland, Ohio: CRC Press, 1973.)

Geologic time scale and biologic evolution

Era	Approx. Age in Millions of Years Before Present	Period	Epoch	Life Form
	Less than 0.01		Recent (Holocene)	
	0.01–2	Quaternary	Pleistocene	Humans
	— 2 —			
	2–5		Pliocene	
	5–24		Miocene	
Cenozoic	24–38	Tertiary	Oligocene	
	38–55		Eocene	Mammals
	55–63		Paleocene	
	— 63 —			
	63–138	Cretaceous		
Mesozoic	138–205	Jurassic		Flying reptiles, birds
	205–240	Triassic		Dinosaurs
	— 240 —			
	240–290	Permian		Reptiles
	290–360	Carboniferous		Insects
				Fossils
Paleozoic	360–410	Devonian		Amphibians
	410–435	Silurian		Land plants
	435–500	Ordovician		Fish
	500–570	Cambrian		
	— 570 —			
	700			Multicelled organisms
	3,400			One-celled organisms
Precambrian	4,000	Approximate age of oldest rocks discovered on earth		
	4,500	Approximate age of the earth and meteorites		

Glossary

A soil horizon Uppermost soil horizon, sometimes referred to as the **zone of leaching.**

Absorption Taking up, incorporating, or assimilating of a material.

Adsorption Attachment of molecules of gas or molecules in solution to the surface of solid materials with which they come in contact.

Aerobic Characterized by the presence of free oxygen.

Aesthetics Originally a branch of philosophy; defined today by artists and art critics.

Age element Elements that characteristically accumulate in tissue with age.

Aggregate Any hard material such as crushed rock, sand, gravel, or other material that is added to cement to make concrete.

Alkaline soils Soils in arid regions that contain a large amount of soluble mineral salts (primarily sodium) that in the dry season may appear on the surface as a crust or powder.

Alluvium Unconsolidated sediments, including sand, gravel, and silt, deposited by streams.

Anaerobic Characterized by the absence of free oxygen.

Angle of repose The maximum angle that loose material will sustain.

Anhydrite Evaporite mineral ($CaSO_4$) calcium sulfate.

Anthracite A type of coal characterized by a high percentage of carbon and low percentage of volatiles, providing a high heat value. Anthracite often forms as a result of metamorphism of bituminous coal.

Anticline Type of fold characterized by an upfold or arch. The oldest rocks are found in the center of the fold.

Appropriation doctrine, water law Holds that prior usage of water is a significant factor. The first to use the water for beneficial purposes is prior in right.

Aquiclude Earth material that retards the flow of groundwater.

Aquifer Earth material containing sufficient groundwater that the water can be pumped out. Highly fractured rocks and unconsolidated sands and gravels make good aquifers.

Area (strip) mining Type of strip mining practiced on relatively flat areas.

Artesian Referring to a groundwater system in which the groundwater is isolated from the surface by a confining layer and the water is under pressure. Groundwater that is under sufficient pressure will flow freely at the surface from a spring or well.

Asbestos Fibrous mineral material used as insulation. It is suspected of being either a true carcinogen or a carrier of carcinogenic trace elements.

Ash fall Volcanic ash eruption that blows up into the atmosphere and then rains down on the landscape.

Ash flow Mixture of volcanic ash, hot gases, and fragments of rock and glass that flows rapidly down the flank of a volcano. May be an extremely hazardous event.

Ash, volcanic Unconsolidated volcanic debris, less than 44 mm in diameter, physically blown out of a volcano during an eruption.

Avalanche A type of landslide involving a large mass of snow, ice, and rock debris that slides, flows, or falls rapidly down a mountainside.

Azonal soil   Recent surface materials such as floodplain deposits that do not have distinctive soil layering.

B soil horizon   Intermediate soil horizon, sometimes known as the **zone of accumulation.**

Balancing doctrine   Asserts that public benefits and importance of a particular action should be balanced against potential injury to certain individuals. The method of balancing is changing, and courts are considering possible long-range injury caused by certain activities to large numbers of citizens other than the immediate complainants.

Barrier island   Island separated from the mainland by a salt marsh. It generally consists of a multiple system of beach ridges and is separated from other barrier islands by inlets that allow exchange of seawater with lagoon water.

Basalt   A fine-grained extrusive igneous rock. It is one of the most common igneous rock types.

Basaltic   Engineering geology term for all fine-grained igneous rocks.

Bauxite   A rock composed almost entirely of hydrous aluminum oxides. It is a common ore of aluminum.

Bed material   Sediment transported and deposited along the bed of a stream channel.

Bedding plane   The plane that delineates the layers of sedimentary rocks.

Bentonite   A type of clay that is extremely unstable; upon wetting, it expands to many times its original volume.

Biosphere   The zone adjacent to the surface of the earth that includes all living organisms.

Biotite   A common ferromagnesian mineral, a member of the mica family.

Bituminous coal   A common type of coal characterized by relatively high carbon content and low volatiles; sometimes called **soft coal.**

Blowout   Failure of an oil, gas, or disposal well resulting from adverse pressures that can physically blow part of the well casing upward. May be associated with leaks of oil, gas, or, in the case of disposal wells, harmful chemicals.

BOD   Biological oxygen demand, a measure of the amount of oxygen necessary to decompose organic materials in a unit volume of water. As the amount of organic waste in water increases, more oxygen is used, resulting in a higher BOD.

Braided river   A river channel characterized by an abundance of islands that continually divide and subdivide the flow of the river.

Breccia   A rock or zone within a rock composed of angular fragments. Sedimentary, volcanic, and tectonic breccias are recognized.

Breeder reactor   A type of nuclear reactor that actually produces more fissionable (fuel) material than it uses.

Brine   Water that has a high concentration of salt.

British thermal unit (Btu)   A unit of heat defined as the heat required to raise the temperature of one pound of water one degree Fahrenheit.

Brittle   Material that ruptures before any plastic deformation.

Bulk Element   Common elements that make up the bulk of living material.

C soil horizon   Lowest soil horizon, sometimes known as the **zone of partially altered parent material.**

Cadmium   A metallic element with an atomic number of 48 and an atomic weight of 112.4. As a trace element, it has been associated with serious health problems.

Calcite   Calcium carbonate ($CaCO_3$); common carbonate mineral that is the major constituent of the rock limestone. It weathers readily by solutional processes, and large cavities and open weathered fractures are common in rocks that contain the mineral calcite.

Caliche   A white-to-gray irregular accumulation of calcium carbonate in soils of arid regions.

Calorie   The quantity of heat required to raise the temperature of one gram of water from 14.5 to 15.5 degrees Celsius.

Capillary action   The rise of water along narrow passages, facilitated and caused by surface tension.

Carbonate   A compound or mineral containing the radical (CO_3^{--}). The common carbonate is calcite.

Carcinogen   Any material known to produce cancer in humans or other animals.

Cesium-137   A fission product produced from nuclear reactors with a half-life of 33 years.

Channelization   An engineering technique to straighten, widen, deepen, or otherwise modify a natural stream channel.

Circum-Pacific belt   One of the three major zones where earthquakes occur. This belt is essentially the Pacific plate. It is also known as the **ring of fire,** as many active volcanoes are found on the edge of the Pacific plate.

Clay   May refer to a mineral family or to a very fine-grained sediment. It is associated with many environmental problems, such as shrinking and swelling of soils and sediment pollution.

Coal   A sedimentary rock formed from plant material that has been buried, compressed, and changed.

Colluvium   Mixture of weathered rock, soil, and other, usually angular, material on a slope.

Columnar jointing   System of fractures (joints) that break rock into polygons of typically five or six sides; the polygons form columns. This type of fracturing is common in basalt and most likely caused by shrinking during cooling of the lava.

Common excavation   Excavation that can be accomplished with an earth mover, backhoe, or dragline.

Composite volcano   Steep-sided volcanic cone produced by alternating layers of pryoclastic debris and lava flows. It characteristically has magma that is intermediate in silica content (60 percent).

Conchoidal fracture   A shell-like or fan-shaped fracture characteristic of the mineral quartz and natural glass.

Cone of depression A cone-shaped depression in the water table caused by withdrawal of water at rates greater than those at which the water can be replenished by natural groundwater flow.

Conglomerate A detrital sedimentary rock composed of rounded fragments, 10 percent of which are larger than 2 mm in diameter.

Connate water Water that is no longer in circulation or in contact with the present water cycle; generally, saline water trapped during deposition of sediments.

Contact metamorphism Type of metamorphism produced when country rocks are in close contact with a cooling body of magma below the surface of the earth.

Continental drift Movement of continents in response to sea-floor spreading. The most recent episode of continental draft supposedly began about 200 million years ago with the breakup of the supercontinent Pangaea.

Continental shelf Relatively shallow ocean area between the shoreline and the continental slope that extends to approximately a 600-foot water depth surrounding a continent.

Contour (strip) mining Type of strip mining used in hilly terrain.

Convection Transfer of heat involving movement of particles; for example, the boiling of water in which hot water rises to the surface and displaces cooler water which moves toward the bottom.

Convergent plate boundary Boundary between two lithospheric plates in which one plate descends below the other (subduction).

Corrosion A slow chemical weathering or chemical decomposition that proceeds inward from the surface. Objects such as pipes experience corrosion when buried in soil.

Cost-benefits analysis A type of site selection that compares benefits and costs of a particular project. The most desirable projects are those for which the benefits-to-cost ratio is greater than one.

Creep A type of downslope movement characterized by slow flowing, sliding, or slipping of soil and other earth materials.

Crystal settling A sinking of earlier-formed crystals to the bottom of a magma chamber.

Crystalline A material with a definite internal structure such that the atoms are in an orderly, repeating arrangement.

Crystallization Processes of crystal formation.

Debris flow Rapid downslope movement of earth material often involving saturated, unconsolidated material that has become unstable because of torrential rainfall.

Deep-well disposal Method of waste disposal that involves pumping waste into subsurface disposal sites such as fractured or otherwise porous rocks.

Detrital Mineral and rock fragments derived from preexisting rocks.

Diamond Very hard mineral composed of the element carbon.

Dilatancy of rocks Inelastic increase in volume of a rock that begins after stress on the rock has reached one-half the rock's breaking strength.

Discharge The quantity of water flowing past a particular point on a stream, usually measured in cubic feet per second (cfs).

Disseminated mineral deposit Mineral deposit in which ore is scattered throughout the rocks; examples are diamonds in kimberlite and many copper deposits.

Divergent plate boundary Boundary between lithospheric plates characterized by production of new lithosphere; found along oceanic ridges.

Dose dependency Relates to the fact that effects of a certain trace element on a particular organism depend upon the dose or concentration of the element.

Drainage basin Area that contributes surface water to a particular stream network.

Drainage net System of stream channels that coalesce to form a stream system.

Dredge spoils Solid material, such as sand, silt, clay, rock or other material deposited from industrial and municipal discharges, that are removed from the bottom of water bodies to improve navigation.

Driving forces Those forces that tend to make earth material slide.

Ductile Material that ruptures following elastic and plastic deformation.

Earthquake Natural shaking or vibrating of the earth in response to the breaking of rocks along faults. The earthquake zones of the earth generally correlate with lithospheric plate boundaries.

Ecology Branch of biology that treats relationships between organisms and their environments.

Economic geology Application of geology to locating and evaluating mineral materials.

Effluent Any material that flows outward from something; examples include waste water from hydroelectric plants and water discharged into streams from waste-disposal sites.

Effluent stream Type of stream where flow is maintained during the dry season by groundwater seepage into the channel.

Elastic deformation Type of deformation where the material returns to its original shape after the stress is removed.

Engineering geology Application of geologic information to engineering problems.

Environment That which surrounds an individual or a community; both physical and cultural surroundings. **Environment** also sometimes denotes a certain set of circumstances surrounding a particular occurrence, for example, environments of deposition.

Environmental geology Application of geologic information to environmental problems.

Environmental geology map   A map that combines geologic and hydrologic data expressed in nontechnical terms to facilitate general understanding by a large audience.

Environmental-impact statement   A written statement that assesses and explores the possible impacts of a particular project that may affect the human environment. The statement is required by the National Environmental Policy Act of 1969.

Environmental law   A field of law that is growing rapidly and becoming a significant part of our jurisprudence.

Environmental resource unit (ERU)   A portion of the environment with a similar set of physical and biological characteristics, a supposedly natural division characterized by specific patterns or assemblages of structural components, such as rocks, soils, vegetation, etc., and natural processes, such as erosion, runoff, soil processes, etc.

Ephemeral   Temporary or very short-lived. Characteristic of beaches, lakes, and some stream channels that change rapidly (geologically).

Epicenter   The point on the surface of the earth directly above the focus (area of first motion) of an earthquake.

Evaporite   Sediments deposited from water as a result of extensive evaporation of seawater or lake water; dissolved materials left behind following evaporation.

Exponential growth   A type of compound growth in which a total amount or number increases at a certain percentage per year, and each year's rate of growth is added to the total from the previous year; characteristically stated in terms of a particular doubling time, that is, the time in years it will take the original number to double. Commonly used in reference to population growth.

Extrusive igneous rocks   Igneous rock that forms when magma reaches the surface of the earth; a volcanic rock.

Fault   A fracture or fracture system that has experienced movement along opposite sides of the fracture.

Feldspar   The most abundant family of minerals in the crust of the earth; silicates of calcium, sodium, and potassium.

Ferromagnesian mineral   Minerals containing iron and magnesium, characteristically dark in color.

Fertile materials   Materials such as uranium-238, which is not naturally fissionable but upon bombardment by neutrons is converted to plutonium-239, which is fissionable.

Fission   The splitting of an atom into smaller fragments with the release of energy.

Floodplain   Flat topography adjacent to a stream in a river valley, produced by the combination of overbank flow and lateral migration of meander bends.

Floodway district   That portion of a channel and floodplain of a stream designated to provide passage of the 100-year regulatory flood without increasing elevation of the flood by more than one foot.

Floodway fringe district   Land located between the floodway district and the maximum elevation subject to flooding by the 100-year regulatory flood.

Fluorine   Important trace element, essential for nutrition.

Fluvial   Concerning or pertaining to rivers.

Fly ash   Very fine particles (ash) resulting from the burning of fuels such as coal.

Focus, earthquake   The point in the earth where an earthquake originates.

Folds   Bends that develop in stratified rocks because of tectonic forces.

Foliation   Property of metamorphic rock characterized by parallel alignment of the platy or elongated mineral grains; environmentally important because it can affect the strength and hydrologic properties of rock.

Formation   Any rock unit that can be mapped.

Fossil fuels   Fuels such as coal, oil, and gas formed by the alteration and decomposition of plants and animals from a previous geologic time.

Fracture zone   A fracture system that may or may not be active and may or may not have an alteration zone along the fracture planes. Fracture zones are environmentally important because they greatly affect the strength of rocks.

Fumarole   A natural vent from which fumes or vapors are emitted, such as the geysers and hot springs characteristic of volcanic areas.

Fusion, nuclear   Combining of light elements to form heavy elements with the release of energy.

Gaging station   Location at a stream channel where discharge of water is measured.

Gasification   Method of producing gas from coal.

Geochemical cycle   Migratory paths of elements during geologic changes and processes.

Geologic cycle   A group of interrelated cycles known as the hydrologic, rock, tectonic, and geochemical cycles.

Geomorphology   The study of landforms and surface processes.

Geothermal energy   The useful conversion of natural heat from the interior of the earth.

Glacier   A landbound mass of moving ice.

Gneiss   A coarse-grained, foliated metamorphic rock in which there is banding of light and dark minerals.

Gravel   Unconsolidated, generally rounded fragments of rocks and minerals greater than 2 mm in diameter.

Groin   A structure designed to protect shorelines and trap sediment in the zone of littoral drift, generally constructed perpendicular to the shoreline.

Groundwater   Water found beneath the surface of the earth within the zone of saturation.

Grout   A mixture of cement and sediment that is sufficiently fluid to be pumped into open fissures or cracks in rocks, thereby increasing the strength of a foundation for an engineering structure.

Gypsum   An evaporite mineral, $CaSO_4 \cdot 2H_2O$.

Half-life   The amount of time necessary for one-half of the atoms of a particular radioactive element to decay.

Halite A common mineral, NaCL (salt).

Hematite An important ore of iron, a mineral (Fe_2O_3).

High-value resource Materials such as diamonds, copper, gold, and aluminum. These materials are extracted wherever they are found and transported around the world to numerous markets.

Humus Black organic material in soil.

Hydrocarbon Organic compounds consisting of carbon and hydrogen.

Hydroconsolidation Consolidation of earth materials upon wetting.

Hydrofracturing Pumping of water under high pressure into subsurface rocks to fracture the rocks and thereby increasing their permeability.

Hydrograph A graph of the discharge of a stream with time.

Hydrologic cycle Circulation of water from the oceans to the atmosphere and back to the oceans by way of evaporation, runoff from streams and rivers, and groundwater flow.

Hydrology The study of surface and subsurface water.

Hydrothermal ore deposit A mineral deposit derived from hot water solutions of magmatic origin.

Igneous rocks Rocks formed from solidification of magma; *extrusive* if they crystallize on the surface of the earth and *intrusive* if they crystallize beneath the surface.

Impermeable Earth materials that greatly retard or prevent movement of fluids through them.

Infiltration Movement of surface water into rocks or soil.

Influent stream Type of stream that is everywhere above the groundwater table and flows in direct response to precipitation. Water from the channel moves down to the water table, forming a recharge mound.

Intrazonal soil Soil characteristically determined by local conditions such as excess water from flooding or high groundwater table.

Island arc A curved group of volcanic islands associated with a deep-oceanic trench and subduction zone (convergent plate boundary).

Isotopes Forms of the same element having a variable atomic weight.

Itaiitai disease Extremely painful disease that attacks bones, causing them to become very brittle so that they break easily. Associated with heavy metals, especially cadmium, in concentrations of a few parts per million in the soil or in food consumed by victims of the disease.

Juvenile water Water derived from the interior of the earth that has not previously existed as atmospheric or surface water.

Karst topography A type of topography characterized by the presence of sinkholes, caverns, and diversion of surface water to subterranean routes.

Kimberlite pipe An igneous intrusive body that may contain diamond crystals disseminated (scattered) throughout the rock type.

Land ethic Ethic that affirms the right of all resources, including plants, animals, and earth materials, to continued existence and, at least in some locations, continued existence in a natural state.

Landslide Specifically, rapid downslope movement of rock and/or soil; also a general term for all types of downslope movement.

Land-use planning Complex process involving development of a land-use plan to include a statement of land-use issues, goals, and objectives; summary of data collection and analysis; land-classification map; and report describing and indicating appropriate development in areas of special environmental concern. An extremely controversial issue.

Laterite Soil formed from intense chemical weathering in tropical or savanna regions.

Lava Molten material produced from a volcanic eruption, or rock that forms from solidification of molten material.

Leachate Obnoxious liquid material capable of carrying bacteria, produced when surface water or groundwater comes into contact with solid waste.

Leaching Process of dissolving, washing, or draining earth materials by percolation of groundwater or other liquids.

Lignite A type of low-grade coal.

Limestone A sedimentary rock composed almost entirely of the mineral calcite.

Limonite Rust; hydrated iron oxide.

Liquefaction Transformation of water-saturated granular material from the solid state to a liquid state.

Lithosphere Outer layer of the earth approximately 100 kilometers thick of which the plates that contain the ocean basins and continents are composed.

Littoral Pertaining to the near-shore and beach environments.

Loess Angular deposits of windblown silt.

Low-value resource Resources such as sand and gravel that have primarily a place value; economically extracted because they are located close to where they are to be used.

Magma A naturally occurring silica melt, much of which is in a liquid state.

Magma tap Attempt to recover geothermal heat directly from magma. Feasibility of such heat extraction is unknown.

Magnetite A mineral and important ore of iron, Fe_3O_4.

Magnitude, earthquake A number on a logarithmic scale referring to the amount of energy released by an earthquake.

Manganese oxide nodule Nodules of manganese, iron with secondary copper, nickel, and cobalt, that cover vast areas of the deep-ocean floor.

Marble Metamorphosed limestone.

Marl Unconsolidated clays, silts, sands, or mixtures of these materials that contain a variable content of calcareous material.

Meanders Bends in a stream channel that migrate back and forth across the floodplain, depositing sediment on the inside of the bends, forming point bars, and eroding outsides of bends.

Meteoric water Water derived from the atmosphere.

Methane A gas, CH_4, the major constituent of natural gas.

Mica A common rock-forming silicate mineral.

Mudflow A mixture of unconsolidated materials and water that flows rapidly downslope or down a channel.

Myth of superabundance The myth that land and water resources are inexhaustible and management of resources is therefore unnecessary.

National Environmental Policy Act of 1969 (NEPA) Act declaring a national policy that harmony between man and his physical environment be encouraged. Established the Council on Environmental Quality. Established requirements that an environmental-impact statement be completed prior to major federal actions that significantly affect the quality of the human environment.

Neutron A subatomic particle having no electric charge, found in the nuclei of atoms. Neutrons are crucial in sustaining nuclear fission in a reactor.

Nonrenewable resource A resource cycled so slowly by natural earth processes that, once used, will be essentially unavailable during any useful time framework.

Nuclear reactor Device in which controlled nuclear fission is maintained; the major component of a nuclear power plant.

Oil shale Organic-rich shale containing substantial quantities of oil that can be extracted by conventional methods of destructive distillation.

Ore Earth material from which a useful commodity can be extracted profitably.

Osteoporosis Disease characterized by reduction in bone mass.

Outcrop A naturally occurring or man-caused exposure of rock at the surface of the earth.

Overburden Earth materials (spoil) that overlie an ore deposit, particularly material overlying or extracted from a surface (strip) mine.

Oxidation Chemical process of combining with oxygen.

P **wave** One of the seismic waves produced by an earthquake; the fastest of the seismic waves, it can move through liquid and solid materials.

Pathogen Any material that can cause disease; for example, microorganisms, including bacteria and fungi.

Pebble A rock fragment between 4 and 64 mm in diameter.

Pedology Study of soils.

Pegmatite A coarse-grained igneous rock that may contain rare minerals rich in elements such as lithium, boron, fluorine, uranium, and others.

Percolation test A standard test for determining rate at which water will infiltrate into the soil. Primarily used to determine feasibility of a septic-tank disposal system.

Permafrost Permanently frozen ground.

Permeabilty A measure of the ability of an earth material to transmit fluids such as water or oil.

Petrology Study of rocks and minerals.

Physiographic determinism Site selection based on the philosophy of designing with nature.

Physiographic province Region characterized by a particular assemblage of landforms, climate, and geomorphic history.

Placer deposit Type of ore deposit found in material transported and deposited by agents such as running water, ice, or wind; for example, gold and diamonds found in stream deposits.

Plastic deformation Type of deformation involving permanent change of shape without rupture.

Plate tectonics A model of global tectonics that suggests that the outer layer of the earth known as the **lithosphere** is composed of several large plates that move relative to one another; continents and ocean basins are passive riders on these plates.

Plutonium-239 A radioactive element produced in a nuclear reactor; has a half-life of approximately 24,000 years.

Point bar Accumulation of sand and other sediments on the inside of meander bends in stream channels.

Pollution Any substance, biological or chemical, in which an identified excess is known to be detrimental to desirable living organisms.

Pool Common bed form produced by scour in meandering and straight channels with relatively low channel slope; characterized at low flow by slow-moving, deep water. Generally, but not exclusively, found on the outside of meander bends.

Porosity The percentage of void (empty space) in earth material such as soil or rock.

Potable water Water that may be safely drunk.

Pyrite Iron sulfide, a mineral, commonly known as fool's gold. Environmentally important because, in contact with oxygen-rich water, it produces a weak acid that can pollute water or dissolve other minerals.

Pyroclastic activity Type of volcanic activity characterized by eruptive or explosive activity in which all types of volcanic debris, from ash to very large particles, are physically blown from a volcanic vent.

Quartz Silicon oxide, a common rock-forming mineral.

Quartzite Metamorphosed sandstone.

Quick clay Type of clay which when disturbed, as by seismic shaking, may experience spontaneous liquefaction and lose all shear strength.

Radioactive waste Type of waste produced in the nuclear fuel cycle, generally classified as *high-level* or *low-level*.

Radon A colorless, radioactive, gaseous element.

Reclamation, mining Restoring land used for mining to other useful purposes, such as agriculture or recreation, after mining operations are concluded.

Recycling The reuse of resources reclaimed from waste.

Regional metamorphism Wide-scale metamorphism of deeply buried rocks by regional stress accompanied by elevated temperatures and pressures.

Renewable resource A resource such as timber, water, or air that is naturally recycled or recycled by man-induced processes within a useful time framework.

Reserves Known and identified deposits of earth materials from which useful materials can be extracted profitably with existing technology under present economic and legal conditions.

Resisting forces Forces that tend to oppose downslope movement of earth materials.

Resistivity A measure of an earth material's ability to retard the flow of electricity; the opposite of conductivity.

Resources Includes reserves plus other deposits of useful earth materials that may eventually become available.

Riffle A section of stream channel characterized at low flow by fast, shallow flow; generally contains relatively coarse bed-load particles.

Riparian rights, water law Right of the landowner to make reasonable use of water on his land, provided the water is returned to the natural stream channel before it leaves his property; the property owner has the right to receive the full flow of the stream undiminished in quantity and quality.

Rippable excavation Type of excavation that requires breaking up soil before it can be removed.

Riprap Layer or assemblage of broken stones placed to protect an embankment against erosion by running water or breaking waves.

Riverine environment Land area adjacent to and influenced by a river.

Rock, Geologic An aggregate of a mineral or minerals; **Engineering**—Any earth material that must be blasted to be removed.

Rock cycle Group of processes that produce igneous, metamorphic, and sedimentary rocks.

Rocksalt Rock composed of the mineral halite.

Rotational landslide Type of landslide that develops in homogeneous material; movement is likely to be rotational along a potential slide plane.

S wave Secondary wave, one of the waves produced by earthquakes.

Saline Salty; characterized by high salinity.

Salinity A measure of the total amount of dissolved solids in water.

Salt dome A structure produced by upward movement of a mass of salt; frequently associated with oil and gas deposits on the flanks of a dome.

Sand Grains of sediment with a size between 1/16 and 2 mm in diameter. Often, sediment composed of quartz particles of this size.

Sand dune Ridge or hill of sand formed by wind action.

Sandstone Detrital sedimentary rock composed of sand grains that have been cemented together.

Sanitary landfill Method of solid-waste disposal that does not produce a public health problem or nuisance; confines and compresses waste and covers it at the end of each day with a layer of compacted, relatively impermeable material, such as clay.

Scarp Steep slope or cliff commonly associated with landslides or earthquakes.

Scenic resources The visual portion of an aesthetic experience; scenery is now recognized as a natural resource with varying values.

Schist Coarse-grained metamorphic rock characterized by foliated texture of the platy or elongated mineral grains.

Schistosomiasis Snail fever, a debilitating and sometimes fatal tropical disease.

Sea wall Engineering structure constructed at the water's edge to minimize coastal erosion by wave activity.

Secondary enrichment Weathering process of sulfide ore deposits that may concentrate the desired minerals.

Sedimentology Study of environments of deposition of sediments.

Seismic Referring to vibrations in the earth produced by earthquakes.

Seismograph Instrument that records earthquakes.

Selenium Important nonmetallic trace element with an atomic number of 34.

Septic tank Tank that receives and temporarily holds solid and liquid waste; anaerobic bacterial activity breaks down the waste, solid wastes are separated out, and liquid waste from the tank overflows into a drainage system.

Serpentine A family of ferromagnesian minerals; environmentally important because they form very weak rocks.

Sewage sludge Solid material that remains after municipal waste-water treatment.

Shale Sedimentary rock composed of silt- and clay-sized particles; the most common sedimentary rock.

Shield volcano A broad, convex volcano built up by successive lava flows; the largest of the volcanoes.

Silicate minerals The most important group of rock-forming minerals.

Silt Sediment between 1/16 and 1/256 mm in diameter.

Sinkhole Surface depression formed by solution of limestone or collapse over a subterranean void such as a cave.

Sinuous channel Type of stream channel (not braided).

Slate A fine-grained, foliated metamorphic rock.

Slump Type of landslide characterized by downward slip of a mass of rock, generally along a curved slide plane.

Soil **Soil science**—Earth material modified by biological, chemical, and physical processes such that the material will support rooted plants. **Engineering**—Earth material that can be removed without blasting.

Soil horizons Layers in soil (A, B, C) that differ from one another in chemical, physical, and biological properties.

Soil survey A survey consisting of a detailed soil map and descriptions of soils and land-use limitations; usually prepared by the Soil Conservation Service in cooperation with local government.

Solar energy Collecting and using energy directly from the sun.

Solid waste Material such as refuse, garbage, and trash.

Spoils, mining Banks or piles that are accumulations of overburden removed during mining processes and discarded on the surface.

Storm surge Wind-driven oceanic waves.

Strain Change in shape or size of a material as a result of applied stress; the result of stress.

Stress Force per unit area; may be compression, tension, or shear.

Strip mining A method of surface mining.

Subduction Process in which one lithospheric plate descends beneath another.

Subsidence Sinking, settling, or other lowering of parts of the crust of the earth.

Subsurface water All the waters within the lithosphere.

Surface water Waters above the surface of the lithosphere.

Surface wave One of the types of waves produced by earthquakes; these waves generally cause most of the damage to structures on the surface of the earth.

Suspended load Sediment in a stream or river carried off the bottom by the fluid.

Syncline Type of fold in which younger rocks are found in the core of the fold; rocks in the limbs of the fold dip inward toward a common axis.

System Any part of the universe that is isolated in thought or in fact for the purpose of studying or observing changes that occur under various imposed conditions.

Tar sands Naturally occurring sand, sandstone, or limestone that contains an extremely viscous petroleum.

Tectonic Referring to rock deformation.

Tectonic creep Slow, more or less continuous movement along a fault.

Tectonic cycle Group of processes that collectively produce external forms on the earth, such as ocean basins, continents, and mountains.

Tephra Any material ejected and physically blown out of a volcano; mostly ash.

Texture, rock The size, shape, and arrangement of mineral grains in rocks.

Tidal energy Electricity generated by tidal power.

Till Unstratified, heterogeneous material deposited directly by glacial ice.

Toxic Harmful, deadly, or poisonous.

Transform fault Type of fault associated with oceanic ridges; may form a plate boundary, such as the San Andreas fault in California.

Translation (slab) landslide Type of landslide in which movement takes place along a definite fracture plane, such as a weak clay layer or bedding plane.

Tropical cyclone Severe storm generated from a tropical disturbance; called **typhoons** in most of the Pacific Ocean and **hurricanes** in the Western Hemisphere.

Tsunami Seismic sea wave generated by submarine volcanic or earthquake activity; characteristically has very long wave length and moves rapidly in the open sea; incorrectly referred to as *tidal wave*.

Tuff Volcanic ash that is compacted, cemented, or welded together.

Unconfined aquifer Type of aquifer in which there is no impermeable layer restricting the upper surface of the zone of saturation.

Unconformity A buried surface of erosion representing a time of nondeposition; a gap in the geologic record.

Unified soil classification system Classification of soils, widely used in engineering practice, based on amount of coarse particles, fine particles, or organic material.

Uniformitarianism Concept that the present is the key to the past; that is, we can read the geologic record by studying present processes.

Volcanic breccia, agglomerate Large rock fragments mixed with ash and other volcanic materials cemented together.

Volcanic dome Type of volcano characterized by very viscous magma with high silica content; activity is generally explosive.

Water table Surface that divides the zone of aeration from the zone of saturation; the surface below which all the pore space in rocks is saturated with water.

Weathering Changes that take place in rocks and minerals at or near the surface of the earth in response to physical, chemical, and biological changes; the physical, chemical, and biological breakdown of rocks and minerals.

Zinc An important trace element necessary in life processes.

Zonal soil Soil in which the profile is in adjustment to the climatic zone.

Zone of aeration Zone or layer above the water table in which some water may be suspended or moving in a downward migration toward the water table or laterally toward a discharge point.

Zone of saturation Zone or layer below the water table in which all the pore space of rock or soil is saturated.

Index